Asheville-Buncombe
Technical Community College
Learning Resources Center
340 Victoria Rd.
Asheville, NC 28801

Discarded
Date SEP 17 2024

Fundamentals of Metal-Matrix Composites

Fundamentals of Metal-Matrix Composites

EDITED BY

Subra Suresh
PROFESSOR OF ENGINEERING, DIVISION OF ENGINEERING, BROWN UNIVERSITY, PROVIDENCE, RHODE ISLAND
PRESENTLY, RICHARD P. SIMMONS PROFESSOR, DEPARTMENT OF MATERIALS SCIENCE AND ENGINEERING, MASSACHUSETTS INSTITUTE OF TECHNOLOGY, CAMBRIDGE, MASSACHUSETTS

Andreas Mortensen
ASSOCIATE PROFESSOR OF MECHANICAL METALLURGY, DEPARTMENT OF MATERIALS SCIENCE AND ENGINEERING, MASSACHUSETTS INSTITUTE OF TECHNOLOGY, CAMBRIDGE, MASSACHUSETTS

Alan Needleman
PROFESSOR OF ENGINEERING, DIVISION OF ENGINEERING, BROWN UNIVERSITY, PROVIDENCE, RHODE ISLAND

WITH 30 CONTRIBUTING AUTHORS

Butterworth–Heinemann
Boston   London   Oxford   Singapore   Sydney   Toronto   Wellington

Copyright © 1993 by Butterworth–Heinemann,

⩇ A member of the Reed Elsevier group
All rights reserved

No part of this publication may be reproduced, stored in a retrieval system, or transmitted, in any form or by any means, electronic, mechanical, photocopying, recording, or otherwise, without the prior written permission of the publisher.

Recognizing the importance of preserving what has been written, it is the policy of Butterworth-Heinemann to have the books it publishes printed on acid-free paper, and we exert our best efforts to that end.

Library of Congress Cataloging-in-Publication Data

Fundamentals of metal-matrix composites / edited by S. Suresh, A. Mortensen, and A. Needleman.
    p. cm.
  Includes bibliographical references and index.
  ISBN 0-7506-9321-5
  1. Metallic composites. I. Suresh, S. (Subra) II. Mortensen, A. (Andreas) III. Needleman, A. (Alan)
TA481.F87   1993
620.1'6—dc20   93-4727
                                                     CIP

British Library Cataloguing-in-Publication Data

A catalogue record for this book is available from the British Library.

Butterworth–Heinemann
80 Montvale Avenue
Stoneham, MA 02180

10 9 8 7 6 5 4 3 2 1

Printed in the United States of America

Contents

Preface	vii
Contributing Authors	ix

Part I: Processing

Chapter 1: Liquid-State Processing VÉRONIQUE J. MICHAUD	3
Chapter 2: Solid-State Processing AMIT K. GHOSH	23
Chapter 3: Capillary Phenomena, Interfacial Bonding, and Reactivity NICOLAS EUSTATHOPOULOS AND ANDREAS MORTENSEN	42

Part II: Microstructure Characterization

Chapter 4: Characterization of Residual Stresses in Composites MARK A.M. BOURKE, JOYCE A. GOLDSTONE, MICHAEL G. STOUT, AND ALAN NEEDLEMAN	61
Chapter 5: Structure and Chemistry of Metal/Ceramic Interfaces MANFRED RÜHLE	81
Chapter 6: Microstructural Evolution in Whisker- and Particle-Containing Materials NIELS HANSEN AND CLAIRE Y. BARLOW	109
Chapter 7: Aging Characteristics of Reinforced Metals SUBRA SURESH AND KRISHAN K. CHAWLA	119

Part III: Micromechanics and Mechanics of Deformation

Chapter 8: Crystal Plasticity Models PETER E. MCHUGH, ROBERT J. ASARO AND C. FONG SHIH	139
Chapter 9: Continuum Models for Deformation: Discontinuous Reinforcements JOHN W. HUTCHINSON AND ROBERT M. MCMEEKING	158
Chapter 10: Continuum Models for Deformation: Metals Reinforced with Continuous Fibers SUBRA SURESH AND JOHN R. BROCKENBROUGH	174
Chapter 11: Creep and Thermal Cycling DAVID C. DUNAND AND BRIAN DERBY	191

Part IV: Damage Micromechanisms and Mechanics of Failure

Chapter 12: Models for Metal/Ceramic Interface Fracture ZHIGANG SUO AND C. FONG SHIH	217
Chapter 13: Matrix, Reinforcement, and Interfacial Failure ALAN NEEDLEMAN, STEVE R. NUTT, SUBRA SURESH AND VIGGO TVERGAARD	233
Chapter 14: Fracture Behavior BRIAN DERBY AND PAUL M. MUMMERY	251
Chapter 15: Fatigue Behavior of Discontinuously Reinforced Metal-Matrix Composites JOHN E. ALLISON AND J. WAYNE JONES	269

Part V: Applications

Chapter 16: Metal-Matrix Composites for Ground Vehicle, Aerospace and Industrial Applications MICHAEL J. KOCZAK, SUBHASH C. KHATRI, JOHN E. ALLISON AND MICHAEL G. BADER	297
Index	327

Preface

Metal-matrix composite materials have been active subjects of scientific investigation and applied research for three decades. However, only in the past few years have these composites become realistic contenders as engineering materials. Mass market products, like automobiles, now contain metal-matrix composite components. The use of these materials is also being explored in many other applications, including aerospace and sporting goods. At the same time, scientific interest in metal-matrix composites has grown substantially. This stems from the realization that a deeper understanding of fundamentals is needed for metal-matrix composites to become more broadly useful.

A significant volume fraction of a stiff nonmetallic phase in a ductile metal matrix results in phenomena that are specific to reinforced metals and that were relatively unexplored until recently. Examples include issues of interfacial bonding between the reinforcement and the matrix, residual stresses, matrix dislocations generated by the thermal mismatch between phases, and reinforcement-induced alterations in matrix precipitation kinetics. In the past few years, we have seen significant advances in our understanding of these materials and of phenomena specific to their fabrication and behavior. Scientific investigations have addressed the governing principles of their processing and general laws have been identified for the influence exerted by the reinforcement on the microstructural evolution of the matrix. Advances in computational mechanics have brought to light practically important micromechanical phenomena that were often ignored in analytic treatments.

The combination of growing engineering importance and scientific understanding provides the motivation for this volume. Our goal is to provide a summary of the current state of the engineering and science pertinent to metal-matrix composites. To contain the coverage within one reasonably sized volume, while maintaining depth and coherence, we have emphasized processing, microstructural development, and mechanical response. Within this scope, we have sought to concentrate on fundamentals, for brevity and generality, and with the hope of promoting interdisciplinary interaction within the community. Coverage is at the level of current research. Relevant background literature can be found in the references provided in each chapter.

The book has four main sections. In the first section, two chapters present the principal processing methods and their governing scientific principles. An additional chapter focuses on interfacial phenomena that are of general importance in these materials, beginning with their processing. The second section focuses on microstructural development and characterization. It covers metallurgical issues that are brought about or that are strongly influenced by the presence of the reinforcement, such as residual stresses, interface characterization, matrix substructure and matrix aging.

The third section covers the micromechanics of deformation. The first chapter of this section deals with the influence of the crystallographic nature of matrix slip. The next two chapters present the micromechanics of short and long fiber composites, treating the matrix as an elastic-plastic continuum. The last chapter of this section provides coverage of the high-temperature mechanical behavior of reinforced metals.

Composite failure is covered in the fourth section. The first two chapters of this section focus on basic phenomena of great relevance to these materials: interfacial failure and localized failure arising from highly constrained matrix flow and particle fracture. The following two chapters then specifically address monotonic fracture and fatigue.

The volume ends with a summary of one of the most important areas of recent progress: the engineering application of these materials. Although not a topic of fundamental investigation, this issue was deemed important in the context of this volume because applications play a prominent role in defining and directing the focus of fundamental research on metal-matrix composites.

We are grateful to the authors for taking time from their active research programs to provide summaries of some of the most exciting and relevant research areas in this field. Our goal is to present a picture of the present state of knowledge and to point out fruitful areas for further investigation. We are resigned to the fact that the utility of such a volume is ephemeral; nonetheless, we hope that these chapters will provide valuable tools for further progress.

Subra Suresh
Andreas Mortensen
Alan Needleman

Contributing Authors

JOHN E. ALLISON
Staff Scientist, Materials Science Department, Ford Research Laboratory, Ford Motor Company, Dearborn, Michigan

ROBERT J. ASARO
Professor, Department of AMES, University of California, San Diego, La Jolla, California.

MICHAEL G. BADER
Lecturer, Materials Department, University of Surrey, Surrey, England.

CLAIRE Y. BARLOW
Senior Research Fellow, Department of Engineering, University of Cambridge, Cambridge, England.

MARK A.M. BOURKE
LANSCE, Los Alamos National Lab, Los Alamos, New Mexico.

JOHN R. BROCKENBROUGH
Engineer, Product Design and Mechanics Division, Alcoa Laboratories, Alcoa Center, Pennsylvania.

KRISHAN K. CHAWLA
Department of Materials and Metals Engineering, New Mexico Tech, Socorro, New Mexico

BRIAN DERBY
University Lecturer in Materials Science, Department of Materials, University of Oxford, Oxford, England.

DAVID C. DUNAND
Assistant Professor, Department of Materials Science and Engineering, Massachusetts Institute of Technology, Cambridge, Massachussetts.

NICHOLAS EUSTATHOPOULOS
Directeur de Recherches, Laboratoire de Thermodynamique et Physico-Chimie Métallurgique, C.N.R.S., I.N.P.G., Grenoble, France.

AMIT K. GHOSH
Professor, Department of Materials Science and Engineering, University of Michigan, Ann Arbor, Michigan.

JOYCE A. GOLDSTONE
LANSCE, Los Alamos National Laboratory, Los Alamos, New Mexico.

NIELS HANSEN
Head of Materials Department, Risø National Laboratory, Denmark.

JOHN W. HUTCHINSON
Professor, Division of Applied Sciences, Harvard University, Cambridge, Massachussetts.

J. WAYNE JONES
Professor, Department of Materials Science and Engineering, The University of Michigan, Ann Arbor, Michigan.

SUBHASH C. KHATRI
Department of Materials Engineering, Drexel University, Philadelphia, Pennsylvania.

MICHAEL J. KOCZAK
Department of Materials Engineering, Drexel University, Philadelphia, Pennsylvania.

PETER E. MCHUGH
Mechanical Engineering Department, University College, Galway, Ireland.

ROBERT M. MCMEEKING
Professor, Departments of Mechanical Engineering and Materials, University of California, Santa Barbara, Santa Barbara, California.

VÉRONIQUE J. MICHAUD
Post-Doctoral Research Associate, Department of Materials Science and Engineering, Massachussetts Institute of Technology, Cambridge, Massachussetts.

PAUL M. MUMMERY
Research Fellow, Department of Materials, University of Oxford, Oxford, England.

STEVE NUTT
Associate Professor, Brown University, Division of Engineering, Providence, Rhode Island.

MANFRED RÜHLE
Professor, Institut für Werkstoffwissenschaft, Max-Planck-Institut (für Metallforschung), Stuttgart, Germany.

C. FONG SHIH
Professor, Division of Engineering, Brown University, Providence, Rhode Island.

MICHAEL G. STOUT
Staff Scientist, Los Alamos National Laboratory, Los Alamos, New Mexico.

ZHIGANG SUO
Associate Professor, Department of Mechanical and Environmental Engineering, University of California, Santa Barbara, California.

VIGGO TVERGAARD
Professor, Technical University of Denmark, Lyngby, Denmark.

PART I
PROCESSING

Chapter 1
Liquid-State Processing

VÉRONIQUE J. MICHAUD

Despite their highly promising mechanical and thermal properties, metal-matrix composites have, for a long time, been afforded only limited use in very specific applications. Shortcomings such as complex processing requirements and the high cost of the final product have presented the greatest barriers to their proliferation. Improvements in the reinforcement fabrication and composite processing techniques are therefore pivotal for increasing their commercial applicability. Significant efforts have been, and continue to be, devoted to this end with encouraging results; reinforced metals have begun to show their presence in large-scale commercial applications. Notable examples include the alumina fiber-reinforced aluminum-alloy pistons for diesel engines introduced by Toyota Motor Corporation in 1982 (Donomoto et al. 1983) and, more recently, the alumina and carbon fiber-reinforced cylinder liners of the Honda Prelude (Ebisawa et al. 1991; Hayashi et al. 1989).

Processing of metal-matrix composites can be broadly divided into two categories of fabrication technique: (1) solid state (including powder metallurgy and diffusion bonding) and (2) liquid state.* A majority of the commercially viable applications are now produced by liquid-state processing because of inherent advantages of this processing technique over solid-state techniques. That is, the liquid metal is generally less expensive and easier to handle than are powders, and the composite material can be produced in a wide variety of shapes, using methods already developed in the casting industry for unreinforced metals. Conversely, liquid-state processes often suffer from a lack of reproducibility as a result of incomplete control of the processing parameters, and of undesirable chemical reactions at the interface between molten metal and reinforcement (see Chapter 3). Also, they are often limited to low-melting-point alloys, although some reinforced intermetallics have now been produced by liquid-state processes.

Liquid-state processing technologies currently being investigated and developed utilize a variety of methods to physically combine the matrix and the reinforcement. On this basis, we can sort them into four major categories: (1) Infiltration, (2) Dispersion, (3) Spraying, and (4) In-situ fabrication. For each category, the underlying method and the specific processes are described herein, together with a discussion of the governing physical phenomena, because their identification and study is crucial for understanding and improving the processes. Two of these phenomena, wettability and interfacial reactivity, are covered more extensively in Chapter 3 and so are touched upon only briefly here. Finally, some recent innovations and strategies for optimization of the processing of these materials are underlined and some possible future trends are briefly discussed.

1.1 Infiltration Processes

1.1.1 Description

Infiltration processes involve holding a porous body of the reinforcing phase within a mold and infiltrating it with molten metal that flows through interstices to fill the pores and produce a composite. Two examples of infiltration processes are given in Figure 1.1. Such

---

*It should be noted that for this monograph, the latter category is defined to include techniques that involve fully or partially molten matrix material during the fabrication of the composite.

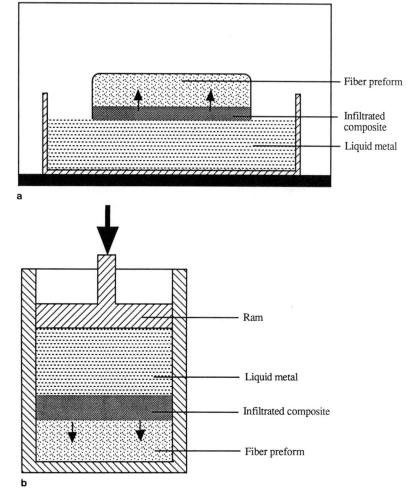

Figure 1.1. Schematic representations. (a) Spontaneous infiltration and (b) pressure-driven infiltration.

processes are not new; liquid-metal infiltration has been used for many years in powder metallurgy to strengthen porous metal parts of iron and steel with copper (Lenel 1980). The major difference between infiltration of solid metal by liquid metal and infiltration of ceramic by liquid metal regards wetting. In the latter case, liquid metal generally does not spontaneously wet the reinforcement. Therefore, it is often forced into the preform by application of an external force that overcomes the capillary and fluid-drag forces.

The main parameters in infiltration processes are the initial composition, morphology, volume fraction and temperature of the reinforcement, the initial composition and temperature of the infiltrating metal, and the nature and magnitude of the external force applied to the metal, if any. In the discussion that follows, we classify the several infiltration processes according to this last parameter.

1.1.1.1 No External Force

In specific cases, the metal may spontaneously infiltrate the reinforcement. Cermets such as titanium-carbide-reinforced steel or nickel-base alloys have been produced by spontaneous infiltration (Lenel 1980). Systems in which specifically tailored chemistry or processing conditions are used to induce wetting also allow spontaneous infiltration. Practical examples include the Ti-B process, in which Ti and B are deposited by chemical vapor deposition on fibers prior to infiltration by aluminum (Harrigan and Flowers 1976; Harrigan and Flowers 1977; Harrigan et al. 1980; Kendall and Pepper 1978; Lachman et al. 1975; Meyerer et al. 1977). Another example is the PRIMEX pressureless metal infiltration process, developed by Lanxide Corporation, Newark, Delaware (Kennedy 1991), whereby Al-Mg alloys infiltrate ceramic preforms at temperatures between 750°C and 1050°C in a nitrogen-rich atmosphere. The infiltration rates are quite low however, at up to 25 cm/hour (Kennedy 1991).

1.1.1.2 Vacuum-Driven Infiltration

For some matrix-reinforcement systems, creating a vacuum around the reinforcement provides a sufficiently large pressure difference to drive infiltration. An example of this is the vacuum infiltration of alumina fiber FP (E.I. du Pont de Nemours, Wilmington, DE) by Al-Li alloys (Champion et al. 1978; Dhingra 1980; Folgar et al. 1987; Hunt 1986). Vacuum infiltration occurs in this case because lithium can reduce alumina; however, because this reaction can also degrade the fiber, the amount of lithium has to be well controlled. Another interesting technique of vacuum infiltration involves the upward infiltration of magnesium into encased preforms of Al_2O_3 or SiC fibers positioned at the metal surface. Either the molten metal or its vapor reacts with the air above the preform to form solid products such as MgO. The vacuum thus created provides the driving force for infiltration (Donomoto 1982; Lawrence 1972; Westengen et al. 1990).

1.1.1.3 Pressure-Driven Infiltration

A strategy to overcome poor wetting is to supply mechanical work to force the metal into a preform that it does not wet. Although the primary purpose of an externally applied pressure is to overcome the capillary forces, higher pressures can provide additional benefits such as increased processing speed, control over chemical reactions, refined matrix microstructures, and better soundness of the product through feeding of solidification shrinkage. Pressure is generally applied by a gas or mechanically: In the case of pressure application by gas, the metal is forced into the preform of reinforcing phase by an inert gas, such as Ar, typically pressurized in the 1 to 10 MPa range. The process was first patented in 1970 (Cochran and Ray 1970), and has been widely used since then to produce both reinforced aluminum alloys (Isaacs 1991; Masur et al. 1989; Michaud 1991) and intermetallic composites (Nourbakhsh et al. 1988; Nourbakhsh and Margolin 1991). The use of higher pressures (up to 17 MPa) has also been recently investigated for aluminum-matrix composites (Yang and Chung 1989), but safety issues then become a major limiting factor.

The technique utilizing mechanically applied pressure involves a force that is exerted on the molten metal by the piston of a hydraulic press, and subsequently maintained during solidification. The pressures involved are generally higher than those in gas-driven processes, ranging from about 10 to 100 MPa. This type of infiltration process is currently the most widely investigated for commercial applications, and has been directly adapted from established processes designed to cast unreinforced metals, such as squeeze-casting, in which the piston constitutes part of the mold (Das et al. 1988; Fukunaga 1988; Sample et al. 1989; Verma and Dorcic 1988) and die-casting (Charbonnier et al. 1988; Girot et al. 1990; Jarry et al. 1992). Composites produced by this method generally feature a pore-free matrix. However, application of pressure may induce preform deformation or breakage during infiltration, as is often observed experimentally (Fukunaga 1988; Imai et al. 1987; Jarry et al. 1992; Kloucek and Singer 1986; Nishida et al. 1988; Rasmussen et al. 1991a; Rasmussen et al. 1991b).

1.1.1.4 Other Forces

Alumina preforms have been infiltrated by Al-Si alloys under low pressure with the assistance of vibrations at a frequency up to 3 kHz (Pennander and Anderson 1991; Pennander et al. 1991). Centrifugal casting methods have been adapted to the production of tubular reinforced metal (Tsunekawa et al. 1988; Vaidyanathan and Rohrmayer 1991). Also, a new infiltration process has recently been developed using electromagnetic body forces to drive molten metal into a preform (Andrews and Mortensen 1991).

A major advantage of infiltration processes is that they allow for near-net shape production of parts fully or selectively reinforced with a variety of materials. If cold dies and reinforcements are used, or if high pressures are maintained during solidification, matrix-reinforcement chemical reactions can be minimized and attractive, defect-free matrix microstructures can be achieved. A limitation of infiltration processes is the need for the reinforcement to be self-supporting, either as a bound preform or as a dense pack of particles or fibers. Heterogeneity of the final product may result from preform deformation during infiltration or from clustering of fibers that are detrimental to the composite mechanical properties. Also, tooling may be expensive if high pressures are used.

1.1.2 Governing Transport Phenomena

Several approaches have been adopted in the past decade to model infiltration processing. Simple analyses similar to those developed at an earlier date for metal/metal systems and cermets (Lenel 1980; Semlak and Rhines 1958; Shaler 1965) model the preform as a bundle of cylindrical capillary tubes (Martins et al. 1988; Maxwell et al. 1990). Other analyses, initially proposed in Japan and the United Kingdom, model the preform as a continuum (Clyne et al. 1985; Fukunaga and Goda 1985a; Fukunaga and Goda 1984; Fukunaga and Goda 1985b; Fukunaga et al. 1983; Fukunaga and Kuriyama 1982; Fukunaga and Ohde 1982; Nagata and Matsuda 1981). In this approach, analysis is based on consideration of a differential volume element ΔV, which contains pores, metal, and reinforcement elements such as fibers or particles, and is schematically represented in Figure 1.2. Within ΔV, parameters such as fraction solid metal g_s, fiber volume fraction V_f or temperature T are uniform. Velocities of the fluid phase v_l and solid phase v_s are also averaged over ΔV. In most cases, preform deformation is neglected, so the solid phase is assumed to be stationary ($v_s = 0$). However, it is necessary to take the solid velocity into account when preform deformation occurs during infiltration (Sommer 1992). The continuum approach is generally legitimate given the small size of the reinforcements (typically 1 to 20 μm in diameter) compared to preform dimensions.

Using this continuum approach, Nagata and Matsuda illustrated the possibility of infiltrating preforms having initial temperatures below the metal liquidus (Nagata and Matsuda 1981). Fukunaga and Goda proposed an explanation for the case of pure aluminum: a layer of solidified metal forms around the fibers, allowing for flow of liquid metal around it (Fukunaga and Goda 1984). More recently, the role and interaction of transport phenomena influencing infiltration kinetics and composite microstructure have been investigated and analyzed quantitatively (Lacoste and Danis 1991; Lacoste et al. 1991; Mortensen et al. 1989; Mortensen and Michaud 1991; Mortensen and Wong 1990).

The basic governing equations of these continuum models in general form will now be discussed. V_f, V_p and V_m are the local volume fraction of reinforcement, pores, and metal, respectively, such that $V_f + V_p + V_m = 1$. If a fraction of the metal g_s solidifies during infiltration, and V_{sf} represents the local fraction solid in the composite, including reinforcement and solidified metal, then $V_{sf} = V_f + g_s V_m$. For a given alloy and reinforcement, the following laws govern.

1.1.2.1 Fluid Flow

Flow of liquid metal through ΔV generally follows the Forscheimer equation:

$$f - \nabla P = [\mu V_m (1 - g_s) K^{-1} + B \rho_m \sqrt{(v_l - v_s)^2}](v_l - v_s) \quad (1.1)$$

where f is the local volumetric value of the gravitational, centrifugal, or electromagnetic force applied to ΔV, P the average pressure in ΔV, and μ the viscosity of the liquid metal, which is a function of the metal temperature and composition. K, the local permeability tensor of the preform, and B are functions of V_m and of the morphology and volume fraction of the solidified metal (Beavers and Sparrow 1969; Coulaud et al. 1988). ρ_m is the metal density, assumed here to be the same in the solid and liquid state. When the relevant Reynolds number is below a critical value near one:

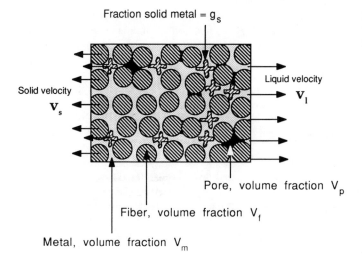

Figure 1.2. Schematic representation of a volume element ΔV, containing liquid metal, solid metal, fibers, and pores.

$$R_e = \frac{d\rho_m \sqrt{(v_l-v_s)^2}}{\mu V_f} \leq R_{e,\text{critical}} \quad (1.2)$$

where d is a characteristic length of the reinforcement (for example the fiber diameter), the Forscheimer equation reduces to D'Arcy's law:

$$v_l - v_s = \frac{K}{\mu V_m (1-g_s)}(f - \nabla P) \quad (1.3)$$

This law applies for analysis of most infiltration processes. There are, however, processes in which R_e is higher than R_{ec} (Andrews and Mortensen 1991).

Continuity of matter dictates:

$$\frac{\partial V_f}{\partial t} = -\nabla \cdot V_f v_s \quad (1.4)$$

$$g_s \frac{\partial (V_m)}{\partial t} = -\nabla \cdot (g_s V_m v_s) \quad (1.5)$$

$$(1-g_s)\frac{\partial (V_m)}{\partial t} = -\nabla \cdot ((1-g_s)V_m v_l) \quad (1.6)$$

The amount of metal V_m in the volume element is a function of current and previous values of pressure P in the metal, and of pressure P_g in the pores, if any. P_g is generally negligible, but, in squeeze-casting processes, build-up of gases ahead of the preform during infiltration can lead to non-negligible values of P_g, causing large defects when pressure is released (Chadwick 1991; Young 1991). It has been observed that the infiltration front is generally not sharp: a nonwetting metal fills the bigger pores first, and infiltration occurs over a range of pressures. The relationship between V_m and the pressure can be measured with drainage/imbibition curves of the preform relevant to the local value of V_f, as in soil science or reservoir engineering (Kaufmann and Mortensen 1992; Mortensen 1991).

If compression of the preform occurs during infiltration, V_f, V_m, and K become functions of both the local stress and the prior stress-strain history of the preform. If inertial terms are neglected, the pressure in the liquid P and the preform stress σ are linked by equations of mechanical equilibrium (Biot 1955; Sommer 1992). The presence of solid metal during infiltration and the presence of friction between the composite and the mold wall may also influence the stress-strain behavior of the porous medium (Jarry et al. 1990; Jarry et al. 1992; Sommer 1992).

1.1.2.2 Heat Transfer

As mentioned previously, it is possible to infiltrate a preform initially at a temperature below the matrix liquidus temperature. This is desirable for several reasons: (1) to minimize matrix-reinforcement chemical interaction, (2) to avoid metal leakage from the mold parting lines, or (3) to obtain smaller matrix grain sizes. A fraction of the metal then solidifies when it contacts the preform, thus releasing heat that brings the system temperature to a value where solid and liquid metal coexist, and where metal can flow within the porous medium constituted by the reinforcement and the solid metal. It follows that for an adiabatic system, flow could continue indefinitely, and that in practical cases, flow stops because of cooling from the mold walls, or eventually stops when the metal has filled the mold. Heat transfer within the volume element is thus governed by conduction and convection, as well as by local changes in fraction solid metal g_s and the amount of heat released by chemical reaction

$$\nabla \cdot (k_c \nabla T) = \rho_c c_c \frac{\partial T}{\partial t} + (\rho_m c_m V_m (1-g_s))v_l \cdot \nabla T$$
$$+ (\rho_f c_f V_f + \rho_m c_m V_m g_s)v_s \cdot \nabla T$$
$$- \rho_m \Delta H \frac{\partial (g_s V_m)}{\partial t} - \dot{Q} \quad (1.7)$$

where k_c is the thermal conductivity tensor of the composite, ΔH the latent heat of solidification of the alloy, and $\dot{Q}$ the rate of heat generation from chemical reaction. ρ and c refer to density and heat capacity respectively; the subscripts m, f, and c refer to metal, reinforcement, and composite, respectively. The heat capacity of the gas phase is assumed to be zero.

1.1.2.3 Mass Transfer

Mass flow in and out of ΔV by diffusion is neglected by comparison to convection. Mass transfer can occur from two different effects: (1) as some metal solidifies, solute may be rejected and entrained by convection, or (2) if a chemical reaction takes place, solute may be consumed or rejected. The governing equation for mass transport is then

$$\frac{\partial \overline{C}}{\partial t} = -\nabla \cdot [(1-g_s)C_L v_l + g_s C_s v] + r_A \quad (1.8)$$

where $\overline{C}$ is the local average matrix content in solute A, C_L and C_s are respectively the average solute content of the liquid and the solid phases, and r_A is the rate of change in solute resulting from chemical reaction. C_L and C_s are linked by the phase diagram and are functions of the local temperature T, and r_A is a function of the reaction kinetics, C_L and/or C_s.

1.1.2.4 Initial and Boundary Conditions

In addition to these equations, the infiltration process is governed by initial and boundary conditions. The initial conditions are imposed by the process parameters, which include the preform, mold and metal initial temperatures, the initial volume fraction of reinforcement, and the applied pressure or the infiltration velocity. Conditions at the boundary between the composite system and its surroundings, or between various regions that may appear within the composite, are derived from physical considerations including continuity, and heat and mass conservation (Mortensen et al. 1989; Mortensen and Michaud 1990).

Modeling of these overall transport processes requires a simultaneous solution of the above equations, taking into account the initial and boundary conditions for each specific case. The solutions for a generalized case are highly complex and necessitate the use of numerical methods. However, it is possible to gain insight on physical phenomena through analytical or numerical studies of simplified systems and geometries. Such studies have contributed to an increased understanding of features of significant practical importance such as macrosegregation resulting from solidification effects (Jarry et al. 1992; Michaud and Mortensen 1992; Mortensen and Wong 1990), chemical reaction between reinforcement and matrix (LePetitcorps et al. 1991; Takenaka and Fukazawa 1988; Westengen et al. 1990), composite microstructure (Masur et al. 1989; Michaud and Mortensen 1992; Mortensen 1989; Mortensen and Wong 1990), and preform compression (Sommer 1992). In this manner, not only can investigators correlate practical observations with theory, they can also predict and manipulate possible system behavior.

1.1.3 Solidification

The mechanical properties of any given reinforcement-matrix system are highly dependent on the matrix microstructure. The latter, in turn, is dictated by solidification in the presence of a fixed reinforcement, which may initiate during infiltration or after its completion. The rules developed for microstructural control in the solidification of unreinforced metal are generally not directly applicable to metal-matrix composites because the reinforcing phase influences matrix solidification. Several factors need to be considered.

1.1.3.1 Reinforcement Nature and Initial Temperature

If the reinforcement is a nucleation catalyst for the metal, the matrix grain size may be reduced. Examples of matrix primary-phase grain refinement include aluminum in the presence of porous TiC (Mortensen et al. 1986), or hypereutectic Al-Si alloys in the presence of SiC, silica, and alumina (Mortensen 1990).

However, in most cases of engineering interest, the fibers do not catalyze matrix nucleation. Larger columnar structures may then develop because liquid convection is reduced by the presence of a reinforcement (Cole and Bolling 1965; Jarry et al. 1992). When the preform initial temperature is below the metal liquidus, regions where solid metal nucleated as a result of rapid cooling by the fibers without remelting feature a fine grain size—on the order of 50 μm for Al-4.5%Cu in the Saffil (Imperial Chemical Industries, Runcorn, U.K.) alumina preforms presented in Figure 1.3a—even if the fibers do not act as nucleating agents. Conversely, large grains are found in alumina-fiber-reinforced aluminum alloys in regions where the solid phase was remelted by incoming superheated metal and solidified after infiltration ceased (see Figure 1.3b) (Clyne and Mason 1987; Jarry et al. 1992; Michaud and Mortensen 1992).

1.1.3.2 Interfiber Spacing

The reinforcement acts as a barrier to mass transfer during solidification. This causes an alloyed matrix to grow in avoidance of the fibers because the fibers present a barrier to solute diffusion. Thus, in most composites, the last phase to solidify is found in the vicinity of the reinforcement surface, as illustrated in Figure 1.4. Fundamental studies have used steady-state solidification experiments, in which alloys reinforced with continuous, parallel fibers are melted and resolidified in a Bridgman furnace along the fiber direction. The growth parameters are then simply defined as temperature gradient G and growth rate V, which are independently controlled in the experiments. If the interfiber spacing is smaller or comparable to the size of the solidification structure found in the unreinforced metal, the transitions from plane front to cellular growth, and from cellular to dendritic growth, are shifted (Dean et al. 1991; Sekhar and Trivedi 1989). Plane-front stabilization, observed in reinforced Al-Cu alloys, arises because the allowed critical perturbation wavelengths λ_c are limited by the interfiber spacing. Near the cellular/dendritic transition regime of the unreinforced alloy, complicated morphologies arise in the

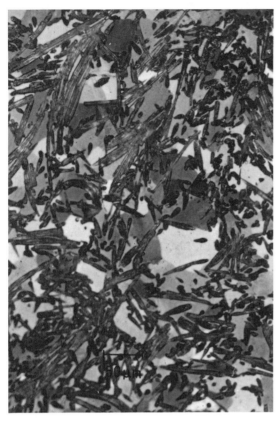
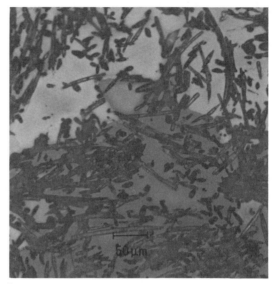

Figure 1.3. Saffil fiber (Imperial Chemical Industries, Runcorn, U.K.) reinforced Al-4.5wt%Cu that was produced by infiltration of preforms initially at a temperature below the matrix liquidus temperature. (a) By cooling the metal by the fibers during infiltration, a fine matrix grain size resulted, as revealed by Barker's etch. (b) Near the infiltration gate, where the solid metal remelted because of incoming superheated metal, large grains are found *(Courtesy of V. Michaud and A. Mortensen, MIT)*.

composite as a result of the presence of the fibers (Sekhar and Trivedi 1989; Shangguan and Hunt 1991). When solidification is dendritic in the unreinforced alloy, ripening of dendrite arms ceases in the composite when the arm spacing reaches the fiber spacing; from then on, further coarsening of the structure occurs by coalescence. Then, if the solidification time exceeds the time for full coalescence of secondary dendrite arms, the fully solidified matrix appears cellular, having concentric coring patterns parallel to the matrix-reinforcement interface (Mortensen et al. 1988). Matrix microsegregation is also strongly affected by the reinforcement, which places an upper limit on diffusion distances in the solid phase during solidification. Microsegregation can thus be reduced by solid-state diffusion during solidification to a much greater extent than that which occurs in unreinforced metals (Dean et al. 1991; Mortensen et al. 1988).

1.1.3.3 Applied Pressure

If high pressure is applied during solidification, the matrix exhibits a finer grain size than would be obtained at atmospheric pressure. This is because of an increase in the heat-transfer coefficient between mold and composite, and because of an increase in the matrix liquidus temperature (up to 10°C for Al at 100 MPa), which results in a larger undercooling (Bhagat 1991; Das et al. 1988).

1.2 Dispersion Processes

1.2.1 Description

In dispersion processes, schematically represented in Figure 1.5, the reinforcement is incorporated in loose

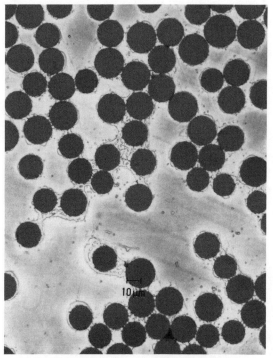

Figure 1.4. Alumina FP fibers (E.I. duPont de Nemours, Wilmington, DE) infiltrated by Al-4.5wt% Cu. Keller's etch reveals the second phase, located preferentially in the vicinity of the fibers *(Courtesy of P. Bystricky and A. Mortensen, MIT)*.

form into the metal matrix. Because most metal-reinforcement systems exhibit poor wetting, mechanical force is required to combine the phases, generally through stirring. This method is currently the most inexpensive manner in which to produce MMCs, and lends itself to production of large quantities of material, which can be further processed via casting or extrusion. The simplest dispersion process in current use is the Vortex method, which consists of vigorous stirring of the liquid metal and the addition of particles in the vortex (Rohatgi et al. 1986). Skibo and Schuster have patented a process for mixing SiC particulates in molten aluminum under vacuum, with a specially designed impeller (Skibo and Schuster 1988a; Skibo and Schuster 1988b) that has the advantage of limiting the incorporation of impurities, oxides, or gases because of the vacuum and reduced vortexing. Ingots of such composites are now commercially produced in large quantities by Duralcan, San Diego, CA (Hoover 1991). Other methods being investigated include the bottom-mixing process, where a rotating blade is progressively lowered into an evacuated bed of particles covered with molten aluminum (Caron and Masounave 1990), and the injection of particles below the surface of the melt using a carrier gas (Hansen et al. 1989).

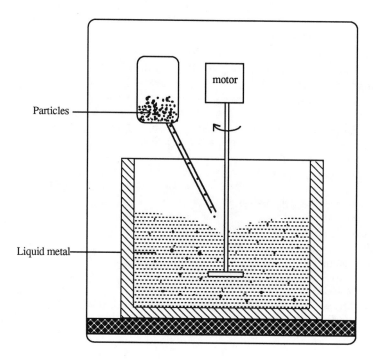

Figure 1.5. Schematic representation of a dispersion process.

Mixing of particles and metal can also be achieved while the alloyed metal is kept between solidus and liquidus temperatures. This process, known as Compocasting, is an outgrowth of the Rheocasting processes for unreinforced semisolid slurries (Balasubramanian et al. 1990; Bayoumi and Suéry 1988; Bayoumi and Suéry 1987; Cornie et al. 1990; Girot et al. 1987; Hosking et al. 1982; Levi et al. 1977; Mehrabian et al. 1973; Millière and Suéry 1985). The advantage of using semisolid metal is the increase in the apparent viscosity of the slurry, and the prevention of the buoyant migration of particles by the solid metal. Another proposed method to mix semisolid metal and particulates is the Thixomolding process, whereby metal pellets and particles are extruded through an injection-molding apparatus (Carnahan et al. 1990).

Critical to the success of dispersion processes is the control over generally undesirable features such as porosity resulting from gas entrapment during mixing, oxide inclusions, reaction between reinforcement and metal favored by long contact times, as well as particle migration and clustering during and after mixing. Furthermore, such processes are not suited for long fibers or oriented reinforcement because of the difficulties in stirring and the need for secondary deformation processing to improve the reinforcement distribution and close any pores.

1.2.2 Governing Phenomena

The behavior of particles suspended in a liquid is relevant to many branches of engineering and has been analyzed by several authors (for example, Papanicolaou 1987). Studies directly concerned with the processing of MMCs have treated the incorporation of particles into the melt, the fluid-flow properties of the solid/liquid mix, and finally the issue of particle interaction and migration.

1.2.2.1 Particle Incorporation

Generally, it is not energetically favorable for a ceramic particle to replace the particle/air interface with a particle/molten-metal interface, because of poor wetting. As a result, it is necessary to apply force on the particle to overcome the surface tension barrier as well as, in some cases, to break the oxide layer on the surface of the metal. A number of authors have addressed this issue for the case of particle engulfment in a stationary fluid (Cornie et al. 1990; Rohatgi et al. 1990a). Adding the important effect of melt velocity, the force balance in the vertical direction on a particle partially immersed in a stirred melt is given by Equation 1.9 (Ilegbusi and Szekely 1988).

$$\Sigma F = F_B + F_G + F_H + F_S + F_E \quad (1.9)$$

where F_B is the buoyancy force on the immersed part, F_G is the gravity on the whole particle, F_H is the hydrostatic pressure of the liquid directly above the contact area, F_S is the vertical component of the surface tension at the solid liquid interface, and F_E is a drag force caused by motion in the fluid. The resulting sign of this algebraic sum indicates whether the particle will sink or float. This analysis is simplified because it does not take into account the interaction between particles, the effect of their shape (which is often angular), or the presence of a thick oxide layer at the surface. Results for electromagnetically stirred melts do agree though with experimental observations that small particles are more difficult to engulf, and that when the ceramic does not wet the liquid, impractically high melt velocities would be required to engulf the particles.

According to Skibo and Schuster, particles can be engulfed in the metal and wetted if high shear is applied at the interface between particles and liquid metal. The shearing action helps break oxides present near the particle surface and spreads the metal onto the particle surface, while the global mixing effects a homogeneous distribution of particles in the melt (Skibo and Schuster 1988a; Skibo and Schuster 1988b). Best results are obtained when no gases are introduced into the melt with the particles, that is, when the vortex at the surface of the melt is minimized, and/or when a vacuum is maintained above the melt.

1.2.2.2 Fluid Flow

Unreinforced liquid metals generally have a viscosity of about 0.01 Poise, similar to water, and behave like newtonian fluids, that is, their viscosity is independent of shear rate, and decreases with temperature. Adding particles to a liquid metal increases its apparent viscosity, because the particles interact with the matrix and with each other, thus creating more resistance to shear stress. This has been reported in a number of experimental studies (Cornie et al. 1990; Flemings 1991). Typical values of viscosity are in the range of 10 to 20 Poise for Al alloy reinforced with 15 vol % SiC particles, as seen in Figure 1.6 (Lloyd 1991). Evidently, the viscosity is a function of the reinforcement-volume fraction, shape, and size. An increase in volume fraction or a decrease in size will increase the viscosity of the slurry. This limits the practically achievable amount of reinforcement to about 30 vol %. The metal temperature is also a factor, and dispersion mixing is effected in some processes while the alloy temperature is between liquidus and solidus (Flemings 1991). In both cases, that is, above liquidus temperature or between liquidus and solidus temperature, the metal-particulate slurries exhibit a non-newtonian behavior. Their viscosity decreases up to an order of magnitude as the shear rate increases (see Figure 1.6). Moreover, it has been experimentally observed that composite slurries are thixotropic, meaning that when the shear rate is abruptly changed, the viscosity of the slurry changes only progressively to reach the steady-state value of the new shear rate. Such behavior has been reported by Moon on SiC/Al-6.5 wt % Si (Moon et al. 1991) in both semisolid and liquid metal cases, and by Mada and Ajersch (1990) on similar systems in the semisolid state only, for a smaller range of shear rates. The mechanisms underlying the pseudoplastic behavior of composite slurries are not yet clearly understood. The existence of such behavior in slurries above the metal liquidus suggests that clustering and declustering of particles plays a major role. Interfacial chemical reaction between reinforcement and metal has also been reported to increase the slurry viscosity, apparently because the reaction products have a different density, break loose, or change the morphology of the reinforcement (Lloyd 1991).

In the semisolid metal temperature range, the behavior of composite slurries is generally explained on the same basis as is their unreinforced counterparts. That is, there is a change in morphology of the solid phase under shear as a result of dendrite fragmentation, ripening and abrasion, and collision and coalescence of solid particles (Flemings 1991). Mada and Ajersch (1990) developed an analytical model of thixotropic behavior in semisolid composite slurries, based on an analogy between rate constants for chemical reactions and the rates of dissociation and formation of solid aggregates. From their analysis and experimental results, they concluded that particle addition has little influence on the thixotropic nature of the melt, and that the primary metal solid phase is the major factor. Yet, an effect of particle addition on the pseudoplastic behavior has been observed by Moon (Moon et al. 1991); at a given shear rate, the viscosity of slurries composed of SiC particles combined with Al-6.5 wt % Si in the semisolid range was lower than that of unreinforced semisolid slurry of the same total volume fraction solid. This result is tentatively explained by the preferential location of reinforcement particles between dendrites arms, limiting contact and agglomeration of dendritic solid particles.

1.2.2.3 Particle Migration

After mixing, the semisolid composite may be at rest before complete solidification. The issue of particle migration because of gravity differences between metal and reinforcement then arises. For high volume fractions, the sedimentation is influenced by interactions between particles and by their size distribution (Lloyd 1991).

When solidification takes place, particle migration caused by solidification effects competes with migration caused by gravity. Understanding and controlling the interaction between growing solid metal and the particles is crucial to producing homogeneously reinforced composites. Thus, when encountering a moving liquid/solid interface, particles may be engulfed in the solid metal, or they may be pushed by the interface and consequently migrate into areas that solidified last, for example, interdendritic regions as illustrated in Figure 1.7. Work on this issue has recently been reviewed by several authors (Mortensen and Jin 1992; Rohatgi et al. 1990b; Stefanescu and Dhindaw 1988). From experimental observations, some particle/metal systems feature a critical interface velocity, V_c, below which the particles are pushed and above which they are trapped. The value of V_c depends on a number of factors, including particle size, shape and composition, metal composition, differences in thermal conductivity between particle and metal, and cooling rate. Several models have been developed to quantify particle-pushing phenomena, but they are somewhat difficult to compare with experiments because the parameters do not all vary independently, and the castings generally do not solidify at steady-state.

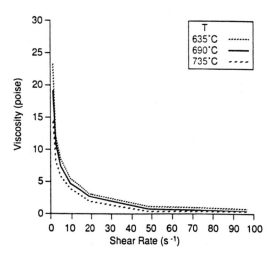

Figure 1.6. Influence of shear rate on the viscosity of A-356–15vol% SiC. *(Reprinted by permission from D. J. Lloyd, 1991.)*

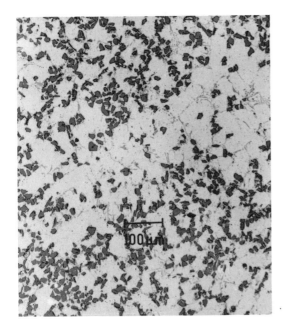

Figure 1.7. Silicon carbide particle reinforced A356, $V_f = 15\%$, produced by Duralcan (San Diego, CA). The SiC particles have been pushed during solidification and have accumulated in the last pools of liquid metal, rich in eutectic phase.

1.3 Spray Processes

1.3.1 Description

In these processes, schematically represented in Figure 1.8, droplets of molten metal are sprayed together with the reinforcing phase and collected on a substrate where metal solidification is completed. Alternatively, the reinforcement may be placed on the substrate, and molten metal may be sprayed onto it. Spray-forming processes have been utilized for unreinforced metal for more than 20 years because the high solidification rate of the droplets yields materials with little segregation and a refined grain structure (Mathur et al. 1989b; Singer 1991). The various composite spray processes differ in the method of spraying the metal, and differ in the (often proprietary) way the reinforcement is mixed with the metal. The critical parameters in spray processing are the initial temperature, size distribution and velocity of the metal drops, the velocity, temperature, and feeding rate of the reinforcement (if it is simultaneously injected), and the position, nature, and temperature of the substrate collecting the material. Most spray-deposition processes use gases to atomize the molten metal into fine droplets (generally up to 300 μm diameter). The particles can be injected within the droplet stream or between the liquid stream and the atomizing gas, as in the Osprey process, developed by Alcan International Ltd, Banbury, U.K. (Singer 1991; White et al. 1988; Willis et al. 1987). The Osprey process has been successfully scaled-up for the production of 100 kg ingots of SiC reinforced Al alloys. Similar techniques are under investigation to incorporate Al_2O_3, SiC, or graphite particles into Al alloys (Gupta et al. 1991; Gupta et al. 1990; Jensen and Kahl 1991; Namai et al. 1986; Nayim and Baram 1991; Zhang et al. 1991). Other methods to produce molten-metal drops from aluminum wires have also been developed. These include arc melting (Buhrmaster et al. 1988), flame spraying (Tsunekawa et al. 1987), or a combination of both (Tiwari and Herman 1991). For high-temperature materials, plasma torches are used to melt and spray metal powders: Ni_3Al-reinforced TiB_2 particles have been produced by this method (Tiwari et al. 1991), as well as various low- and high-temperature matrices such as, Al alloys and Ti_3Al or $MoSi_2$, reinforced with SiC, TiC, or TiB_2 particles (Gungor et al. 1991). Monotapes of continuous fibers and Ti-based matrices are also prepared by plasma spraying, to be further processed in the solid state, for example by diffusion bonding (Backman 1990).

One advantage of spray-deposition techniques resides in the resulting matrix microstructures that feature fine grain size and low segregation. Also, because liquid metal and reinforcement contact only briefly, interfacial reaction is minimized, allowing the production of thermodynamically metastable two-phase materials such as iron particles in aluminum alloys (Singer 1991). Simple forms, such as ingots or tubes, can be produced by these processes, although they lack in part-shape versatility. Other drawbacks include the amount of residual porosity (at least a few percent by volume) and the resulting need to further process the materials. Also, this process is not as economical as dispersion or infiltration processes, because of the high cost of the gases used and the large amounts of waste powder to be collected and disposed.

1.3.2 Governing Phenomena

The spray-deposition process for unreinforced metals has recently been analyzed by various authors (Gutierrez-Miravete et al. 1989; Mathur et al. 1989b). Analysis is generally performed in two steps that correspond to the two phases of the process: flight of the atomized metal droplets and deposition on the substrate. The addition of particles sprayed with the metal considerably increases the complexity of the analysis, and current models are based on modifications of models for unreinforced sprays, taking into account the

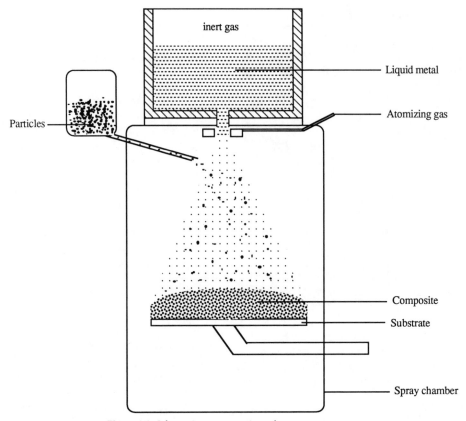

Figure 1.8. Schematic representation of a spray process.

global enthalpy changes caused by the particles (Gupta et al. 1992; Gupta et al. 1991).

1.3.2.1 Transport Phenomena During Flight

The first step in the analysis of such a process is to determine the metal drop size distribution which is either measured experimentally (Mathur et al. 1989b) or calculated using semiempirical correlations (Gutierrez-Miravete et al. 1989). Then, assuming a linear trajectory and no interactions between drops, the acceleration of a given drop of diameter d can be calculated using a simple force balance,

$$F = m\frac{dV_d}{dt}$$
$$= mg + C_d \rho_g (V_g - V_d)|V_g - V_d|\frac{\pi d^2}{8} \quad (1.10)$$

where V_d is the drop velocity, V_g the gas velocity, m the drop mass, g the acceleration due to gravity, C_d the drag coefficient, ρ_g the gas density. The drop velocity is then numerically calculated as a function of position from this equation. Heat transfer to the surrounding gas during flight is mainly by convection and radiation, and is generally assumed to be interface controlled (Biot number < 0.1), allowing the assignment of a uniform temperature to each drop. Depending on the degree of undercooling upon nucleation in the drops, a temperature rise may result from recalescence.

When particles are incorporated, heat transfer between the drops and the particles needs to be taken into account. This is rather difficult because it is necessary to know if the particles penetrate into the metal drops, stick to their surface, or do not contact them at all. The particle sizes, the position from which they are injected, and surface characteristics of both metal drops and particles are key factors. This problem has been raised by several authors, and experimental evidence indicates that particles may surround the droplets during flight (Gupta et al. 1991; Namai et al. 1986).

1.3.2.2 Transport Phenomena After Deposition

Once the thermal history for each drop size and the size distribution are established, droplet/substrate interactions are considered. During deposition, the temperature in the composite is not uniform. The temperature distribution can be estimated from a global heat balance taking into account the average heat content of the spray at the deposit surface, the deposition rate, the rate of heat exchange with the gas at the top surface of the deposit, the rate of heat extraction at the substrate/deposit interface, and the thermal properties of the deposited substrate. If the deposition rate is higher than the solidification rate, liquid metal may be present at the surface of the substrate.

1.3.2.3 Solidification

Results of calculations for the unreinforced metal are in agreement with the temperature profiles in the deposit, and provide useful tools for prediction of the composite microstructure. Important to determine is whether the surface of the substrate is fully solid, or if a partially or totally liquid layer develops. It seems that, in most cases, a partially liquid layer would be favorable to minimize porosity, but may result in partial loss of the rapidly solidified structures that make this process attractive. However, grain sizes observed on the final product are generally lower than predicted by theory, possibly because dendrites that broke during impact on the substrate create new nucleation sites (Gutierrez-Miravete et al. 1989). A further reduction in grain size is observed in composites; the grain size is about 30% less than that of unreinforced alloys produced in the same conditions (from 30 μm to about 20 μm for 6061 matrix when reinforced with about 10% of 3 μm diameter SiC particles). This additional effect is in part explained by the influence of thermal transfer from the particles to the metal on the droplets' solidification behavior (Gupta et al. 1992).

1.4 In-Situ Processes

1.4.1 Description

The term *in-situ composites* was first used for materials produced by solidification of polyphase alloys. When polyphase alloys solidify directionally with a plane front, they may exhibit a fine lamellar or rod-like structure of β phase in an α phase matrix, and the interphase spacing is a function of the growth rate (Flemings 1974). Extensive work has been done in this field since the 1960s, and is reviewed in depth by Elliott (1983). Applications have been proposed for these materials in the areas of optics and electronics, but their generally low growth rates, and problems resulting from the gradual coarsening of the structure at elevated temperatures, restricted their possible use. The recent emphasis on processing of high-temperature materials, however, has given a new impetus to research on in-situ composites, because reinforced intermetallic alloys may be produced by controlled solidification or by other in-situ processes such as chemical reaction between a melt and solid or gaseous phases. Schematic examples of in-situ processes are given in Figure 1.9.

Some examples of composites obtained from alloy solidification include: eutectic systems based on Cr and Ta (Mazdiyasni and Miracle 1990), Fe-TiC composites produced from solidification of Fe-Ti-C melts (Chambers et al. 1987; Kattamis and Suganuma 1990; Terry and Chinyamakobvu 1991), TiB rods in TiAl matrices from solidification of melts containing γ-TiAl, Ta, and B (Valencia et al. 1991), as well as TiC/Ti composites from mixtures of Ti, and C with Al additions (Chen et al. 1989). A chemical reaction between a liquid metal and a solid metal or ceramic can also yield suitable composites such as: TiB_2-reinforced Al alloys made by the XD process, patented by Martin Marietta Laboratories, Baltimore, MD, whereby Ti, B, and Al powders heated to 800°C react to form TiB_2 (Christodoulou et al. 1986; Westwood 1988), TiB whiskers obtained after laser melting of Ti and ZrB_2 powders (Thompson and Nardone 1991), or TiAl matrix obtained after squeeze-casting of molten aluminum into TiO_2 powders or short fibers (Fukunaga et al. 1990).

Another way to produce in-situ composites is to react molten metal with a gas. Using a process similar to Lanxide's (Newark, DE) DIMOX process for ceramic matrix composites, Al_2O_3/Al composites have been formed by oxidation of Al (Xiao and Derby 1991), and TiC-reinforced Al-Cu alloys have been obtained by bubbling CH_4 and Ar gas through a melt of Al-Cu-Ti. This gas injection method can be used to produce a wide array of carbide- and nitride-reinforced alloys (Koczak and Sahoo 1991; Sahoo and Koczak 1991).

A major advantage of in-situ composite materials is that the reinforcing phase is generally homogeneously distributed, and spacing or size of the reinforcement may be adjusted in several cases by the solidification or reaction time (Kattamis and Suganuma 1990). However, the choice of systems and the orientation of the reinforcement are limited, and the kinetics of the processes (in the case of reactions), or the shape of the reinforcing phases, are sometimes difficult to control.

1.4.2 Governing Phenomena

In-situ composites can be produced using a wide range of different processes, as described above, and it is difficult to underline governing phenomena that are

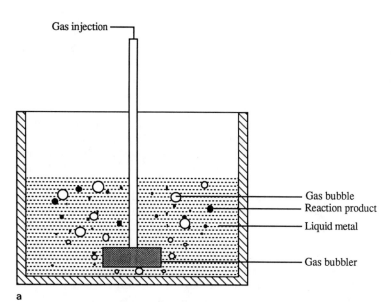

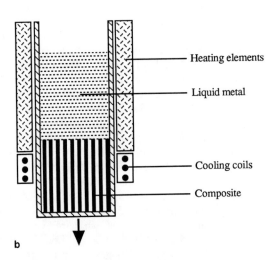

Figure 1.9. Schematic representation of in-situ processes. (a) Reaction between a gas and a liquid metal (*from Koczak and Sahoo 1991*). (b) Directional solidification of a eutectic alloy.

common to all these processes. We can, however, distinguish between composites obtained from solidification of a melt and those obtained from chemical reaction between phases.

1.4.2.1 Solidification

Complete treatment of the mechanisms governing polyphase alloys solidification can be found in Elliott 1983; Flemings 1974; and Kurz and Fisher 1989. Generally, for a binary eutectic alloy, the spacing λ between rods or lamellae is proportional to $R^{-1/2}$, where R is the cooling rate. The volume fraction of the reinforcing phase is given by the alloy composition. From surface energy considerations, it appears that rod-like morphology is favored for low-volume fraction of one phase (theoretically $V_f < 1/\pi$ if the surface energy is isotropic).

For alloys containing three or more constituents, similar basic rules hold, for example, the reinforcement spacing decreases with increasing cooling rate. However, a wide array of reinforcement shapes and compositions can be obtained (Kattamis and Suganuma 1990), and additional alloying elements may favor the formation of one reinforcement species over another; for example, solidification of a γ-TiAl/B melt produces TiB_2 equiaxed particles, whereas additions of Nb or Ta to the same melt favor the formation of TiB rods (Valencia et al. 1991).

The aging of in-situ composites produced by solidification from a melt is a critical issue; these materials are intrinsically stable because they are produced under near-equilibrium conditions and generally feature low-energy interphase boundaries. However, when exposed to temperature changes or gradients, coarsening occurs, that is, the rods tend to thicken and shorten, thus affecting the mechanical properties of the material. The mechanisms responsible for coarsening include oxidation and interfacial or bulk diffusion caused by a change in the solubility of the phases, and occur preferentially at grain boundaries or faults in the structure (Elliott 1983).

1.4.2.2 Chemical Reactions

In the case of in-situ composites produced by reaction between a liquid and other phases, such as gas or solid, the governing mechanisms are those that are encountered in chemical reactions. Their elucidation involves identification of the possible reactions taking place and evaluation of the driving force derived from thermodynamic considerations and reaction kinetics. These are functions of temperature and alloy, gas or solid compositions and concentrations, as well as diffusion mechanisms across reaction or boundary layers, as detailed for example, by Sahoo and Koczak (1991). Some of these reactions are highly exothermic, including the XD process (Marietta Laboratories, Baltimore, MD) (Westwood 1988; Merzhanov 1990), and present the advantage of being fast and self-propagating.

1.5 Optimization of Liquid-Metal Processing

1.5.1 Reinforcement

A number of problems associated with liquid-state processes arise from the reinforcement; consequently, attempts to improve processes need to include optimization of the reinforcing-phase architecture and its behavior during combination with the metal. (Modification of the reinforcement surface to improve wetting or alter reaction mechanisms are addressed in Chapter 3.)

In infiltration processes, a preform of the reinforcing phase is often prepared prior to infiltration, for example by dispersing the reinforcement in a solution containing an inorganic binder, pressing to the desired volume fraction, and drying the cake composed of fibers and residual binder material (Baty et al. 1987; Ebisawa et al. 1991). When short fibers are used, fiber alignment is difficult to achieve, unless the composite is extruded, which generally damages the fibers. An innovative approach has been proposed by Itoh and coworkers to solve this problem: a nonconductive fluid, containing the fibers and a surfactant, is subjected to a DC electric field; the fibers become polarized and thus orient themselves along the field lines (Itoh et al. 1988; Masuda and Itoh 1989). For in-situ composites, remelting and directional solidification allows for alignment of the reinforcement in initially randomly oriented XD material (Kampe et al. 1990).

Gradients in volume fraction reinforcement resulting from preform compression during infiltration, as mentioned in Section 1.1.1.3., can be minimized based on an understanding of the process governing phenomena. For example, if the compressive strength of the preform is known, pressure (or infiltration velocity) can be adjusted during the infiltration phase in order to avoid preform compression, and then raised again later during solidification to obtain the desired microstructure (Booth et al. 1983).

To alter the often time-consuming and delicate stage of preform fabrication, several approaches have been investigated. Foams constituting a three-dimensional network of ceramic have recently been used for liquid-metal infiltration (Figure 1.10) (Battelle Advanced Materials 1990; Fitzgerald and Mortensen 1992). Another type of three-dimensional network was proposed by Lange in which slurries of alumina particles are infiltrated into a pyrolizable phase and subsequently sintered after pyrolyzation of the initial phase (Lange et al. 1988). If loose fibers or particles are used, hybridization allows the minimization of contact points and a reduction of the volume fraction. For example, coating continuous SiC fibers with a small amount of fine SiC particulates or whiskers prevents the loose fibers from touching during infiltration. The resulting hybrid composites feature increased longitudinal and transverse strength (Feest et al. 1989; Towata and Yamada 1986). In a similar fashion, adding some fibers when compacting a particle preform allows for reduction of the volume-fraction reinforcement because fibers do not pack as closely as particles (Friend et al. 1991). Another way to produce a particle-reinforced composite with a lower volume fraction combines infiltration and dispersion processes. A dense compact is first infiltrated under pressure, then diluted into liquid metal to achieve the desired volume-fraction reinforcement (Klier et al. 1991). If well controlled, the inhomogeneity of the reinforcement distribution can also be used to advantage: Al-reinforced alumina fibers or ceramic particles composites have been produced by centrifugal casting and feature increased wear resistance at the surface, where it is most needed (Suéry and LaJoye 1990; Tsunekawa et al. 1988).

It seems, however, that a main obstacle to a wider use of many metal-matrix composite materials is the price of the reinforcement. Continuous fiber-reinforced composites successfully compete with other materials only

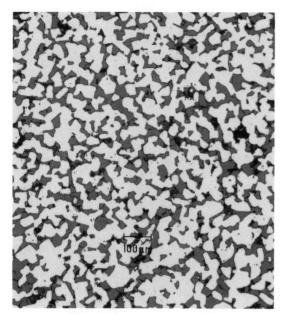

Figure 1.10. Tridimensional network of silicon carbide foam infiltrated with copper. *(Courtesy of T. Fitzgerald and A. Mortensen, MIT.)*

when cost is not relevant, whereas particles or chopped-fiber reinforced alloys have found larger-scale commercial applications (Donomoto et al. 1983; Ebisawa et al. 1991; Hoover 1991; Hoover 1990; Klimovicz 1990). Efforts are still needed to improve the fabrication of high-quality fibers and preforms. Conversely, inexpensive reinforcements that may not produce excellent structural materials, but feature low weight, good wear resistance, or noise damping, have been proposed. An example of such material is red mud, a by-product of aluminum fabrication (Sing Solanki et al. 1991).

1.5.2 Matrix

Porosity may be found in the matrix of materials produced by any of the liquid-metal processes, but can be minimized in several cases by a better understanding of the processes. Application of a vacuum and reduction of the vortex during mixing, for example, reduces the amount of residual porosity in dispersion processes. In infiltration, heterogeneity of the matrix composition or microstructure in infiltrated materials may also be reduced through control of the processing parameters. For example, high-speed infiltration of a low-temperature Saffil (Imperial Chemical Industries, RUNCORN, U.K.) δ-alumina preform by molten Al-Cu alloys with very little superheat results in a fine-grained structure and concentration of macro-segregation in a small region (Michaud and Mortensen 1992; Mortensen and Michaud 1990). New alloys specifically designed for use in MMC fabrication have also been developed to minimize chemical reaction between matrix and reinforcement or to modify the matrix microstructure.

1.5.3 Process Development

The four basic methods presented in this chapter have been modified and combined to create new processes. Plasma spray and in-situ processes have been combined to form metal-ceramic composite by reaction of the metal with the surrounding gas during flight (Mathur et al. 1989a). Combinations of infiltration and in-situ processing have also been investigated (Fukunaga et al. 1990), as well as combination of infiltration and dispersion (Klier et al. 1990; Tank 1990). Other liquid-metal routes have been investigated, including laser melting of a metal surface to locally incorporate a reinforcing phase, or methods closely related to powder metallurgy processes, such as liquid-phase hot pressing and sintering. The processes investigated for MMC production are generally batch processes, which limit the products' size and fabrication speed. Various methods for continuous casting of MMCs have been proposed, adapting squeeze casting (Atsushei 1985) or pressure casting (Clifford and Cook 1989), although both are often impractical because of the need for a pressure vessel. Infiltration using electromagnetic body forces might be used for continuous casting (Andrews and Mortensen 1991). Spray processes could conceivably also be adapted for semicontinuous processes (Mathur et al. 1989a).

In order to minimize the composites' cost, emphasis has been put on the development of net-shape or near-net-shape processes such as squeeze casting to reinforce metals with fibers or whiskers. Conversely, low-cost, particle-reinforced aluminum alloy ingots have become commercially available; they can be remelted and recast using investment-casting or die-casting processes. An advantage of these materials during die casting is their increased viscosity in the molten state as compared with unreinforced alloys, which reduces the turbulence during mold filling and thus the porosity in the casting (Hoover 1991).

1.6 Summary

During the last 10 years, a large research effort has focused on developing and understanding various liquid-state processes for the production of MMCs. As a result, these processes are becoming more reproducible and controllable, and the materials produced feature improved quality. Particle-reinforced composite ingots are now commercially available for casting in a manner similar to that for unreinforced alloys, and fiber-reinforced metal parts are finding more applications. The increasing need to develop structural materials for high-temperature applications creates high hopes for the development of ceramic-reinforced, intermetallic composite produced by infiltration, plasma spraying, or in-situ processes. But, there is still a need to increase the competitiveness of MMCs vis-à-vis more traditional materials. Innovative thinking to develop new processes is essential for achieving this goal, and so is the work still lying ahead to gain a more thorough understanding of the available processes, allowing for consistent production of defect-free composites featuring optimized microstructures.

References

Andrews, R.M., and A. Mortensen. 1991. *Metall. Trans.* 22A:2903–2915.
Atsushei, K. 1985. Japan Patent No. 60-29433, February 14, 1985.
Backman, D. 1990. *J. of Metals.* 7:17–20.
Balasubramanian, P.K., Rao, P.S., B.C. Pai et al. 1990. *Composites Sci. and Techn.* 39:245–259.
Battelle Advanced Materials 1990. Columbus, OH: Battelle Advanced Materials, Advanced Materials Technical Notes.
Baty, D.L., Price, J.P., and B.G. Coleman. 1987. Society of Manufacturing Engineers, EM87-573, SME Technical Paper, 25 pp.
Bayoumi, M.A., and M. Suéry. 1987. *In:* Sixth International Conference on Composite Materials, ICCM 6 (F.L. Matthews, N.C.R. Buskell, J.M. Hodginson and J. Morton, eds.), 2.481-2.490, London: Elsevier Applied Science.
Bayoumi, M.A., and M. Suéry. 1988. *In:* International Symposium on Advances in Cast Reinforced Metal Composites (S.G. Fishman and A.K. Dhingra, eds.), 167–172, Materials Park, OH: ASM International.
Beavers, G.S., and E.M. Sparrow. 1969. *J. Appl. Mech.* 12:711–714.
Bhagat, R.B. 1991. *In:* Metal Matrix Composites: Processing and Interfaces. (R.K. Everett and R.J. Arsenault, eds.), 43–82, Boston: Academic Press.
Biot, M.A. 1955. *J. of Appl. Phys.* 26:182–185.
Booth, S.E., Clifford, A.W., and N.J. Parratt. 1983. U.K. Patent No. GB 2,115,327A, September 7, 1983.
Buhrmaster, C.L., Clark, D.E., and H.B. Smartt. 1988. *J. of Metals.* 11:44–45.
Carnahan, R.D., Decker, R.F., Bradley, N., and P. Frederick. 1990. *In:* Fabrication of Particulates Reinforced Metal Composites (J. Masounave and F.G. Hamel, eds.), 101–105, Materials Park, OH: ASM International.

Caron, S., and J. Masounave. 1990. *In:* Fabrication of Particulates Reinforced Metal Composites (J. Masounave and F.G. Hamel, eds.), 107–113, Materials Park, OH: ASM International.
Chadwick, G.A. 1991. *Mater. Sci. & Eng.* A 135:23–28.
Chambers, B.V., Kattamis, T.Z., Cornie, J.A., and M.C. Flemings. 1987. *In:* Third International Conference on Solidification Processing, 453, London: The Institute of Metals.
Champion, A.R., Krueger, W.H., Hartmann, H.S., and A.K. Dhingra. 1978. *In:* 1978 International Conference on Composite Materials ICCM 2 (B. Noton, R.A. Signorelli, K.N. Street and L.N. Phillips, eds.), 883–904, Warrendale, PA: The Metallurgical Society.
Charbonnier, J., Dermarkar, S., M. Santarini et al. 1988. *In:* International Symposium on Advances in Cast Reinforced Metal Composites (S.G. Fishman and A.K. Dhingra, eds.), 127–132, Materials Park, OH: ASM International.
Chen, J., Geng, Z., and B.A. Chin. 1989. *In:* Materials Research Society Symposium (C.T. Liu, A.I. Taub, N.S. Stoloff and C.C. Koch, eds.), 447–452, Pittsburgh, PA: Materials Research Society.
Christodoulou, L., Nagle, D.C., and J.M. Brupbacher. 1986. International Patent No. WO 86/06366, November 6, 1986.
Clifford, A.W., and W.J. Cook. 1989. European Application Patent No. 0,304,167, July 20, 1988.
Clyne, T.W., Bader, M.G., Cappleman, G.R., and P.A. Hubert. 1985. *J. Mater. Sci.* 10:85–96.
Clyne, T.W., and J.F. Mason. 1987. *Metall. Trans.* 18A:1519–1530.
Cochran, C.N., and R.C. Ray. 1970. U.S. Patent No. 3,547,180, December 15, 1970.
Cole, G.S., and G.F. Bolling. 1965. *Transactions of the Metallurgical Society of AIME.* 233:1568–1572.
Cornie, J.A., Moon, H.K., and M.C. Flemings. 1990. *In:* Fabrication of Particulates Reinforced Metal Composites (J. Masounave and F.G. Hamel, eds.), 63–78, Materials Park, OH: ASM International.
Coulaud, O., Morel, P., and J.P. Caltagirone. 1988. *J. Fluid Mech.* 190:393–407.
Das, A.A., Clegg, A.J., Zantout, B., and M.M. Yakoub. 1988. *In:* International Symposium on Advances in Cast Reinforced Metal Composites (S.G. Fishman and A.K. Dhingra, eds.), 139–147, Materials Park, OH: ASM International.
Dean, N.F., Mortensen, A., and M.C. Flemings. 1991. Steady State Plane Front and Cellular Solidification in Fiber Composites, Work in Progress.
Dhingra, A.K. 1980. *Phil. Trans. R. Soc. Lond.* 294:559–564.
Donomoto, T. 1982. European Patent Application Patent No. 0 045 002 A1, 03-02-82.
Donomoto, T., Miura, N., Funatani, K., and N. Miyake. 1983. SAE Technical Paper Series, Paper No. 830252, 10 pp.
Ebisawa, M., Hara, T., Hayashi, T., and H. Ushio. 1991. SAE Technical Paper Series, Paper No. 910835, 13 pp.
Elliott, R. 1983. *Eutectic Solidification Processing; Crystalline and Glassy Alloys,* London: Butterworth & Co.
Feest, E.A., Young, R.M.K., Yamada, S.I., and S.I. Towata. 1989. *In:* Developments in the Science and Technology of Composite Materials, ECCM 3 (A.R. Bunsell, P. Lamicq and A. Massiah, eds.), 165–170, London and New York: Elsevier.
Fitzgerald, T., and A. Mortensen. 1992. Work in progress, Massachusetts Institute of Technology, Cambridge, MA.

Flemings, M.C. 1974. *Solidification Processing*, New York: McGraw-Hill.
Flemings, M.C. 1991. *Metall. Trans.* 22A:957–981.
Folgar, F., Widrig, J.E., and J.W. Hunt. 1987. SAE Technical Paper Series, Paper No. 870406, 9 pp.
Friend, C.M., Horsfall, I., and C.L. Burrows. 1991. *J. Mater. Sci.* 26:225–231.
Fukunaga, H. 1988. *In:* International Symposium on Advances in Cast Reinforced Metal Composites (S.G. Fishman and A.K. Dhingra, eds.), 101–107, Materials Park, OH: ASM International.
Fukunaga, H., and K. Goda. 1984. *Bull. Jpn. Soc. Mech. Engrs.* 27:1245–1250.
Fukunaga, H., and K. Goda. 1985a. *Bulletin of the JSME.* 28:1–6.
Fukunaga, H., and K. Goda. 1985b. *J. Jap. Inst. Met.* 49:78–83.
Fukunaga, H., Komatsu, S., and Y. Kanoh. 1983. *Bull. Jpn. Soc. Mech. Engrs.* 26:1814–1819.
Fukunaga, H., and M. Kuriyama. 1982. *Bull. Jpn. Soc. Mech. Engrs.* 25:842–847.
Fukunaga, H., and T. Ohde. 1982. *In:* Fourth International Conference on Composite Materials, ICCM-IV (T. Hayashi, K. Kawata and S. Umakawa, eds.), 1443–1450, Tokyo: Japan Society for Composite Materials.
Fukunaga, H., Wang, X., and Y. Aramaki. 1990. *J. of Mater. Sci. Lett.* 9:23–25.
Girot, F.A., Albingre, L., Quenisset, J.M., and R. Naslain. 1987. *J. of Metals.* 39:18–21.
Girot, F.A., Fédou, R., Quenisset, J.M., and R. Naslain. 1990. *J. of Reinforced Plastics and Composites.* 9:456–469.
Gungor, M.N., Roidt, R.M., and M.G. Burke. 1991. *Mater. Sci. & Eng.* A 144:111–119.
Gupta, M., Ibrahim, I.A., Mohamed, F.A., and E.J. Lavernia. 1991. *J. Mater. Sci.* 26:6673–6684.
Gupta, M., Mohamed, F., and E. Lavernia. 1990. *In:* Metal & Ceramic Matrix Composites: Processing, Modeling & Mechanical Behavior (R.B. Bhagat, A.H. Clauer, P. Kumar and A.M. Ritter, eds.), 91–106, Warrendale, PA: The Minerals, Metals & Materials Society.
Gupta, M., Mohamed, F.A., and E. Lavernia. 1992. *Metall. Trans.* 23 A:831–843.
Gupta, M., Mohamed, F.A., and E.J. Lavernia. 1991. *Mater. Sci. & Eng.* A144:99–110.
Gutierrez-Miravete, E., Lavernia, E.J., G.M. Trapaga et al. 1989. *Metall. Trans.* 20A:71–85.
Hansen, N.L., Engh, T.A., and O. Lohne. 1989. *In:* Interfaces in Metal-Ceramic Composites (R.Y. Lin, R.J. Arsenault, G.P. Martins and S.G. Fishman, eds.), 241–257, Warrendale, PA: The Minerals, Metals & Materials Society.
Harrigan, W.C., and R.H. Flowers. 1976. *In:* Failure Modes in Composites III (T.T. Chiao and D.M. Schuster, eds.), 212–225, New York: The Metallurgical Society of AIME.
Harrigan, W.C., and R.H. Flowers. 1977. *In:* Failure Modes in Composites IV (J.A. Cornie and F.W. Crossman, eds.), 319–335, New York: The Metallurgical Society of AIME.
Harrigan, W.C., Flowers, R.H., and S.P. Hudson. 1980. U.S. Patent No. 4,223,075, September 16, 1980.
Hayashi, T., Ushio, H., and M. Ebisawa. 1989. SAE Technical Paper Series, Paper No. 890557, 10 pp.

Hoover, W.R. 1990. *In:* Fabrication of Particulates Reinforced Metal Composites (J. Masounave and F.G. Hamel, eds.), 115–123, Materials Park, OH: ASM International.
Hoover, W.R. 1991. *In:* Metal Matrix Composites—Processing, Microstructure and Properties, 12th Risø International Symposium on Materials Science (N. Hansen, D. Juul-Jensen, T. Leffers, et al. eds.), 387–392, Roskilde, Denmark: Risø National Laboratory.
Hosking, F.M., Folgar-Portillo, F., Wunderlin, R., and R. Mehrabian. 1982. *J. Mater. Sci.* 17:477–498.
Hunt, W.H. 1986. In: *Interfaces in Metal Matrix Composites* (A.K. Dhingra and S.G. Fishman, eds.), 3–25, Warrendale, PA: The Metallurgical Society.
Ilegbusi, O., and J. Szekely. 1988. *J. Colloid Interface Sci.* 125:567–574.
Imai, T., Nishida, Y., M. Yamada et al. 1987. *J. of Mater. Sci. Lett.* 6:343–345.
Isaacs, J.A. 1991. Ph.D. Thesis, Department of Materials Science and Engineering, Massachusetts Institute of Technology, Cambridge, MA.
Itoh, T., Hirai, H., and R.-I. Isomura. 1988. *J. of Jpn Inst. of Light Metals.* 38:620–625.
Jarry, P., Dubus, A., G. Regazzoni et al. 1990. *In:* F. Weinberg International Symposium on Solidification Processing (J.E. Lait and I.V. Samarasekera, eds.), 195–204, Oxford, England: Pergamon Press.
Jarry, P., Michaud, V.J., A. Mortensen et al. 1992. *Metall. Trans.* 23A:2281–2289.
Jensen, P.S., and W. Kahl. 1991. *In:* 12th Risø International Symposium on Metallurgy and Materials Science: Metal Matrix Composites—Processing, Microstructure and Properties (N. Hansen, D. Juul-Jensen, T. Leffers et al., eds.), 405–410, Roskilde, Denmark: Risø National Laboratory.
Kampe, S.L., Swope, G.H., and L. Christodoulou. 1990. *In:* Materials Research Society Symposium (D.L. Anton, P.L. Martin, D.B. Miracle and R. McMeeking, eds.), 97–103, Pittsburgh, PA: Materials Research Society.
Kattamis, T.Z., and T. Suganuma. 1990. *Mater. Sci. & Eng.* A128:241–252.
Kaufmann, H., and A. Mortensen. 1992. *Metall. Trans.* 23A:2017–2073.
Kendall, E.G., and R.T. Pepper. 1978. U.S. Patent No. 4,082,864, April 4, 1978.
Kennedy, C.R. 1991. *In:* Satellite Symposium 2 on Advanced Structural Inorganic Composites of the 7th International Meeting on Modern Ceramics Technologies (7th CIMTEC—World Ceramics Congress) (P. Vincenzini, ed.), 691–700, London and New York: Elsevier Science.
Klier, E.M., Mortensen, A., Cornie, J.A., and M.C. Flemings. 1990. US Patent No. US 4,961,461, October 9, 1990.
Klier, E.M., Mortensen, A., Cornie, J.A., and M.C. Flemings. 1991. *J. Mater. Sci.* 26:2519–2526.
Klimovicz, T.F. 1990. SAE Technical Paper Series, Paper No: 900532, pp. 589–596.
Kloucek, F., and R.F. Singer. 1986. *In:* Materials Science for the Future: 31st International SAMPE Symposium and Exhibition (J.L. Bauer and R. Dunaetz, eds.), 1701–1712, Covina, CA: SAMPE.

Koczak, M.J., and P. Sahoo. 1991. *In:* Second Japan International SAMPE Symposium and Exhibition (I. Kimpara, K. Kageyama and Y. Kagawa, eds.), 713–722, International Convention Management, Inc, Tokyo, Japan.

Kurz, W., and D.J. Fisher. 1989. *Fundamentals of Solidification*, Aedermannsdorf, Switzerland: Trans Tech Publications.

Lachman, W.L., Penty, R.A., and A.F. Jahn. 1975. U.S. Patent No. 3,894,863, July 15, 1975.

Lacoste, E., and M. Danis. 1991. *Compte Rendus de l'Académie des Sciences de Paris* 313:15–20.

Lacoste, E., Danis, M., Girot, F., and J.M. Quenisset. 1991. *Mater. Sci. & Eng.* A135:45–49.

Lange, F.F., Velamakanni, B.V., and A.G. Evans. 1988. *In:* 41st Pacific Coast Regional Meeting of the American Ceramic Society, Paper No. 39-B-88P.

Lawrence, G.D. 1972. Columbus, OH: *AFS Transactions* 80:283–286.

Lenel, F.V. 1980. Powder Metallurgy, Princeton, NJ: Metal Powder Industries Federation.

LePetitcorps, Y., Quenisset, J.M., Borgne, G.L., and M. Barthole. 1991. *Mater. Sci. & Eng.* A135:37–40.

Levi, C.G., Abbashian, G.J., and R. Mehrabian. 1977. *In:* Failure Modes in Composites IV (J.A. Cornie and F.W. Crossman, eds.), 370–391, The Minerals, Metals and Materials Society, Warrendale, PA: TMS/AIME.

Lloyd, D. 1991. *In:* Metal Matrix Composites—Processing, Microstructure and Properties, 12th Risø International Symposium on Materials Science (N. Hansen, D. Juul-Jensen, T. Leffers et al., eds.), 81–99, Roskilde, Denmark: Risø National Laboratory.

Mada, M., and F. Ajersch. 1990. *In:* Metal & Ceramic Matrix Composites: Processing, Modeling and Mechanical Behavior (R.B. Bhagat, A.H. CLauer, P. Kumar and A.M. Ritter, eds.), 337–350, Warrendale, PA: The Metallurgical Society.

Martins, G.P., Olson, D.L., and G.R. Edwards. 1988. *Metall. Trans.* 19B:95–101.

Masuda, S.-I., and T. Itoh. 1989. *IEEE Trans. on Ind. Applic.* 25:1989.

Masur, L.J., Mortensen, A., Cornie, J.A., and M.C. Flemings. 1989. *Metall. Trans.* 20A:2549–2557.

Mathur, P., Annavarapu, S., Apelian, D., and A. Lawley. (1989a). *J. of Metals*. 41:23–28.

Mathur, P., Apelian, D., and A. Lawley. (1989b). *Acta Metall.* 37:429–443.

Maxwell, P.B., Martins, G.P., Olson, D.L., and G.R. Edwards. 1990. *Metall. Trans.* 21B:475–485.

Mazdiyasni, S., and D.B. Miracle. 1990. *In:* Materials Research Society Symposium (D.L. Anton, P.L. Martin, D.B. Miracle and R. McMeeking, eds.), 155–162, Pittsburgh, PA: Materials Research Society.

Mehrabian, R., Riek, R.G., and M.C. Flemings. 1973. *Metall. Trans.* 5:1899–1905.

Merzhanov, A.G. 1990. *In:* Combustion and Plasma Synthesis of High-Temperature Materials. (Z.A. Munir and J.B. Holt, eds.) , 1–53, New York: VCH.

Meyerer, W., Kizer, D., and S. Paprocki. 1977. *In:* Failure Modes in Composites IV (J.A. Cornie and F.W. Crossman, eds.), 297–307, Warrendale, PA: TMS/AIME.

Michaud, V.J. 1991. Ph.D. Thesis, Department of Materials Science and Engineering, Massachusetts Institute of Technology, Cambridge, MA.

Michaud, V.J., and A. Mortensen. 1992. *Metall. Trans.* 23A:2263–2280.

Millière, C., and M. Suéry. 1985. *In:* Advanced Materials Research and Developments for Transport—Composites (P. Lamiq, W.J.G. Bunk and J.G. Wurm, eds.), 241–248, MRS-Europe.

Moon, H.K., Cornie, J.A., and M.C. Flemings. 1991. *Mater. Sci. & Eng.* A144:253–265.

Mortensen, A. 1989. *In:* Japan-U.S. Cooperative Science Program Seminar on "Solidification Processing of Advanced Materials," Tokyo: Japan Society for Promotion of Science and National Science Foundation.

Mortensen, A. 1990. *In:* Solidification of Metal Matrix Composites (P.K. Rohatgi, ed.), 1–21, Warrendale, PA: The Minerals, Metals & Materials Society.

Mortensen, A. 1991. *Mater. Sci. & Eng.* A135:1–11.

Mortensen, A., Cornie, J.A., and M.C. Flemings. 1988. *Metall. Trans.* 19A:709–721.

Mortensen, A., Gungor, M.N., Cornie, J.A., and M.C. Flemings. 1986. *J. of Metals*. 38:30–35.

Mortensen, A., and I. Jin. 1992. *Intern. Mater. Rev.* 37(3):101–128.

Mortensen, A., Masur, L.J., Cornie, J.A., and M.C. Flemings. 1989. *Metall. Trans.* 20A:2535–2547.

Mortensen, A., and V.J. Michaud. 1990. *Metall. Trans.* 21A:2059–2072.

Mortensen, A., and V.J. Michaud. 1991. *In:* Advanced Materials for Future Industries: Needs and Seeds (I. Kimpara, K. Kagayama and Y. Kagawa, eds.), 1011–1018, Tokyo: International Convention Management, Inc.

Mortensen, A., and T. Wong. 1990. *Metall. Trans.* 21A:2257–2263.

Nagata, S., and K. Matsuda. 1981. IMONO, *Trans. of the Jpn Foundrymen's Soc.* 53:300–304.

Namai, T., Osawa, Y., and M. Kikuchi. 1986. *Trans. of the Jpn Foundrymen's Soc.* 5:29–32.

Nayim, S., and J.C. Baram. 1991. *In:* Advanced Composite Materials: New Developments and Applications, 267–273, Materials Park, OH: ASM International.

Nishida, Y., Matsubara, H., Yamada, M., and T. Imai. 1988. *In:* Fourth Japan-U.S. Conference on Composite Materials ed.), 429–438, Lancaster-Basel: Technomic Publishing Co.

Nourbakhsh, S., Liang, F.L., and H. Margolin. 1988. *Journal of Physics E: Scientific Instruments*. 21:898–902.

Nourbakhssh, S., and H. Margolin. 1991. *Mater. Sci. & Eng.* A144:133–141.

Papanicolaou, G. 1987. Hydrodynamic Behavior and Interacting Particle Systems, vol. 9, New York: Springer-Verlag.

Pennander, L., and C.-H. Andersson. 1991. *In:* Metal Matrix Composites—Processing, Microstructure and Properties, 12th Risø International Symposium on Materials Science (N. Hansen, D. Juul-Jensen, T. Leffers et al., eds.), 575–580, Roskilde, Denmark: Risø National Laboratory.

Pennander, L., Ståhl, J.E., and C.H. Andersson. 1991. *In:* Eighth International Conference on Composite Materials,

ICCM 8 (S.W. Tsai and G.S. Springer, eds.), 17B1–17B10, Covina, CA: SAMPE.

Rasmussen, N.W., Hansen, P.N., and S.F. Hansen. 1991a. *Mater. Sci. & Eng.* A135:41–43.

Rasmussen, N.W., Hansen, P.N., and S.F. Hansen. 1991b. *In:* Metal Matrix Composites—Processing, Microstructure and Properties, 12th Risø International Symposium on Materials Science (N. Hansen, D. Juul-Jensen, T. Leffers et al., eds.), 617–620, Roskilde, Denmark: Risø National Laboratory.

Rohatgi, P.K., Asthana, R., and S. Das. 1986. *Intern. Metals Rev.* 31:115–139.

Rohatgi, P.K., Asthana, R., Yadav, R.N., and S. Ray. 1990a. *Metall. Trans.* 21A:2073–2082.

Rohatgi, P.K., Asthana, R., and F. Yarandi. 1990b. *In:* Solidification of Metal Matrix Composites (P.K. Rohatgi, ed.), 51–75, Warrendale, PA: The Minerals, Metals & Materials Society.

Sahoo, P., and M.J. Koczak. 1991. *Mater. Sci. & Eng.* A144:37–44.

Sample, R.J., Bhagat, R.B., and M.F. Amateau. 1989. *J. of Comp. Mater.* 23:1021–1028.

Sekhar, J.A., and R. Trivedi. 1989. *Mater. Sci. & Eng.* A114:133–146.

Semlak, K.A., and F.N. Rhines. 1958. *Transactions of The Metallurgical Society of AIME.* 212:325–331.

Shaler, A.J. 1965. *Intern. J. of Powder Metall.* 1:3–14.

Sing Solanki, R., Sing, A.K., Basu, K., and C.B. Raju. 1991. *In:* Satellite Symposium 2 on Advanced Structural Inorganic Composites of the 7th International Meeting on Modern Ceramics Technologies (7th CIMTEC—World Ceramics Congress) (P. Vincenzini, ed.), 747–756, London and New York: Elsevier Science.

Singer, A.R.E. 1991. *Mater. Sci. & Eng.* A135:13–17.

Skibo, M.D., and D.M. Schuster. 1988a. U.S. Patent No. 4,786,467, November 22, 1988.

Skibo, M.D., and D.M. Schuster. 1988b. U.S. Patent No. 4,759,995, July, 26, 1988.

Sommer, J.L. 1992. Ph.D. Thesis, Dept. of Materials Science and Engineering, Massachusetts Institute of Technology, Cambridge, MA.

Stefanescu, D.M., and B.K. Dhindaw. 1988. *In:* Metals Handbook. 15 (A.I.H. Committee, ed.), 142–147, Metals Park, OH: American Society of Metals.

Suéry, M., and L. LaJoye. 1990. *In:* Solidification of Metal Matrix Composites (P. Rohatgi, ed.), 171–179, Warrendale, PA: The Minerals, Metals & Materials Society.

Takenaka, T., and M. Fukazawa. 1988. SAE Technical Paper Series, Paper No. 881189, pp. 47–55.

Tank, E. 1990. U.S. Patent No. 4,943,413, July 24, 1990.

Terry, B.S., and O.S. Chinyamakobvu. 1991. *J. of Mater. Sci. Lett.* 10:628–629.

Thompson, M.S., and V.C. Nardone. 1991. *Mater. Sci. & Eng.* A144:121–126.

Tiwari, R., and H. Herman. 1991. *Scripta Metall. et Mater.* 25:1103–1107.

Tiwari, R., Herman, H., Sampath, S., and B. Gudmundsson. 1991. *Mater. Sci. & Eng.* A144:127–131.

Towata, S.-I., and S.-I. Yamada. 1986. *In:* Composites '86: Recent Advances in Japan and the United States, CCM-III (K. Kawata, S. Umekawa and A. Kobayashi, eds.), 497–503, Tokyo: The Japan Society for Composite Materials.

Tsunekawa, Y., Okumiya, M., Niimi, I., and K. Okumura. 1987. *J. of Mater. Sci. Lett.* 6:191–193.

Tsunekawa, Y., Okumiya, M., Niimi, I., and K. Yoneyama. 1988. *J. of Mater. Sci. Lett.* 7:830–832.

Vaidyanathan, C., and R. Rohrmayer. 1991. *In:* Eighth International Conference on Composite Materials, ICCM 8 (S.W. Tsai and G.S. Springer, eds.), 17D1–17D10, Covina, CA: SAMPE.

Valencia, J.J., Löfvander, J.P.A., C. McCullough et al. 1991. *Mater. Sci. & Eng.* A144:25–36.

Verma, S.K., and J.L. Dorcic. 1988. *In:* International Symposium on Advances in Cast Reinforced Metal Composites (S.G. Fishman and A.K. Dhingra, eds.), 115–126, Materials Park, OH: ASM International.

Westengen, H., Albright, D.L., and A. Nygard. 1990. SAE Technical Paper Series, Paper No. 900534, pp. 606–612.

Westwood, A.R.C. 1988. *Metall. Trans.* 19A:740–758.

White, J., Willis, T.C., Hughes, I.R., and R.M. Jordan. 1988. *In:* Dispersion Strengthened Aluminum Alloys (Y.-W. Kim and W.M. Griffith, eds.), 693–708, Warrendale, PA: The Minerals, Metals & Materials Society.

Willis, T.C., White, J., Jordan, R.M., and I.R. Hughes. 1987. *In:* Third International Conference on Solidification Processing ed.), 476–478, London: The Institute of Metals.

Xiao, P., and B. Derby. 1991. *In:* Surfaces and Interfaces, British Ceramics Proceedings (R. Morrell and G. Partridge, eds.), 153–159, Stoke-on-Trent, U.K.: The Institute of Ceramics.

Yang, J., and D.D.L. Chung. 1989. *J. Mater. Sci.* 24:3605–3612.

Young, R.M.K. 1991. *Mater. Sci. & Eng.* A135:19–22.

Zhang, J., Gungor, M.N., and E.J. Lavernia. 1991. *In:* Eighth International Conference on Composite Materials (ICCM/8) (S.W. Tsai and G.S. Springer, eds.), 17H1–17H12, Covina, CA: SAMPE.

Chapter 2
Solid-State Processing

AMIT K. GHOSH

Significant improvements in the mechanical properties of a metal matrix can be achieved by incorporating a stronger and stiffer phase as reinforcement in such matrix (Cooper and Kelly 1969; Piggott 1980). In addition, the physical properties of these metal-matrix composites (MMCs) are also improved as a result of the incorporation of reinforcements. As evidenced by the increased demand within the aerospace and commercial industries for materials with mechanical and physical properties superior to those of advanced monolithic alloys, it is becoming increasingly clear that composites will have many actual applications (McDaniels et al. 1986; Keyworth 1985). The specific benefits of MMCs include (1) increased strength and stiffness per unit density, which reduces structural weight and increases performance, (2) a decrease in the coefficient of thermal expansion (CTE), which reduces thermal strains in structures undergoing thermal cycles, (3) increased creep strength, which is necessary for improving operating temperatures and performance of turbine engines, and (4) other specialized attributes, such as wear resistance and high conductivity.

Continuous fiber MMCs such as graphite-aluminum and boron-aluminum provide the highest strength and modulus levels, and are currently utilized in tubular struts for lightweight applications. These composites are made by methods such as foil-fiber-foil lay-up, solid-state bonding, and centrifugal casting. Some of these methods are discussed later, or are covered in other literature (Grierson and Krock 1972; Zweben and Dvorak 1982; Ghosh 1989). The majority of MMCs today are discontinuously reinforced with particulate or whisker forms of SiC or Al_2O_3 in aluminum, magnesium, and other matrix alloys. These materials are processed using methods that are similar to conventional deformation processing or machining of metallic alloys, and consequently are considerably less expensive than continuous fiber reinforced composites because of the lower costs of the reinforcement material and the lower cost of processing. In this chapter, a detailed discussion about solid-state processing of metal-matrix composites is presented with the focus on particulate-reinforced composites, primarily because the knowledge base and industrial interest are wider for these materials.

2.1 Solid-State Methods of Fabrication of MMCs

There are many different methods available for fabricating composites (Grierson and Krock 1972; Ghosh 1989; Geiger and Walker 1991; Wu and Lavernia 1991; Karmakar and Divecha 1992; Backman et al. 1977). However, these are generally based on (1) liquid metal processes, (2) solid-state processes (with or without some liquid present), and (3) deposition processes, often followed by solid-state bonding processes. Solid-state processes are generally used to obtain the highest mechanical properties in MMCs, particularly in discontinuous MMCs. This is because segregation effects and brittle reaction product formation are at a minimum for these processes, especially when compared with liquid-state processes.

2.1.1 Powder Consolidation

Powder metallurgy is the most common method for fabricating metal–ceramic and metal–metal composites.

With the advent of rapid solidification technology, the matrix alloy is produced in a prealloyed powder form rather than starting from elemental blends (Mahoney and Ghosh 1987). After blending the powder with ceramic reinforcement particulates (or whiskers), cold isostatic pressing is utilized to obtain a green compact that is then thoroughly outgassed and forged or extruded. In some cases, hot isostatic pressing of the powder blend is required, prior to which complete outgassing is essential. Consolidation of matrix powder with ceramic fibers has also been achieved, but the difficulties encountered when attempting to maintain uniform fiber spacing dictate that powder slurry be sprayed onto the fiber mat (parallel, woven row of fibers) or that powder slurry be tape-cast to produce the initial lay-up for press consolidation (Brindley 1987; Reed 1988). The chief difficulty in these processes is the removal of the binder used to hold the powder particles together. The organic binders often leave residual contamination that causes deterioration of the mechanical properties of the composite.

Although most powder consolidation and processing work is carried out below the matrix solidus temperature, sometimes it becomes necessary to maintain the consolidation temperature slightly above the solidus to minimize deformation stresses and to avoid whisker damage (an example of this is discussed in detail in Section 2.2). In liquid-phase sintering, powder consolidation may be achieved without the use of any external pressure because a low melting phase pulls solid particles together via the force of surface tension (Kingery et al. 1975). The higher melting temperature phase in this process should be slightly soluble in this liquid. An example is the Ni–W system. Liquid-phase sintering has been discussed in detail in many books and is not covered in this chapter (Grierson and Krock 1972; Kingery et al. 1975; Reed 1988).

2.1.2 Diffusion Bonding

Diffusion bonding is used for consolidating alternate layers of foils and fibers to create single- or multiple-ply composites (Schoutens 1982). This is a solid-state creep-deformation process and is schematically illustrated in Figure 2.1. After creep flow of the matrix between the fibers to make complete metal-to-metal contact, diffusion across the foil interfaces completes the process. Pressure-time requirements for consolidation can be determined from matrix flow stress, taking into consideration the above mentioned matrix flow processes (Ghosh 1989). To avoid fiber degradation, care must be exercised to maintain low pressure during consolidation. Furthermore, obtaining complete flow in the interstices between the fiber midplane and the foil segments on either side is extremely difficult. This is accomplished by using a fine-grained foil and by consolidating at a temperature where the matrix is in either a soft or a superplastic state. Examples of this type of composite currently in use are aluminum or titanium alloys reinforced with SiC, B, or other fibers. Diffusion bonding is performed either by loading a vacuum pack in a hot press or by a hot isostatic pressing operation. In either case, thorough outgassing of the pack prior to consolidation is required. Figure 2.2 shows cross-sectional microstructures of Ti-6Al-4V alloy reinforced with SCS-6 SiC fiber and Borsic fiber. Because of their high stiffness and strength at elevated temperatures, such composites find applications in gas turbine engines. Fiber-matrix interfacial reactions during diffusion bonding can cause degradation of the fiber interface region and reduce its load-carrying ability (Ghosh 1989). Thermal expansion mismatch between the fiber and the matrix can often cause tensile stresses and matrix cracking upon cooling from the diffusion bonding temperature. Thus, steps may be taken during prebonding to provide fibers with a diffusion barrier coating as well as a coating of a ductile material. Coatings of Ta, Nb, or other metals have been used on the interface of titanium aluminide matrices and SCS-6 SiC fibers to meet these needs and simultaneously provide a refractory metal layer for enhanced creep resistance (Ghosh 1988).

Roll bonding and coextrusion have also been used for solid-state bonding of metallic foils and fibers and ceramic particulates in metallic matrix. Powder blends packed and evacuated in a container can be subjected to these consolidation methods. Laminated composites are ideally produced by high-temperature roll-bonding operations starting from either foils of individual metals or from alloys. During roll bonding, both surface deformation and diffusion actively cause asperity deformation and interdiffusion to produce strong interfaces. Figure 2.3 shows the microstructure of a multilayer Cu–Nb composite produced by roll bonding. Rods of Cu and Ni (or W, Nb) can also be coextruded so that both metallic phases deform and lead to the fine microfibers of the harder phase (Spitzig et al. 1987).

Deposition processes for fiber composites generally yield a monotape that can be stacked in layers and diffusion bonded into a composite. Some degree of porosity is usually present in the monotape but this can be eliminated during diffusion bonding or during hot isostatic pressing. W-fiber reinforced Ni-base alloy monotapes produced by this method have been used for the fabrication of lightweight, hollow turbine blades for a high-performance engine. The process involves placing monotapes around a bent steel core and diffusion bonding several monotape layers. Near-net shape blades are

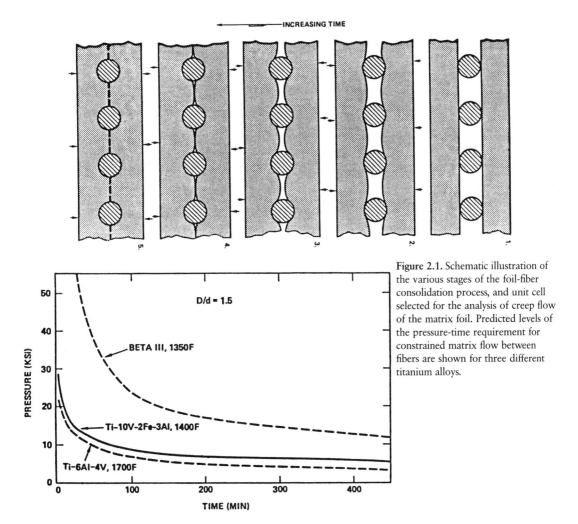

Figure 2.1. Schematic illustration of the various stages of the foil-fiber consolidation process, and unit cell selected for the analysis of creep flow of the matrix foil. Predicted levels of the pressure-time requirement for constrained matrix flow between fibers are shown for three different titanium alloys.

produced after the steel core is removed by acid. Figure 2.4 is a schematic illustration of the process and cross-sections of the blade illustrated in Petrasek et al. (1986) indicate excellent consolidation with uniform fiber spacing and no detrimental interfacial reaction. This method is currently being applied to fiber-reinforced titanium matrix composites to improve fiber distribution compared to that shown in Figure 2.2 (Textron Specialty Materials 1990), and enhance composite mechanical properties.

2.1.3 High-Rate Consolidation

This consolidation process of powder blends is most suitable for rapidly solidified (RS) metals and hard-to-deform metals. Frictional heating at the powder-particle interface causes local melting and consolidation, and rapid heat extraction by the cooler particle interior causes rapid solidification (Murr et al. 1986). Thus, an RS microstructure can be better preserved by this method. High-rate consolidation leads to strengthening of alloys (due to high dislocation density) but reduced ductility. Composites produced by this method often contain cracks. It has yet to be shown if reasonable composite properties can be achieved by this method.

Mechanical alloying is truly a solid-state process that can be applied to particulate composites. In this method, a high-energy impact mill is used to continuously fragment and reweld the powder particles as fresh internal surfaces are exposed. The continuous fragmentation leads to thorough mixing of the constituents, and subsequent processes such as hot pressing and extrusion are used to consolidate and/or synthesize the alloy or composite (Jatkar et al. 1986).

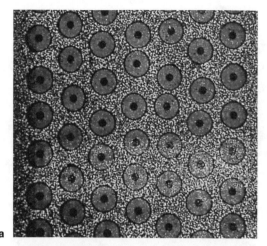

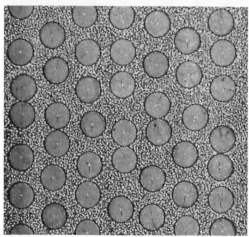

Figure 2.2. Cross-sectional micrographs of multi-ply, unidirectional composites of (a) SCS-6 SiC fiber reinforced Ti-6Al-4V alloy and (b) Borsic fiber reinforced Ti-6Al-4V alloy produced by diffusion bonding of fiber-foil. Notice the cracks present within Borsic fibers from the consolidation process (fiber diameter equal to approximately 140 μm).

2.2 Primary Solid-State Processing of Discontinuously Reinforced Composites

As mentioned in the last section, primary solid-state synthesis of particulate reinforced composites is performed by a variety of techniques, such as powder blending and consolidation, mechanical alloying, diffusion bonding or roll bonding, high-rate consolidation, and powder coating by electrochemical or vapor processes followed by solid-state consolidation methods.

Coated particles have an advantage in that composites made from these have a reasonable amount of matrix spacing maintained between reinforcement particles. Because a chief cause of poor mechanical properties in composites stems from particle contact or too small an interparticle spacing, composites with coated particles have a great potential advantage. Coating methods are not, however, so advanced at this time as to provide contamination-free uniform thickness coating, and thus few applications are using these methods. Hollow microspheres of Al_2O_3 and quartz, coated with a variety of metallic materials, have been consolidated into lightweight composites (Rawal and Misra 1990). These have excellent stiffness and sound-absorbing qualities.

Blending and consolidation is the primary solid-state synthesis process used to produce both particle- or whisker-reinforced metals. Prealloyed atomized powder of the matrix alloy (or elemental powder) is mixed with the powder or whisker of the ceramic phase (for example, SiC or Al_2O_3) and thoroughly blended. The prealloyed powder is commonly a -325 mesh size (the average particle size is 25–30 μm), but ceramic particle size needed for obtaining reasonable mechanical properties is in the 1 to 5 μm range. This large size difference creates agglomeration of the ceramic particles in the blend. Figure 2.5 shows such agglomeration in extruded 6013 Al/SiC composites fabricated with different SiC particle sizes. Such agglomerates can be broken up during powder mixing by ultrasonic agitation in a liquid slurry. The addition of surfactants, which causes a repulsive force between the ceramic particles, can also be used to improve the distribution of constituent phases although this is normally avoided for metal matrices because contamination from impurity atoms adversely affects the mechanical properties of most MMCs. The most commonly used blending practice is, therefore, dry blending in a V-blender or other agitation devices, preferably carried out in an inert environment (glove box) to avoid contamination. For short fiber-reinforced composites, the difference in powder size ratio is not as critical a factor in controlling the distribution of short fibers because both fiber diameter (10 to 25 μm) and length (100 to 400 μm) are substantially

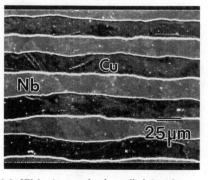

Figure 2.3. SEM micrograph of as-rolled Cu-Nb microlaminate produced by the roll-bonding process.

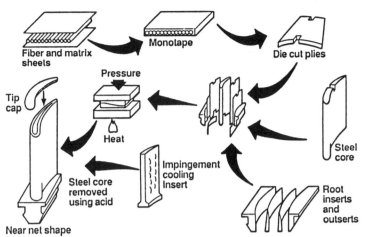

Figure 2.4. Schematic illustration of the fabrication of hollow turbine blades by diffusion bonding of arc-sprayed fiber monotapes *(Courtesy of Petrasek et al. 1986)*.

larger (as for carbon, Nicalon-SiC, and Al_2O_3 fibers). For whisker-reinforced composites (d = 0.2 to 2 μm), agglomeration can be a significant problem.

Currently, a great deal of interest has developed regarding the use of mechanical alloying as a means of dispersing a ceramic phase more evenly within the metallic matrix in the blend. By repeated impact of the powder particles and their fragmentation, a fresh metallic surface is constantly exposed, within which the ceramic particles become embedded. Particle distribution has been shown to be vastly improved by this method (Jatkar et al. 1986), but difficulty arises from contamination introduced from the vessel, balls, hammers, or surfactants used in the pulverizing mill, and from the chemical reaction that occurs with the atmosphere within the mill. In certain instances, this reaction problem can be judiciously utilized. For example, cryomilling with liquid nitrogen has been carried out to intentionally produce ultrafine AlN particles in an NiAl matrix (Whittenberger et al. 1990). In most cases, however, contamination leads to the undesirable presence of oxides, carbides, Fe, or W in the product. With proper

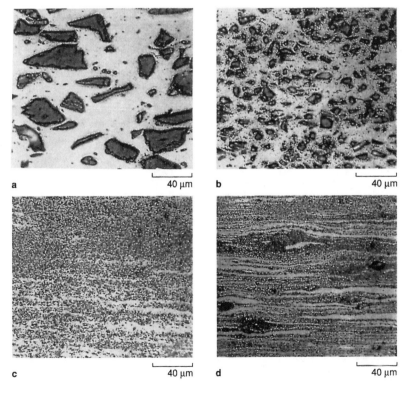

Figure 2.5. Optical micrographs of longitudinal sections of 6013 Al/20%SiC$_p$ DRA composites with different SiC particle diameters: (a) 28 μm, (b) 10 μm, (c) 2 μm, and (d) 0.7 μm *(Courtesy of Geiger and Walker, 1991)*.

care, it is possible to keep such problems to a minimum. As demonstrated recently by Japanese researchers, an extremely fine grain size (< 1 μm) stabilized by a fine dispersion of ceramic particles and a high degree of superplastic ductility can be achieved in 6061 Al/SiC and 6061 Al/Si$_3$N$_4$ particulate composites prepared by mechanical alloying (Mabuchi et al 1991; Higashi et al. 1992).

Prior to consolidation, the powder blend is loaded into a can of a compatible material (such as the predominant constituent of the matrix alloy), and evacuated for a long period of time. This process is typically carried out at 400°C to 500°C, where most volatile constituents are driven out. Because the path of exiting gas molecules is rather tortuous in a large column of fine powder particles, required outgassing times, which increase by as much as the square of the length dimension of the can in diffusion control, can easily extend to 10 to 30 hours. Typically, the consolidation and subsequent processing of discontinuous MMCs are carried out via a powder metallurgy method similar to that used for the metal-matrix alloys. After outgassing, the can is sealed and loaded into a hot press for consolidation. When a vacuum hot press is used, the canning and sealing steps can be avoided, and the powder can be loaded directly into a die (Geiger and Walker 1991).

Powder consolidation of MMCs is typically not a liquid-phase sintering process, although this method has been used for WC-Co composites and other high-temperature composite materials. Typically, hot pressing or hot isostatic pressing (HIP) is carried out at as high a temperature as possible to render the matrix in its softest solid-state condition. Generation of liquid phase can adversely influence the mechanical properties of the product due to grain boundary segregation and the formation of brittle intermetallic phases. Even in the production of aluminum matrix composites, certain manufacturers hot press powder at temperatures above the solidus. The presence of a small amount of liquid aids in lowering the consolidation pressure for achieving full densification in hot pressing (Geiger and Walker 1991). This is helpful in the consolidation of whisker-reinforced composites in which fragmentation of whiskers is an issue. An alternative practiced by others is to consolidate powder in a can to about 95% density via hot pressing, subsequently HIPing this billet to achieve full density (Center for Technology, Kaiser Aluminum Co., 1983). Because a hot isostatic press can generate much higher pressures (20 to 30 Ksi) than do hot pressing systems, consolidation in the solid state is not difficult using such equipment.

When consolidation is carried out in the solid state, the process is similar to diffusion bonding, in which there is considerable creep flow of the softer matrix material into the interstices between the ceramic particles or whiskers. This is the primary process during hot pressing or HIPing. Such a creep process is a combination of dislocation and diffusion creep. When all interconnected porosity is eliminated (density > 90%), and only triple points and grain corners contain the remaining porosity, the transport of matter via pressure-assisted diffusion becomes the dominant mechanism. The pressure-time requirements for complete consolidation can be calculated from the creep properties of the matrix powder in a relatively simple manner. The HIPing diagrams produced by Arzt and co-workers (Arzt et al. 1983) and the consolidation analysis by Besson and Evans (1992) are helpful in this respect. The presence of nondeformable ceramic inclusions actually decreases the time requirements for initial consolidation because their jagged edges and corners locally enhance the stresses in the matrix during the early part of the consolidation process. Agglomeration of ceramic particles can however provide significant resistance to completion of the consolidation process, and increase the overall HIPing time.

After full consolidation is achieved, most discontinuously reinforced MMCs are subjected to additional deformation processing to improve their microstructure and mechanical properties as well as to create a product of useful shape. Unbroken prior particle boundaries and interfaces existing in a consolidated MMC do not provide the best toughness and strength. Extrusion is a commonly used process to create substantial amounts of shear within the material and to generate new grain boundaries and strongly bonded interfaces. Because an extrusion press is used for this step, improved process economics can be achieved by combining the consolidation process with this step. Certain manufacturers thus hot consolidate the can containing blended powder against a blind die in the extrusion press (similar to hot pressing), and subsequently replace the blind die by an extrusion die to force the billet through as in a normal extrusion operation. In such a process, some remaining porosity from the first step undergoes complete removal due to shear flow and hydrostatic compression present during the extrusion operation.

2.3 Deformation Processing of Metal-Matrix Composites

Secondary processing of the discontinuously reinforced composites leads to break up of particle (or whisker) agglomerates, reduction or elimination of porosity, and improved particle-to-particle bonding, all of which tend to improve the mechanical properties of these materials.

2.3.1 Extrusion

The most common secondary processing method is extrusion, which is performed at a temperature that enables strain-rate sensitivity ($m = d\log\sigma/d\log\dot{\varepsilon}$) to reach a relatively high value. Typically, the process is carried out at a high strain rate (1–100s^{-1}) and primarily involves dislocation creep deformation of the matrix. Thus, the value of m does not exceed 0.3 during such an operation and can be as low as 0.1 for the particulate composites. Nevertheless, the highest possible values of m are sought in order to obtain the most uniform flow and minimize tendencies for cracking. Apart from improving the homogeneity of the product, extrusion can produce net-shape product forms in large lengths. Because of the presence of 15 to 25% nondeformable particulates (or whiskers) as in Al/SiC or Al/Al$_2$O$_3$ composites, fracture during extrusion can occur unless appropriate process design is employed. It has been found that the use of streamlined dies can produce many complex section shapes without cracks, which would be impossible to do with conventional extrusion die shapes (Gunasekera and Hoshino 1980).

The microstructure of the as-extruded material is generally heterogeneous, as seen in Figure 2.6(a). The local volume fraction of SiC particulates in the cluster is as high as 50 to 60%. Many of these clusters are nearly round in shape, and they are randomly dispersed in the matrix without any apparent orientation preference from the consolidation step (or even a low level of extrusion 5:1). The typical size of the clusters varies from 10 to 50 mm in diameter. There are also areas that contain virtually no particles, shown as the white areas in the photograph. Figure 2.6(b) shows the same structure by using an SEM at a higher magnification; from this, the inhomogeneity of the microstructure can be easily visualized.

Strain-rate effects during the deformation process have a strong effect on the break-up of particle clusters. Because the particle clusters act as harder regions than particle-free regions, strain partitioning takes place during flow. Thus, at low strain rates, the imposed deformation is accommodated primarily by the flow of the particle-free regions. Because the strain-rate sensitivity of the matrix alloy exceeds that of the clustered region, a high-applied strain rate raises matrix flow stress preferentially thereby forcing deformation to take place in the particle clustered regions and leads to shearing of clustered regions. Thus, at high strain rates, a reasonable amount of strain within the cluster can be expected, as structural homogenization occurs at 3 to 5 times the cluster strain of about 0.35. Conversely, the stresses and strains in the cluster are so small at low strain rates that its break-up is insufficient even when the overall strain on the composite is 100 times that of the cluster.

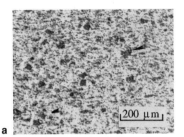

Figure 2.6. Microstructures of as-extruded 7064Al/SiC particulate composite. (a) Lower magnification optical micrograph showing particulate clusters (arrows). (b) SEM photograph showing details within clustered region (extrusion direction is horizontal).

In whisker- (or short fiber) reinforced composites, rotation of the fibers takes place as shear deformation of the composite occurs within the extrusion die. Figure 2.7 illustrates the manner in which the motion of internal grid lines during the extrusion process is affected by the presence of the short fibers (Stanford-Beale and Clyne 1989). Note that the local turbulence that is set up around these fibers perturbs the uniform shear that tends to occur during this process. The aspect ratio of the fibers is reduced during this process due to fiber fracture, although they continue to be realigned in the extrusion direction. Typically, an extrusion ratio of at least 16:1 is required to cause a complete realignment of all whiskers (and/or fibers), as well as a break-up of prior particle boundaries to achieve improved mechanical properties. Although extensive shear can improve atomic registry of grain boundaries and interfaces and thereby improve composite mechanical properties, particle fracture and particle-matrix debonding can also take place, both leading to void formation in these materials, that could adversely affect their mechanical properties. It is essential, therefore, that the state of stress existing during extrusion is significantly compressive in nature. Hydrostatic extrusion is thus a key process for minimizing extrusion-induced damage in these materials. Currently, effort is being directed toward hydrostatic extrusion of high-temperature composite materials such as composites of Nb and Mo silicides and titanium aluminide.

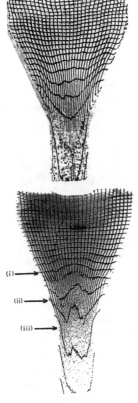

Figure 2.7. Distortion of internal grid marks in a 6061 Al/18% short fiber (SiC) reinforcement, showing irregular shearing of grid lines. *(From: Stanford-Beale & Clyne 1989.)*

2.3.2 Rolling

When composite sheet or plate products are desired, extrusion is followed by rolling. Because compressive stresses are lower in the rolling operation than they are in extrusion, edge cracking is a serious problem with these materials. Thus, rolling from more than 40-mm thick sections often leads to a significant amount of cracked and therefore scrap material. It is found that rolling of discontinuously reinforced composites is most successful in the range of 0.5 T_m using relatively low rolling speeds. Warm, isothermal rolling using smaller passes and a large roll diameter can produce rolled sheets with minimum edge-cracking problems. When the rolling temperature is lower, reductions per pass must be minimized and intermediate annealing steps must be utilized to produce successful products. Near isothermal conditions during rolling often require pack-rolling procedures, involving encapsulation of the workpiece in a tight pack of a stronger material. During rolling, the pack undergoes chilling by the rolls while maintaining a sufficient temperature in the workpiece for rolling. The temperature and the strain rate in pack rolling can be adjusted to minimize void formation during the rolling operation.

As in the case of extrusion, further break-up of particulate agglomerates takes place during rolling. In heavily rolled sheet materials, particle clusters are completely broken up and one finds that the matrix has flowed between individual particles that were previously in contact. Such a well-distributed microstructure in a thin rolled foil of 2014 Al/15v%Al$_2$O$_3$ particulate composite is shown in Figure 2.8.

Figure 2.9 illustrates the change in distribution of SiC clusters as a function of the rolling reduction in a 7064 Al/SiC composite. Because deformation of the harder cluster regions is considerably less than that of the matrix, the matrix flows past the clusters during rolling, causing shear flow. The particles adjacent to the cluster-matrix interface are less constrained than those in the interior of the cluster and appear to flow by tumbling along the cluster interface because of the matrix shearing effect. With increasing strain, the clusters continue to change their shapes as more particulates are removed from the cluster surface and are transported to the end of the elongated clusters, seen as tails in the photographs. After a certain amount of deformation, the clusters show an orientation preference, that is, they string themselves out along the rolling direction. This helps to embed the particles in the matrix where the particle density originally was low. At about 90% thickness reduction, the number of clusters is substantially reduced, and the microstructure becomes reasonably homogeneous, which continues to improve all the way to 99% reduction in thickness.

Figures 2.10(a and b) show variations in the area fractions of SiC clusters and that of particle-free regions

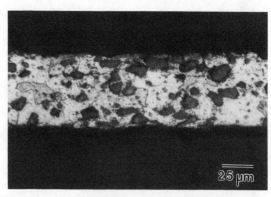

Figure 2.8. Microstructures of rolled foil of 2014 Al/15% Al$_2$O$_3$ particles showing reasonably good particle distribution.

sharply at the early stage of deformation, reaches its peak at about 25% rolling reduction, and then decreases substantially before leveling off.

2.3.3 Thermomechanical Processing

Thermomechanical processing is often utilized for particulate composites for the purpose of grain refinement and recrystallization. Microstructures produced by TMT are influenced by the presence of ceramic particulates. A grain refinement method practiced for heat-treatable aluminum matrix composites involves (1) overaging to produce coarse precipitates, (2) rolling to

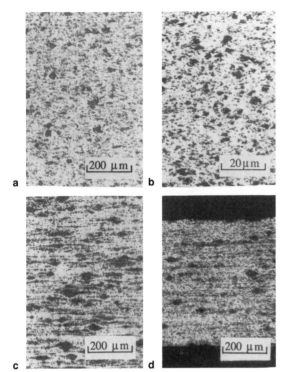

Figure 2.9. Optical microstructures of 7064 Al/15v%SiC at various stages of rolling at 250°C. Note the progression of clump break-up via interfacial shear; (a) as extruded, (b) rolled 25% at 250°C, (c) rolled 53% at 250°C, (d) rolled 92% at 250°C.

as a function of applied strain. Also shown is the density of SiC particles in nonclustered regions from the samples rolled at 250°C and 350°C, respectively. The area fraction values are to be considered only in a qualitative manner. For a given rolling temperature, while the area fraction of SiC clusters and that of particle-free regions decrease as the effective strain increases, the area fraction occupied by SiC particles in the nonclustered regions continuously increases. The continuous changes in these parameters as a function of strain reflect the homogenization process for particle distribution.

The rolling temperature has a strong effect on the flow pattern of the particles. Figure 2.11 shows the change in aspect ratio of SiC clusters with effective strain. For both rolling temperatures, the aspect ratio increases rapidly at lower strains until it reaches its maximum. With increasing strain, the aspect ratio decreases first and then remains unchanged. Plastic flow of the matrix is more difficult at lower temperatures (250°C), and higher stresses are required to break up the agglomerates. Thus, the peak of the aspect ratio plot shifts toward higher strains in comparison to that for 350°C. In contrast, at 350°C, the aspect ratio increases

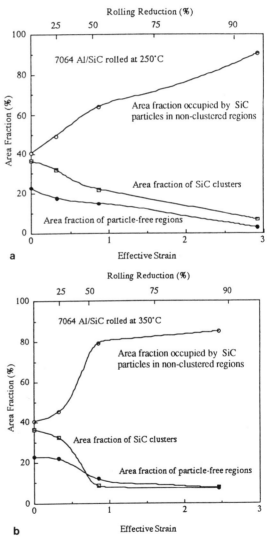

Figure 2.10. Area fraction of SiC particles in 7064 Al/15%SiC as a function of rolling strain for isothermal rolling at (a) 250°C and (b) 350°C.

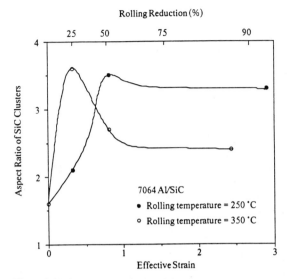

Figure 2.11. Average aspect ratio of SiC clusters as a function of rolling reduction at two different isothermal rolling temperatures.

95% reduction at the lowest possible temperatures, and (3) static annealing to develop a recrystallized microstructure (Paton and Hamilton 1978). The object of the rolling deformation step is to introduce as much cold work as possible within the matrix to produce nuclei for recrystallized grains. The ceramic particles cause additional strain concentration during this step over and above that caused by the aging precipitates. However, because strain localization and the tendency for internal fracture are greater in the case of the composite, the rolling process for the composite, to be successful, must be carried out at a higher temperature (350°C instead of 250°C) than that necessary for the matrix. This reduces the density of defects stored within the composite, and thus minimizes the density of the recrystallization nuclei during the subsequent annealing step. Therefore, contrary to expectations, a composite containing 15% to 20% ceramic particles does not exhibit a significantly finer grain size in comparison to the matrix alloy processed by a similar method.

Even though the storage of dislocations is not significantly enhanced because of the higher processing temperature, the spacing between particulates does influence the pinning of migrating grain boundaries during the recrystallization step. For example, the particle-clustered regions in the composites, which have small interparticle spacing, show finer grain sizes, dictated by the particle spacing itself. Thus, a bimodal grain structure develops within the composite. This kind of structure also forms easily by dynamic recrystallization during high-temperature deformation of these materials. Under the appropriate temperature and strain-rate con-

ditions, these particles can cause both nucleation of new grains as a result of strain concentration as well as pinning of their grain boundaries.

A bimodal grain structure is observed in the samples of 7064 Al/SiC$_p$ composites isothermally rolled at 250°C and 350°C. Figure 2.12 shows such a microstructure for the 350°C rolled material. The structure is typified as coarse matrix grains existing between SiC clusters and fine grains near and/or within the SiC clusters. The nonuniformity of grain size is rather large for the 350°C rolled structure (average grain size ~ 8.2 ± 5.3 μm) in comparison to the 250°C rolled structure (average grain size ~ 4.0 ± 1.3 μm).

Grain coarsening taking place during rolling and subsequent reheating is believed to be partly caused by the existence of SiC particle clusters. Progressive rolling reductions with intermediate reheating steps can cause grain coarsening and even exaggerated grain growth in the particulate composites. This result is possibly because of a large difference in the accumulated strain level between the regions containing ceramic particles and the remaining matrix. Because recrystallization nuclei can be produced easily in this area with only small levels of rolling reduction, the remaining matrix with lower strain levels is a clear path through which migrating grain boundaries can sweep. This can lead to the development of a coarse-grain microstructure. Under these conditions, the ceramic particles become embedded within the grains rather than being located on the grain boundaries. In particle-clustered regions, the local deformation within the matrix results in the formation of dislocation cells in the deformation zone around the

Figure 2.12. Bimodal grain size distribution in an isothermally (350°C) rolled 7064 Al/SiC composite, showing coarse grains between regions of particle clusters and fine grains within the clusters.

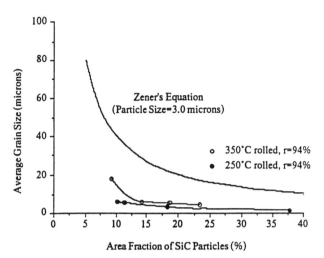

Figure 2.13. Average grain size of 7064 Al/SiC composite as a function of local SiC density.

particles (Wert 1985), which frequently have high misorientation with their neighboring cells. In addition, the differential thermal expansion between the matrix and the particles causes additional local deformation at the interfaces (Christman et al. 1989). During annealing, these cells rearrange themselves into subgrain boundaries and subsequently convert to recrystallized grains. Because of the larger area percentage of interfaces, and therefore the larger number of potential nucleation sites, the rate of nucleation in the high-particle-density areas (cluster areas) is considerably greater, and a substantial number of new grains are nucleated before their mutual impingement process, resulting in an ultrafine structure.

The effect of the SiC particle content on the matrix grain size is shown in Figure 2.13. As the local particle density increases, the average grain size for both rolling temperatures decreases. The dependence of grain size as a function of area fraction of SiC particles is based on Zener's pinning effect theory (Smith 1948), such that

$$D_{crit} = \frac{4r}{3f}$$

where D_{crit} = the limiting grain diameter, r = particle size, and f = volume fraction of particles. Although discrepancy exists between the theoretical prediction and the experimental data due to the fact that the model is based on the assumption that the migrating boundary is purely under the influence of its own interfacial tension, the general trend is clearly obeyed.

Microstructures developed by rolling at lower temperatures can become rather unstable at higher temperatures because particle-pinning effects are small. A typical example of the bimodal structure is shown in Figure 2.14 in the 350°C rolled specimen tested under uniaxial tension at 480°C and $0.0001 s^{-1}$. This microstructure showed the greatest instability in the grip region as grain

Figure 2.14. (a) Grain refinement caused by the presence of SiC particles in a 7064/SiC composite deformed at 480°C at a strain rate of $10^{-4} s^{-1}$. It is believed that this structure undergoes dynamic recrystallization. (b) Grip area from the same specimen ($\varepsilon = 0$) shows grain coarsening between particle clustered regions.

growth caused the grain size to increase from 8.2 μm to 10.1 μm, whereas, with concurrent deformation, the same material underwent significant grain refinement, decreasing the grain size to 4.8 μm.

Depending on their size, the SiC particulates generally play different roles during processing: accelerating or retarding recrystallization and grain growth. In this case, it is apparent that the SiC particles facilitated nucleation of new grains (aided by straining and the formation of high-angle grain boundaries around the particles). However, as far as the particle-pinning effect is concerned, they provided little resistance to boundary migration simply because their sizes were too large to allow them to exert any high pressure on dislocations.

Attempts to store large amounts of lattice defects in particulate composites by deformation processing at ambient temperatures require avoidance of cavitation and fracture through the use of superimposed hydrostatic pressure. Figure 2.15 shows a TEM microstructure of a 7064Al/SiC$_p$ composite that has been severely deformed and given a short-time annealing treatment to nucleate new grains. A distinct band of a defect-free zone is seen at the particle interface that does not show any cell structure. It is believed that the high levels of deformation experienced by matrix shearing in this zone is recovered during the deformation step, and that there is inadequate strain left for the process to generate fine grain sizes in this region. With the appropriate thermomechanical processing method, however, fine recrystallized grain size can be obtained in these composites. A discussion of such methods and utilization of the resulting product are summarized below.

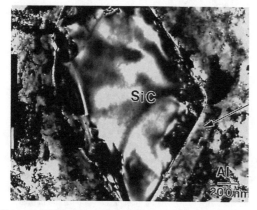

Figure 2.15. Heavily cold-worked 7064 Al/SiC composite exhibiting regions of high shear and a band of recovered material near the particle interface (arrow).

2.3.4 Superplastic Deformation

Particulate composites with micrograin superplasticity have been produced by both foil processing (7475 Al) (Mahoney and Ghosh 1987; Mahoney et al. 1987) and by a P/M process (7064 Al). In early work, high-strength aluminum alloys of the Al-Zn-Mg-Cu type with Cr-containing dispersoids were used. The latter alloy had Cr, Zr, and Co as dispersoid formers. The foil method has the advantage of low oxide content because of decreased surface area in comparison to the powder alloy. The starting point for the foil work was a fine grain 7475 Al sheet produced via a thermomechanical process (Paton and Hamilton 1978). When this process is performed a second time on this sheet, grain sizes of the order of 6 μm can be obtained. The foils (0.10 mm thick) so produced are then sprayed with 1200 mesh β SiC particulate (or whiskers) in a volatile binder to obtain a uniformly thick coating. A schematic illustration is shown in Figure 2.16 (Ghosh et al. 1984). Multiple foil layers are then stacked, outgassed, and forged. The extensive deformation exposes a new Al surface that causes interlayer bonding and forms a billet. The fine grain size allows consolidation to take place at relatively low pressure (2.85 MPa). The billet is then thermomechanically processed into a sheet. This sheet is superplastic at 516°C, with elongations in the 400% to 500% range for 10 to 15 vol % reinforcement.

In the P/M process, 7064 Al prealloyed powder (produced by Kaiser Alumina, Pleasantown, CA) was blended with SiC particulates and cold isostatically pressed to make a green compact. The inert gas (N$_2$) atomized powder (150 μm average particle diameter) is found to be of superior quality and produced excellent superplasticity as an unreinforced alloy (Mahoney and Ghosh 1987). The compact is then hot outgassed in a can with intermediate flushing of an inert gas and then hot pressed to 100% density. After removal of the can, the billet is hot extruded with at least an 18:1 extrusion ratio. This step is very important to cause breakup of the powder surface oxide as well as to achieve good mechanical properties in the P/M composite. The extruded billet is then thermomechanically processed into a sheet as before (Ghosh and Mahoney 1988).

A schematic illustration of the P/M processing steps is shown in Figure 2.17. The initial microstructure in the as-hot-pressed billet shows finer SiC particles in the interstices between coarser Al powder particles. Upon extrusion, the Al powder particles become elongated and thinner with continued tumbling of the SiC particles between them. As the elongated Al particles stretch and shear, mixing with finer SiC occurs in a reasonably homogeneous manner. The thermomechanically processed microstructure is well recrystallized with a high

Solid-State Processing 35

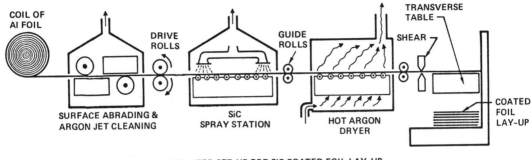

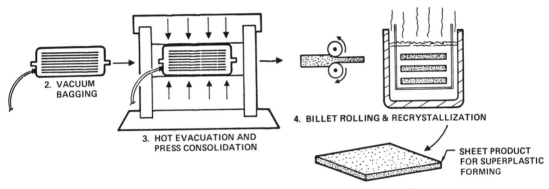

Figure 2.16. Schematic illustration of the steps involved in the foil-particulate composite fabrication method *(From: Ghosh, Hamilton and Paton, 1984).*

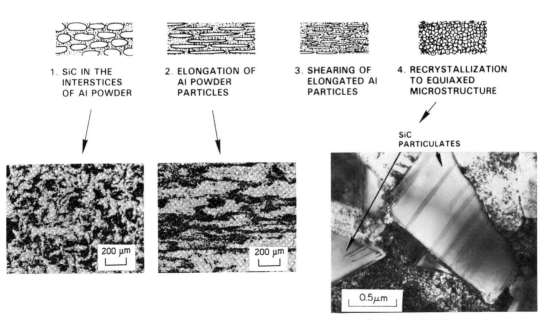

Figure 2.17. Schematic model (with microstructures) at various stages of P/M composite fabrication.

density of dislocations present between SiC reinforcements as shown in the transmission electron micrograph on the bottom corner of Figure 2.17. The method for thermomechanical processing involves solution treatment (at 500°C) followed by overaging (at 400°C for 8 hours) to generate many coarse precipitates, then warm rolling (at 250°C to 300°C) the alloy to produce a highly deformed (large local density of dislocations) structure. Upon heating the composite alloy to its recrystallization temperature (500°C), a fine, equiaxed microstructure results (Figure 2.18). This material is superplastic at 500°C at a strain rate of approximately $5 \times 10^{-4} s^{-1}$ with elongations in the 400% to 500% range for 10 to 15 vol % SiC reinforcement. To prevent void formation during superplastic forming, hydrostatic pressure of the order of material flow stress may be superimposed, thereby providing the large tensile elongations in this material.

Figure 2.19 shows stress versus strain-rate data for the P/M composite for several levels of reinforcement content. It is clear that there is greater composite strengthening at lower strain rates, and the high m (= $d\log\sigma/\log\dot{\epsilon}$) of the unreinforced matrix is lost with increasing reinforcement-volume fraction. This leads to an adverse effect on the uniformity of deformation with increasing SiC content. It has been found that void formation during deformation cannot be suppressed at higher strain rates because it is difficult to superimpose more than 6 MPa of hydrostatic pressure. The lack of adequate composite strengthening at higher strain rates is believed to be due to particle-matrix interface decohesion and excessive cavitation. At very low strain rates, the extent of composite strengthening has been found

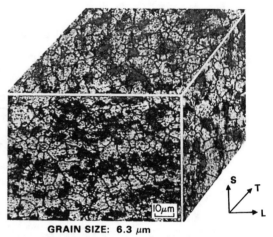

GRAIN SIZE: 6.3 μm

Figure 2.18. Microstructure of P/M 7064 Al/SiC composite after thermomechanical processing.

to be greater than that predicted by continuum models. It is believed (and quantitatively agreed) that the retardation of grain boundary sliding caused by the SiC particles (invariably situated on grain boundaries) may be responsible for the extent of composite strengthening at lower strain rates.

Figure 2.20 shows the effect of superimposed hydrostatic pressure on the stress-strain curve of the P/M composite under superplastic conditions. One can see that cavitation (atmospheric pressure data) can cause significant weakening in comparison to the case where superimposed pressure of 300 psi (2 MPa) was used. Correspondingly, considerably larger tensile elongation is achievable when cavitation is avoided. The degree of

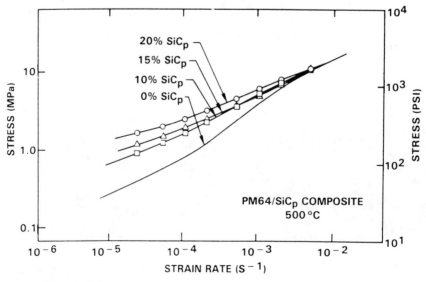

Figure 2.19. Stress-strain rate data for P/M 7064 Al/SiC composite at 500°C.

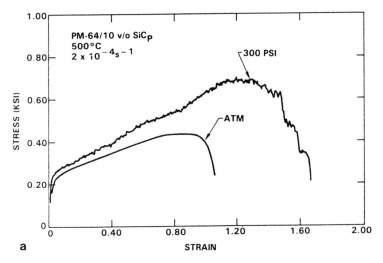

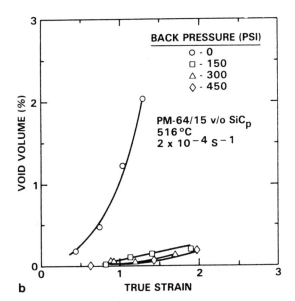

Figure 2.20. The effect of superimposed hydrostatic pressure on (a) the stress-strain curve of 7064 Al/SiC (top) and (b) the cavitation void volume (bottom).

cavitation (as measured by the density change technique) shown in Figure 2.20 also indicates that 300 to 450 psi (2 to 3.2 MPa) is sufficient to suppress cavitation up to fairly large strain levels.

Although complicated shapes can be fabricated using both types of composites discussed above, the mechanical properties after superplastic forming also appear to be quite respectable (Table 2.1). High tensile strength and room temperature tensile elongation of 5% to 7% with a 30% to 40% increase in elastic modulus over base-aluminum alloys offer an advanced capability in aerospace materials.

Nieh and others (Nieh 1984; Wu and Sherby 1984) have examined two approaches to obtaining superplastic elongations in these types of composites other than the micrograin superplasticity effects discussed so far. In the former work, large CTE differences between the matrix and SiC induce large dislocation density during thermal cycling of the composites. Because the matrix has a much higher CTE than the reinforcement, cooling cycles induce tensile hoop strain and compressive radial strain in the matrix (proportional to $\Delta\alpha\Delta T$), which lead to plastic shear. Dislocations are stored in the matrix during the cool-down cycle while during the heat-up cycle, these regions undergo dynamic recovery (and recrystallization) and contribute to large amounts of accumulated strain under a small applied load. Chen et al. have recently taken advantage of this mechanism to blow form (by using gas pressure) sheet samples of 6061 Al/10/SiCp into hemispherical shapes (Chen et al.

Table 2.1. Characteristics of fine grain high strength Al/SiC composites

Material	Grain Size (μm)	Yield Strength		U.T.S.		Elastic Modulus		Tensile Elongation	Superplastic Elongation PCT
		KSI	MPa	KSI	MPa	MSI	GPa		
7475 Al/ 15V% SiC$_p$	6	70	483	81	558	14	97	7%	500 (at 516°C, $2 \times 10^{-4} s^{-1}$)
7064 Al/ 15V% SiC$_p$	6	82	565	88	607	13	90	5%	460 (at 500°C $5 \times 10^{-4} s^{-1}$)

1990). It was found that Al composites become much softer as a result of thermal cycling and that they become formable. It is claimed that strain-rate sensitivity during this process approaches unity because strain accumulated during each cycle is proportional to stress. However, this kind of stress does not represent instantaneous strain-rate sensitivity (m) value that controls necking resistance. It is well known that when m approaches unity, the distribution of strain within the part should also become uniform (Ghosh 1977; Cornfield and Johnson 1970). As demonstrated by the strain distribution plots in the work by Chen et al., newtonian viscous behavior is not exhibited by these materials. Although this type of deformation is an interesting phenomenon to study, it is not yet a practical method for forming.

In the work by Nieh (1984), an aligned SiC whisker reinforced with 2124 Al alloy was deformed at a high strain rate near its solidus temperature. This type of deformation is likely to cause continuous recrystallization of the alloy with associated viscous flow. Recently, the work on high-rate superplasticity was greatly expanded by the work of Japanese researchers (Mabuchi et al. 1991; Higashi et al. 1992). It was found that, in a 6061 Al/Si$_3$N$_4$ particulate (or whisker) reinforced alloy, 400% to 600% tensile elongations are possible by deforming the alloy at an extremely high strain rate of 1 to 100s^{-1} at a relatively high forming temperature (833 K). The alloys are specially prepared from a clean starting powder and a 100:1 extrusion ratio to develop an extremely fine grain size (0.2 μm) in the rolled sheets. At these high strain rates, the mechanism for deformation is expected to be dislocation creep, although it is possible that adiabatic heating and the high rate of defect production can lead to enhanced diffusivities, particularly near grain boundaries and interface boundaries. The strength of the particle-matrix interface and the nature of chemical interaction between nitride and the nearby molten aluminum may also have a strong influence on this process. If such a chemical interaction is possible, it is likely to contribute to a viscous interface process and preclude strain localization and fracture.

2.3.5 Forging

Net-shape fabrication of complex composite shapes by hot forging or superplastic forging is of significant recent interest. Automotive connecting rods, missile components, navigational systems, and structures for space applications are being considered (or made) as composite forged products. Forging is often limited by cracking on the outer surface during the process of forging. In the closed-die forging process, it has been found that the workpiece can develop an incipiently cracked surface at an intermediate stage of forging, but surface cracks may no longer be visible at the end of the process as the workpiece is firmly pressed against the die. Closed-up surface cracks are not acceptable for fatigue-loaded structures and so a knowledge of forging limit for the appearance of outer surface cracks is essential.

Syu and Ghosh have investigated the limiting strains during forging of particulate-reinforced MMCs (Syu and Ghosh 1992). The forging-limit diagrams are plots of tensile surface strain at the point of incipient surface cracking as a function of compressive strain applied to a series of samples of varying geometry and lubrication. This was popularized in the 1970s by Kuhn et al. 1973, Kobayashi 1970, and others. Figure 2.21 shows forging limit diagrams for 6061 Al/20v%SiC$_p$ at 400°C produced via the P/M route. The SiC particulates are 1 to 5 μm in size and jagged. This compares with results on 6061 Al/20v%Al$_2$O$_3$ (microsphere) reinforced material produced by the ingot metallurgy route. The microspheres are considerably larger (10 to 30 μm) in size. 6061 Al alloy is known to have good high-temperature ductility. The grain size of the extruded P/M alloy composite matrix is considerably finer (< 3 μm) as

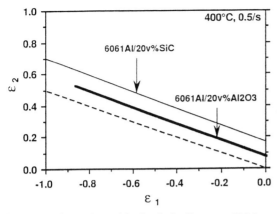

Figure 2.21. Comparison of forging limits diagrams of P/M 6061 Al/20v%SiC and 6061 Al/20v%Al$_2$O$_3$ (microspheres), cast and extruded.

compared with the microsphere reinforced material (~ 15 μm). During elevated temperature forging, the finer grain matrix does maintain a lower flow stress and minimizes cracking tendencies. Thus, matrix grain size and ductility are the most important factors in enhancing the forgeability of particulate composites.

The fracture path for these composites is typically through the matrix and so the strength of particle-matrix interface may not be a critical factor for such fracture. Conversely, the perturbation of flow around the large spherical particles is so significant that a combination of high local shear strain and hydrostatic tension is generated between these particles, and may lead to early fracture. Polycrystalline microspheres also exhibit more damage during forging than do fine SiC particles. However, the jaggedness of particles may not be as critical a factor in terms of fracture at these high temperatures as it is at the lower temperature. Typically, for the 6061 Al matrix, the composite forgeability is found to increase with increasing test temperature up to 500°C.

Figure 2.22 shows a comparison of forging limits for 6061 Al/Al$_2$O$_3$ with that of 2014 Al/Al$_2$O$_3$, the latter having been produced by the ingot metallurgy route using Al$_2$O$_3$ platelets. Clearly, even with smaller-size platelets, the forging limit for the 2014 Al composite is lower and it drops even further at higher temperatures. Again lower matrix forgeability has a strong bearing on the overall forgeability of the composite.

2.3.6 Deformation Processing of Fiber Composites

Deformation processing is rarely used with fiber composites, except when the reinforcement is metallic, because of the extensive fiber fracture that would result. With metal-metal fiber composites, a system that has attracted great interest for combustion chamber liner applications is that of Cu-Nb composites produced by co-drawing or roll bonding. Cu and Nb have very limited solid solubility, which is instrumental in developing a strong interfacial bond without excessive chemical reaction. Cast and rapidly cooled alloys of Cu-12% Nb and Cu-20% Nb have been studied by Spitzig et al. 1987. Because both phases are highly ductile, with an increasing degree of deformation, the Nb precipitates adopt a filamentary morphology with the filament cross-section being curled. Figure 2.23 shows the increase in the strength of the composites as a function of drawing strain, compared with that of the Cu matrix. The strengthening is considerably greater than rule of mixtures strength and increases as the extent of deformation increases. Thus, while a portion of the strength increase comes from strain hardening of Cu, the additional strengthening cannot be explained by strain hardening of Nb. When the fiber spacing enters the micron range, deformation of the Cu phase becomes highly restricted. This can be shown from dislocation slip distance arguments as well as through continuum analysis. This difficulty of yielding and plastic flow in Cu is responsible for the extremely high strengths that can be developed in these composites. Recrystallization effects can reduce this strength somewhat, but not to a significant degree. Electrodeposited Cu-Ni layered microcomposites have also shown extremely high strength as the layer thickness falls below 1μm (Tench and White 1984).

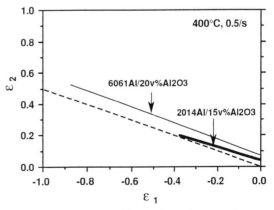

Figure 2.22. Comparison of forging-limits diagrams of 6061 Al/20v%Al$_2$O$_3$ microspheres with 2014 Al/15v%Al$_2$O$_3$ particulates, both cast and extruded.

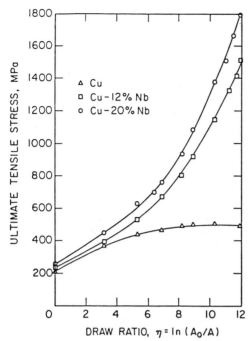

Figure 2.23. Strength increase as a function of drawing strain in Cu-Nb composites. *(From: Spitzig et al. 1987.)*

2.4 Summary

State-of-the-art solid-state processes for fabrication of MMCs cover a wide range of materials, including light-alloy matrices, high-temperature matrices, special application cases such as high-conductivity matrices, and a variety of particulates, whiskers, and fibers as reinforcements. Processing methods involving several solid-state processes, deposition, and deformation have been discussed briefly. The following points are highlighted from the text.

1. The metal-reinforcement interface plays a critical role in MMC properties and so of prime concern during consolidation is the prevention of degradation of the interface region. A high interfacial bond strength is desirable for load transfer in MMCs, but ductility and toughness may be adversely affected.
2. Thermal mismatch strains near the interface after cooling from the fabrication are typically in the plastic regime. Without a ductile layer near the interface, the matrix can often exhibit cracking as a result of the large tensile residual stresses. Fiber-foil consolidation processes often use interface treatments to impart ductilizing effects and barriers for further diffusion.
3. Complicated net shape parts can be fabricated by monotape development (deposition processes) and diffusion bonding. High-temperature composites typically require a diffusion barrier coating on the fiber surface, particularly when a reactive matrix such as titanium is used. Without this, excessive reaction leading to brittle phase formation and fiber notching can take place.
4. For fabrication of discontinuously reinforced MMCs and for achieving the most uniform reinforcement distribution, elimination of agglomerates poses the greatest challenge. A significant amount of deformation during primary and secondary processing is required to achieve this goal. A physical model of agglomerate break-up during processing has been presented.
5. Matrix grain structure is affected by the presence of nondeformable reinforcement particles during rolling and other secondary processing operations. It is possible to have dynamic recrystallization triggered by these particles at high temperatures although grain coarsening may occur in undeformed areas.
6. Inert processing of prealloyed powder/reinforcement blends and a significant amount of deformation imparted to the composite are key to obtaining high mechanical properties in discontinuous reinforced composites. This helps the break-up of both the powder surface oxides and the agglomerates.
7. By refining the grain size of the matrix, a high degree of superplasticity can be achieved in Al/Sic (discontinuous) composites. Diffusional creep is believed to be operative as a flow mechanism; however, added resistance to initiating grain boundary sliding is present because of SiC particulates.
8. In discontinuous reinforced composites, significant triaxial tensile stresses can develop during deformation processing. It is possible to suppress cavitation in reinforcement clusters by applying superimposed hydrostatic pressure during superplastic forming at low strain rates. Cavitation is more difficult to suppress at high strain rates.
9. By reducing the spacing between the hard phases in a metal-metal composite (Cu-Nb) to within a few microns via deformation processing, significant strengthening in excess of rule-of-mixtures is possible. This is because of the difficulty of slip in a highly constrained soft phase (Cu).

Acknowledgments

The author wishes to acknowledge the assistance of Dr. Yun Zhu and Dr. Charles Syu, former members of the University of Michigan Materials Engineering Department, for providing some of their research results for this paper. Their work was performed under AFOSR Contract No. F49620-86-C-0058, Dr. A.H. Rosenstein,

Program Monitor, and also partly supported by General Motors-AES, Dr. Brian Taylor, Program Monitor. Figure 2.2 was obtained from the Master's thesis of Mr. A. Sunderrajan under AFOSR Grant No. DOD-G-AFOSR-90-0141. Contributions of test materials and/or funding from DWA Composites (now BP) and Comalco are also gratefully acknowledged. Finally, the author is indebted to Dr. Subra Suresh and Dr. Andreas Mortensen for their extreme patience during the period of writing this paper. Thanks also go to Ms. Bonni Viets for typing the manuscript.

References

Arzt, E., Ashby, M.F., and K.E. Easterling. 1983. *Metall. Trans. A.* 14A:211.
Backman, D.G., Mehrabian R., and M.C. Flemings. 1977. *Met. Trans. B.* 8B:471.
Besson, J., and A.G. Evans. 1992. *Acta Met. et Mat.* 40:2247.
Brindley, P.K. 1987. *Mat. Res. Soc. Symp. Proc.* 81:419.
Chen, Y-C., Daehn, G.S., and R.H. Wagoner. 1990. *Scripta Met.* 24:2157.
Christman, T., Needleman, A., and S. Suresh. 1989. *Mat. Sci. and Eng.* 107A:49.
Cooper, G.A., and A. Kelly. 1969. *ASTM STP452*. Philadelphia, PA: ASTM.
Cornfield, J., and W. Johnson. 1970. *Int. J. Mech. Sci.* 12:479.
Geiger, A.L., and J.A. Walker. 1991. *J. of Metals.* 43:8.
Ghosh, A.K., U.S. Patent, Ser. No. 217,253, July 11, 1988.
Ghosh, A.K. 1977. *Acta Met.* 25:162.
Ghosh, A.K. 1989. *In:* Principles of Solidification and Materials Processing. Vol. 2. (R. Trivedi, J.A. Sekhar, and J. Mazumdar, eds.), 585, New Delhi, India: Oxford & IBH Publishing Co. Pvt. Ltd.
Ghosh, A.K., Hamilton, C.H., and N.E. Paton. 1984. U.S. Patent No. 4,469,757, September 4, 1984.
Ghosh, A.K., and M.W. Mahoney. 1988. U.S. Patent No. 4,722,754, February 2, 1988.
Grierson, R., and R.H. Krock. 1972. Techniques for Fabrication of Composite Materials, Techniques of Metals Research, Vol. VII, Part 1 (R.F. Bunshah, ed.), Interscience Publishers.
Gunasekera, J.S., and S. Hoshino. 1980. *Annals Int. Inst. Prod. Eng. Res. (CIRP)*, 29:141.
Higashi, K., Okada, T., T. Mukai, et al. 1992. *Scripta Metall. et Mater.* 26:185.
Jatkar, A.D., Schelling, R.D., and S.J. Donachie. 1986. *J. M.* 38:74.
Karmakar, S.D., and A.P. Divecha. 1992. NSWC Report No. 753654.
Keyworth II, G.A. 1985. National Aeronautical R & D Goals, Executive Office of the President, Office of Science and Technology Policy, Washington, D.C.
Kingery, W.D., Bowen, H.K., and D.R. Uhlmann. 1975. Introduction to Ceramics, 2nd Ed. 498. New York: J. Wiley & Sons.

Kobayashi, S. 1970. *J of Eng. Ind.* May:391.
Kuhn, H.A. 1978. Formability—Analysis, Modeling and Experimentation. *In:* (S.S. Hecker, A.K. Ghosh, and H.L. Gegel, eds.), Cleveland, OH: American Society of Metals.
Kuhn, H.A., and P.W. Lee. 1973. *Metal Trans. A.* 4:969.
Mabuchi, M., Higashi, K., S. Okada et al. 1991. *Scripta Met. and Mat.* 25:2517.
Mahoney, M.W., and A.K. Ghosh. 1987. *Metal. Trans. A.* 18A:653.
Mahoney, M.W., Ghosh, A.K., and C.C. Bampton. 1987. *In:* ICCM and ECCM Proceedings. Vol. 2. (F.L. Mathews, N.C.R. Buskell, J.M. Hodgkinson, and J. Morton, eds.), Elsevier, Applied Science Publishers, 2.372.
McDaniels, D.L., Serafini, T.T., and J.A. Dicarlo. 1986. *J. Materials for Energy Systems.* 8:80.
Metals Handbook, vol. 15. 1992. *Centrifugal Casting.* Cleveland, OH: American Society of Metals.
Murr, L.E., Staudhammer, K.P., and M.A. Meyers. 1986. Metallurgical Applications of Shock-Wave and High-Strain-Rate Phenomena, 129, New York: Marcel Dekker, Inc.
Nieh, T.G. 1984. *Metall. Trans. A.* 15A:139.
Paton, N.E., and C.H. Hamilton. 1978. U.S. Patent No. 4,092,181, May 15, 1978.
Petrasek, D.W., McDaniels, D.L., Westfall, L.J., and J.R. Stephens. 1986. *Metal Progress.* 136:26.
Piggott, M.A. 1980. Load Bearing Fiber Composites. London: Pergamon Press.
Rawal, S., and M. Misra. 1990. *Sampe Conf. Proceedings*, 317–326.
Reed, J.S. Introduction to the Principles of Ceramic Processing. 1988, 23, 397. New York: J. Wiley & Sons.
Schoutens, J.E. 1982. *MMCIAC Tutorial Series*, Introduction to MMC Materials, 272, Santa Barbara, CA: Kaman Tempo.
Smith, C.S. 1948. *Trans. AIME*, 175:15.
Spitzig, W.A., Pelton, A.R., and F.C. Laabs. 1987. *Acta Met.* 35:2427.
Stanford-Beale, C.A., and T.W. Clyne. 1989. *Composite Science and Technology.* 35:121.
Syu, D-G.C., and A.K. Ghosh. 1992. A comparison of forging limits of several aluminum matrix composites. Submitted to *Metall. Trans. A.*
Tench, D., and J. White. 1984. *Metall. Trans. A.* 15A:2039.
Textron Specialty Materials. 1990. Databook for Silicon Carbide Materials. Lowell, MA.
Wert, J.A. 1985. *Deformation, Recovery and Recrystallization.* (E.H. Chia, and H.J. McQueen, eds.), 67, Warrendale, PA: TMS-AIME.
Whittenberger, J.D., Artz, E., and M.J. Lutton. 1990. *J. Mater. Res.* 5:271.
Wu, Y., and E.J. Lavernia. 1991. *JOM.* 16. Vol. 43, No. 8.
Wu, M.Y., and O.D. Sherby. 1984. *Scripta Met.* 18:773.
Zweben, C., and G. Dvorak. 1985. Metal Matrix Composites, UCLA Short Course Book. Los Angeles, CA: Lifelong Learning, UCLA extension.

Chapter 3
Capillary Phenomena, Interfacial Bonding, and Reactivity

NICOLAS EUSTATHOPOULOS
ANDREAS MORTENSEN

Metal-matrix composites are generally produced by combining a metallic matrix with a pre-existing reinforcement. This processing feature allows for added freedom when conducting microstructural engineering because the bond that unites the two constituent materials of the composite can be altered by means of matrix alloying, reinforcement surface coating, or alteration of composite fabrication process temperature cycles. However, many of the difficulties encountered in the processing of metal-matrix composites, such as poor wetting of the reinforcement by molten metal, are related to the interface. The ability to engineer the interface thus represents both a challenge in the processing of these materials, as well as an additional opportunity for optimization of their properties.

The question of interface design and optimization is complex and not fully understood. There are many parameters that either affect the structure of the interface as it evolves during processing or that relate interface structure with composite performance. Desirable interfacial features are, however, often contradictory. As an example based on a very elementary level of analysis, high chemical affinity between matrix and reinforcement is desired so that the two phases combine spontaneously, yet low chemical affinity is desired to avoid unwanted chemical reactions at the interface during processing of the composite.

Although metal-matrix composite engineering has mostly preceded scientific explanations of the phenomena it involves, sufficient scientific knowledge now exists to guide the materials engineer when seeking to improve both the processing and the properties of the composite. This scientific base addresses two essential classes of interfacial phenomena influencing the materials' processing: (1) physicochemical phenomena, which govern the nature of the bond or the interfacial zone that unites matrix and reinforcement, and (2) mechanical phenomena, which depend largely on interface geometry and on the nature of the composite fabrication process. These phenomena are the focus of this chapter, which begins with a presentation of governing parameters and a brief description of their measurement. The following section places focus on the principles that govern the physicochemistry of interfaces between metals and ceramics. The third and last main section is a succinct presentation of methods currently used to tailor the structure of the interface and the resulting performance of the composite.

3.1 Governing Parameters

3.1.1 Thermodynamics

The interfacial zone that links the reinforcement with the matrix can take a variety of forms, ranging from a single surface of atomic bonds (which we call a *simple interface*), to one, or even several, new reaction phases and simple interfaces located between the matrix and the reinforcement. These result from the chemical reaction between matrix and reinforcement. In all cases, an elementary quantitative characteristic of the interface that is relevant to composite processing is the chemical free energy change that accompanies its formation.

In processes such as infiltration and dispersion casting, in which the reinforcement is combined with bulk metal, this free energy change is ΔG_i, the change in free energy upon replacement of a square meter of reinforcement surface, of surface energy σ_{CV}, with a square meter of interfacial zone, of energy G^s_{CM} per square meter of interfacial zone:

$$\Delta G_i = G^s_{CM} - \sigma_{CV} \quad (3.1)$$

In solid-state processes or spray casting, the metal is in finely divided form before it is combined with the reinforcement. The interface is then created at the expense of both a ceramic surface and a metal surface, so that the relevant free energy change per square meter of interface is ΔG_a, defined as the change in free energy upon replacement of a square meter of reinforcement surface, of surface energy σ_{CV}, plus a square meter of metal surface, of surface energy σ_{MV}, with a square meter of interfacial zone, of energy G^s_{CM} per square meter of interfacial zone:

$$\Delta G_a = G^s_{CM} - \sigma_{CV} - \sigma_{MV} \quad (3.2)$$

When there is no formation of interfacial chemical reaction products during processing, the interfacial zone is a simple interface and G^s_{CM} is the matrix-reinforcement surface energy, σ_{MC}. ΔG_i and ΔG_a then respectively become the work of immersion W_i:

$$W_i = \sigma_{CM} - \sigma_{CV} \quad (3.3)$$

or the negative of the work of adhesion W_a, which is conventionally defined as:

$$W_a = \sigma_{MV} + \sigma_{CV} - \sigma_{CM} \quad (3.4)$$

When there is no interfacial reaction, the sessile-drop experiment and other capillary tests provide unambiguous methods for characterization of interfacial bond energy.

In the sessile-drop experiment, a drop of liquid metal is caused to rest in a controlled atmosphere on a flat substrate that is representative of the reinforcing phase(s). The shape of the drop is recorded and this measurement is used to compute the surface tension of the liquid metal, σ_{MV}, together with the contact angle of the drop on the flat substrate, θ, which is defined in Figure 3.1. These parameters are related to W_i and W_a via the Young-Dupré equation shown in Equation 3.5.

$$\begin{aligned}\sigma_{CV} - \sigma_{CM} &= \sigma_{MV}\cos(\theta)\\ &= -W_i = W_a - \sigma_{MV}\end{aligned} \quad (3.5)$$

Some variants of the sessile-drop experiment have also been used to measure contact angles in other configurations such as experiments in which the substrate is a fiber (Nogi et al. 1991).

In measuring the contact angle θ, great care must be exerted to control several important parameters. Because deviations of the substrate surface geometry from a plane alter θ, the surface roughness of the substrate must be characterized (Chatain et al. 1986). The chemical purity of all phases present must be tightly controlled and documented. The composition and pressure of the vapor phase can exert a significant influence on θ. A good example of this can be seen in metals that form a strong oxide layer at their surface. In cases such as this, the oxygen partial pressure has been shown to strongly affect the measured value of θ because oxide formation at the metal drop surface prevents proper contact between metal and substrate (Brennan and Pask 1968; Laurent et al. 1988; Weirauch 1988b). For these reasons, published values for the contact angle θ must be used with caution.

The sessile-drop experiment has a very significant advantage in that, for simple interfaces, it yields directly the surface energy terms defined above and, because σ_{CM} and σ_{MV} vary little with solidification of the matrix (provided matrix solidification brings no significant changes in matrix composition along the interface by segregation of alloying elements) (Eustathopoulos et al. 1988; Miedema 1978; Pilliar and Nutting 1967), resulting measurements of W_a or W_i are applicable to both solid-state and liquid-metal processes.

When interfacial reactions take place, data generated by the sessile-drop experiment will bear a less direct relation to the thermochemistry of the interfacial zone

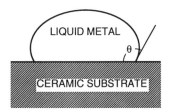

Figure 3.1. Schematic of a metal drop resting on a flat substrate in a sessile-drop experiment, to define the contact angle θ.

(see Section 3.2.2). Microstructural investigation, described in Chapter 5, and indirect methods of monitoring interfacial reactions, such as X-ray analysis (Lloyd 1989), measurement of changes in matrix composition (Zhang et al. 1991), matrix property variations (Salvo et al. 1991; Thanh and Suéry 1991), or differential thermal analysis (Kindl et al. 1992; Lloyd and Jin 1988), have mostly been used for characterization of reactive interfacial zones.

3.1.2 Mechanics

Depending on the process, W_i or $-W_a$ quantifies the free energy change upon reversible creation of the composite from its constituent phases: if W_i or $-W_a$ is negative, spontaneous composite formation is thermodynamically possible by the relevant process.

In composite fabrication processes, there are additional irreversible energy loss mechanisms, related to the mechanics of creating the interface. For instance, in solid-state processes the solid matrix must be deformed by plastic flow or by creep before it can come into contact with the reinforcement. This represents a considerable energy expenditure, and limits the rate of bond formation in processing. With molten-metal processes, deformation of the matrix is much easier, but still consumes energy by viscous friction or by inertial losses in the liquid metal. These expenditures of energy are important, often being greater than ΔG_i or ΔG_a and, consequently, they frequently dictate process parameters such as applied load or reinforcement preheating temperatures (see Chapters 1 and 2). An important consequence of these phenomena is that a wetting angle smaller than 90 degrees will generally not be a sufficient condition for spontaneous wetting of the reinforcement by the metal during composite processing (this is paralleled in brazing, where θ is quoted as preferably being less than 20 degrees for the molten brazing metal to penetrate joints [Nicholas 1986]).

Mechanical phenomena are also important on a more microscopic scale, in that they influence the formation and structure of the interfacial zone in the composite. In diffusion bonding of an aligned fiber composite, the microscopic process of creating the interface at the point farthest removed from the plane of the metal foils requires considerable metal deformation, far in excess of what is required to create most of the interface of the composite (see Figure 2.1, page 25). This may leave interfacial voids at these locations in the composite (e.g., Hall et al. 1987). Likewise, in infiltration with a nonwetting metal, very high (theoretically infinite) pressures are required to force the metal to come into contact with the fiber surface near the contact line between two fibers (Mortensen and Cornie 1987), which also may result in matrix voids along the interface at these locations, Figure 3.2. These examples show that microstructural features of the interfacial zone also depend on the microscopic mechanics of composite processing.

These mechanical phenomena, which cause irreversible energy losses in processing and govern microscopic features of the composite, are strong functions of the internal microscopic geometry of the reinforcement. For example, preventing fiber-to-fiber contact lines in preforms to be infiltrated will reduce the pressure required for full preform infiltration and will also eliminate interfacial voids.

To account for mechanical features of interface creation, and to remedy some other drawbacks of the sessile-drop technique in its applicability to the processing of metal-matrix composites (the experiment is relatively static, whereas processes such as infiltration involve rapidly moving liquid/solid/vapor contact lines), alternative methods have been devised to characterize interface creation as relevant to metal-matrix composite fabrication processes. These process-specific tests seek to account for the interplay between the physicochemistry and the micromechanics of interface creation. In infiltration, relevant measurement techniques generally involve measurement of the pressure required to fully infiltrate a preform of the reinforcement (Fletcher et al. 1991; Fletcher et al. 1988; Maxwell et al. 1990; Mortensen and Wong 1990; Oh et al. 1989a; Oh et al. 1989b; Seitz et al. 1989). Alternatively, so-called "drainage/imbibition curves" have been measured in one system (Kaufmann and Mortensen 1992). Such curves

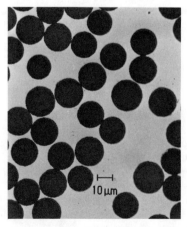

Figure 3.2. Microstructure of PRD166 (Wilmington, DE: DuPont de Nemours) alumina-zirconia fiber reinforced aluminum produced by pressure infiltration. Interfacial voids are apparent at several fiber contact lines.

have been used for decades in other branches of engineering such as hydrogeology and reservoir engineering, and are plots of volume-fraction matrix as a function of applied pressure (Mortensen 1991). In stir-casting processes, several authors have characterized wetting with various measures of ease of particle incorporation into stirred metal (Agarwala and Dixit 1981; Ghosh and Ray 1988; Hanumanth and Irons 1990). In other processes, no such characterization techniques seem to have been developed, either because the mechanics of wetting are still largely unknown (for example in spray casting) or because mechanical energies involved are so much higher than capillary energy terms that the latter exert no discernible influence (solid-state processes).

Various mechanical tests have been devised to characterize the strength of the bond uniting the solid metal with its reinforcement (Argon et al. 1989; Clough et al. 1990; Clyne and Watson 1991; Eldridge and Brindley 1989; Evans et al. 1990; Gupta et al. 1990; Oh et al. 1988; Reimanis et al. 1991; Roman and Aharonov 1992). If the interface were perfectly brittle, W_a would be the interfacial fracture energy according to the Griffith criterion for crack propagation, and could thus be measured by mechanical means. With a metal matrix, however, crack propagation along a metal/reinforcement interface always involves significant additional energy dissipation. Furthermore, the interfacial cracking mode also strongly influences the fracture energy (Evans et al. 1990), and thus W_a is not simply related to the mechanical behavior of the interface. Thus, results from mechanical characterization of the interface reflect the influence of many parameters, such as the mechanical characteristics of the matrix and the reinforcement as well as the morphology of the interface (location and shape of second phases and roughness of the interface being examples).

3.2 Physicochemistry of Wetting and Bonding

3.2.1 Nonreactive Systems

3.2.1.1 Oxide Reinforcements

Many reinforcements used in metal-matrix composites are made of an oxide, Al_2O_3 being an example. A useful but approximate criterion for nonreactivity at the matrix/oxide interface under vacuum or inert gas atmosphere is that the standard Gibbs free-energy change ΔG^o_R of the reaction

$$n\ Me + n'\ MO_n = n\ MeO_{n'} + n'\ M \quad (3.6)$$

be strongly positive (Me is the matrix and MO_n the reinforcing oxide). A more exact criterion takes into account the fact that some dissolution of M in Me will always occur, and therefore incorporates interactions between dissolved M and the matrix Me,

$$\Delta G^*_R = \Delta G^o_R + \Delta H^\infty_{M(Me)} \gg 0 \quad (3.7)$$

where $\Delta H^\infty_{M(Me)}$ is the partial enthalpy of M in Me at infinite dilution. $\Delta H^\infty_{M(Me)}$ has been measured for a great number of binary metal alloys (Hultgren et al. 1973; Kubaschewski and Alcock 1979), and estimated using Miedema's model (Miedema et al. 1977). When the criterion (Equation 3.7) is satisfied, some dissolution of M into liquid Me can still occur; however, the maximum amount of dissolution is in the range of a few parts per million (ppm), that is, on the order of the metallic impurity concentrations typically found in metals.

In nonreactive metal/oxide systems, wetting is generally marginal or poor (Table 3.1). The kinetics of wetting are rapid, the time needed to reach equilibrium being on the order of 10^{-3}s (Laurent 1988; Naidich et al. 1972). W_a typically represents from 15% to 40% of the work of cohesion W_c of Me ($W_c \approx 2\ \sigma_{MV}$, Table 3.1). The absolute values of $d\theta/dT$ and dW_a/dT, where T is temperature, are small, being negative for the former and positive for the latter. For example, a 1000 K rise in temperature causes a decrease in the contact angle of Ga on sapphire or quartz substrates of only 5 to 10 degrees and an increase in W_a of only 50 to 100 mJ · m^{-2} (Naidich and Chuvashov 1983). It is also observed that θ and W_a depend little on the orientation of the substrate oxide when it is a single-crystal (Chatain et al. 1987; Ownby et al. 1991; Ownby and Liu 1988).

There is, at the time of this writing, no model capable of describing satisfactorily the bond between a metal and an ionocovalent oxide (Chatain et al. 1988; Russell et al. 1991). It has been proposed that the only possible type of interaction between metal and oxide is in the

Table 3.1. Experimental values of the contact angle θ, the work of adhesion W_a and the ratio of W_a and metal cohesive energy W_c (approximately equal to $2\sigma_{MV}$), for nonreactive metals on monocrystalline alumina under high vacuum or neutral gas.

Metal	T (K)	θ (degrees)	W_a (mJ · m^{-2})	W_a/W_c
Pb	1,173	117	215	0.27
Sn	1,373	125	205	0.21
Ag	1,373	130	325	0.17
Cu	1,373	128	490	0.19
Ni	1,773	109	1,200	0.34

From a review by Chatain et al. 1986.

form of physical, van der Waals (vdW) interactions resulting from dispersion forces (Naidich 1981; Pask 1987). Calculated values of W_a for various metal/oxide pairs by Naidich and MacDonald and Eberhardt vary by less than 50%, while experimentally observed variations from pair to pair are as large as 500% (Naidich 1981; MacDonald and Eberhart 1965). It also seems, from theoretic considerations (Gubanov and Dunaevskii 1977; Hicter et al. 1988; Johnson and Pepper 1982) and thermodynamic considerations (Chatain et al. 1986), that even for nonreactive metal/oxide couples, chemical interactions localized at the interface are involved in the interfacial bond. On this basis, the following equation has been proposed for the work of adhesion of metal/oxide couples (Chatain et al. 1987):

$$W_a = -\frac{C}{\mathcal{N}_a^{1/3} V_{Me}^{2/3}} \left[\overline{\Delta H}_O^\infty (Me) + \frac{1}{S} \overline{\Delta H}_M^\infty (Me) \right] \quad (3.8)$$

where $\Delta H^\infty_{O(Me)}$ and $\Delta H^\infty_{M(Me)}$ are the partial enthalpy of mixing at infinite dilution of oxygen and oxide metal M in the metal Me, respectively, $\mathcal{N}_a$ is Avogadro's number and V_{Me} is the molar volume of the liquid metal (the quantity $\mathcal{N}_a^1 V_{Me}^{2/3} \equiv \Omega_{Me}$ is roughly the area occupied by a monolayer of one mole of Me atoms). C and S are empirical constants that depend only on the nature of the oxide substrate. Equation 3.8 gives good results for both alumina (Chatain et al. 1987) and silica (Sangiorgi et al. 1988), with $C \approx 0.2$ and $s \approx n$ ($n = 1.5$ for alumina and 2 for silica).

This equation has two limitations. In some refractory oxides, including rare-earth oxides such as Y_2O_3, the ratio of the radius of the cation to that of the oxygen anion can be as much as twice that found in alumina or silica (Naidich et al. 1990). In this case, S would be lower than n, so that values of W_a estimated by Equation 3.8 would give only a lower limit of this quantity. Also, Equation 3.8 is only valid for oxides that are electrical insulators. Some oxides, such as some titanium oxides, are partly metallic, and it has been proposed by Ramqvist (Ramqvist 1965) and Naidich (Naidich 1981) that metallicity increases metal/ceramic adhesion. This seems to be confirmed by experiment (Table 3.2), which shows that when the metallicity of the oxide increases, W_a in contact with pure copper increases much more than is predicted by Equation 3.8.

3.2.1.2 Other Reinforcements

As with oxide ceramics, wetting of nonreactive covalent ceramic such as boron carbide and silicon carbide is generally poor (Table 3.3). In some cases, contact angles as large as 150 degrees, corresponding to very low values of W_a/W_c, have been obtained and explained by

Table 3.2. Contact angle θ and work of adhesion W_a of copper on different oxides at $T = 1423$ K. The metallic character of the oxides increases from top to bottom.

Oxide	θ (degrees)	W_a (experimental) (mJ · m^{-2})	W_a (calculated) (mJ · m^{-2})
Al_2O_3	128	460	550
Ti_2O_3	113	740	670
$TiO_{1.14}$	82	1,460	750
$TiO_{0.86}$	72	1,650	860

Experimental data are from Naidich 1981. Calculated values were derived from Equation 3.8.

assuming that interface bonding is entirely due to vdW interactions (Naidich 1981). Conversely, with carbides of metallic character, such as those of some d-metals, W_a increases by almost an order of magnitude and good wetting is observed (Naidich 1981; Ramqvist 1965).

In the absence of interfacial reactivity, the wetting of liquid metals on carbon substrates, of both graphite and diamond, is similar to that of covalent carbides and is nearly independent of temperature. Calculations of vdW interactions between metals and C by Naidich give the right order of magnitude for W_a and account for significant variations (of a factor of five or less) between different metals, in agreement with experiment (Naidich 1981).

3.2.1.3 Influence of Alloying Additions

The addition of an alloying element to the melt metal Me can influence W_a and θ directly by adsorption of the alloying addition to the metal/ceramic interface or to the metal/atmosphere interface, respectively leading to a decrease of σ_{CM} and σ_{MV}. It can be seen from Equation 3.5 that interfacial adsorption improves both adhesion and wetting. Surface adsorption, however, always deteriorates adhesion and only improves wetting if θ is originally below 90 degrees (Figure 3.3) (Li et al. 1988a). These effects have been modeled by Li et al. using the classical approximation of a monolayer surface, in agreement with experimental results on binary alloy/alumina systems (Li et al. 1989).

To obtain greater improvements in wetting, a two-solute approach can be used (Li et al. 1988b). This approach requires the addition of (1) a first solute element to produce a transition from nonwetting to wetting (θ < 90 degrees) by adsorption at the metal/ceramic interface, and (2) a second tensioactive solute that segregates to the melt free surface to reduce its surface tension and thus decreases θ further. Quantitative criteria for effectiveness of this approach are that

Table 3.3. Contact angle θ, the work of adhesion W_a and the ratio of W_a and metal cohesive energy W_c (approximately equal to $2\sigma_{MV}$) for two nonreactive metals on various carbides and carbon substrates at 1323 K for Sn and 1373 K for Cu.

Metal	Ceramic	θ (degrees)	W_a (mJ·m^{-2})	W_a/W_c	References
Sn	BN (hex.)	150	60	0.07	(Naidich 1981)
Sn	C (gr.)	149	65	0.07	(Naidich 1981)
Sn	SiC	135	145	0.15	(Naidich 1981)
Sn	B$_4$C	135	145	0.15	(Naidich 1981)
Cu	BN (hex.)	146	225	0.085	(Naidich 1981)
Cu	C (gr.)	140	315	0.12	(Naidich 1981)
Cu	B$_4$C	136	380	0.14	(Naidich 1981)
Cu	Cr$_3$C$_2$	45–47	≈2200	0.85	(Naidich 1981; Ramqvist 1965)
Cu	WC	20	2500	0.97	(Ramqvist 1965)
Cu	Mo$_2$C	18	2500	0.97	(Ramqvist 1965)

Hex. = hexagonal; gr. = graphite.

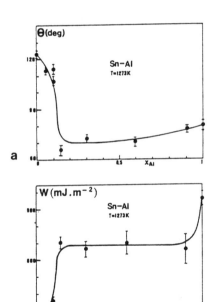

Figure 3.3. (a) Contact angle θ and (b) work of adhesion W_a isotherms of Sn-Al alloys in contact with monocrystalline alumina at $T = 1273$ K. When Al is added to Sn, θ decreases because the solid-liquid interfacial energy σ_{CM} decreases. When Sn is added to Al, θ also decreases, but this is because the liquid-vapor surface energy σ_{MV} is decreased. As a result, the curve of contact angle versus composition passes through a minimum. (From: Li, J.G., et al. Journal of Materials Science, vol. 24, 1109–1116. © Copyright 1989 Chapman & Hall.)

the adsorption energy of the solute at the metal/ceramic interface E_{CM} and at the metal/vapor interface E_{MV} be strongly negative. These parameters are defined (Li et al. 1989) as

$$E_{CM} = E_{MV} + (W_a^{Me} - W_a^A)\Omega_{Me} \quad (3.9)$$

$$E_{MV} = (\sigma_{MV}^A - \sigma_{MV}^{Me})\Omega_{Me} - m\lambda \quad (3.10)$$

where λ is the exchange energy of the Me-A solution modeled as a regular solution (a quantity that is evaluated from enthalpy of mixing data) and m is a structural parameter roughly equal to ¼.

An example of this approach is found in Cu-Al-Sn alloys on alumina: adding Al and Sn to Cu respectively decreases Cu/Al$_2$O$_3$ and Cu-vapor interfacial energies (Li et al. 1988b). There are two limitations to this approach. First, significant improvements in wettability require alloying additions in rather high quantities (typically around 10 at. % or more), which may produce undesirable changes in matrix bulk properties. Secondly, even in the most favorable cases, θ cannot be reduced below about 60 degrees.

3.2.1.4 Influence of Oxygen

In all metal/ceramic systems, the role of oxygen is of particular interest, as this element is present in most processes and because it influences the surface properties of many metals at partial pressures as low as 10^{-15} atm. (Eustathopoulos and Joud 1980). The interaction of liquid metals with oxygen results either in the formation of oxides, a tendency that predominates at low temperature, or, alternatively, yields a single metallic phase containing dissolved oxygen.

A typical example for the former case is aluminum. Near its melting point, the surface of liquid aluminum is covered by a film of aluminum oxide that inhibits the formation of Al/ceramic and Al/atmosphere interfaces, leading, in sessile-drop experiments performed under moderate vacuum, to a high value for θ of about 160 degrees, which is nearly independent of the nature of the substrate (Al_2O_3, SiC, C, . . .). This value changes little until about 1150 K to 1200 K, at which point a sudden decrease in θ to a value that depends on the ceramic occurs as the oxide layer is degraded by evaporation (Laurent et al. 1988) (Figure 3.4). The use of static atmospheres of neutral gas slows down evaporation of the oxide layer and displaces this transition to higher temperatures (Coudurier et al. 1984; Weirauch 1988a), while Mg (Weirauch 1988a) or Ca (Mori et al. 1983) additions to the aluminum melt interact with the film to reduce its effect.

Dissolution of oxygen increases the wettability of oxides by liquid metals even though the oxygen concentration may be as low as a few ppm or tens of ppm (Ownby and Liu 1988). An explanation of this phenomenon was given by Naidich, who proposed that oxygen in solution in the metal associates with metal atoms to form clusters having a partially ionic character, which results from charge transfer from the metal to the oxygen atoms (Naidich 1981). These clusters can therefore develop coulombian interactions with ionocovalent ceramics and, as a consequence, adsorb strongly at the metal/oxide interfaces. It has recently been shown that the beneficial effect of oxygen can be enhanced by adding to the metal Me a solute A capable of developing strong solute-solute interactions with dissolved oxygen (Kritsalis et al. 1990). The thermodynamic requirement for this is that $\varepsilon O^A \ll 0$, εO^A being Wagner's first order interaction parameter, defined by:

$$\ln(\gamma O) = \ln(\gamma O(Me)) + \varepsilon_o^A X_A + \ldots \quad (3.11)$$

where γO(Me) and γO are the activity coefficients of oxygen in pure Me and the Me-A alloy, respectively, and X_A is the molar fraction of A in Me. The more highly negative εO^A is, the stronger the interaction is between A and O, and the greater is the interfacial activity of a cluster of O and A atoms. Examples of this effect are found in O and Cr clusters in Cu/Al_2O_3 (Kritsalis et al. 1990), or O and Si clusters in Au/Al_2O_3 (Drevet et al. 1990).

3.2.2 Reactive Systems

Despite substantial progress reviewed by Laurent (Laurent 1988), a complete description of reactive wetting is still not available. Following Laurent, the smallest contact angle possible in a reactive system is given by

$$\cos(\theta_{min}) = \cos(\theta_o) - \frac{\Delta\sigma_r}{\sigma_{MV}} - \frac{\Delta G_r}{\sigma_{MV}} \quad (3.12)$$

where θ_o is the contact angle of the liquid on the substrate in the absence of any reaction. $\Delta\sigma_r$ is the change in interfacial energies brought about by the interfacial reaction. This term essentially accounts for the change in interface nature that results from the interfacial reaction. The last term on the right-hand side of Equation 3.12 was proposed by Naidich and Aksay et al. on the reasoning that the reaction between the liquid and a fresh, unreacted, solid surface at the periphery of the drop increases the driving force for wetting (Naidich 1968; Naidich 1981; Naidich and Perevertajlo 1971; Aksay et al. 1974). ΔG_r is the change in free energy per unit area released by reaction of material contained in the immediate vicinity of the metal/substrate interface.

One of the major difficulties associated with reactive wetting is the coupling of time-dependent interfacial reactivity with the kinetics of wetting. It was argued by Aksay et al. that because the rate of interfacial reaction is maximum at the very first moments of contact between the liquid metal and the substrate, the effect of this last term of Equation 3.12 is strongest in these early moments of wetting (Aksay et al. 1974). Thereafter, as the reaction kinetics slow down because the interface becomes saturated with reaction products and because diffusion becomes rate limiting, the same authors argued that the contact angle will increase and gradually

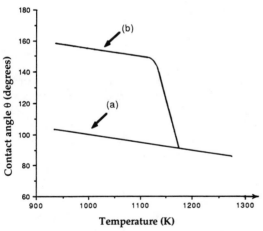

Figure 3.4. Schematic variation with temperature of the contact angle measured on a flat substrate for pure aluminum on alumina for two vacuum levels. (a) $P \approx 10^{-5}$ Pa and (b) $P \approx 10^{-3}$ Pa. The difference between the two curves arises because of an oxide layer covering the metal at higher pressures. At 1150 K, the oxide layer evaporates, which causes curve (a) to join curve (b). *(Data from work by Laurent et al. 1988.)*

approach the equilibrium contact angle of the metal on the product of interfacial reaction.

For interfacial reactivity to drive wetting according to Equation 3.12, the interfacial reaction must be localized at the triple line. For this to be, because of the short characteristic time for spreading of liquid metal (apparent from the rapid spreading observed in nonreactive systems), the rate of interfacial reaction must be very high. When the rate of interfacial reaction is low, the interfacial reaction will tend to become delocalized and precede or follow the triple line. In this case, the observed angle will not be influenced by ΔG_r. Examples are provided in the following two practically relevant systems:

(1) In the wetting of silica by aluminum under high vacuum, the interfacial reaction is the reduction of silica to form alumina, an oxide of higher stability. In a sessile-drop experiment of Al on SiO_2 near the melting point of Al, the reaction progresses ahead of the triple line because of atom transport by surface diffusion or transport through the atmosphere (Laurent 1988; Laurent et al. 1991). The contact angle that is observed is, therefore, close to that of aluminum on alumina (Laurent et al. 1991; Mori et al. 1983). (Deviations of a few degrees arise because of slight metal enrichment in Si and the fact that the oxide formed at the interface differs somewhat from Al_2O_3.)

(2) The case of aluminum on SiC provides an example in which the triple line precedes interfacial reaction product formation. In this system, aluminum carbide forms as discrete plate-like particles that grow into the melt from the interface. Because of the low rate of growth of the carbide platelets, the triple line remains, for long periods, one of aluminum, slightly enriched in Si and C, on SiC, and there is no influence of ΔG_r on θ (Laurent et al. 1987; Laurent et al. 1988). This is most eloquently illustrated by the fact that with sufficiently high additions of Si to the melt, SiC becomes thermodynamically stable and aluminum carbide formation is inhibited; neither the kinetics of wetting nor the value of the stationary contact angle are changed by such addition of Si.

In the case of interface reactions that produce no new phase at the interface and involve only dissolution of the substrate into the melt, divergent opinions exist regarding the influence of ΔG_r on wetting (Naidich 1981; Pask 1987).

In addition to the kinetic complexities involved in the theory of reactive wetting, an added difficulty exists when evaluating ΔG_r. Calculated values of ΔG_r are very sensitive to the choice of the thickness of the effective interface, that is, the number of atomic layers constituting the zone in which the diffusionless reaction occurs. At the time of this writing, the authors must rely on somewhat arbitrary assumptions with regard to this quantity, and can only provide order-of-magnitude estimations.

Experimental data showing an influence of ΔG_r, the free-energy change that accompanies the interfacial reaction, on wetting by a metal are relatively scarce (these are cited in Naidich 1981 and Pask 1987). The majority of analyzed experimental data from reactive sessile-drop experiments have been explained with no account of ΔG_r, but solely in terms of changes in the nature of the metal/substrate interface, together with other effects such as shape of reaction products and kinetics of reaction that accompany the reaction (Chatain et al. 1991; Nicholas 1986; Nicholas 1988). These include the studies of wetting of silica and silicon carbide by aluminum cited above, in which the reaction takes place between the metal solvent and the substrate.

To exploit interfacial reactions as a means of promoting wetting without causing massive reactions between matrix and reinforcement, a preferable approach is to alloy a nonreactive base metal with controlled quantities of reactive solute additions. This approach is used in the design of brazing alloys for ceramics (Kritsalis et al. 1991a; Naidich 1981; Nicholas 1986; Pask 1987) and is well illustrated by the case of Al_2O_3 and SiO_2 wetted by nickel containing a reactive alloying addition A (Merlin 1992; Merlin et al. 1992). The governing parameter in this category of systems is ε_O^A, Wagner's first-order interaction parameter, defined earlier in Equation 3.11. Two types of reaction may occur, depending on the alloying addition and its concentration: (1) dissolution of the substrate oxide into the alloy, favored by a moderately negative ε_O^A and (2) precipitation of a new oxide by reaction between dissolved oxygen and A atoms, favored by a strongly negative ε_O^A. In both cases, the influence of these reactions on wetting can be explained using the framework developed for nonreactive wetting, applied to the reacted system. Namely, in reaction 1, wetting is improved by the resulting increased oxygen concentration and the increased adsorption capability of O-A clusters compared to O-Me clusters at the interface. This is illustrated by Ni-Cr alloys on alumina ($\varepsilon_O^{Cr} = -25$), in Figure 3.5. In reaction 2, particularly good wetting results when the reaction product oxide features high metallicity. An example is the Ni-Ti/Al_2O_3 system shown in Figure 3.5. Because of strong O-Ti interactions ($\varepsilon_O^{Ti} \approx -100$), O-Ti clusters must be very tensioactive at the Ni/oxide interface, which is transformed by formation of a continuous layer of semimetallic Ti_2O_3. The beneficial influence of Ti on wettability is even more dramatic in the case of Cu-Ti/Al_2O_3 because Ti-Cu interactions are weaker than Ti-Ni interactions, leading to a higher thermodynamic activity of Ti. In this case, TiO is formed at the interface which, being highly metallic in its bonding character, provides for a much smoother chemical and electronic transition across the interface than does the sharp Cu/Al$_2$O$_3$ (Kritsalis et al. 1991a). These features explain the near-perfect wetting observed in this system and illustrate the considerable improvements in wetting that can be achieved via alloy design.

These effects, investigated and analyzed for metal-oxide systems, also seem to be valid for other families of ceramic reinforcements. For example, variations in the contact angle of Ni-Pd alloys on vitreous carbon with chromium additions show two wetting transitions with increasing chromium atom fraction X_{Cr} (see Figure 3.6) (Kritsalis et al. 1991b). The first transition is caused by a simple interfacial adsorption, which can be explained in terms of Cr-C cluster formation from chromium atoms segregating from the bulk liquid and C atoms resulting from substrate dissolution ($\varepsilon_c^{Cr} \ll 0$). The second transition is caused by the formation of Cr$_3$C$_2$, a highly metallic carbide. Other examples of improved wetting of liquid alloys on carbon or covalent carbides by the interfacial formation of metallic carbides are to be found in extensive work by Nicholas et al. (Standing and Nicholas 1978) and by Naidich (Naidich 1981; Naidich and Chuvashov 1983).

Significant differences thus exist, both fundamentally and practically, between reactive and nonreactive wetting. One important difference is that, in reactive wetting, very low contact angles can be obtained with small quantities of additional alloying elements, as low as 1 at. pct. Conversely, in nonreactive systems, additions on

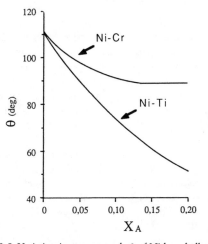

Figure 3.5. Variation in contact angle θ of Ni-based alloys on alumina with molar fraction X of Cr and Ti at 1773 K. The effect of Cr results from adsorption only, whereas the effect of Ti results from a combination of adsorption and the formation of a new phase at the interface *(Data from work reported in Merlin et al. 1992).*

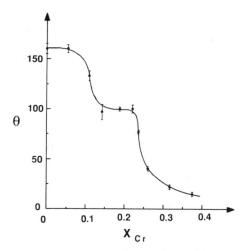

Figure 3.6. Influence of Cr molar fraction X_{Cr} in Ni-Pd-Cr alloys on their contact angle on vitreous carbon substrates at 1523 K. The first transition from 160 degrees to about 90 degrees is due to an adsorption process. The second transition results from the formation of metal-like chromium carbide Cr$_3$C$_2$ at the interface *(Data from work reported in Kritsalis et al. 1991b).*

the order of 10 at. pct. are required to obtain more limited improvements in wetting. A second difference arises from the influence exerted by the kinetics of interfacial reaction. In sessile-drop experiments, the attainment of stationary contact angle values takes much longer than in nonreactive systems (in which case equilibrium is attained in about 10^{-3} s): for example, in the Cu-Ti/Al$_2$O$_3$ and Ni-Pd-Cr/C systems, the contact angle stabilizes after 100 and 1,000 seconds, respectively. Even higher times are required for a drop of Al-Si alloy to stabilize on the basal face of hexagonal SiC monocrystals. In this system, the kinetics of wetting are limited by the rate of dissociation of SiC and the stationary contact-angle value is only attained once the liquid aluminum drop is saturated with carbon and silicon (Figure 3.7) (Laurent et al. 1988). Similarly slow kinetics have been observed in the case of Al on graphite (Mori et al. 1983; Weirauch and Krafick 1990). Finally (and unlike nonreactive wetting), in reactive wetting, temperature can exert a strong influence on wetting, by altering both the nature and the kinetics of interfacial reactions.

3.3 Interface Engineering

3.3.1 Interface Tailoring for Processing

Two primary concerns guide interface engineering in processing: (1) preventing composite microstructural degradation by interfacial reactions and (2) promoting interface bond formation (the latter concern being

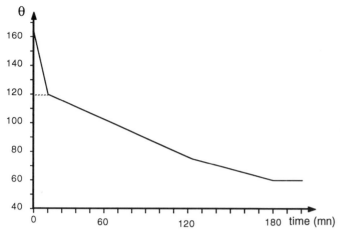

Figure 3.7. Variation of the contact angle θ with time t for a drop of Al-18wt%Si at 1073 K under high vacuum ($P = 4 \cdot 10^{-5}$ Pa). The initial angle (about 160 degrees) is that of Al covered by a film of its oxide. The first true contact angle of Al on SiC is about 120 degrees. A nearly stationary value of θ (about 60 degrees) is obtained after three hours *(Data from work by Laurent et al. 1988)*.

mostly present in solidification processes). As shown in Section 3.2, the energetics and kinetics of bond formation from the initial constituents of the composite are highly dependent on the nature of the chemical interaction that takes place beween the two phases, and can be tailored to a significant extent via compositional control of the composite. Mechanical aspects of the process of bond creation, presented in Section 3.1.2, can also be tailored to some extent to ease processing and to improve properties of the composite.

Chemical means of easing bond formation are means that improve wetting of the reinforcement by the molten metal. These are reviewed in more detail by Mortensen and Jin (1992), and can be divided into several broad categories.

Some processes rely, entirely or in part, on tailoring the matrix alloy chemistry to promote wetting. Examples include addition of Li to aluminum, which has been used to improve wetting of Al_2O_3 fibers during vacuum infiltration (Champion et al. 1978; Dhingra 1980; Folgar et al. 1987; Hunt 1986), addition of 2% Ca to Al and 0.5% Ti to Cu for improvement of wetting of various particles during spray-casting (Namai et al. 1986), or the addition of magnesium to aluminum, which forms an ingredient in the more complicated Primex process (Lanxide, Newark, DE) for spontaneous infiltration in the presence of nitrogen-containing atmospheres (Aghajanian et al. 1989; Aghajanian et al. 1991; Burke et al. 1989; Halverson and Landingham 1988; Halverson et al. 1989; Pyzik and Aksay 1987; White et al. 1989).

A different and frequently used strategy is to modify the reinforcement. Simple heat treatment of the reinforcement has been used (Banerji et al. 1984; Caron and Masounave 1990; Krishnan et al. 1981; Rohatgi et al. 1986), presumably to eliminate deleterious adsorbed species; however, the most generally accepted technique is to coat the reinforcement with a material that is well wetted by the matrix. By coating the reinforcement, the volume of chemical additives to the composite and, hence, their incidence on composite properties, may be minimized. In particular, when interfacial reaction comes with improved wetting, the volume of resulting reaction products can potentially be restricted by exhaustion of the reactants. Examples of reinforcement coatings for wetting improvement are numerous, and include Ti-B coatings for infiltration of C fibers by aluminum in the absence of oxygen (Harrigan and Flowers 1977), SiO_2 coatings developed for infiltration of carbon fiber preforms by Mg (Katzman 1983; Katzman 1987), SiO_2 coatings produced on SiC by roasting the reinforcement in air (Skibo and Schuster 1989), and K_2ZrF_6 crystal coatings for wetting by aluminum alloys (Schamm et al. 1991).

Although many wettability enhancement techniques used in metal-matrix composites processing were designed prior to their scientific rationalization, explanations for several of the techniques described above can be given in light of work summarized in Section 3.2. The effectiveness of Ti-B coatings, which promote wetting and react at most to a limited extent with Al (Tsai et al. 1981), can be explained by the metallic character of TiB_2, which is known to result in low wetting angles with aluminum under high vacuum (Coudurier et al. 1984; Rhee 1970). In the case of K_2ZrF_6 coatings, wetting may be promoted by reaction and/or by disruption of the oxide film that covers molten aluminum. The beneficial influence of adding small quantities of Ca to Al and Ti to Cu, documented for spray-casting with alumina, SiC, and graphite reinforcements, is also in agreement with experiment and theory summarized in the previous section.

Mechanical means of promoting bond formation are less numerous, but include an elegant method of controlling the microscopic geometry of the reinforcement distribution by artificial separation of fibers with small particles prior to infiltration to prevent fiber contact lines (Ikuno et al. 1989). More generally, because bond formation is seldom spontaneous, the process of interface creation is primarily driven by mechanical means, by application of pressure, as in solid-state consolidation and infiltration, or by using other methods recently designed for infiltration processing (Andrews and Mortensen 1991; Pennander et al. 1991; Sui et al. 1991).

There is, as shown in the preceding section, considerable latitude for optimization of interfacial chemistry to improve wetting; however, the range of possible strategies is limited by other requirements. In particular, interface reactions that may drive wetting are generally detrimental because they degrade reinforcement strength (this is, for example, the case with the Al-Li/Al$_2$O$_3$ system).

Interfacial reactivity has been studied by many researchers, producing a wide variety of results for the kinetics of the reaction, the morphology of the reaction products, the latitude for control of the reaction and its incidence on composite properties. The appendix at the end of this chapter provides a partial listing of such studies to illustrate the variety and, at times, the extent to which interfacial reactions in metal-matrix composite fabrication are being studied. It can easily be reasoned, and it is confirmed by experiment, that the results from so many studies on so wide a variety of systems lend themselves to little, if any, systematization. The shape of the reaction products (discrete plate-like particles in the case of Al$_4$C$_3$, recrystallized fibers in the case of Ni/C composites, or a continuous layer in the case of Al-Li/Al$_2$O$_3$ composites), their rate-limiting mechanism (diffusion in the solid reinforcement, diffusion in the liquid metal, diffusion through the reaction layer when it coats the reinforcement) and their relation to processing parameters such as preform temperature (Isaacs et al. 1991), vary significantly from case to case.

Control of the extent of interfacial reaction is of great concern in the processing of metal-matrix composites. This concern imposes restrictions on allowable process parameters, as is the case in infiltration which, in reactive systems, may need to be conducted with low initial reinforcement and low die temperatures as well as high pressures (Fukunaga and Goda 1985). In cases of very high interfacial reactivity, found for example with the majority of titanium-matrix composites, solid-state or rapid-solidification processes must be used to minimize matrix/reinforcement temperatures and contact times at elevated temperatures.

An alternative (but sometimes costly) method for control of interfacial reactions is to coat the reinforcement with an inert diffusion barrier to protect the reinforcement during processing. This approach has been implemented for a large number of systems, using coatings produced by sol-gel techniques (Kindl et al. 1992), by chemical vapor deposition (Bouix et al. 1987), by sputter deposition (Kieschke et al. 1991), or by surface reaction of a fraction of the reinforcement (Himbeault et al. 1989; Skibo and Schuster 1989).

3.3.2 Interface Tailoring for Composite Performance

Beyond concerns associated with producing a chosen composite efficiently and with minimal deviation from the desired structure, the structure and properties of the interfacial zone can be tailored to optimize the performance of the material, particularly with regard to its mechanical strength. Quantitative criteria, backed with characterization methods generally based on flat replications of composite interfaces, exist to measure relevant interface attributes such as the interfacial fracture energy or strength (see Section 3.1.2 and Chapter 12), although none are established as a standard test.

Because of the large number of interfacial reaction configurations observed, the incidence of interfacial reaction on composite properties is system-dependent. Generally, massive reaction of most reinforcements is undesirable because the reinforcement strength is degraded. Formation of a brittle reaction product in the matrix (such as Al$_4$C$_3$ in carbon fiber reinforced aluminum, or Al$_3$Ni in infiltrated aluminum-matrix composites produced via reinforcement coating with Ni) is also generally undesirable because this provides a site for initiation and propagation of internal damage across the matrix. However, a thin and smooth layer of interfacial reaction product may be compatible with preservation of reinforcement strength (Metcalfe and Klein 1974; Ochiai et al. 1982; Shorshorov et al. 1979).

At the time of this writing there are no established criteria for interface optimization from the standpoint of composite properties, although much progress has been made recently on the mechanics of interface fracture and the incidence of interface properties on those of the composite. These topics are addressed in detail elsewhere (Chapter 12); for the present discussion, we shall simplify the argument by assuming that regardless of the particular interface fracture configuration, energy dissipative mechanisms along an interfacial crack path depend primarily on the nature and geometry of the surrounding material, to act roughly as a constant amplification factor for the intrinsic fracture energy of interface cracks. The interfacial fracture energy then scales with the strength of atomic bonds ruptured by a crack propagating at the interface, and, hence, with the interfacial work of adhesion W_a.

In these simplified terms, a strong interface comes with a high value of W_a, corresponding to the establishment of a strong chemical bond across the interface. With the important exception of aligned, continuous fiber reinforced metals stressed parallel to the fibers and having too small a fiber diameter for the matrix to toughen the composite, a strong interface is generally desirable because a weak interface would create a path for easy crack propagation in the composite. Some degree of chemical affinity between matrix and reinforcement is then required, which can be brought about by the use of alloying additives, including oxygen, as described in Section 3.2. Limited reaction may also be used to tailor interface properties. There are examples showing that the apparent interface strength can be increased by a factor on the order of four via interfacial reaction including Cu-Ti alloys brazed on Al_2O_3 (Nicholas 1986), and the Pt/NiO system, for which the interfacial reaction layer was as thin as 1 nm (Shieu et al. 1990).

With aligned fiber composites, a weak interface is desirable if the matrix is not capable of absorbing sufficient plastic deformation energy along the path of a crack propagating across the fiber direction. In this case, the fracture energy of the interface must be significantly smaller than that of the matrix, by a factor that varies between 4 and 10 (depending on the analysis) (Evans et al. 1990; Kelly and Macmillan 1986; Kendall 1976; Kendall 1975). This implies that large differences must exist between the interface bond strength and that of atomic bonds in the matrix. If it suffices that the interfacial bond strength be about one-fourth the matrix cohesive strength, interfaces may be designed with some degree of chemical bonding and yet induce crack deflection by debonding. The more stringent criterion that W_a be no more than about one-tenth of the matrix cohesive energy W_c is much more difficult to fulfill because atomic chemical bond strengths seldom vary by a factor as large as 10 (Mortensen 1988). Sufficiently weak interfacial zones are then essentially restricted to ones that (1) solely feature physical bonding at the interface or (2) contain a layer of material with strongly anisotropic bonds properly oriented with regard to the interface plane. This is apparent in Tables 3.1 and 3.3, where it is shown that the measured values of W_a/W_c only fall below 0.1 with graphite and boron nitride. These low ratios result from the nature of bonding in these two solids, which is highly concentrated within a plane and very weak across this plane. Graphite and boron nitride thus provide a means of satisfying even the more stringent criterion for interfacial debonding by either of the effects (1) and (2) presented above. Indeed, the three principal inorganic composite systems successfully toughened by crack deflection at fiber/matrix interfaces presently being produced comprise a properly oriented layer of graphite at their interface (Mortensen 1988). The design of strategies for interface optimization in fiber-reinforced metals to be toughened by interfacial crack deflection is nonetheless still open to question, and future research may lead to the definition of alternate strategies for interfacial toughening via control of interface bond strength and interfacial zone geometry.

Resistance to corrosion of metal-matrix composites may also be influenced by the interfacial zone. Wet corrosion is particularly critical in carbon fiber aluminum or magnesium because of galvanic coupling of fiber and matrix (Hall 1987; Hihara and Latanision 1988). It has been proposed to reduce the (generally high) wet corrosion rate of these composites by electrically insulating the fibers from the matrix with a resistive layer, which may be deliberately inserted by coating the reinforcement, or formed by chemical reaction between matrix and reinforcement. Although there is one reference documenting such an effect (Isaikin et al. 1980), Hihara predicts that for continuous carbon fiber reinforced aluminum composites, coatings of reasonable thickness (less than about 1 μm) would need to have a resistivity higher than about 10^{16} $\Omega \cdot$ cm; this would be difficult to achieve in practice (Hihara 1989).

3.4 Summary

The wide variety of engineering approaches that have been proposed to improve the interface in metal-matrix composites have mostly been designed empirically. The vast number of parameters and issues that must be taken into account largely explains this state of affairs, as does the fact that several fundamental questions remain to be fully understood (such as reactive wetting or the relation of interface energetics and structure with metal-matrix composite performance).

Research on the energetics of interface formation, and its relation to composite processing and performance, is nonetheless progressing at a relatively fast pace, and it may not be long before such research guides the design and the processing of these materials to a much greater extent than now. Two particularly critical areas being addressed in this regard are the identification of the main parameters linking composite performance with interface structure and energetics, and the need for understanding of the full range of strategies available for controlling these parameters, including the role of limited interfacial chemical reaction. As the range of materials systems and composite attributes considered for engineering applications of metal-matrix composites increases, the benefits to be derived from a scientific approach to interface engineering will increase—not only because of the increased number of parameters and systems to be addressed, but also because of the increased palette of strategies made available for optimization of the interface.

References

Agarwala, V., and D. Dixit. 1981. *Trans. of the Jpn. Inst. of Met.* 22:521-526.

Aghajanian, M.K., Burke, J.T., White, D.R., and A.S. Nagelberg. 1989. *SAMPE Quart.* 20:43-46.

Aghajanian, M.K., Rocazella, M.A., Burke, J.T., and S.D. Keck. 1991. *J. Mater. Sci.* 26:447-454.

Aksay, L.A., Hoge, C.E., and J.A. Pask. 1974. *J. of Phys. Chem.* 78:1178-1183.

Andrews, R.M., and A. Mortensen. 1991. *Metall. Trans.* 22A:2903-2915.

Argon, A.S., Gupta, V., Landis, H., and J.A. Cornie. 1989. *Mater. Sci. & Eng.* A107:41-47.

Banerji, A., Rohatgi, P.K., and W. Reif. 1984. *Metallwiss. & Technik* 38:656-661.

Bouix, J., Cromer, M., J. Dazord et al. 1987. *Rev. Int. Hautes Tempér. Réfract.* 24:5-26.

Brennan, J.J., and J.A. Pask. 1968. *J. Am. Cer. Soc.* 51:569-573.

Burke, J.T., Aghajanian, M.K., and M.A. Rocazella. 1989. *In:* 34th International SAMPE Symposium and Exhibition 2440-2454, Covina, CA: SAMPE.

Caron, S., and J. Masounave. 1990. *In:* Fabrication of Particulates Reinforced Metal Composites (J. Masounave and F.G. Hamel, eds.), 107-113, Materials Park, OH: ASM International.

Champion, A.R., Krueger, W.H., Hartmann, H.S., and A.K. Dhingra. 1978. *In:* 1978 International Conference on Composite Materials ICCM 2. (B. Noton, R.A. Signorelli, K.N. Street and L.N. Phillips, eds.), 883-904, Warrendale, PA: The Metallurgical Society.

Chatain, D., Coudurier, L., and N. Eustathopoulos. 1988. *Rev. Phys. Appl.* 23:1055-1064.

Chatain, D., Coudurier, L., Steinchen, A., and N. Eustathopoulos. 1991. *In:* Interfaces in New Materials (P. Grange and B. Belmon, eds.), 210-218, London: Elsevier Applied Science.

Chatain, D., Rivollet, I., and N. Eustathopoulos. 1986. *J. de Chim. Phys.* 83:561-567.

Chatain, D., Rivollet, I., and N. Eustathopoulos. 1987. *J. de Chim. Phys.* 84:201-203.

Clough, R.B., Biancaniello, F.S., Wadley, H.N.G., and U.R. Kattner. 1990. *Metall. Trans.* 21A:2747-2757.

Clyne, T.W., and M.C. Watson. 1991. *Composites Sci. and Techn.* 42:25-55.

Coudurier, L., Adorian, J., Pique, D., and N. Eustathopoulos. 1984. *Rev. Int. des Hautes Temp. et Réfract.* 21:81-90.

Dhingra, A.K. 1980. *Phil. Trans. R. Soc. Lond.* 294:559-564.

Drevet, B., Chatain, D., and N. Eustathopoulos. 1990. *J. de Chim. Phys.* 87:117-126.

Eldridge, J.I., and P.K. Brindley. 1989. *J. of Mater. Sci. Lett.* 8:1451-1454.

Eustathopoulos, N., Camel, D., and J.J. Favier. 1988. *In:* Solidification des Alliages (F. Durand, ed.), 133-168, France: Les Editions de Physique.

Eustathopoulos, N., and J.C. Joud. 1980. *In:* Current Topics in Materials Science. (E. Kaldis, ed.), 4:281-360, Amsterdam: North Holland.

Evans, A.G., Rühle, M., Dalgleish, B.J., and P.G. Charalambides. 1990. *Mater. Sci. & Eng.* A126:53-64.

Fletcher, T.R., Cornie, J.A., and K.C. Russell. 1988. *In:* Cast Reinforced Metal Composites (S.G.Fishman and A.K. Dhingra, eds.), 21-25, Metals Park, OH: ASM International.

Fletcher, T.R., Cornie, J.A., and K.C. Russell. 1991. *Mater. Sci. & Eng.* A144:159-163.

Folgar, F., Widrig, J.E., and J.W. Hunt. 1987. SAE Technical Paper Series, Paper No. 870406, 9 pp.

Fukunaga, H., and K. Goda. 1985. *J. Jap. Inst. Met.* 49:78-83.

Ghosh, P.K., and S. Ray. 1988. *Trans. of the Jpn. Inst. of Met.* 29:509-519.

Gubanov, A.I., and S.M. Dunaevskii. 1977. *Soviet Physics Solid State* 19:795-797.

Gupta, V., Argon, A.S., Cornie, J.A., and D.M. Parks. 1990. *Mater. Sci. & Eng.* A126:105-117.

Hall, I.W. 1987. *Scripta Metall.* 21:1717-1721.

Hall, I.W., Kyono, T., and A. Diwanji. 1987. *J. Mater. Sci.* 22:1743-1748.

Halverson, D.C., and R.L. Landingham. 1988. U.S. Patent No. 4,718,941, January 12, 1988.

Halverson, D.C., Pyzik, A.J., Aksay, I.A., and W.E. Snowden. 1989. *J. Am. Cer. Soc.* 72:775-780.

Hanumanth, G.S., and G.A. Irons. 1990. *In:* Fabrication of Particulates Reinforced Metal Composites (J. Masounave and F.G. Hamel, eds.), 41-47, Materials Park, OH: ASM International.

Harrigan, W.C., and R.H. Flowers. 1977. *In:* Failure Modes in Composites IV (J.A. Cornie and F.W. Crossman, eds.), 319-335, Warrendale, PA: The Metallurgical Society of AIME.

Hicter, P., Chatain, D., Pasturel, A., and N. Eustathopoulos. 1988. *J. de Chim. Phys.* 85:941-945.

Hihara, L.H. 1989. Ph.D. Thesis, Department of Materials Science and Engineering, Massachusetts Institute of Technology, Cambridge, MA.

Hihara, L.H., and R.M. Latanision. 1988. *Scripta Metall.* 22:413-418.

Himbeault, D.D., Varin, R.A., and K. Piekarski. 1989. *J. Mater. Sci.* 24:2746-2750.

Hultgren, R., Desai, P.D., and D.T. Hawkins et al. 1973. Selected Values of the Thermodynamic Properties of Binary Alloys, Metals Park, OH: American Society for Metals.

Hunt, W.H. 1986. *In:* Interfaces in Metal Matrix Composites (A.K. Dhingra and S.G. Fishman, eds.), 3-25, Warrendale, PA: The Metallurgical Society.

Ikuno, H., Towata, S.-I., and S.-I. Yamada. 1989. *J. Jap. Inst. Met.* 53:327-332.

Isaacs, J.A., Taricco, F., Michaud, V.J., and A. Mortensen. 1991. *Metall. Trans.* 22A:2855-2862.

Isaikin, A.S., Chubarov, V.M., and B.F. Trefilov et al. 1980. *Metal Sci. and Heat Treat.* 22:815-817.

Johnson, K.H., and S.V. Pepper. 1982. *J. Appl. Phys.* 53:6634-6637.

Katzman, H.A. 1983. U.S. Patent No. 4,376,803, March 15, 1983.
Katzman, H.A. 1987. *J. Mater. Sci.* 22:144–148.
Kaufmann, H., and A. Mortensen. 1992. *Metall. Trans.* 23A:2071–2073.
Kelly, A., and N.H. Macmillan. 1986. *Strong Solids*. Oxford: Clarendon Press.
Kendall, K. 1975. *Proc. Roy. Soc. Lond.* A344:287–302.
Kendall, K. 1976. *J. Mater. Sci.* 11:638–644.
Kieschke, R.R., Somekh, R.E., and T.W. Clyne. 1991. *Acta Metall. et Mater.* 39:427–435.
Kindl, B., Liu, Y.L., Nyberg, E., and N. Hansen. 1992. *Composites Sci. and Techn.* 43:85–93.
Krishnan, B.P., Surappa, M.K., and P.K. Rohatgi. 1981. *J. Mater. Sci.* 16:1209–1216.
Kritsalis, P., Coudurier, L., and N. Eustathopoulos. 1991a. *J. Mater. Sci.* 26:3400–3408.
Kritsalis, P., Coudurier, L., Parayre, C., and N. Eustathopoulos. 1991b. *Journal of the Less-Common Metals* 175:13–27.
Kritsalis, P., Li, J.G., Coudurier, L., and N. Eustathopoulos. 1990. *J. of Mater. Sci. Lett.* 9:1332–1335.
Kubaschewski, O., and C.B. Alcock. 1979. *Metallurgical Thermochemistry*. Oxford: Pergamon Press.
Laurent, V. 1988. D.Sc. Thesis, Science des Matériaux-Métallurgie, Institut National Polytechnique de Grenoble, France.
Laurent, V., Chatain, D., Chatillon, C., and N. Eustathopoulos. 1988. *Acta Metall.* 36:1797–1803.
Laurent, V., Chatain, D., and N. Eustathopoulos. 1987. *J. Mater. Sci.* 22:244–250.
Laurent, V., Chatain, D., and N. Eustathopoulos. 1991. *Mater. Sci. & Eng.* A135:89–94.
Laurent, V., Chatain, D., Eustathopoulos, N., and X. Dumant. 1988. *In:* Cast Reinforced Metal Composites (S.G. Fishman and A.K. Dhingra, eds.), 27–31, Metals Park, OH: ASM International.
Li, J.G., Chatain, D., Coudurier, L., and N. Eustathopoulos. 1988a. *J. Mater. Sci. Lett.* 7:961–963.
Li, J.G., Coudurier, L., Ansara, I., and N. Eustathopoulos. 1988b. *Ann. Chim. Franc.* 143:145–153.
Li, J.G., Coudurier, L., and N. Eustathopoulos. 1989. *J. Mater. Sci.* 24:1109–1116.
Lloyd, D.J. 1989. *Composites Sci. and Techn.* 35:159–179.
Lloyd, D.J., and I. Jin. 1988. *Metall. Trans.* 19A:3107–3110.
MacDonald, J.E., and J.G. Eberhart. 1965. *Trans. Metall. Soc. AIME.* 233:512–517.
Maxwell, P.B., Martins, G.P., Olson, D.L., and G.R. Edwards. 1990. *Metall. Trans.* 21B:475–485.
Merlin, V. 1992. D.Sc. Thesis, Science des Matériaux-Métallurgie, Institut Polytechnique National de Grenoble, France.
Merlin, V., Kritsalis, P., Coudurier, L., and N. Eustathopoulos. 1992. *In:* Proc. MRS Fall Meeting, Boston, MA, Pittsburgh, PA: Materials Research Society. 238:511–516.
Metcalfe, A.G., and M.J. Klein. 1974. *In:* Interfaces in Metal Matrix Composites. (A.G. Metcalfe, ed.), 1:127–168, New York: Acadamic Press.
Miedema, A.R. 1978. *Z. Metallk.* 69:287–292.
Miedema, A.R., Boer, F.R. den, Boom, R., and J.W.F. Dorleijn. 1977. *Calphad* 1:353–359.
Mori, N., Sorano, H., and A. Kitahara et al. 1983. *J. Jap. Inst. Met.* 47:1132–1139.
Mortensen, A. 1988. *In:* 9th Risø International Symposium on Metallurgy and Materials Science (S.I. Andersen, H. Lilholt, and O.B. Pedersen, eds.), 141–155, Roskilde, Denmark: Risø National Laboratory.
Mortensen, A. 1991. *Mater. Sci. & Eng.* A135:1–11.
Mortensen, A., and J.A. Cornie. 1987. *Metall. Trans.* 18A:1160–1163.
Mortensen, A., and I. Jin. 1992. *Intern. Mater. Rev.* 37:101–128.
Mortensen, A., and T. Wong. 1990. *Metall. Trans.* 21A:2257–2263.
Naidich, J.V. 1968. *Russian Journal of Physical Chemistry* 42:1023–1026.
Naidich, J.V. 1981. *In:* Progress in Surface and Membrane Science. 14 (D.A. Cadenhead and J.F. Danielli, eds.), 353–484, New York: Academic Press.
Naidich, J.V., and J.N. Chuvashov. 1983. *J. Mater. Sci.* 18:2071–2080.
Naidich, J.V., and V.M. Perevertajlo. 1971. *Russian Journal Phys. Chem.* 45:1025–1027.
Naidich, J.V., Perevertajlo, V.M., and G.M. Nevodnik. 1972. *Poroshk. Metallurg.* 7:51–55.
Naidich, J.V., Zhuravljov, V.S., and N.I. Frumina. 1990. *J. Mater. Sci.* 25:1895–1901.
Namai, T., Osawa, Y., and M. Kikuchi. 1986. *Trans. of the Jpn Foundrymen's Soc.* 5:29–32.
Nicholas, M.G. 1986. *British Ceramic Transactions Journal* 85:144–146.
Nicholas, M.G. 1988. *Materials Forum* 29:127–150.
Nogi, K., Ogino, K., and N. Iwamoto. 1991. *In:* Metal Matrix Composites—Processing, Microstructure and Properties, 12th Risø International Symposium on Materials Science (N. Hansen, D. Juul-Jensen, T. Leffers et al., eds.), 559–564, Roskilde, Denmark: Risø National Laboratory.
Ochiai, S., Osamura, K., and Y. Murakami. 1982. *In:* Fourth International Conference on Composite Materials, ICCM-IV (T. Hayashi, K. Kawata and S. Umekawa, eds.), 1331–1338, Tokyo: Japan Society for Composite Materials.
Oh, S.-Y., Cornie, J.A., and K.C. Russell. 1989a. *Metall. Trans.* 20A:533–541.
Oh, S.-Y., Cornie, J.A., and K.C. Russell. 1989b. *Metall. Trans.* 20A:527–532.
Oh, T.S., Rödel, J., Cannon, R.M., and R.O. Ritchie. 1988. *Acta Metall.* 36:2083–2093.
Ownby, P.D., Li, W.K., and D.A. Weirauch. 1991. *J. Am. Cer. Soc.* 74:1275–1281.
Ownby, P.D., and J. Liu. 1988. *J. of Adhes. Sci. and Technol.* 2:255–269.
Pask, J.A. 1987. *Ceram. Bull.* 66:1587–1592.
Pennander, L.O., Ståhl, J.E., and C.H. Andersson. 1991. *In:* Eighth International Conference on Composite Materials, ICCM 8 (S.W. Tsai and G.S. Springer, eds.), 17B1–17B10, Covina, CA: SAMPE.
Pilliar, R.M., and J. Nutting. 1967. *Phil. Mag.* 16:181–188.
Pyzik, A.J., and I.A. Aksay. 1987. U.S. Patent No. 4,702,770, October 27, 1987.
Ramqvist, L. 1965. *Intern. J. Powder Metall.* 1:2–34.

Reimanis, I.E., Dalgleisch, B.J., and A.G. Evans. 1991. *Acta Metall. et Mater.* 39:3133–3141.

Rhee, S.K. 1970. *J. Am. Cer. Soc.* 53:386–389.

Rohatgi, P.K., Asthana, R., and S. Das. 1986. *Intern. Metals Rev.* 31:115–139.

Roman, I., and R. Aharonov. 1992. *Acta Metall. et Mater.* 40:477–485.

Russell, K.C., Oh, S.-Y., and A. Figueredo. 1991. *MRS Bull.* 16:46–52.

Salvo, L., Suéry, M., Legoux, J.G., and G. l'Espérance. 1991. *Mater. Sci. & Eng.* A126:129–133.

Sangiorgi, R., Muolo, M.L., Chatain, D., and N. Eustathopoulos. 1988. *J. Am. Cer. Soc.* 71:742–748.

Schamm, S., Fedou, R., and J.P. Rocher et al. 1991. *Metall. Trans.* 22A:2133–2139.

Seitz, J.D., Edwards, G.R., Martins, G.P., and P.Q. Campbell. 1989. *In:* Interfaces in Metal-Ceramic Composites (R.Y. Lin, R.J. Arsenault, G.P. Martins, and S.G. Fishman, eds.), 197–212, Warrendale, PA: The Minerals, Metals & Materials Society.

Shieu, F.-S., Raj, R., and S.L. Sass. 1990. *Acta Metall. et Mater.* 38:2215–2224.

Shorshorov, M.K., Ustinov, L.M., and A.M. Zirlin et al. 1979. *J. Mater. Sci.* 14:1850–1861.

Skibo, M.D., and D.M. Schuster. 1989. U.S. Patent No. 4,865,806, September 12, 1989.

Standing, R., and M. Nicholas. 1978. *J. Mater. Sci.* 13:1509–1514.

Sui, Q., Guo, S., and F. Tang. 1991. *In:* Eighth International Conference on Composite Materials, ICCM 8 (S.W. Tsai and G.S. Springer, eds.), 21E1–21E6, Covina, CA: SAMPE.

Thanh, L.N., and M. Suéry. 1991. *Scripta Metall. et Mater.* 25:2781–2786.

Tsai, S.D., Schmerling, M., and H.L. Marcus. 1981. *In:* Ceramic Engineering and Science Proceeding (J.W. McCauley, ed.), 798–808, Columbus, OH: American Ceramic Society.

Weirauch, D.A. 1988a. *J. of Mater. Res.* 3:729–739.

Weirauch, D.A. 1988b. *In:* Ceramic Microstructures '86 (J.A. Pask and A.G. Evans, eds.), 329–339, New York: Plenum Publishing Corp.

Weirauch, D.A., and W.F. Krafick. 1990. *Metall. Trans.* 21A:1745–1751.

White, D.R., Urquhart, A.W., Aghajanian, M.K., and D.K. Creber. 1989. U.S. Patent No. 4,828,008, May 9, 1989.

Zhang, G.D., Tsai, H.W., and R.J. Wu. 1991. *In:* Advanced Materials for Future Industries: Needs and Seeds (I. Kimpara, K. Kagayama and Y. Kagawa, eds.), 723–730, Tokyo: International Convention Management, Inc.

Appendix A
Illustrative List of Interfacial Reaction Studies in Selected Metal-Matrix Composite Systems

An example of a material system that has attracted considerable attention from the standpoint of interface microstructure and reactivity is provided by the Si-C-Al system, of interest for both SiC- and C-reinforced Al. The following references are relevant to interfacial structure and reactivity in this system:

Allard, L.F., Rawal, S.P., and M.S. Misra. 1986. *J. of Metals.* (October), 38:40–43.

Andersson, C.-H., and R. Warren. 1984. *Composites* 15:16–24.

Arsenault, R.J., and C.S. Pande. 1984. *Scripta Metall.* 18:1131–1134.

Baker, S.J., and W. Bonfield. 1978. *J. Mater. Sci.* 13:1329–1334.

Bermudez, V.M. 1983. *Appl. Phys. Lett.* 42:70–72.

Bienvenu, Y., Le Flour, J.C., and Y. Favry et al. 1989. *In:* The Materials Revolution through the 90's: Powders, M.M.C., Magnetics (Oxford: BNF Metals Technology Centre), Paper No. 34.

Blankenburgs, G. 1969. *J. of the Austr. Inst. of Metals.* 14:236–241.

Chen, X.-Q., and G.-X. Hu. 1990. *In:* Controlled Interphases in Composite Materials, Proceedings of the Third International Conference on Composite Interfaces (ICCI-III) (H. Ishida, ed.), 381–388, New York: Elsevier.

Cheng, H.M., Akiyama, S., and A. Kitahara et al. 1991. *Scripta Metall. et Mater.* 25:1951–1956.

Chernyshova, T.A., Kobeleva, L.I., Tylkina, M.I., and A.V. Rebrov. 1984. *Metal Sci. and Heat Treat.* 26:592–595.

Clough, R.B., Biancaniello, F.S., Wadley, H.N.G., and U.R. Kattner. 1990. *Metall. Trans.* 21A:2747–2757.

Diwanji, A.P., and I.W. Hall. 1987. In: Sixth International Conference on Composite Materials, ICCM (F.L. Matthews, N.C.R. Buskell, J.M. Hodgkinson and J. Morton, eds.), 2.265–2.274, London: Elsevier Applied Science.

Everett, R.K., and C.J. Skowronek. 1984. In: Failure Mechanisms in High Performance Materials (J.G. Early, T.R. Shives and J.H. Smith, eds.), 128–137, Cambridge, U.K.: Cambridge University Press.

Favry, Y., and A.R. Bunsell. 1987. *Composites Sci. and Techn.* 30:85–97.

Fukunaga, H., and K. Goda. 1985. *J. Jap. Inst. Met.* 49:78–83.

Girot, F.A., Albingre, L., Quenisset, J.M., and R. Naslain. 1987. *J. of Metals* 39:18–21.

Hall, I.W. 1991. *J. Mater. Sci.* 26:776–781.

Handwerker, C.A., Cahn, J.W., and J.R. Manning. 1990. *Mater. Sci. & Eng.* A126:173–189.

Henriksen, B.R., and T.E. Johnsen. 1990. *Mater. Sci. and Technol.* 6:857–861.

Iseki, T., Kameda, T., and T. Maruyama. 1984. *J. Mater. Sci.* 19:1692–1698.

Islam, M.U., and W. Wallace. 1988. *Adv. Mater. and Proc.* 3:1–35.

Kendall, E.G. 1974. In: Metallic Matrix Composites. Vol. 4 (K.G. Kreider, ed.), 319–397, New York: Academic Press.

Khan, I.H. 1976. *Metall. Trans.* 7A:1281–1289.

Kindl, B., Liu, Y.L., Nyberg, E., and N. Hansen. 1992. *Composites Sci. and Techn.* 43:85–93.

Kohara, S. 1981. In: Composite Materials, Proc. Japan-U.S. Conference. (K. Kawata and T. Akasaka, eds.), 224–229, Tokyo: Japan Society of Composite Materials.

Lee, D.J., Vaudin, M.D., Handwerker, C.A., and U.R. Kattner. 1988. In: High Temperature/High Performance Composites, MRS Symp. Proc. Vol. 120 (F.D. Lemkey, S.G. Fishman, A.G. Evans and J.R. Strife, eds.), 357–365, Pittsburgh, PA: Materials Research Society.

Legoux, J.G., L'Espérance, G., Salvo, L., and M. Suéry. 1990. In: Fabrication of Particulates Reinforced Metal Composite (J. Masounave and F.G. Hamel, eds.), 31–39, Materials Park, OH: ASM International.

Li, Q., Megusar, J., Masur, L.J., and J.A. Cornie. 1989. *Mater. Sci. & Eng.* A117:199–206.

Lin, R.L., and K. Kannikeswaran. 1989. In: Interfaces in Metal-Ceramic Composites (R.Y. Lin, R.J. Arsenault, G.P. Martins and S.G. Fishman, eds.), 153–164, Warrendale, PA: The Minerals, Metals & Materials Society.

Lloyd, D.J. 1989. *Composites Sci. and Techn.* 35:159–179.

Lloyd, D.J., and I. Jin. 1988. *Metall. Trans.* 19A:3107–3110.

Maruyama, B., Ohuchi, F., and L. Rabenberg. 1990a. In: Interfaces in Composites, MRS Symp. Proc. Vol. 170 (C.G. Pantano and E.J.H. Chen, eds.), 167–172, Pittsburgh, PA: Materials Research Society.

Maruyama, B., Ohuchi, F.S., and L. Rabenberg. 1990b. *J. of Mater. Sci. Lett.* 9:864–866.

Maruyama, B., and L. Rabenberg. 1986. In: Interfaces in Metal Matrix Composites (A.K. Dhingra and S.G. Fishman, eds.), 233–238, Warrendale, PA: The Metallurgical Society.

Masson, J.J., Schulte, K., Girot, F., and Y.L. Petitcorps. 1991. *Mater. Sci. & Eng.* A135:59–63.

Nakata, E., and Y. Kagawa. 1985. *J. of Mater. Sci. Lett.* 4:61–62.

Nutt, S.R. 1986. In: Interfaces in Metal Matrix Composites (A.K. Dhingra and S.G. Fishman, eds.), 157–167, Warrendale, PA: The Metallurgical Society.

Nutt, S.R., and R.W. Carpenter. 1985. *Mater. Sci. & Eng.* 75:169–177.

Portnoi, K.I., Timofeeva, N.I., A.A. Zabolotskii et al. 1981. *Soviet Powder Met. and Metal Cer.* 20:116–119.

Portnoi, K.I., Zabolotskii, A.A., and N.I. Timofeeva. 1980. *Metal Sci. and Heat Treat.* 22:813–815.

Pu, T., and W. Peng. 1991. In: Advanced Structural Inorganic Composites (P. Vicenzini, ed.), 151–159, Amsterdam: Elsevier Science Publishers.

Rawal, S.P., Allard, L.F., and M.S. Misra. 1987. In: Sixth International Conference on Composite Materials, ICCM 6 (F.L. Matthews, N.C.R. Buskell, J.M. Hodgkinson and J. Morton, eds.), 2.169–2.182, London: Elsevier Applied Science.

Ribes, H., Suéry, M., L'Espérance, G., and J.G. Legoux. 1990. *Metall. Trans.* 21A:2489–2496.

Roman, I., and R. Aharonov. 1992. *Acta Metall. et Mater.* 40:477–485.

Salvo, L., Suéry, M., Legoux, J.G., and G. l'Espérance. 1991. *Mater. Sci. & Eng.* A126:129–133.

Sawada, Y., and M.G. Bader. 1985. In: Fifth International Conference on Composite Materials, ICCM-V (W.C. Harrigan, J. Strife, and A.K. Dhingra, eds.), 785–794, Warrendale, PA: The Metallurgical Society.

Schamm, S., Lepetitcorps, Y., and R. Naslain. 1991. *Composites Sci. and Techn.* 40:193–211.

Shorshorov, M.K., Chernyshova, T.A., and L.I. Kobeleva. 1982. In: Fourth International Conference on Composite Materials, ICCM-IV (T. Hayashi, K. Kawata, and S. Umekawa, eds.), 1273–1279, Tokyo: Japan Society for Composite Materials.

Si, W., Li, P., and G. Li. 1991. In: Eighth International Conference on Composite Materials, ICCM 8 (S.W. Tsai and G.S. Springer, eds.) 19K1–19K8, Covina, CA: SAMPE.

Smyth, I. 1991. Ph.D. Thesis. Department of Materials Science and Engineering, Massachusetts Institute of Technology, Cambridge, MA.

Sui, Q., Guo, S., and F. Tang. 1991. In: Eighth International Conference on Composite Materials, ICCM 8 (S.W. Tsai and G.S. Springer, eds.) 21E1–21E6, Covina, CA: SAMPE.

Thanh, L.N., and M. Suéry. 1991. *Scripta Metall. et Mater.* 25:2781–2786.

Tsai, H.W., Wang, W.L., and G.D. Zhang. 1991. In: Eighth International Conference on Composite Materials, ICCM 8 (S.W. Tsai and G.S. Springer, eds.) 19C1–19C9, Covina, CA: SAMPE.

Viala, J.C., Fortier, P., and J. Bouix. 1990. *J. Mater. Sci.* 25:1842–1850.

Wadley, H.N.G., Biancaniello, F.S., and R.B. Clough. 1988. In: High Temperature/High Performance Composites, MRS Symp. Proc. Vol. 120 (F.D. Lemkey, S.G. Fishman, A.G. Evans and J.R. Strife, eds.), 35–44, Pittsburgh, PA: Materials Research Society.

Warner, T.J., Withers, P.J., and J. White et al. 1988. In: Second International Conference on Composite Interfaces (ICCI-II) (H. Ishida, ed.), 537–551, New York: Elsevier.

Warren, R., and C.-H. Andersson. 1984. *Composites* 15:101–111.

Wilkinson, B.W., Holm, C., and Y.P. Lin et al. 1990. *In:* Interfaces in Composites, MRS Symp. Proc. Vol. 170 (C.G. Pantano and E.J.H. Chen, eds.), 105–110, Pittsburgh, PA: Materials Society.

Wu, R. 1990. *In:* Controlled Interphases in Composite Materials, Proceedings of the Third International Conference on Composite Interfaces (ICCI-III) (H. Ishida, ed.), 43–56, New York: Elsevier.

Yang, M., and V.D. Scott. 1991. *J. Mater. Sci.* 26:1609–1617.

Yoon, H.-S., Ojura, A., and H. Ichinose. 1989. *J. of the Iron and Steel Inst. of Jap.* 75:1455–1462.

Yumoto, H., Takahashi, A., and N. Igata. 1991. *In:* Advanced Materials for Future Industries: Needs and Seeds (I. Kimpara, K. Kagayama, and Y. Kagawa, eds.), 737–743, Tokyo: International Convention Management, Inc.

Zhang, G.D., Tsai, H.W., and R.J. Wu. 1991. *In:* Advanced Materials for Future Industries: Needs and Seeds (I. Kimpara, K. Kagayama, and Y. Kagawa, eds.), 723–730, Tokyo: International Convention Management, Inc.

Zhuo, Y., Yin, X.F., and D.M. Yang. 1990. *In:* Controlled Interphases in Composite Materials, Proceedings of the Third International Conference on Composite Interfaces (ICCI-III) (H. Ishida, ed.), 277–283, New York: Elsevier.

Another system that has received considerable attention is the Al-Mg-Si-O system, of interest for Al_2O_3 reinforced Al, Mg, and their alloys. Recent references include a detailed review of earlier work by Warren et al., given last in the following list:

Fishkis, M. 1991. *J. Mater. Sci.* 26:2651–2661.

Ghosh, P.K., and S. Ray. 1990. *In:* Fabrication of Particulates Reinforced Metal Composite (J. Masounave and F.G. Hamel, eds.), 23–29, Materials Park, OH: ASM International.

Molins, R., Bartout, J.D., and Y. Bienvenu. 1991. *Mater. Sci. & Eng.* A135:111–117.

Pfeifer, M., Rigsbee, J.M., and K.K. Chawla. 1990. *J. Mater. Sci.* 25:1563–1567.

Warren, R., and C. Li. 1990. *In:* Controlled Interphases in Composite Materials, Proceedings of the Third International Conference on Composite Interfaces (ICCI-III) (H. Ishida, ed.), 583–598, New York: Elsevier.

Other systems having attracted equally considerable attention include those based on the Ti-C-Si system, a system that is of interest for use with Si-C reinforced titanium composites. The variety of systems that have been investigated is illustrated by the following list of systems and relevant references:

1. Al-Li alloys with Al_2O_3

Champion, A.R., Krueger, W.H., Hartmann, H.S., and A.K. Dhingra. 1978. *In:* 1978 International Conference on Composite Materials ICCM 2 (B. Noton, R.A. Signorelli, K.N. Street and L.N. Phillips, eds.), 883–904, Warrendale, PA: The Metallurgical Society.

Hunt, W.H. 1986. *In:* Interfaces in Metal Matrix Composites (A.K. Dhingra and S.G. Fishman, eds.), 3–25, Warrendale, PA: The Metallurgical Society

2. the Ni/C-fiber system

Warren, R., Anderson, C.H., and M. Carlsson. 1978. *J. Mater. Sci.* 13:178–188.

3. Pt with NiO

Shieu, F.-S., Raj, R., and S.L. Sass. 1990. *Acta Metall. et Mater.* 38:2215–2224.

4. Al with Al_2O_3/ZrO_2 fibers

Isaacs, J.A., Taricco, F., Michaud, V.J., and A. Mortensen. 1991. *Metall. Trans.* 22A:2855–2862.

5. Fe_3Al with Al_2O_3/ZrO_2 fibers:

Nourbakhsh, S., Margolin, H., and F.L. Liang. 1990. *Metall. Trans.* 21A:2881–2889.

6. Ti-Al with Al_2O_3/ZrO_2 fibers:

Nourbakhsh, S., Liang, F.L., and H. Margolin. 1990. *Metall. Trans.* 21A:213–219.

7. Ni and Ni_3Al with Al_2O_3:

Povirk, G.L., Horton, J.A., and C.G. McKamey et al. 1988. *J. Mater. Sci.* 23:3945–3950.

Nourbakhsh, S., Margolin, H., and F.L. Liang. 1989. *Metall. Trans.* 20A:2159–2166.

Trumble, K.P., and M. Rühle. 1991. *Acta Metall. et Mater.* 39:1915–1924.

8. Cr with Al_2O_3-Cr_2O_3:

Handwerker, C.A., Cahn, J.W., and J.R. Manning. 1990. *Mater. Sci. & Eng.* A126:173–189.

PART II
MICROSTRUCTURE CHARACTERIZATION

Chapter 4
Characterization of Residual Stresses in Composites

MARK A.M. BOURKE, JOYCE A. GOLDSTONE,
MICHAEL G. STOUT, AND ALAN NEEDLEMAN

4.1 Definition

Materials can maintain elastic strains in their bulk even though they are not subject to any applied tractions. If a material is in equilibrium, the elastic strains or stresses that exist within it, in the absence of an external load, are called residual stresses. Classically, residual stresses are categorized into three types according to the length scale over which they act (Hauk 1986; James and Buck 1980). "Type-I" stresses act over distances measured in millimeters and are often referred to as macroscopic. They can arise because of different cooling rates within a specimen following welding or machining. Stresses that act over lengths of the microstructure (typically 1 to 100 μm) are called "type-II" residual stresses. Intergranular interactions between the reinforcement and matrix of a metal matrix composite (MMC) typically result in type-II stresses. Finally, stresses that exist at an atomic scale and vary over individual grains are described as type III, resulting, for example, from dislocation pile-ups at nonshearable precipitates. Types II and III are often collectively referred to as microstrain effects.

4.2 Effects

The importance of residual stresses in the manufacture of metals and ceramics has been recognized for many years. Annealing of glass by heat treatment was practiced by ancient civilizations and is among the first documented uses of an engineering procedure to control residual stresses (Cooper 1981). Typical undesirable effects associated with residual stresses include diminished fatigue resistance, ply delamination, shape distortion and accelerated stress corrosion. In addition, mechanical properties such as yield strength or fracture toughness can be affected by the presence, sign, and magnitude of residual stresses (Hauk 1986; James and Buck 1980; Matzkanin 1987; Pedersen 1985; Metcalfe and Klein 1974). Even more than in conventional materials, the control of residual stress in composites during fabrication and processing is crucial to their integrity. Fabrication processing, welding, machining, and thermal treatments can all produce substantial residual stress in composites. Because composites are multiphase materials with an extremely anisotropic structure, the detrimental effects of residual stresses on material shape and properties are compounded.

4.3 Origins

Residual stresses are inherent in most metal-matrix, organic-matrix and ceramic composites (Cheskis and Heckel 1968; French and MacDonald 1969; Cox 1987; Krawitz et al. 1981; Peck 1987; Pedersen 1985; Predecki and Barrett 1981; Ruud et al. 1987; Tsai et al.

1981b; Krawitz 1985; Matzkanin 1987; James 1989; Abuhasan et al. 1990; Cox et al. 1990; Stepanova 1990; James 1991). Type-II stresses that result from differences in the coefficients of thermal expansion (CTE) of the constituents are frequently dominant. However, phase transformations (during cooling from fabrication temperatures), quenching, or heat treatment also lead to type-II stresses. In addition, even small amounts of plastic deformation can significantly alter the residual stress state produced during fabrication. In laminated composites, type-I and type-II stresses can coexist (Nairan and Zoller 1985b). Individual plies have thermal expansion properties with orthotropic symmetry. Thus, the stacking order and orientation of individual plies during fabrication or filament winding affects the residual stress state. Simultaneously, individual plies may experience type-II stresses because of incompatibility between the matrix and fibers (Nairan and Zoller 1985a).

The state of residual stress is rarely static. Many relaxation processes exist, for example interface sliding, decohesion, plastic strain, microcracking, and buckling of fibers. Some of the relaxation processes can be active at room temperature, but it is rarely possible to identify those that are active from measurements alone. The same is true for organic-base composites in which water absorption can swell the matrix but have little effect on the fibers (Predecki and Barrett 1981; Matzkanin 1987). Thus, residual stresses in the matrix that were originally tensile are generally decreased as moisture content increases.

In laminated composites, most measurements have addressed the macroscopic stress resulting from anisotropic expansion of plies laid up in different orientations. The measurements have generally been made using surface displacement techniques. Few studies have been undertaken to assess the stresses between the fibers and organic matrix. Both photoelasticity techniques applied to a single-fiber-thickness ply (Nairan and Zoller 1985a) and X-ray diffraction from metallic flakes imbedded between plies (Predecki and Barrett 1981) have been successfully used for this purpose.

4.4 Techniques

Both destructive and nondestructive techniques have been developed to measure residual stresses. Measurements based on surface relaxation or diffraction are the most common approaches but recently a variety of other techniques, for example ultrasonic or magnetic methods, are seeing application to MMCs. The selection of a particular technique depends not only on the material but also on the type of residual stress of interest. Thus, for conventional engineering materials, surface-relaxation techniques are appropriate for determining macroscopic (type-I) stress distributions. However, for MMCs, type-II stresses are important, and diffraction is the technique of choice because it distinguishes between the strains in the matrix and in the reinforcement. It is also nondestructive and measures both type-I and type-II residual stresses. Both X-ray (Hauk 1981; Noyan and Cohen 1987; Peck 1987) and neutron (Krawitz et al. 1981; Prask and Choi 1987; Krawitz 1990a; Krawitz 1990b) diffraction have been applied to composites. Although experimental techniques can generally distinguish type-I and type-II residual stresses, they cannot reliably determine atomic-length scale, type-III, stresses.

In recent experiments, flakes of crystalline material have been embedded in organic matrix composites to enable diffraction measurements (Predecki and Barrett 1981). However, for most amorphous materials, techniques other than diffraction are applied. Of these, destructive or semidestructive procedures coupled with the concomitant measurement of surface displacements are most used. Hole drilling and coring with strain-gauge rosettes are widely used (Hauk 1986; Boag et al. 1987; ASTM Standards 1989). As alternatives to mechanical gauge measurements of surface displacement, photoelastic techniques are applied (Nairan and Zoller 1985a). These usually use photoelastic coatings but, in special cases, observations are taken directly from the matrix of organic composites. In another mechanical approach, the radius of curvature of one surface is measured while removing material from the opposite side (Doi and Kataoka 1982; Nairan and Zoller 1985b; Vikar et al. 1988). Surface removal techniques are limited in their application to composites; they address only type-I stresses, and depth profiling usually involves destruction of the specimen.

Other techniques have been employed in specific situations. They include Raman spectroscopy (Cheong et al. 1986; Cheong et al. 1987) acoustic emission (Armstrong et al. 1990), ultrasonics (James and Buck 1980), scanning electron microscopy (with an acoustic transducer) (Cantrell et al. 1990), Moiré fringe displacement measurement (Lee and Czarnek 1991; McDonach et al. 1983; Nicoletto 1991), stress-pattern analysis by measurement of thermal emission (SPATE) (Leaity and Smith 1989), and magnetic measurements (Barkhausen noise and magnetostriction) (James and Buck 1980; Boag et al. 1987; Abuku and Cullity 1971). In some cases these techniques are still being developed or have seen limited application to composites and so will not be discussed here.

4.5 Reasons for Measuring Residual Stresses

Metal-matrix composites often experience fabrication temperatures in excess of 1000°C, and large stresses can be induced on cooling. These stresses can lead to plastic deformation of the matrix or, in severe cases, to interfacial debonding of the reinforcement from the matrix. The small-length scales associated with the interfaces of MMCs preclude experimental techniques from accurately resolving the interface behavior. However, the interface is important, and numerical descriptions have been attempted. Finite element or Eshelby approaches predict interface behavior, and although it is impossible to resolve strains at matrix-reinforcement interfaces, average strains can be measured in each phase and correlated with the numerical predictions. Because of the complex stress and strain state in multiphase materials it is particularly important to verify numerical predictions by experiment.

Before a metal-matrix composite can be used, it must be characterized and its residual stress must be related to variables that include: the type of reinforcement (particulate or fibers [chopped or continuous]); volume fraction; diameter; aspect ratio; and shape. The residual stress state of aligned fiber composites is often anisotropic at both the macroscopic and microscopic level and can evolve during thermomechanical loading. Indeed, changes can occur in the absence of loading because of room temperature relaxation.

At this point we will focus on examples of experimental and analytical approaches applicable to metal-matrix composites. Particular attention will be given to diffraction. Surface displacement measurements will also be discussed. These can be applied to metal-matrix composites, although they have generally been applied to organic-matrix composites. Both the elastic Eshelby and elastic-plastic finite element method (FEM) analyses will be examined in context with experimental results.

4.6 Measuring Techniques

4.6.1 Diffraction

Although the average macroscopic stress in a composite may be zero, adjacent grains of the matrix and reinforcement are often in opposing stress states. Thus, both the macroscopic and microstructural behaviors must be considered, and techniques that address only type-I stresses are limited in their application to composites, in which type-II stresses are prevalent. The ability to separately measure the residual elastic strains in multiple crystalline phases makes diffraction a powerful probe. Strains can be determined because residual (or applied) stresses alter the interatomic spacings in crystalline materials (Noyan and Cohen 1987).

Intergranular residual elastic strains in metal-matrix composites are often less than 0.002 in the matrix or 0.0005 for the stiff refractory reinforcement. Thus, changes in the angle or wavelength of diffracted reflections are small and can only be measured with high-resolution instruments. Strains can result from applied or residual stresses (or a mixture of both), but only the elastic component is measured by diffraction. However, qualitative information concerning plastic strain or the distribution of microstrain can sometimes be inferred (Krawitz et al. 1988; Krawitz et al. 1989).

Diffraction from crystalline materials is possible using X-rays or neutrons. The two radiations differ primarily in their relative penetration into materials. X-rays are limited to surfaces and typically address the first 100μm in materials. By contrast, neutrons penetrate many millimeters and give measurements averaged over the interior of a specimen (although it is rare that specimens thicker than a few centimeters are examined). Table 4.1 gives a few examples of the relative penetration depth of X-rays and neutrons. The application of X-rays or neutrons to a problem depends on the material and region of interest.

Much of the discussion about diffraction is pertinent to both types of radiation. Before commenting on X-rays and neutrons in particular, problems common to both approaches are noted. The first concerns the identification of a value for the stress-free response of the material. This is needed if absolute strain values are required. Stress-free values are also always needed for neutron measurements and for X-rays, when a triaxial stress analysis is employed. Unfortunately if changes in microstructure or stoichiometry occur (during fabrication) stress-free values are not easily identifiable for individual phases. This is more likely to be a problem for the soft matrix than for the hard reinforcement. However, the strains in the reinforcement are smaller and harder to measure than the strains in the matrix.

Identifying the compliances corresponding to individual lattice reflections is another problem. To calculate stresses from measured strains the compliance (or effective modulus) for grains in specific orientations in a polycrystal must be known. Single crystals have anisotropic compliances which can be measured experimentally. However, grains in a polycrystal are constrained by their neighbors and do not deform like isolated crystals. Thus, neither the bulk modulus nor the measured (or calculated) single-crystal elastic-constants are appropriate. Experimental determination of the effective compliance for sets of lattice planes is recommended (Dolle 1979; Hauk 1981) and is the only way to account for strong texture variations.

Table 4.1. Penetration Depths of X-rays and Neutrons for Different Materials

Material	90% Absorption Depth (cm)		Coefficient of Thermal Expansion (K^{-1})	Elastic Modulus (GPa)
	X-rays*	Neutrons†		
Al	0.018	288	24	$(68.5)^1$
Ti	0.002	12	10	$(90-125)^2$
Mg	0.034	1355	30	$(44.2)^3$
SiC	0.029	1023	4	$(470-483)^{1,4}$
Graphite	0.850	4606	≈ 0, (axial) 25, (transverse)	$(218-380)^3$

*For 1.54Å Cu Kα radiation
†Neutron absorption data is for 1.08Å neutrons. From Bacon (1975).
[1]Povirk et al. (1992). [2]Cox et al. (1990). [3]Armstrong et al. (1990). [4]Shi et al. (1992).

4.6.1.1 X-ray Diffraction

X-rays were first employed by researchers for measuring type-I residual stresses in the 1920s. However, not until the late 1960s was some of the earliest work on composites performed by Cheskis and Heckel who studied Al-B and Cu-W composites under load (Cheskis and Heckel 1968; Cheskis and Heckel 1970). Since then, diffractometers have become widely available and X-ray diffraction has become well established for measuring stress. Many authors (French and MacDonald 1969; Dolle 1979; James and Cohen 1980; Hauk 1981; Tsai et al. 1981b) have discussed this technique. Noyan and Cohen (1987) provide a particularly comprehensive review. Because the subject has been covered in detail elsewhere, only a brief description is given here.

In the simplest form of X-ray stress analysis, the stress normal to a free surface is assumed to be zero. This assumption reduces the measurement to determining two principal stress components lying in the plane of the surface. This is justifiable because of the shallow penetration of X-rays in most materials. By recording the position of Bragg reflections (and thus the lattice spacings) at a variety of tilt angles, it is possible to obtain the well known d versus $\sin^2\psi$ plot. From these data, the stress state at the surface can be inferred. Variations in approach include limiting the number of tilt angles or the number of points to define a Bragg reflection (Peck 1987). In the more general analysis, the stress state is not assumed to be biaxial and a triaxial analysis is incorporated for which measurement of the stress/strain response is necessary (James 1989). In either case, Lorentz polarization and absorption corrections are applied after recording the Bragg reflections (Tsai et al. 1981b).

The heterogeneous nature of composites often invalidates the plane stress assumption of simple X-ray stress analysis. For single-phase materials, the biaxial assumption is reasonable, but in a composite, if the penetration depth of the X-rays is comparable to the dimensions of one dispersed phase, the stress state must be treated as triaxial. In uniformly dispersed particulate metal-matrix composites, this is likely to be true. For example, 90% of the X-rays produced by a Co source are diffracted from within 54μm of the surface of an aluminum alloy containing SiC particles less than 10μm in diameter (Tsai et al. 1981b). Triaxial analyses are used but are less accurate than biaxial analyses because stress-free values of the interplanar spacings are needed. As already mentioned, it is often difficult to identify values for the stress-free material response because of differences in texture, microstructure, or stoichiometry between monolithic and composite specimens. Other complicating features of X-ray measurements are differences in the effective sampling depths associated with different ψ tilts, surface preparation effects (ASTM Standards 1989), protective coatings, laminates, grain size effects, and the need to identify the effective X-ray elastic constants (which may vary with deformation). Despite the problems, X-rays have been successfully applied to composites (French and MacDonald 1969; Predecki and Barrett 1981; Tsai et al. 1981; James 1989; Abuhasan et al. 1990). In practice, X-ray and neutron measurements can be complimentary as shown by James et al. (1992) in a study of Ti_3Al containing continuous SiC fibers (a material that may have applications in high temperature hypersonic aerospace design). By removing surface layers of the Ti_3AlSiC by progressive electropolishing, the longitudinal stress was shown to increase below the surface in material containing a high volume-fraction of fibers (Cox et al. 1990). The neutron diffraction measurements on similar material determined the average bulk longitudinal stress and supported the existence of a stress gradient

(James et al. 1992). In other depth profiling examinations by electropolishing, Tsai et al. (1981a) and Tsai et al. (1981b) reported X-ray measurements of the strains near fiber-matrix interfaces in Al-graphite composites.

4.6.1.2 Neutron Diffraction

Neutrons provide a special and powerful probe of lattice dimensions (Prask 1991; Axe 1991). In contrast to X-rays, their use for measuring strains is very recent, with references first appearing in the early 1980s (Krawitz and Schmank 1982; Allen et al. 1985). Since then, their application to problems of macroscopic residual stresses has increased dramatically, and more recently they have been applied to metal-matrix composites (Kupperman et al. 1989; Krawitz 1990b; Kupperman et al. 1990; Majumdar et al. 1991). Apart from surface studies, the application of neutrons to MMCs is generally superior to using X-rays because the strain measurement is an average over the bulk of a specimen.

However, neutrons suffer from limited accessibility to researchers because of the need for an intense source of thermal neutrons. Specimens must be taken to a facility dedicated to neutron production, which means either a research reactor or a spallation (pulsed) neutron source. Few research neutron sources exist. In North America some of the more well known include reactors at the Oak Ridge National Laboratory, the University of Missouri, and at Chalk River (AECL-Canada). In Europe, well-known research reactors exist at Risø (Denmark), Petten (Holland), and the Institut Laue Langevin (France). Spallation neutron sources in North America exist at Los Alamos National Laboratory (LANSCE) and at Argonne National Laboratory (IPNS), in Europe at ISIS (England), and in Japan (KENS). Despite the limited number of sources, most facilities operate user programs that make instruments available for research on the basis of scientific merit (though neutron beam time can often be purchased for proprietary research). Thus, for fundamental work or pilot projects, neutrons are available.

The principal advantage of using neutrons for residual strain measurements is their ability to penetrate deeper into materials than X-rays. For example, the 90% absorption depth of thermal neutrons in aluminum is 288 cm compared with 0.018 cm for Cu-Kα X-rays (Table 4.1). The greater penetration eliminates surface effects and gives values of the lattice spacing averaged over many grains. Neutron diffraction has been widely used to make spatially resolved measurements of type-I stresses in engineering components (Allen et al. 1985). However, for studies of type-II residual stresses, spatial resolution is not usually required.

Measurements of residual strain can be made using monochromatic (single wavelength) or polychromatic (multiple wavelength) neutrons (see Appendix A of this chapter). These alternatives are usually associated with reactor and pulsed sources, respectively. Using monochromatic neutrons, strains are inferred from changes in angle of the interplanar reflections (Allen et al. 1985). With polychromatic neutrons, strains are determined from changes in wavelength for diffracted neutrons at a constant angle (Bourke, Goldstone, and Holden 1990). In either case, the changes are measured relative to stress-free reference material. The strain is measured in the direction of the scattering vector that bisects the incident and the diffracted beams (Figure 4.1).

Typical residual strains result in changes in the angle of Bragg reflections (for monochromatic neutrons) of less than 0.01 degree. Similarly, changes in wavelength of diffracted reflections at a pulsed source can be less than 0.002Å. To measure these small changes, high-resolution powder diffractometers, commonly used for crystal structure analysis, are required. The intensities of neutron sources are usually less than those of X-ray sources, and count times are longer to record single Bragg reflections. It is often necessary to compromise between count times and the statistical accuracy of the measurement. Sampling volumes examined by neutron diffraction are typically between 50 mm^3 and 2000 mm^3. Thus, they are much larger than the volumes examined during an X-ray measurement. Samples of less than a few cubic centimeters in volume can be completely immersed in a neutron beam that will average out any macroscopic effects. For most studies of composites, the sampling volume is maximized to improve the statistical accuracy of the results.

If the stress state in a composite is close to isotropic, then a single measurement may be sufficient. More often strains must be measured in different directions in a specimen, for example, parallel and transverse to a fiber axis. In this respect, a big advantage of a pulsed source is that strains can be measured simultaneously in two or more directions by using multiple detector banks arrayed around the specimen. At pulsed neutron sources, detectors are placed at fixed angles and specimens are irradiated with neutrons of all wavelengths. The wavelength of the detected neutrons is determined from their time of flight from the production target to the detector. The scattering vectors of all the reflections recorded in one detector bank all lie in the same direction in the specimen. Each diffraction pattern includes all the Bragg reflections. Thus, multiple phases can be examined simultaneously. Collection of all the lattice reflections means that measurements are less vulnerable to texture effects (than when using monochromatic neutrons), and that comprehensive assessment of the

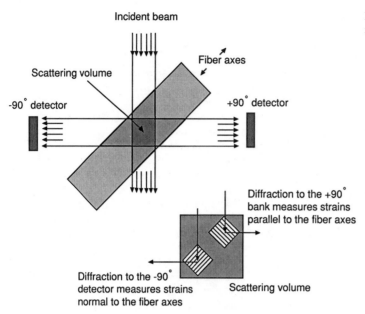

Figure 4.1. Sampling volume in a neutron measurement.

elastic-plastic anisotropy of different reflections is possible. The application of a pulsed source for strain measurement in deformed zircalloy was demonstrated by MacEwen et al. (1983).

Two recent studies on Al/SiC have compared residual strains measured by neutron diffraction with numerical predictions. Allen et al. (1992) and Povirk et al. (1991b) and (1992), applied Eshelby (equivalent inclusion) and finite element analyses respectively to predict thermal residual stresses and the effects of plastic deformation. Neutron diffraction provided a method for validating the predictions because it discerned the average behavior of each phase. Both studies demonstrated the ability of neutron diffraction to measure strains in different directions, which is important in cases where the stresses are directionally anisotropic, for example, in aligned fiber composites. Other examples of the application of neutrons to composites are found in studies of cermet materials (Krawitz 1985; Krawitz et al. 1988; Krawitz et al. 1989; James et al. 1992 and of a variety of systems by Kupperman et al. 1989; Krawitz 1990b; Kupperman et al. 1990; Wright et al. 1991; Majumdar et al. 1991).

The development and evolution of residual stresses in composites is intimately related to the ability of their constituents to share a load. By virtue of their nondestructive penetration into materials, neutrons have been used to measure elastic strains induced in each phase by the application of an applied load. With small stress rigs that can fit into the limited space available to neutron diffractometers, the anisotropic strains that result from elastic anisotropy of grains in different orientations or in individual phases can be measured (Hutchings 1990; Allen et al. 1988, Windsor 1990). For example, shared-load measurements in Al/SiC have been made by Allen et al. (1987) and Withers et al. (1989). Neutrons can penetrate heat shields and enable creep and relaxation studies. Recently there has been considerable interest in making in situ measurements of strain in composites under load at elevated temperatures.

4.6.2 Hole Drilling

Semidestructive techniques have been developed over many years to assess type-I residual stresses in metals by measuring the displacements that result when material is removed from the component. Material is removed by hole drilling, coring, or grinding. In hole drilling measurements (Soete and Vancrombrugge 1950; Mather 1934; Boag et al. 1987), a strain-gauge rosette, (manufactured specifically for this application) is attached to the specimen. A hole is drilled at the center of the rosette (Figure 4.2), and as material is removed the surface residual strain in the vicinity of the hole is relaxed and strain displacements result. The technique, which assumes isotropic elasticity, has been developed for metals and is described in the ASTM Standards (1989). The standard was written for hole drilling in the center of a rosette containing three strain gauges. Two of the gauges are orthogonal to one another, and the other lies along the bisector. A tool guide, with a fixture for a microscope, is used to assure that the

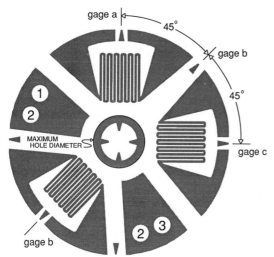

Figure 4.2. A commercial strain-gauge rosette used to measure residual stresses by the hole drilling technique.

directions of principal stress and strain are often not known and will not necessarily coincide with one another. However, if a composite is orthotropic, the directions of the material symmetry and the values of the elastic constants in those directions are likely to be known. The elastic constants, E_{11}, E_{22} and E_{12}, then appear in the following equations, which relate the surface strains in the material-symmetry directions to the corresponding stress components:

$$\varepsilon_{11} = (1/E_{11})\sigma_{11} - (\nu_{21}/E_{22})\sigma_{22} \quad (4.1)$$

$$\varepsilon_{22} = -(\nu_{12}/E_{11})\sigma_{11} + (1/E_{22})\sigma_{22} \quad (4.2)$$

$$\varepsilon_{12} = \sigma_{12}/G. \quad (4.3)$$

hole is drilled or abraded in the exact center of the rosette (Figure 4.3).

The hole drilling technique has principally been applied to organic-based composites in which macroscopic stresses result from differences in the thermal expansion of the longitudinal and transverse directions in individual laminate plies and the "lay-up" of these plies in different orientations (Doi and Kataoka 1982; Nairan and Zoller 1985b; Kataoka et al. 1986; White and Hahn 1990; Aleong and Munro 1991). However, the technique is applicable to metal-matrix composites, with unidirectional reinforcement if macroscopic residual stress gradients exist between the surface and the interior of a bulk component.

Because of the orthotropic symmetry of some composites, it has been necessary to modify the equations around which the ASTM standard has been written (Bert and Thompson 1968). In real composites, the

Bert and Thompson (1968) considered an orthotropic material in a generalized plane-stress field, the free-surface results in the plane-stress condition, represented by the stress components σ_{11}, σ_{22}, and σ_{12}. They derived the equations necessary to obtain the residual stresses for such a condition in terms of three coefficients that contain the elastic properties of the material. A summary of their results and their proposed method for determining the calibration coefficients are given in Appendix B of this chapter. In addition, Prasad et al. (1987) have experimentally determined the coefficients for a 20-ply unidirectionally reinforced graphite-polyimide composite. Their procedures used a uniaxially loaded specimen with the tensile axis parallel to the ply axes and a four-point asymmetric bend-bar geometry for producing a pure shear loading. The calibrations are valid only for the hole depth from which they were obtained. In the actual measurement, the holes in the component must be drilled to this depth as well.

The standard techniques of using strain gauges to measure the residual surface strains have recently been

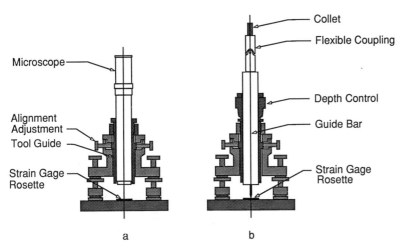

Figure 4.3. The tool guide used to drill the hole for a residual stress measurement.
(a) Configured for alignment.
(b) Configured to drill the hole.

modified to include strain measurement using Moiré fringe (McDonach et al. 1983; Lee and Czarnek 1991; Nicoletto 1991) and laser holography. The advantage of optical techniques is that they address the full-field displacements rather than just the displacements under the strain gauges. The reader should consult the references for details regarding utilization of these techniques.

4.7 Experimental Results

Because of the influence of residual stresses, it is important to determine their extent and origins. Their magnitude depends on processing, the yield strengths of the constituents, and the volume fraction of the reinforcement. They influence the bulk mechanical behavior by affecting the bond between the matrix and the reinforcement, usually with degradation of the bulk properties. For example, tensile matrix residual stresses produced by thermal cycling can degrade the composite integrity. Thermal effects may be compounded by plastic deformation. Thus, the origins of residual stresses can be difficult to separate.

In practice, the importance of residual stresses is related to their effects on a material's mechanical properties. In particular, type-II residual stresses are implied by a strength differential between tension and compression (Cheskis and Heckel 1970). The thermal and mechanical environment of components is rarely static, and the residual stress state often evolves. Thermally activated recovery or shape distortions will characteristically relieve residual stresses. In the following sections, selected results are presented concerning the characterization of residual stresses that are induced by thermal treatments or originate from plastic deformation.

4.7.1 Thermal Effects

Thermal stresses in metal-matrix composites result during cooling either from temperature gradients within the specimen (leading to type-I stresses) or from a mismatch of coefficients of thermal expansion (CTE) between the constituents (introducing type-II stresses). The former effect can be controlled by slow cooling, but the latter is inherent and will always give interaction stresses unless the CTEs of the constituents are identical. Typically, the difference in CTEs in metal-matrix composites is large, for example, Al $\approx 25 \times 10^{-6}$/K and SiC $\approx 4 \times 10^{-6}$/K. Processing temperatures of metal-matrix composites are usually several hundred degrees above ambient. Thus, on cooling, the larger contraction of the matrix compared to the reinforcement leads to tensile and compressive residual stresses in the matrix and the reinforcement, respectively. In continuous fiber or aligned whisker composites, the anisotropic shape of the reinforcement induces larger axial than transverse residual stresses. Thermal residual stresses have been studied in several metal-matrix systems: Al/SiC (Krawitz 1990b; Majumdar et al. 1990), Al/Graphite (Hauk 1981; Predecki and Barrett 1981; Tsai et al. 1981b; Tsai et al. 1981a; Cheong and Marcus 1987; Vedula et al. 1988a, 1988b; Zong and Marcus 1991), Al/B (Stepanova 1990), Mg/Graphite (Predecki and Barrett 1981; Zong and Marcus 1991), Cu/W (Cheskis and Heckel 1970) and Ti-alloys/SiC (Cox et al. 1990; Kupperman et al. 1990; James 1991; Wright et al. 1991; Saigal et al. 1992) by a variety of techniques that include X-ray (Cox 1987; Tsai et al. 1981; Tsai et al. 1981b; Ledbetter and Austin 1987; Cox et al. 1990; Stepanova 1990; Wright et al. 1991; Saigal et al. 1992) and neutron diffraction (Ledbetter and Austin 1987; Krawitz 1990b, Kupperman et al. 1990; Majumdar et al. 1991; Wright et al. 1991), ultrasonics (Wright et al. 1991), high-energy electron diffraction (Khan 1976), scanning-electron microscopy (Khan 1976; Cheong and Marcus 1987), acoustic emission (Predecki and Barrett 1981), and microbuckling (Cox et al. 1990; Zong and Marcus 1991).

Experiments confirm that, during cooling, tensile residual stresses are generated in the matrix material in both the axial and transverse directions, although the magnitudes vary with the material. Continuous fiber and aligned whisker composites (Krawitz 1990b; Kupperman et al. 1990) exhibit a large directional anisotropy, but particulate or spheroid reinforcements result in nearly isotropic conditions (Ledbetter and Austin 1987; Krawitz 1990b). When researchers use neutron diffraction, which measures both the matrix and fiber stresses, the fibers are found to be in compression in the transverse direction. Experimental limitations in resolving micron-scale stress gradients can lead to an apparent lack of equilibrium between the stresses in the fiber and the average matrix stresses (Ledbetter and Austin 1987; Povirk et al. 1992). The steep stress gradients in the interfacial region cannot be adequately resolved by any experimental techniques, although attempts to measure interfacial stresses using intense synchrotron radiation have been made. For an isolated fiber (Wright et al. 1991), these results show sharply increasing stresses near an alumina fiber embedded in titanium, but a much shallower gradient for a SiC fiber in titanium.

Relaxation of residual stress during heating has been measured with neutrons (Krawitz 1990b; Kupperman et al. 1990; Majumdar et al. 1991; Wright et al. 1991;

Saigal et al. 1992). Establishing the temperature at which stresses in a composite are completely relaxed is important for guiding the development of constitutive models and for verification of analytic and numerical calculations (Figure 4.4). Conversely, this is also the temperature at which stresses begin to be introduced during cooling. In general, this temperature is less than the fabrication temperature, and the intervening range corresponds to a regimen in which dislocation relaxation (recovery) processes prevent the development of residual stress. One should note that the heating rate used in such experiments is of critical importance. Relaxation of residual stresses in Al/SiC, over long periods of time, has even been observed at room temperature (Withers et al. 1987). Thermal cycling, between high or low temperatures and room temperature, can also modify the residual stresses. In general, repeated cooling to liquid nitrogen temperatures (77K) from room temperature reduces the residual stresses because of plastic deformation in the matrix induced by stresses generated from the different coefficients of thermal expansion (Tsai et al. 1981b; Tsai et al. 1981a; Kupperman et al. 1990). Similarly, thermal cycling to elevated temperatures also reduces residual stresses, but the reduction can vary from significant (Kupperman et al. 1990) to very little (Stepanova 1990).

4.7.2 Plastic Deformation

In designing metal-matrix composites, it is usually not possible to exploit the strength of the reinforcement without incurring some plastic flow in the matrix. During mechanical deformation, the presence of hard reinforcement particles promotes heterogeneous plastic deformation in the matrix. This results in residual stresses after the applied loads are removed. Alternatively, plastic deformation from thermal cycling can reduce the residual stresses present at room temperature, and it is clear that thermal and plasticity effects are intimately related. For example, cooling to cryogenic temperatures produces plastic flow in the matrix of an Al/graphite composite. On returning to room temperature, the original residual stresses are reduced by 30% (Tsai et al. 1981a). These effects are even more pronounced in an Al/B composite. The plastic deformation on cooling can be great enough for the initial matrix tensile residual stresses to become compressive at sufficiently low temperatures (Stepanova 1990). In practice, separating the origin of residual stresses can be difficult, and studies of plastic deformation must be accompanied by analytic or numerical calculations. The techniques that have been applied to investigate deformed metal-matrix composites are neutron and X-ray diffraction. They offer complementary information by examining the bulk and surface of the composite. Observations of the effects of plastic deformation have largely been made in conjunction with observations of other effects, such as machining or loading.

James (1989) reported on stresses induced by machining Al/SiC whisker composites. Milling with a carbide tool produced large biaxial compressive stresses (approximately −300MPa) near the surface in the matrix material. In contrast, when researchers used a diamond grinding wheel, they produced nearly stress-free (−10 to −20MPa) specimens. In the as-fabricated condition, compressive stresses near the surface were observed by X-ray diffraction that were not significantly altered by careful grinding and mechanical polishing. The same samples were examined using neutron diffraction. The Al matrix was found to be in tension and the SiC reinforcement in compression. This stress state arose from thermal effects during fabrication, and was clearly not modified by the subsequent machining.

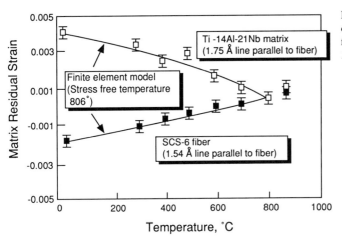

Figure 4.4. Neutron diffraction measurements of residual strain in SCS-6/Ti-24Al-11Nb as a function of temperature. *(From: Saigal et al. 1992.)*

Many studies of the effects of deformation on residual stress states in composite materials have been made. Cheskis and Heckel (1968) and (1970) used X-ray diffraction to look at deformation effects in Al/B composites with a four-point bending specimen geometry. Only the Al matrix was examined because the reinforcement was noncrystalline. Some specimens received a thermal treatment after fabrication. The treatment produced initial axial tensile residual stresses in the matrix of approximately 55MPa. Cheskis and Heckel found that yielding and unloading these specimens in tension significantly altered the initial stress state. The measured residual stresses were approximately zero after this deformation cycle. Upon compressing these specimens past the yield point and unloading, tensile residual stresses, ≈ 55MPa in the matrix, developed again from the plastic deformation.

Allen et al. (1987) and Allen et al. (1992) used in situ neutron diffraction to examine the strain in Al/SiC whisker and particulate composites under a uniaxial tensile load. The initial thermal stress state of the composite was not considered. After loading, the axial lattice displacements returned to an approximately strain-free state. However, there were compressive lattice strains in the matrix transverse to the loading axis. Correspondingly, the SiC reinforcement was in tension. These data are compared with predictions based on an average-field theory using inclusion solutions developed by Eshelby (Withers et al. 1989; Withers et al. 1989).

Shi and Arsenault (1991) and Shi et al. (1992a) have also presented neutron diffraction strain measurements on Al/SiC whisker material before and after tensile and compressive loading. The axial residual stresses in the Al matrix are shown in Figure 4.5. The data indicate an initial reduction in the matrix axial residual tensile stresses upon either tensile or compressive loading. With continued tensile loading, the matrix axial stresses decreased. However, with continued compressive loading, matrix axial residual tensile stresses increased. In this example, the matrix remained in tension in all cases, but after 2.5% total tensile strain the axial residual stress was reduced from 215MPa to 50MPa (Krawitz 1990b). Transverse stresses showed much less response to axial loading conditions. Shi and Arsenault (1991) compared the average axial and transverse residual stresses with those simulated by two- and three-dimensional FEM calculations (Shi and Arsenault 1991; Shi et al. 1992). The trends in the changes of axial residual stresses with compressive and tensile deformation were correctly predicted.

Povirk et al. (1991b) and (1992) have also studied plastic deformation in Al/SiC whiskers and particulate composites using neutron diffraction. Residual fabrication stresses were considerably less than in the experi-

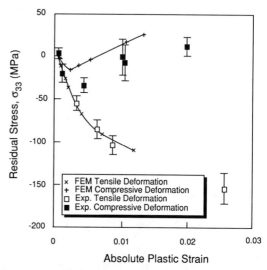

Figure 4.5. Residual stress in a 6061-Al, 20%SiC whisker after axial tensile or compressive plastic deformation. *(From: Shih et al. 1992b.)*

ments of Shi et al. (1991) and (1992b), the highest being 94MPa. Different specimens were plastically deformed both in monotonic tension and compression. Some were given a cycle of deformation, loaded first in tension and then in compression. The maximum plastic strain levels that could be obtained prior to fracture in tension and compression were 3% in the whisker and 4% in the particulate materials respectively. After tensile loading, the measured axial residual stresses changed sign and became compressive in the aluminum matrix and tensile in the SiC. The transverse stresses were considerably smaller and of opposite sign to the axial stresses. Conversely, after compressive loading, the matrix retained tensile residual strains and the reinforcement retained compressive strains. Figures 4.6 and 4.7 show data for the Al/SiC whisker material. Additional samples were loaded in tension and then reverse-loaded in compression. These showed a lower residual axial strain than the monotonically loaded samples, although in the transverse direction the strains were in reasonable agreement with the monotonically loaded specimens. The data were compared to finite element calculations that were developed using standard continuum mechanics based on an axisymmetric unit cell (Povirk et al. 1991b and 1992; Povirk et al. 1990 and 1991a). The complete temperature and loading history of the samples were included in the calculations. In general, experimental data and calculations agree well for the Al matrix in both the whisker and the particulate cases. For the SiC reinforcement, the trends of the axial strains are predicted, but the magnitudes were not. The computed transverse strains in the SiC were typically in poor agreement with the measured strains. The data showed that thermally induced residual stresses are quickly modified by even relatively small amounts of plastic flow (Povirk et al. 1991b and 1992).

The response of the composites to plastic deformation depends on the form of the reinforcement, whether continuous, short fiber, or particulate. In general, however, it appears that the effects of plasticity soon dominate the residual strains resulting from differentials in thermal expansion and the fabrication heat treatments. In metal-matrix composites, type-II residual stresses have a paramount importance. We cannot experimentally measure the very steep residual-stress gradients immediately adjacent to the interface between the reinforcement and the matrix, even with diffraction techniques. Rather, we can only determine the residual stress averaged over a broad cross-section of the composite. Numerical simulation is thus required to estimate gradients and spatially resolve the residual stresses. The study of the kinetics of thermal relaxation of residual stress is a difficult problem. Understanding this

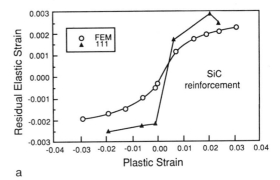

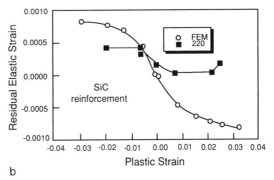

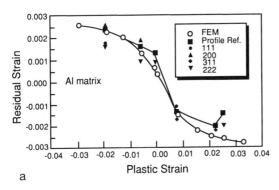

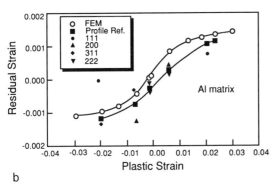

Figure 4.6. Measured and predicted residual strains in the aluminum matrix of a 2009-Al and 15% SiC whisker material following uniaxial plastic deformation in tension and compression. (a) and (b) show data in the axial and transverse directions, respectively. *(From: Povirk et al. 1992.)*

Figure 4.7. Measured and predicted residual strains in the silicon carbide reinforcement of a 2009-Al and 15% SiC whisker composite following uniaxial plastic deformation in tension and compression. (a) and (b) show data in the axial and transverse directions, respectively. *(From: Povirk et al. 1992.)*

phenomenon requires a knowledge of plasticity and thermal and metallurgical effects. Again, an intimate connection between experimental measurement of average residual stresses and numerical simulations is required to successfully address the issue.

4.8 Numerical Simulations

Residual stresses are a consequence of inhomogeneous inelastic deformation and, in composite materials, inhomogeneity is inevitable because of the mismatch in mechanical and thermal properties between the matrix and the reinforcement. Accordingly, the problem of calculating residual stresses is one of determining inhomogeneous equilibrium stress fields. The methods used are almost always based on classical continuum formulations, although the material model may reflect microscopic features of the deformation process.

The discussion in this section is carried out within the context of infinitesmal deformation theory. Within each phase,

$$\sigma = L\epsilon^e \quad (4.4)$$

where L is the tensor of elastic moduli and the elastic strain, ϵ^e, is the total strain minus the (history-dependent) accumulated plastic strain.

After the program of loading and unloading, the residual stress at each point in the material is given by

$$\sigma_{res} = L\epsilon^e_{res} \quad (4.5)$$

For a dual-phase composite material, the tensor of elastic moduli, L in Equation 4.5, takes on two possible values depending on whether the point in question lies in the matrix or in the reinforcement. The experimental methods discussed previously measure ϵ^e_{res}, or its phase average, and the residual stresses are inferred from Equation 4.5. Because the residual stress state is in equilibrium with stress-free surfaces and with body forces neglected, the average of the residual stress is zero. Hence, from Equation 4.4,

$$c_m \hat{\sigma}_{m,res} + c_r \hat{\sigma}_{r,res} = 0 \quad (4.6)$$

where c_m and c_r are the volume fraction of matrix and reinforcement, respectively, and the superposed hat denotes a volume average. Assuming $\widehat{L\epsilon^e} = \hat{L}\hat{\epsilon}^e$ in each phase, which holds if the elastic moduli are uniform, Equations 4.4, 4.5, and 4.6 give

$$c_m L_m \hat{\epsilon}^e_{m,res} + c_r L_r \hat{\epsilon}^e_{r,res} = 0 \quad (4.7)$$

The relation (Equation 4.7) provides a consistency condition for measured, average, residual elastic strains in the two phases, although it rests on the assumption that the average of the product of the elastic moduli and the elastic strains is equal to the product of the averages of these quantities. This does not necessarily hold when both quantities are spatially nonuniform and one source of spatial nonuniformity comes from the different orientations of different grains in the matrix of a composite. For an aluminum-matrix composite, this source of elastic anisotropy is small because aluminum is only slightly anisotropic, and grain-to-grain variations in elastic stiffness are most likely negligible. However, this may not be the case for more elastically anisotropic matrix materials.

Before considering numerical solutions, some general features can be illustrated from an exact solution of a very simple problem—a pressurized spherical shell (see for example, Hill 1950). Assuming spherical symmetry, all field quantities are functions only of the radial coordinate r. Equilibrium is expressed by

$$\frac{d\sigma_r}{dr} + \frac{2(\sigma_r - \sigma_\theta)}{r} = 0 \quad (4.8)$$

where σ_r is the radial stress and σ_θ is the hoop stress.

With the internal radius of the shell denoted by a and the external radius by b, the boundary conditions are

$$\sigma_r(a) = 0 \quad \sigma_r(b) = \Sigma \quad (4.9)$$

Here, Σ is the applied hydrostatic tension.

The material is taken to be isotropic, incompressible, and elastic-perfectly plastic. Incompressiblity implies that

$$\epsilon_r + 2\epsilon_\theta = 0 \quad (4.10)$$

where

$$\epsilon_r = \frac{du}{dr} \quad \epsilon_\theta = \frac{u}{r} \quad (4.11)$$

Regardless of the material response, the displacement field is obtained from Equations 4.10 and 4.11 as $u(r) = Ub^3/r^3$, where U is the expansion of the outer radius.

For an incompressible solid, the stress-strain relation takes the form

$$\sigma_r = \tfrac{2}{3}E\epsilon^e_r + \sigma_m \quad \sigma_\theta = \tfrac{2}{3}E\epsilon^e_\theta + \sigma_m \quad (4.12)$$

where $\sigma_m = \sigma_r + 2\sigma_\theta$ is the mean normal stress, E is Young's modulus and ϵ^e_r and ϵ^e_θ are the elastic strains. The Mises yield criterion is

$$|\sigma_\theta - \sigma_r| = Y \quad (4.13)$$

The loading program is to increase Σ to some specified positive value and then to reduce it to zero. For sufficiently small values of Σ, the deformations remain elastic and the spherical shell is free of residual stresses after unloading. Yielding initiates at $r = a$ when $\Sigma = \Sigma_y$ with

$$\Sigma_y = \tfrac{2}{3}Y\left[1 - \left(\tfrac{a}{b}\right)^3\right] \quad (4.14)$$

The stresses due to loading are

$$\sigma^\ell_r = \begin{cases} 2Y \log\frac{r}{a} & r \leq r_p \\ C\left[1 - \left(\frac{r_p}{r}\right)^3\right] + 2Y \log\frac{r_p}{a} & r > r_p \end{cases} \quad (4.15)$$

$$\sigma^\ell_\theta = \begin{cases} Y + 2Y \log\frac{r}{a} & r \leq r_p; \\ C\left[1 + \tfrac{1}{2}\left(\frac{r_p}{r}\right)^3\right] + 2Y \log\frac{r_p}{a} & r > r_p \end{cases} \quad (4.16)$$

where

$$C = \frac{\Sigma - 2Y\log\frac{r_p}{a}}{1 - \left(\frac{r_p}{b}\right)^3} \quad (4.17)$$

The radius of the plastic zone, r_p, and the applied hydrostatic tension are related by

$$\Sigma = 2Y\log\frac{r_p}{a} + \frac{2}{3}Y\left[1 - \left(\frac{a}{b}\right)^3\right] \quad (4.18)$$

The plastic zone expands as the applied hydrostatic tension, Σ, increases until the sphere becomes fully plastic, $r_p = b$, at which point unrestricted plastic flow occurs at the constant applied hydrostatic tension $2Y \log b/a$.

To unload a pressure of $-\Sigma$ is applied at $r = b$. The elastic unloading solution is

$$\sigma_r^{unl} = -\Sigma\frac{1 - \left(\frac{a}{r}\right)^3}{1 - \left(\frac{a}{b}\right)^3} \quad \sigma_\theta^{unl} = -\Sigma\frac{1 + \frac{1}{2}\left(\frac{a}{r}\right)^3}{1 - \left(\frac{a}{b}\right)^3} \quad (4.19)$$

As long as unloading occurs elastically, the residual stresses are obtained from a superposition of Equations 4.15, 4.16, and 4.19, that is, $\sigma_r^{res} = \sigma_r^\ell + \sigma_r^{unl}$ and $\sigma_\theta^{res} = \sigma_\theta^\ell + \sigma_\theta^{unl}$. However, for a sufficiently large value of Σ, compressive yielding will occur during the unloading process. Since $\sigma_\theta - \sigma_r = Y$ in the plastic region, compressive yielding on unloading requires $\sigma_\theta - \sigma_r = -2Y$. When reverse yielding occurs, the unloading solution is given by Equations 4.15 and 4.16, with Y replaced by $-2Y$.

In any case, the residual elastic strains are obtained from Equations 4.10 and 4.12 as

$$\epsilon_r^{e,res} = -\frac{\sigma_\theta^{res} - \sigma_r^{res}}{E}$$

$$\epsilon_\theta^{e,res} = \frac{1}{2}\frac{\sigma_\theta^{res} - \sigma_r^{res}}{E} \quad (4.20)$$

Figure 4.8 shows the dependence of the residual expansion, $U^{e,res}/b = \epsilon_\theta^{e,res}(b)$ on the expansion during loading, U^ℓ/b. The response for loading in compression as well as for loading in tension is shown. For sufficiently small $|U^\ell|/b$ the response is elastic and there is no residual deformation. Once yielding initiates, the residual deformation increases rapidly with the expansion of the plastic region. Eventually, the limiting value of the applied radial stress, $2Y \log b/a$, is reached and unrestricted plastic flow occurs, with the stress state given by the first of Equations 4.15 and 4.16, $r_p = b$. Subsequently, the stress states during loading and during unloading are independent of the expansion U^ℓ/b, and hence, from Equation 4.20, so is the residual deformation.

Two curves are shown in Figure 4.8; one corresponding to a thick shell and the other corresponding to a thin shell. The strains are more uniform in the thin shell. Therefore, for a given imposed deformation, the residual deformation is less than that for the thick shell. For a strain-hardening solid, the residual deformation would increase somewhat even when there was large-scale plasticity. This example is highly idealized because strain hardening is not accounted for and the loading in the shell is proportional. The calculation of deformation-induced residual strains and stresses in metal-matrix composites is, of course, more complex. In particular, what is measured in Figure 4.6 is an average of the residual elastic strain in one phase of a two-phase material. Nevertheless, this simple model problem reveals the general features of the behavior seen in Figure 4.6.

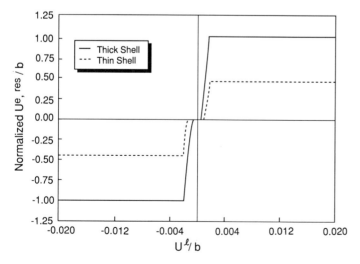

Figure 4.8. Dependence of the residual expansion $U^{e,res}/b$ on the imposed prestrain U^ℓ/b for the shell model.

One approach to the calculation of mechanically induced residual stresses follows the same general procedure as outlined for the simple spherical shell problem. A continuum mechanics boundary value problem is formulated for a model composite morphology. A solution is found for the stress and strain fields, during both loading and unloading, for this idealized composite. Because for the composite the loading is not proportional and the plastic flow is path dependent, an incremental solution procedure is used. The average residual strains are obtained by integrating the computed fields over the matrix volume.

Figure 4.9, from Povirk et al. (1992) uses this approach and shows the effect of particle shape on the average residual strain in the matrix after unloading. The particle shapes, each having the same volume fraction, are shown in Figure 4.9(a). The general features of the variation of residual strain with imposed deformation are much the same as in Figure 4.8, when account is taken of the fact that strain hardening is accounted for in the analyses in Povirk et al. (1992). The cylindrical particle gives rise to highly nonuniform plastic straining near the cylinder corners and, accordingly, to the largest residual strains. The elliptical particle, which has rounded corners, has much smaller residual strains. The conical particle gives rise to residual elastic strains that are much smaller than in the other two cases.

A similar approach was used to compute the thermal residual elastic strains. In general, thermal loading gives rise to a fully coupled thermomechanical boundary value problem, which involves accounting for the nonuniform temperature fields caused by thermal conduction and the heat generated by plastic dissipation. Fortunately, these effects often play a negligible role in determining thermally induced residual stresses in metal-matrix composites. In order to illustrate this, consider a one-dimensional heat conduction problem. There is a characteristic time, given by $L^2/\pi^2\beta^2$, where L is a geometric length and β^2 is the thermal diffusivity of the material, that provides a measure of the time to reach thermal equilibrium. For Al, $\beta^2 = 0.69 \times 10^{-4}$ m²/sec and for SiC $\beta^2 = 0.53 \times 10^{-4}$ m²/sec. This means that for a cell size in Figure 4.9(a) of the order of 10 μm, the characteristic thermal diffusion time is of the order of 10^{-5} sec and for a length of 1 mm increases to $\approx 10^{-3}$ sec. Hence, over times of hundredths of seconds the temperature distribution in a unit cell is effectively uniform.

The development of the residual stresses is driven by the thermal expansion mismatch between the matrix and the reinforcement. If we assume small strains and negligible thermal conduction and heat generated by plastic flow, the rate boundary value problem for unit cell model analyses of thermal processing induced residual stresses involves solution of the equilibrium equation

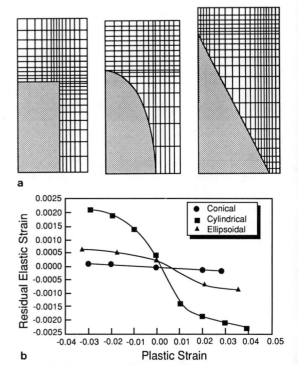

Figure 4.9. (a) Finite element meshes for cylindrical, ellipsoidal, and conical reinforcements. (b) Predicted matrix residual strains parallel to the deformation axis, as a function of particle shape. *(From: Povirk et al. 1992.)*

$$\frac{\partial \dot{\sigma}_{ij}}{\partial x_j} = 0 \qquad (4.21)$$

with

$$\dot{\sigma}_{ij} = L_{ijkl}[\dot{\epsilon}_{kl} - \dot{\epsilon}^p_{kl} - \alpha_{kl}\dot{T}] \qquad (4.22)$$

where $\dot{\epsilon}^p_{kl}$ is the plastic strain rate, L_{ijkl} is the tensor of elastic moduli, α_{kl} is the thermal expansion tensor, and a superposed dot denotes differentiation with respect to time. If the material is isotropic, then

$$L_{ijkl} = \frac{E}{1+\nu}\left[\frac{1}{2}(\delta_{ik}\delta_{jl} + \delta_{il}\delta_{jk}) + \frac{\nu}{1-2\nu}\delta_{ij}\delta_{kl}\right]$$
$$\alpha_{kl} = \alpha\delta_{kl} \qquad (4.23)$$

where δ_{ij} is the Kronecker delta.

For a Mises solid (J_2-flow theory), the flow rule for $\dot{\epsilon}^p_{kl}$ takes the form

$$\dot{\epsilon}^p_{kl} = \frac{3\dot{\epsilon}}{2\sigma_e}\sigma'_{kl} \qquad (4.24)$$

where $\dot{\epsilon}$ is the effective plastic strain rate and σ_e is defined by

$$\sigma'_{ij} = \sigma_{ij} - \tfrac{1}{3}\sigma_{kk}\delta_{ij} \qquad \sigma_e^2 = \tfrac{3}{2}\sigma'_{kl}\sigma'_{lk} \qquad (4.25)$$

The constitutive relation is completed by specifying an evolution equation for $\dot{\bar{\epsilon}}$. For the calculation of thermal-processing-induced residual stresses this means specifying the strain and strain rate hardening properties as a function of temperature. Additionally, the temperature dependencies of the elastic moduli and the thermal expansion coefficients, need to be specified, which, in general, requires large amounts of input data for the model. Typically, simplifying assumptions have been made, e.g., neglecting the temperature variation of Poisson's ratio or of the thermal expansion coefficient. Although such simplifications can be appropriate, and may even be necessary because of limited data, they can lead to significant errors because of the complex interactions that occur. For example, Nakamura and Suresh (1992) have shown the strong effect of the temperature variation of the coefficient of thermal expansion on residual stresses in a 6061 aluminum alloy reinforced with 46 vol % boron continuous fibers.

With $\alpha_{kl} = 0$, and assuming that the heat generated by plastic flow is negligible, the governing Equations (4.21) to (4.25) can be used to analyze deformation-induced residual stresses. In this case, property data are only needed at one temperature.

In any case, periodic boundary conditions are used to complete the cell model problem specification. The specific boundary conditions depend on the cell model used. In the simplest case, the cell consists of a single reinforcement embedded in the surrounding matrix. For short fiber or particle reinforced materials, these unit cells are inherently three dimensional. In order to simplify the calculation procedure, an axisymmetric cell model is often used that involves replacing a hexagonal cylindrical cell with an axisymmetric circular cylindrical cell. Models using a single reinforcement within a cell constrain the distributions that can be analyzed. Nevertheless, some insight into distribution effects has been gained by comparing predictions for fibers with end-to-end alignment with corresponding predictions for staggered alignment. Consideration of a broader range of distributions requires using cell models that contain multiple reinforcements per cell. At this point, such models have only been used in a two-dimensional context. Even for the simplest cell model, determining the residual stress state involves solving a series of nonlinear boundary value problems to determine the residual stresses and strains due to some specified loading/thermal history. Hence, numerical, principally finite element techniques have been used. (Such finite element studies are reported in Shi et al. 1992b; Shi et al. 1992a; Povirk et al. 1991a and 1992; Tvergaard 1991; and Levy and Papazian 1991.) In particular Shi et al. (1992a) and Levy and Papazian (1991) have carried out full, three-dimensional analyses.

In addition to full numerical simulations, residual stresses have been predicted based on mean-field theory approximations, e.g., Withers et al. (1989), which deal with phase-average quantities from the outset. These mean-field theories are simpler than numerical solutions of full boundary value problems. However, mean-field theories have, for analytical reasons, been restricted to consideration of ellipsoidal reinforcements, so that the important shape effects illustrated in Figure 4.9 cannot be revealed. Furthermore, the formulation of mean field theory is such that a nearly linear relation between residual stress (or strain) and imposed deformation is obtained, as in the steeply rising branch of the curve in Figure 4.8 (see, for example, Withers et al. 1989a). The transition to large-scale plastic flow behavior does not occur as seen in Povirk et al. (1991b), where mean field theory and full-field predictions for the dependence of residual elastic strains on imposed plastic deformation in a whisker-reinforced composite are compared.

There are also, of course, significant idealizations that enter into the boundary value problem formulation that forms the basis for the full-field solutions. First, in order to limit the computing time, the composite morphology has been highly idealized. It is worth noting that, for continuous fiber reinforced composites, for which two-dimensional models are appropriate, quite realistic distributions are amenable to computation (Nakamura and Suresh 1992). It is expected that increased computing power will shortly permit computations of realistic distributions in a fully three-dimensional context. Secondly, the material models that have been used to describe the matrix plastic flow have been highly simplified — versions of standard isotropic hardening (Povirk et al. 1991b and 1992) or combined isotropic and kinematic hardening (Levy and Papazian 1991) flow rules. The extent to which more accurate (and more complex) descriptions of plastic flow are needed remains to be assessed. A separate issue concerns limits to the adequacy of any continuum characterization of matrix deformation behavior, that is, the role of discrete dislocation fields in affecting type-II residual stresses.

Appendix A
Strain Measurement by Neutron Diffraction

4.A.1 Constant Wavelength

Neutron diffractometers at reactors generally employ an intense continuous monochromatic (single wavelength) beam of neutrons. It is usual to scan the lattice spacings d_{hkl} at a fixed wavelength, λ_0 and variable 2θ. In this case, from Bragg's law:

$$\frac{\Delta(d_{hkl})}{d_{hkl}} = -\Delta(\theta_{hkl})\cot(\theta_{hkl}) \quad (4.A.1)$$

If a crystal is subjected to a load, its unstressed lattice spacing d_0 will change by $\Delta(d_{hkl}) = d_{hkl} - d_0$. Thus, Equation 4.A.1 defines the elastic strain in a material. From Equation A.1 there is an implied change $\Delta\theta$ in the direction of the scattered maximum (positive for compression, negative for tension). The change is measured as a change in the peak center of a Bragg reflection. Peak shifts as small as five-thousandths of a degree can usually be recorded which, for a scattering angle of 120 degrees, 2θ, corresponds to approximately 50 microstrain. The $\cot(\theta_{hkl})$ term in Equation A.1 means that larger peak shifts (and thus greater strain sensitivity) occur at larger scattering angles. Unfortunately, high-angle reflections are not always available because of instrumental or specimen limitations.

4.A.2 Constant Scattering Angle

A pulsed neutron source produces a continuous spectrum of neutron energies in every pulse (after moderation), and the energy or wavelength of a diffracted neutron is determined by its time of flight from creation to detection. The lattice spacing, d_{hkl} is determined by maintaining a constant scattering angle, θ_0, at the detector and scanning the wavelength. By differentiation of Bragg's law and noting that the wavelength of a neutron is inversely proportional to its energy (de Broglie relationship) we have:

$$\frac{\Delta d_{hkl}}{d_{hkl}} = \frac{\Delta\lambda_{hkl}}{\lambda_{hkl}} = \frac{\Delta t_{hkl}}{t_{hkl}} \quad (4.A.2)$$

Thus, a lattice strain may be determined from the small change in time of flight, Δt_{hkl}, for a diffracted maximum, provided there is no change in the path length between the unstrained and strained material.

Thus, when a load is applied to a crystal lattice, deformation takes place. If the yield stress is not exceeded, the deformation is a reversible change of distance between atomic planes in the unit cell. If the yield stress is exceeded, the deformation then includes a plastic component as a result of the irreversible shear that occurs when planes of atoms slide in suitable directions. However, as strain measurement by neutron diffraction determines the separation of lattice planes, the technique only records the elastic strain in a specimen.

Appendix B
Hole Drilling and Orthotropic Materials

Bert and Thompson (1968) have shown that for a commercial strain-gauge rosette (see Figure 4.2) designed for measuring residual stresses, the residual stresses can be expressed in terms of the measured strains ε_a, ε_b, and ε_c. Note that the subscripts a, b, and c correspond to the rosette-gauge strains at 0, 45, and 90 degrees. The orientation of the rosette must be such that the a and c gauges are parallel to the axes of orthotropic symmetry of the composite. For this particular situation:

$$\varepsilon_a = (A + B)\sigma_{11} + (A - B)\sigma_{22}, \quad (4.B.1)$$

$$\varepsilon_b = A\sigma_{11} + A\sigma_{22} + C\sigma_{12}, \quad (4.B.2)$$

and

$$\varepsilon_c = (A - B)\sigma_{11} + (A + B)\sigma_{22} \quad (4.B.3)$$

If one solves Equations 4.B.1 to 4.B.3 for the corresponding stresses σ_{11}, σ_{22}, and σ_{12} one finds that:

$$\varepsilon_{11} = [(A + B)\varepsilon_a - (A - B)\varepsilon_c]/4AB, \quad (4.B.4)$$

$$\varepsilon_{22} = [(A + B)\varepsilon_c - (A - B)\varepsilon_a]/4AB \quad (4.B.5)$$

and

$$\varepsilon_{12} = 2(\varepsilon_b - \varepsilon_a - \varepsilon_c)/2C \quad (4.B.6)$$

The coefficients A, B, and C reflect the elasticity coefficients of the composite material and can be determined experimentally for a particular composite material and a hole drilled to a specific depth.

The constants A and B can be determined from a uniaxial tensile test of the material in question. The preparation of the tensile calibration sample should be identical to that of the engineering component one wishes to study, but care must be taken to eliminate any stress artifacts that could arise in the calibration specimen, from thermal gradients in heat treating, for example.

The tensile specimen is machined from the composite material such that its axis is parallel to an axis of orthotropic symmetry of the composite. The calibration constants are determined by applying a known stress σ. Noting that $\sigma = \sigma_{11}$ and by orienting the rosette such that $\varepsilon_a = \varepsilon_{11}$, $\varepsilon_c = \varepsilon_{22}$, one can solve Equations 4.B.1 and 4.B.3 for the coefficients A and B,

$$A = (\varepsilon_a + \varepsilon_c)/2\sigma \quad B = (\varepsilon_a - \varepsilon_c)/2\sigma \quad (4.B.7)$$

The coefficient C can also be evaluated with a tensile specimen and a superimposed stress. To obtain C for Equation 4.B.2 requires either superimposing a shear stress, σ_{12}, along one of the planes of orthotropic symmetry or a tensile stress at 45 degrees to the symmetry axes. With metal-matrix composite materials, it is generally possible to produce a tensile coupon with this orientation. In this case, with the rosette orientation still aligned parallel with the material axes (gauge b is now parallel to the 45 degree calibration-specimen tensile axis), we have

$$C = \varepsilon_b/\sigma_{12} = 2\varepsilon_b/\sigma \quad (4.B.8)$$

Bert and Thompson (1968) have mentioned that one should conduct calibration experiments with specimens having different ratios of hole diameter to width to optimize the ratio. A ratio value of one is generally regarded as a reasonable starting point. The constants A, B, and C will vary for a given material depending on this ratio. They also believe that it would be wise, when performing the calibration experiments, to measure strains at other angles as well. These strains can be related to those used in the calibration as a check.

One should note that for organic-matrix composites with continuous fibers, it is not necessarily possible to grip and test a sample at 45 degrees to the material axes. In that case, it is possible to use a shear specimen geometry to obtain C. Prasad et al. (1987) have proposed using an asymmetric four-point bend-bar specimen for that purpose.

References

Abuhasan, A., Balasingh, C., and P. Predecki. 1990. *J. Amer. Ceram. Soc.* 73:2474-2484.

Abuku, S., and B.D. Cullity. 1971. *Experimental Mechanics.* 11:217-223.

Aleong, C., and M. Munro. 1991. *Experimental Techniques.* 15:55-58.

Allen, A.J., Bourke, M.A.M., David, W.I.F. et al. 1988. *In:* Proceedings of the Second International Conference on Residual Stresses I (G. Beck, ed.), 78-83, London: Elsevier.

Allen, A.J., Bourke, M.A.M., Dawes, S. et al. 1992. "The Analysis of Internal Strains Measured by Neutron Diffraction in Al/SiC Metal Matrix Composites," *Acta Metall. Mater.* In press.

Allen, A.J., Bourke, M.A.M., Hutchings, M.T. et al. 1987. *In:* Residual Stresses in Science Technology (E. Macherauch and V. Hauk, eds.), 151-157, Germany: DGM Informationgesellschaft mbH Verlag.

Allen, A.J., Hutchings, M.T., Windsor, C.G., and C. Andreanni. 1985. *Advances in Physics.* 34:445-473.

Armstrong, J.H., Rawal, S.P., and M.S. Misra. 1990. *Mater. Sci. Engin.* A126:119-124.

Arsenault, R.J., and M. Taya. 1987. *Acta Metall.* 35:651-659.

ASTM Standards. 1989. *In:* Annual Book of ASTM Standards, ASTM Designation: E 837-89, 715-720, Philadelphia: American Society for Testing and Materials.

Axe, J.D. 1991. *Science* 252:795-801.

Bert, C.W., and G.L. Thompson. 1968. *J. Composite Materials* 2:244-253.

Boag, J.M., Flaman, M.T., and J.A. Herring. 1987. *In:* Proceedings of ASM's Conference on Residual Stress—in Design, Process and Materials Selection (W.B. Young, ed.), 1-6, Metals Park, OH: ASM International.

Bourke, M.A.M., Goldstone, J.A., and T.M. Holden. 1990. "Residual Stress Measurement Using the Pulsed Neutron Source at LANSCE." To be published in Proceedings of the NATO Workshop on Residual Stresses (M.T. Hutchings, ed.), Amsterdam: Kluwer.

Cantrell, J.H., Qian, M., Ravichandran, M.V., and K.M. Knowles. 1990. *Appl. Phys. Lett.* 57:1870-1872.

Cheong, Y.M., and H.L. Marcus. 1987. *Scripta Metall.* 21:1529-1534.

Cheong, Y.M., Marcus, H.L., and F. Adar. 1986. *In:* Proceedings of a Symposium Sponsored by the AIME—American Society for Metals Composite Committee. (A.K. Dhingra and S.G. Fishman, eds.), 147-154, Warrendale, PA: The Metallurgical Society.

Cheong, Y.M., Marcus, H.L., and F. Adar. 1987. *J. of Mater. Res.* 2:902-909.

Cheskis, H.P., and R.W. Heckel. 1968. *In:* ASTM STP 438. 76-91, Philadelphia: American Society of Testing and Materials.

Cheskis, H.P., and R.W. Heckel. 1970. *Metall. Trans.* 1:1931-1942.

Cooper, A.R. 1981. *In:* Proceedings of the Twenty-Eighth Sagamore Army Materials Research Conference (E. Kula, and V. Weiss, eds.), 439-465, New York: Plenum Press.

Cox, L.C. 1987. *In:* Proceedings of ASM's Conference on Residual Stress—in Design, Process and Materials Selection (W.B. Young, ed.), 109-115, Metals Park, OH: ASM International.

Cox, B.N., James, M.R., Marshall, D.B., and R.C. Addison, Jr. 1990. *Metall. Trans. A* 21A:2701-2707.

Doi, O., and K. Kataoka. 1982. *Bulletin of the JSME* 25:1373-1377.

Dolle, H. 1979. *J. Appl. Cryst.* 12:489-501.

French, D.A., and B.A. MacDonald. 1969. *Experimental Mechanics* 9:456-462.

Hauk, V. 1981. *In:* Proceedings of the Twenty-Eighth Sagamore Army Materials Research Conference (E. Kula and V. Weiss, eds.), 117-139, New York: Plenum Press.

Hauk, V. 1986. *In:* Proceedings of the European Conference on Residual Stresses, 1983 (E. Macherauch, and V. Hauk, eds.), 9-45, Karlsruhe, Germany: DGM Informationgesellschaft mbH Verlag.

Hill, R. 1950. The Mathematical Theory of Plasticity. Oxford: Clarendon Press.

Hutchings, M.T. 1990. *Nondestructive Test. Eval.* 5:395-413.

James, M.R. 1989. *In:* Proceedings of the Second International Conference on Residual Stresses II (G. Beck, ed.), 429-435, London: Elsevier.

James, M.R. 1991. "Behavior of Residual Stresses During Fatigue of Metal Matrix Composites." To be published in Proceedings of the Third International Conference on Residual Stresses, Kyoto, Japan.

James, M.R., and O. Buck. 1980. *In:* CRC Critical Reviews in Solid State and Materials Sciences. 61:105.

James, M.R., Bourke, M.A.M., Lawson, A.C., and J.A. Goldstone. 1992. "Residual Stress Measurement in Continuous Fiber Titanium Matrix Composites." To be published in Proceedings of the Advances in X-Ray Analysis.

James, M.R., and J.B. Cohen. 1980. *Treaties on Material Science and Technology 19A.* New York: Academic Press.

Kataoka, K., Doi, O., and M. Sato. 1986. *Bulletin of JSME.* 29:393-399.

Khan, I.H. 1976. *Metall. Trans. A.* 7A:1281-1289.

Krawitz, A.D. 1985. *Mater. Sci. Engin.* 75:29-36.

Krawitz, A.D. 1990a. *Materials Research Society Symposium Proceedings.* 166:281-292.

Krawitz, A.D. 1990b. "Stress Measurements in Composites Using Neutron Diffraction." To be published in Proceedings of the NATO Workshop on Residual Stresses (M.T. Hutchings, ed.), Amsterdam: Kluwer.

Krawitz, A.D., Brune, J.E., and M.J. Schmank. 1981. *In:* Proceedings of the Twenty-Eighth Sagamore Army Materials

Research Conference (E. Kula and V. Weiss, eds.), 139–155, New York: Plenum Press.

Krawitz, A.D., Crapenhoft, M.L., Reichel, D.G., and R. Warren. 1988. *Mater. Sci. Eng.* A105/106:275–281.

Krawitz, A.D., Reichel, D.G., and R.L. Hitterman. 1989. *Mater. Sci. Eng.* A119:127–134.

Krawitz, A.D., and M.J. Schmank. 1982. *Metall. Trans.* 13A:1069–1075.

Kupperman, D.S., Majumdar, S., Singh, J.P., and A. Saigal. 1990. "Application of Neutron Diffraction Time of Flight Measurement to the Study of Strain in Composites." To be published in *Proceedings of the NATO Workshop on Residual Stresses* (M.T. Hutchings. ed.), Amsterdam: Kluwer.

Kupperman, D.S., Singh, J.P., Faber, J. Jr., and R.L. Hitterman. 1989. *J. Appl. Phys. Comm.* 66:3396–3398.

Leaity, G. P., and R.A. Smith. 1989. *Fatigue Fract. Engng. Mater. Struct.* 12:271–282.

Ledbetter, H. M., and M.W. Austin. 1987. *Mater. Sci. Eng.* 89:53–61.

Lee, J. and R. Czarnek. 1991. *In:* The Proceedings of the 1991 SEM Spring Conference on Experimental Mechanics. 405–415, Bethel: The Society for Experimental Mechanics.

Levy, A. and J.M. Papazian. 1991. *Acta Metall. Mater.* 39:2255–2266.

MacEwen, S.R., Faber, J., and P.L. Turner. 1983. *Acta Metall.* 31:657–676.

Majumdar, S., Singh, J.P., Kupperman, D., and A.D. Krawitz. 1991. *J. Engin. Mater. Tech.* 113:51–59.

Mathar, J. 1934. *Transactions of the ASME.* 56:249–254.

Matzkanin, G.A. 1987. *In:* Residual Stresses in Science and Technology. (E. Macherauch, and V. Hauk, eds.), 101–108, Karlsruhe, Germany: DGM Informationgesellschaft mbH Verlag.

McDonach, A., McKelvie, J., MacKenzie, P., and C.A. Walker. 1983. *Experimental Techniques.* 7:20–24.

Metcalfe, A.G., and M.J. Klein. 1974. *In:* Composite Materials, Vol. I: Interfaces in Metal Matrix Composites (A.G. Metcalfe, ed.), 125–168, New York: Academic Press.

Nairan, J.A., and P. Zoller. 1985a. *J. of Mater. Sci.* 20:355–367.

Nairan, J.A., and P. Zoller. 1985b. *In:* Fifth International Conference on Composite Materials ICCM–V (W.C. Harrigan, Jr., J. Strife, and A.K. Dingra, eds.), 931–946, Warrendale, PA: The Metallurgical Society.

Nakamura, T., and S. Suresh. "Effects of Thermal Residual Stresses and Fiber Packing on Deformation of Metal-Matrix Composites." To be published.

Nicoletto, G. 1991. *Experimental Mechanics.* 31:252–256.

Noyan, I.C., and J.B. Cohen. 1987. Residual Stress—Measurement by Diffraction and Interpretation. Germany: Springer Verlag.

Peck, C.A. 1987. *In:* Proceedings of ASM's Conference on Residual Stress—in Design, Process and Materials Selection (W.B. Young, ed.), 7–9, Metals Park, OH: ASM International.

Pedersen, O.B. 1985. *In:* Proceedings of the Fifth International Conference on Composite Materials (W.C. Harrigan, Jr., J. Strife, and A.K. Dingra, eds.), 1–19, Warrendale, PA: The Metallurgical Society.

Povirk, G.L., Needleman, A., and S.R. Nutt. 1990. *Mater. Sci. Eng.* A125:129–140.

Povirk, G.L., Needleman, A., and S.R. Nutt. 1991a. *Mater. Sci. Eng.* A132:31–38.

Povirk, G.L., Stout, M.G., Bourke, M. et al. 1991b. *Scripta Metall. Mater.* 25:1883–1888.

Povirk, G.L., Stout, M.G., Bourke, M. et al. 1992. *Acta Metall. Mater.* 40:2391–2412.

Prasad, C.B., Prabhakaran, R., and S. Tompkins. 1987. *Composite Structures.* 8:165–172.

Prask, H.J. 1991. *Advanced Materials and Processes.* 9:26–35.

Prask, H.J., and C.S. Choi. 1987. *In:* Proceedings of ASM's Conference on Residual Stress—in Design, Process and Materials Selection (W.B. Young, ed.), 21–26, Metals Park, OH: ASM International.

Predecki, P., and C.S. Barrett. 1981. *In:* Proceedings of the Twenty-Eighth Sagamore Army Materials Research Conference (E. Kula and V. Weiss, eds.), 409–424, New York: Plenum Press.

Ruud, C.O., Snoha, D.J., Gazzara, C.P., and P. Wong. 1987. *In:* Proceedings of ASM's Conference on Residual Stress—in Design, Process and Materials Selection (W.B. Young, ed.), 103–108, Metals Park, OH: ASM International.

Saigal, A., Kuppermann, D.S., and S. Majumdar. 1992. *Mater. Sci. Engin.* A150:59–65.

Seol, K., and A.D. Krawitz. 1990. *Mater. Sci. Eng.* A127:1–5.

Soete, W., and R. Vancrombrugge. 1950. *Proceedings SESA.* 8:17–28.

Shi, N., and R.J. Arsenault. 1991. *J. Composites Tech. and Res.* 13:211–225.

Shi, N., Arsenault, R.J., Krawitz, A.D, and L.F. Smith. 1992a. Deformation–Induced Residual Stress Changes in SiC Whisker Reinforced 6061 Al Composites. *Metall. Trans. A.* 24A:187–196.

Shi, N., Arsenault, R.J., and B. Wilner. 1992b. "A FEM Study of the Plastic Deformation Process of Whisker Reinforced SiC/Al Composites." *Acta Metall. Mater.* 40:2841–2854.

Stepanova, T.R. 1990. *Mechanics of Composite Materials.* 26:319–323.

Tsai, S.-D., Mahulikar, D., Marcus, H.L. et al. 1981a. *Mater. Sci. Eng.* 47:145–149.

Tsai, S.-D., Schmerling, M., and H.L. Marcus. 1981b. *In:* Proceedings of the Twenty-eighth Sagamore Army Materials Research Conference (E. Kula and V. Weiss, eds.), 425–439, New York: Plenum Press.

Tvergaard, V. 1991. *Mech. Matl.* 11:149–161.

Vedula, M., Pangborn, R.N., and R.A. Queeney. 1988a. *Composites.* 19:55–60.

Vedula, M., Pangborn, R.N., and R.A. Queeney. 1988b. *Composites.* 19:133–137.

Vikar, A.V., Jue, J.F., Hansen, J.J., and R.A. Cutler. 1988. *J. Am. Ceram. Soc.* 71:C148–C151.

White, S.R., and H.T. Hahn. 1990. *Polymer Engineering and Science.* 30:1465–1473.

Windsor, C.G. 1990. "The Effects of Crystalline Anisotropy on the Elastic Response of Materials." To be published in *Proceedings of the NATO Workshop on Residual Stresses* (M.T. Hutchings, ed.), Amsterdam: Kluwer.

Withers, P.J., Jensen, D., Jull, D. et al. 1987. *In:* Proceedings of the Sixth International Conference on Composite Materials, 255–264, London: Elsevier Applied Science.

Withers, P.J., Lorentzen, T., and W.M. Stobbs. 1989a. *In:* Proceedings of the Seventh International Conference on Composite Materials. New York: Pergamon Press.

Withers, P., Stobbs, W.M., and O.B. Pedersen. 1989. *Acta Metall. Mater.* 37:3061–3084.

Wright, P.K., Sensmeier, M.D., Kupperman, D., and H. Wadley. 1991. *NASA HITEMP Review, NASA-CP-10082.* 45:1–14.

Zong, G., and H.L. Marcus. 1991. *Scripta Metall. Mater.* 25:277–282.

Chapter 5
Structure and Chemistry of Metal/Ceramic Interfaces

MANFRED RÜHLE

The applications of modern engineering materials often necessitates the bonding of two different materials. The resulting interfaces must typically sustain mechanical and/or electrical forces without failure. Consequently, interfaces exert an important, and sometimes controlling, influence on performance in such applications as composites (Mehrabian 1983; Dhingra and Fishman 1986; Lemkey et al. 1988; Suresh and Needleman 1989), electronic packaging systems used in information processing (Giess et al. 1985; Jackson et al. 1986), thin-film technology (Gibson and Dawson, 1985; Yoo et al. 1988; Huang et al. 1991; Kasper and Parker 1989), and joining (Rühle et al. 1985; Ishida 1987; Doyema et al. 1989; Nicholas 1990; Shimida 1990). Furthermore, interfaces play an important role in the internal and external oxidation or reduction of materials. The importance of systematic studies in this area was emphasized during several international conferences (Rühle et al. 1985; Rühle et al. 1989; Rühle et al. 1992).

An understanding of the properties of interfaces requires a knowledge of both the structure and the chemistry of the interface. The structure should be understood at the atomic level. During bonding of two dissimilar materials, chemical processes can occur that lead to a modification of the properties of the interfaces. Therefore, it is essential to also study the chemistry at the interface. The chemical reactions there will lead to a reaction product and/or gradients in the chemical composition in one or both components adjacent to the interface.

This chapter will report results with respect to structural studies as well as investigations of chemical processes. The structure (to the atomic level) of the interface can be detected either by high-resolution electron microscopy (Section 5.2) or by X-ray scattering (Section 5.3). Both complementary techniques will be described and the results will be reported for different interfaces. Chemical processes at metal interfaces will be discussed in general in Section 5.4, and thermodynamic results will be applied to a group of metal/ceramic interfaces. A case study presented in Section 5.5 regarding structure and bonding at the silver/magnesia (Ag/MgO) interfaces concludes this chapter.

5.1 Fundamentals

5.1.1 Bonding Between Metal and Ceramics

If two different materials are brought into close contact, interaction processes may occur that will reduce the total energy of the system. This amount of energy (the work of adhesion, W_{ad}) represents the driving force for the bonding formation at the interface (Mattox et al. 1988). W_{ad} is the energy that is required for separating the solids at the interfaces. However, dissipative processes that may occur during the separation of these interfaces such as heat, sound, microcracks, and plastic deformation are not taken into account for this basic consideration.

In nature, different interaction processes occur that may be responsible for bonding across dissimilar interfaces. These interaction processes can be either physical or chemical. Characteristic physical bonding represents the electric dipole/dipole type of interaction (Kinloch 1981); these bonds possess very low bonding energies, less than 40 J/mol. Chemical bonding can be distributed in homopolar (covalent) and heteropolar bondings. Heteropolar bondings result in the electrostatic interaction between opposite charges juxtaposing at the interfaces. These chemical bondings exist, for example, in ionic crystals. Covalent bonding exists between nonpolar atoms; those atoms existing in lattices of elements of the same type such as silicon or silicon nitride. In contrast to physical bonding, chemical bondings possess very high bonding energy, which may exeed 1 MJ/mol (Kinloch 1981). Well-defined boundaries do not exist between physical and chemical bonding.

The level of bonding energy between a metal and a ceramic can be strongly influenced by impurities and by chemical processes occurring at the interface. Thus far, no closed theory exists on the nature of a bonding across a heterophase boundary, which can be generalized for all types of metals. There are a few fundamental studies presently available that allow for prediction of type and level of adhesion energy (Section 5.5).

5.1.2 Interface Energies

Young and Dupré (Klomp 1985) detailed the relationship between the surface energy of a liquid and that of a solid, and the relationship of contact angle Θ of a liquid to that of a solid (Figure 5.1). The Young-Dupré equation results in

$$\Gamma_{SL} = \Gamma_{SV} + \Gamma_{LV} \cdot \cos \Theta \quad (5.1)$$

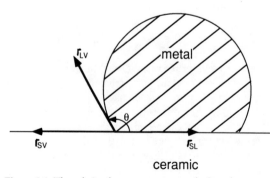

Figure 5.1. The relation between contact angle Θ and interfacial energy Γ_{ij}.

where Γ_{LV} is the surface energy between liquid and vacuum, Γ_{SV} is the corresponding energy between solid and vacuum, and Γ_{SL} is the energy between a solid and a liquid (Ishida 1987). The energy that is required to separate a drop of liquid from a solid can be represented as the energy to form free surfaces reduced by the interfacial energy

$$W_{ad} = \Gamma_{SV} + \Gamma_{LV} - \Gamma_{SL}. \quad (5.2)$$

Combining these equations results in

$$W_{ad} = \Gamma_{LV} \cdot (1 + \cos \Theta). \quad (5.3)$$

This relationship represents a practical way for the determination of work of adhesion W_{ad} between nonreactive metals in both a liquid state and in a nonmetallic substrate. A similar equation is also applied for solids on a ceramic, although it may be difficult to reach an equilibrium situation during the annealing of that material.

5.1.3 Geometric Interface Model

5.1.3.1 The Coincidence Site Lattice (CSL)

A qualitative description of grain boundaries requires theoretical models that relate the energy of grain boundaries with the crystallographic properties of a crystal (Friedel 1926; Kronberg and Wilson 1949; Grimmer et al. 1970). From these studies, Bollmann developed the three-dimensional coincidence site lattice (CSL) (Bollmann 1970). In this model, a stable, low-energy interface is assumed when the crystal possesses a relative orientation that allows the coincidence of as many as possible lattice points of both lattices. The three-dimensional consideration does not take into account the positions of the atoms within the interface, which is often key for an understanding of the evaluation of low-energy interfaces (Balluffi et al. 1982). Therefore, a two-dimensional CSL was also developed that only considers the orientation within the interfaces. The two-dimensional CSL will be called the reciprocal coincidence density, and will be defined as

$$\Sigma_2 = \frac{\text{number of atoms in the coincidence cell}}{\text{number of coincidences}} \quad (5.4)$$

A simple description predicts that the energy of an interface is lower if the two adjacent lattices are oriented so that as many atoms as possible coincide within the lattice interface plane. The coinciding atomic positions may vary by a predetermined misfit τ_i; Figure 5.2 shows

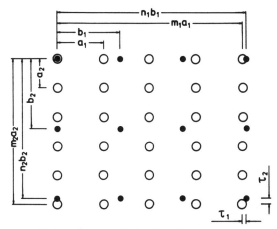

Figure 5.2. Two-dimensional coincidence lattice of two orthogonal lattices with lattice parameters (a_1, a_2) and (b_1, b_2).

the coincidence cell of two cubic lattices with lattice parameters (a_1, a_2) of lattice 1 and lattice parameters (b_1, b_2) of lattice 2. A perfect coincidence requires some strain of ε_i. This strain depends on the misfit ε_i and the lattice constants of lattice i within the interface. For both lattices, these strains can be represented as (Balluffi 1982)

$$\varepsilon_1 = \frac{2\tau_1}{n_1 \cdot a_1 + m_1 \cdot b_1} \quad (5.5)$$

$$\varepsilon_2 = \frac{2\tau_2}{n_2 \cdot a_2 + m_2 \cdot b_2} \quad (5.6)$$

The indices are related to direction 1 and direction 2, respectively (see Figure 5.2). The average strain results in

$$\varepsilon = \left(\frac{\varepsilon_1^2 + e_2^2}{2}\right)^{1/2} \quad (5.7)$$

The size of the coincidence shell Σ_2^* within the interface can be described as the averaged geometric value of the size of the coincidence shell of the two individual lattices

$$\Sigma_2^* = [\Sigma_2(\text{lattice 1}) \cdot \Sigma_2(\text{lattice 2})]^{1/2} \quad (5.8)$$

Certain properties can be predicted with the CSL as a function of misorientation. Low-energy interfaces are expected when the coincidence cell Σ_2 is as small as possible. This model is insufficient though for explaining all possible low-energy interfaces (Sutton and Balluffi 1987). It was shown that no reliable correlation exists between the energy of the interfaces and the inverse coincidence density Σ_2. However, it is important to note that the size of the coincidence cell depends on the residual strain ε. Often, a residual strain of ε_{max} = 1.66% is tolerable. It is assumed that for ε_{max} a coincidence site still occurs. The geometric considerations, however, seem useful as configurations of interface properties or for explaining structural results.

5.2 Structure Determination by HREM

5.2.1 HREM—The Method

Recently, the resolution of commercially available electron microscopes has been significantly improved. The point resolution is now better than 0.17 nm for instruments with an acceleration voltage of 400 kV. A new generation of high-voltage, ultra-high-resolution instruments pushes the resolution limit to about 0.1 nm. Considerable advances have also been made in the analysis of micrographs obtained by high-resolution electron microscopy (HREM) (Spence 1988; Busek et al. 1988; O'Keefe 1985). Methods and programs have been developed that allow the simulation of HREM images of any given atom arrangement (O'Keefe 1985; Stadelmann 1987). These simulated images are used to interpret experimentally obtained micrographs. These recent developments enable us to use HREM as a method for solving problems in materials science. Atomic structures of different lattice defects, such as phase boundaries, grain boundaries, and dislocations, can be determined by HREM. The difficulties in the interpretation of HREM are similar to those encountered in conventional transmission electron microscopy of lattice defects in crystalline materials: the experimentally obtained micrographs do not usually present a direct image of the object.

The extraction of information on the structure of lattice defects from HREM micrographs is complicated (Spence 1988; Busek et al. 1988). Despite this, useful information has been successfully obtained regarding the structure of lattice defects, particularly for semiconductors. A new generation of instruments with better resolution also allows characterization of defects in ceramics and metals by HREM. This chapter summarizes the possibilities of HREM and then discusses applications of HREM in materials science. The various steps needed for working successfully with HREM are described. Specifically, results for a Nb/Al$_2$O$_3$ interface are described in detail.

5.2.2 Quantitative HREM

The different possibilities offered by modern transmission electron microscopy are summarized schematically in Figure 5.3. Conventional TEM techniques (CTEM) involve bright-field (BF) and dark-field (DF) imaging and selected area diffraction (SAD). These are used for

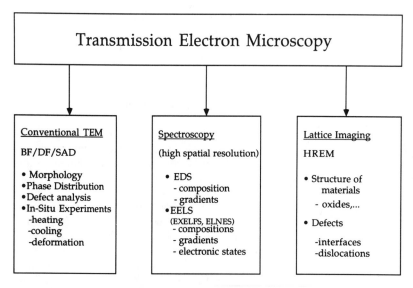

Figure 5.3. Schematic illustrations of different TEM techniques (*BF*, bright-field imaging; *DF*, dark-field imaging; *SAD*, selected area diffraction; *EDS*, energy dispersive spectroscopy; *EELS*, electron energy-loss spectroscopy; *EXELFS*, extended energy-loss fine structure; *ELNES*, electron energy loss near-edge structure; and *HREM*, high-resolution electron microscopy).

morphological analysis as well as for phase and defect identification studies. Spectroscopy can be performed with high spatial resolution. The probe size in scanning transmission electron microscopy ranges from ≤ 1.0 nm to 50 nm. Different signals emitted after inelastic scattering of electrons at atoms of the crystal contain information on chemical composition. Energy dispersive spectroscopy (EDS) uses X-rays emitted from the specimen for a chemical characterization, while characteristic energy losses can be used for identifying qualitatively and quantitatively the light element by energy loss spectroscopy (Hren et al. 1979; Joy et al. 1979, Egerton 1986). The surrounding atoms can be probed by extended energy loss fine structure studies (EXELFS) whereas energy loss near edge structure (ELNES) investigations result in information on the excitation state of atoms or ions. HREM studies result in structure images. This chapter focuses on the latter.

The geometric beam path through the objective lens of a TEM is shown in Figure 5.4(a). Beams from the lower side of the object travel in both the direction of the incoming beam and the direction of the diffracted beam. All beams are focused by the objective lens in the back focal plane to form the diffraction pattern. In the image plane, the image of the object is produced by interference of the transmitted and diffracted beams. Figure 5.4(b) uses wave optics to describe physical processes that contribute to the image formation. From the lower side of the foil, a wave field emerges that can be described by a transmission function $q(x,y)$. For an undistorted lattice, $q(x,y)$ represents a simple periodic amplitude and intensity distribution. The transmission function of a complex lattice with a large periodicity such as a periodic grain boundary, is very complicated and $q(x,y)$ is a nonperiodic function for the distorted region of a crystal.

The intensity distribution in the diffraction pattern is given by the Fourier transformation $Q(u,v)$ of the transmission function $q(x,y)$ where u and v are the coordinates in the diffraction plane (reciprocal space). Because spherical aberration cannot be avoided with rotationally symmetrical electromagnetic lenses (Spence 1988; Busek et al. 1988), the beams emerging from an object at a certain angle (Figure 5.4(a)) undergo a phase shift relative to the direct beam. Imaging under a slightly defocusing mode, Δf, also leads to a phase shift that depends on the sign and magnitude of the defocus value, Δf. The influence of the lens errors and the defocus on the amplitude of the diffraction patterns can be described by the contrast transfer function $\chi(u,v)$,

$$\chi(u,v) = \tfrac{\pi}{\lambda}(\Delta f(u^2+v^2)\lambda^2 - \tfrac{1}{2}C_s(u^2+v^2)^2\lambda^4) \quad (5.9)$$

in which λ is the wavelength of electrons and C_s is the constant of spherical aberration. The dependence of the contrast transfer function as a function of reciprocal distance (space frequency) is shown in Figure 5.5 for a modern 400-kV instrument and for the atomic resolution microscope at the National Center for Electron Microscopy in Berkeley, CA (Hetherington et al. 1989).

The image is formed by a second Fourier transformation of the amplitude distribution in the diffraction pattern multiplied by the contrast-transfer function. Thus, the amplitude in the image plane, $\Psi(x,y)$, is not identical to the wave field in the object plane (transmission function $q[x,y]$). The image is severely modified if scattering to large angle occurs, because the influence of

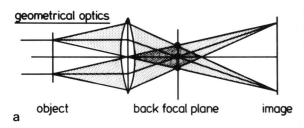

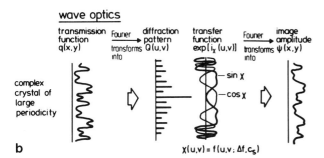

Figure 5.4. Image formation by the objective lens of a transmission electron microscope. (a) Geometric optical path diagram. (b) Wave optical description (see text for explanation).

the spherical aberration increases greatly with increasing scattering angle. The modification is most severe if the components in the diffraction pattern coincide with the oscillating part of the contrast-transfer function (large values of u,v). If, however, the wave vectors lie within the first wide maximum of the contrast-transfer function $\chi(u,v)$, it can be assumed that characteristic features and properties of the object can be directly recognized in the image (Spence 1988; Busek et al. 1988). For good HREM imaging, the first zero value of the contrast-transfer function under the optimum defocusing condition (Scherzer focus) must be at sufficiently large reciprocal lattice spacings so that lattices of many materials can be directly imaged. Good imaging conditions exist when lattices have large lattice parameters (Busek 1985; Mitchell and Davies 1988; Barbier et al. 1985; Merkle 1987). If, however, deviations in the periodicity exist, components of the diffraction pattern appear at large diffraction vectors. It is then most likely that certain Fourier components possess reciprocal lattice distances larger than the Scherzer focus. Lens aberrations and defocusing cannot be neglected.

5.2.2.1 Resolution

The image contrast obtained with a high-resolution electron microscope (see Figure 5.4) is produced by interference of the diffracted beams with the transmitted beam. The interference can also be represented as an impulse response in the form of a Δ function, the location and shape of which is highly dependent on the focusing distance (Spence 1988; Bourret 1985). From this description, it can easily be concluded that the resolution is not uniquely defined and that various distances can be adopted to define the resolution: d is defined as the Scherzer resolution, which is determined from the focusing distance Δf, at which the influence of negative defocusing compensates as far as possible for the influence of the spherical aberration (Spence 1988; Busek et al. 1988). The information limit, d_2, represents the smallest resolvable distance on an image at an optimum defocusing distance.

For HREM analysis of crystalline materials, the number and location of the positions of atom rows must be established. It is assumed that the lattice to be investigated is periodic in the direction of the incoming beam so that the projected potential on the lower surface of the object is formed by the superposition of identical atoms that lie upon each other in the z direction (Figure 5.6). The determination of the atomic structure of distorted regions (or regions of large lattice periodicity), requires a certain resolution of the TEM. The resolution of the TEM can be described by at least two distances, d_1 the Scherzer resolution and d_2 the information limit (Spence 1988). The evaluation of the number of atomic rows and their exact positions are discussed in the next section. Different zone axes of crystalline materials can be used for the evaluation of crystal structures and for the evaluation of defects if the lattice plane distances are larger than the Scherzer resolution d_1 of the instrument (Spence 1988).

If lattice defects must be analyzed by HREM, one must first establish the number of atomic rows that exist in the surroundings of a lattice defect. Naturally, no information can be obtained in any HREM investigation on defects that are not parallel to the direction of

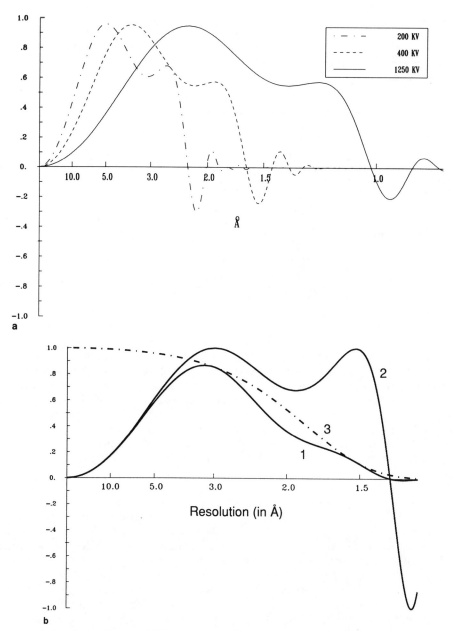

Figure 5.5. Contrast transfer function of high-resolution electron microscopy. (a) (-.-.- 200 kV; ------ 400 kV; ———1250 kV). (b) Atomic resolution microscopy. (1) Actual calculated contrast transfer functions (CTF), (2) undamped CTF, and (3) damping envelope of the atomic resolution electron microscope (ARM) at the Center for Electron Microscopy, Berkeley, CA *(Hetherington et al. 1989)*. Operating conditions: Voltage 800 kV, defocus −55 nm). *(From: Hetherington et al. Mat Res. Soc. Symp. Proc. (1989), 139, 277.)*

the incoming electron beam. HREM images always represent a two-dimensional projection of electron distribution on the exit surface of the transmitted foil. For the image evaluation, it is implied that the atomic rows are exactly parallel to the incoming electron beam.

In order to discuss the possibilities of determining the atomic structure, it is necessary to introduce a parameter d_0, which represents the minimum distance of two neighboring atom columns in the region of the lattice defect. Comparing d_0 with d_1 and d_2 enables us

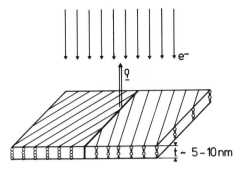

Figure 5.6. Direct lattice imaging by HREM. The crystalline specimen must be adjusted so that the direction of the incoming electron beam coincides directly with the orientation of atomic rows. The schematic drawing includes a grain boundary. HREM can be successfully performed for pure-tilt boundaries with the tilt axis parallel to the direction of the incoming beam.

to decide whether the distorted structure (or structural unit) in the lattice defects can be resolved in all details or not. Three cases can be distinguished:

1. $d_0 > d_1$ In this case, all atom columns can be observed; artifacts are not introduced into the Scherzer image. For thin foils (typical thicknesses 5 nm to 10 nm), channels in the lattice structure appear as white spots on the image. If certain a priori information on the atom configuration is available, the observed HREM micrographs can be directly compared with the atomic structure (Spence 1988; Busek et al. 1988).
2. $d_1 > d_0 > d_2$ The determination of the number of atom columns of a distorted structure requires that micrographs be taken under a series of defocusing distances that typically differ by small values, ~ 5 nm. A quantitative evaluation requires the comparison of micrographs taken under specific focusing conditions with computer-simulated images. This enables one to differentiate between interference produced by artifacts and interference produced by real lattice distortions such as additional atom columns.
3. $d_0 < d_2$ The existing atom rows can no longer be resolved as such, thus, only an absolute measurement of the intensity distribution would give the exact number of atom rows that produce a light (or dark) contrast spot on an HREM image. This evaluation would be very difficult, especially because any surface defects introduced during specimen preparation might cause phase shifts that would drastically modify the intensity distribution on the HREM images.

From this brief summary of the optimum conditions for observing structures in the vicinity of lattice distortions (grain boundary, dislocations), the following conclusions can be drawn:

1. d_1, the point resolution in the Scherzer focus must be as small as possible.
2. Analyses of structures surrounding lattice defects are confined to those cases in which strict symmetry exists in the direction of the incoming electron beam.
3. The specimen thickness must be as small as possible and less than the effective extinction length of the selected zone axes. In most cases, the foil thickness should be between 5 nm and 15 nm. The exact values of the foil thickness t and the defocusing value Δf are important parameters that must be precisely determined. Computer simulations for which the experimentally determined parameters t and Δf are required are thus indispensable if details are to be resolved that are at or below the magnitude d_1.
4. Extremely careful adjustment of the electron microscope has to be made to obtain a good correlation between the image and the projected structure.

5.2.2.2 Determination of Atom Column Locations

After determining the number of atom columns in the vicinity of the lattice defect, the coordinates of the individual atom columns must be determined with the best possible accuracy. This evaluation demands a comparison between experimentally obtained micrographs and computer-simulated images. A trial-and-error method is usually applied. The comparison is made by superimposing two images and visually comparing the location of the individual contrast features. It is important that the location of all contrast maxima and minima coincide. Very reliable data of atomic coalescence coordinates are obtained for $d_0 < d_1$ the accuracy is ± $0.1\,d_1$ (Busek 1985; Bourret 1985). The accuracy of the atom coordinate determination is thus substantially (by a factor of ~ 10) better than the actual resolution. This can easily be explained by the fact that very small changes of the phases at different waves may result in an observable change in the intensity distribution and in the location of the intensity maxima during formation of a phase-contrast image (Spence 1988; Busek et al. 1988). The relative displacement of perfect lattices against each other (in the neighborhood of phase boundaries or even at grain boundaries) can be analyzed from HREM micrographs. The values obtained by HREM are often sufficiently accurate for a differentiation between different grain boundary models. The resolution $d_1 = 0.16$ nm (400 kV HREM) is not sufficient to determine small displacements in the vicinity of ordered grain boundaries (such as twins) in metals. Figure 5.7 shows a flow chart for quantitative HREM.

88   MICROSTRUCTURE CHARACTERIZATION

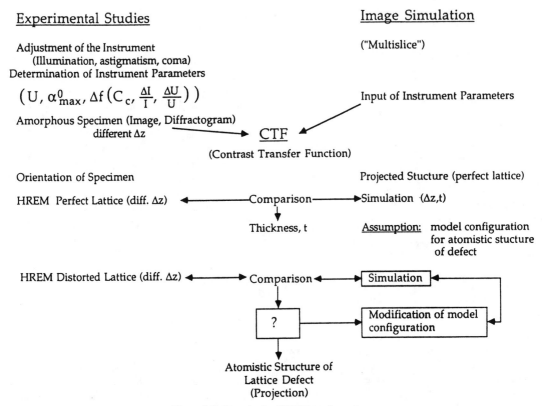

Figure 5.7. Quantitative HREM—a flow chart.

5.2.2.3 Complementary Diffraction Investigations

Selected area diffraction investigations provide important information on the periodicity along the common axis in the case of periodic defects (such as periodic twin boundaries). This is the only possible technique that permits any type of reconstruction along the atom columns with regard to the common axis. This method was recently used to complement high-resolution electron microscopy for determining the structure of grain boundaries (Spence 1988; Busek et al. 1988; Hetherington et al. 1989).

5.2.3 Example: MBE Grown Nb Films on Sapphire (α-Al_2O_3)

Several bonding methods are capable of generating well-defined metal/ceramic interfaces. At the most fundamental level, ultra-clean, flat surfaces readily bond at moderate temperatures and pressures (Fischmeister et al. 1988). Interfaces can also be produced by internal oxidation of metallic alloys (Mader 1989), where small oxide particles in different metals (Nb, Pd, Ag, . . .) are formed by oxidation of a less noble alloying component such as Al or Cd. Interfaces produced by the internal oxidation process usually show a well-defined, low-energy crystallographic orientation relationship between the two components and so were used as model systems. A third method for manufacturing metal/ceramic interfaces is by evaporation of metals onto clean, ceramic surfaces. This method allows control over both substrate material/orientation and overlayer composition. Well-defined interfaces can be obtained by this method (Flynn 1990).

Nb/Al_2O_3 serves as an excellent model system because Nb and Al_2O_3 possess nearly the same thermal expansion coefficient and because thermodynamic quantities, such as solubility and diffusion data, are well established for both components. Nb/Al_2O_3 composites are used in different applications such as for Josephson junctions and as components for structural materials.

To date, only a few detailed studies have been reported concerning the atomistic structure of Nb/Al_2O_3 interfaces formed after diffusion bonding (Florjancic et al. 1985; Mader and Rühle 1989; Mayer et al. 1990a), internal oxidation (Mader 1989; Kuwabara et al. 1989), and after thin-film deposition (Knowles et al. 1987; Mayer et al. 1989; Mayer et al. 1990b). The studies were all performed by high-resolution electron microscopy (HREM). Orientation relationships (OR) were evaluated from diffraction studies: either by X-rays or by selected area diffraction (SAD) patterns obtained in a transmission electron microscope (TEM). The OR between Nb and Al_2O_3 is determined by the manufacturing route (Mayer et al. 1989). Although OR is preset for interfaces prepared by diffusion bonding, topotaxial or epitaxial OR develops during internal oxidation and epitaxial growth, respectively. During internal oxidation, a topotaxial relationship forms between Nb and Al_2O_3 (Mader 1989; Kuwabara et al. 1989), so that close-packed planes of both systems are parallel to each other.

$$(0001)_S \parallel (110)_{Nb} \text{ and } [01\bar{1}1]_S \parallel [001]_{Nb}$$
$$(S = \text{sapphire})^* \qquad (5.10)$$

Epitaxial growth of very-high-quality single-crystalline overlayers of Nb on sapphire has been a subject of recent experiments. There exists experimental evidence (Mayer et al. 1990b) that for most sapphire surfaces a unique three-dimensional *epitaxial* relationship between Nb and Al_2O_3 develops, which is given by two sets of zone axes as shown in Equation 5.11.

$$(0001)_S \parallel (111)_{Nb} \text{ and } [10\bar{1}0]_S \parallel [1\bar{2}1]_{Nb} \qquad (5.11)$$

The Nb layers were fabricated in an MBE growth chamber equipped with electron beam sources for evaporating refractory metals. The sapphire substrates were parallel to $(0001)_S$, $(1\bar{2}10)$, and $(1\bar{1}10)$, respectively. The substrates were preheated and cleaned by annealing (Arbab et al. 1989). A special technique was used to obtain transmission electron microscopy (TEM) cross-section samples of the Nb/Al_2O_3 interface (Mayer et al. 1990b).

The HREM studies were performed at the Atomic Resolution Microscope (ARM) at the National Center for Electron Microscopy (NCEM) at Berkeley, CA. The microscope was operated at 800 kV (Hetherington et al. 1989). The actual contrast-transfer function (CTF) of the instrument is shown in Figure 5.5(b). The point-to-point resolution of the instrument is in the range of 0.16 nm. The CTF is mostly determined by the energy spread of the instrument (~15 nm). As a result of the chromatic aberration the CTF is damped as shown in Figure 5.5(b).

The OR between Nb and Al_2O_3 was evaluated from selected area diffraction (SAD) patterns, which were taken from different Nb films on sapphire substrates in three different orientations. For all films, a unique OR was evaluated and is characterized by the following coinciding planes

$$(0001)_S \parallel (111)_{Nb} \qquad (5.12)$$

and coinciding directions.

$$[2\bar{1}\bar{1}0]_S \parallel [1\bar{1}0]_{Nb} \qquad \text{(direction A)} \qquad (5.13)$$

or

$$[10\bar{1}0]_S \parallel [1\bar{2}1]_{Nb} \qquad \text{(direction B)}. \qquad (5.14)$$

From crystallographic symmetry arguments the angle between the directions A and B is 30 degrees. Figures 5.8(a) and (b) illustrate the relative OR of the two crystals. From Equation 5.12, it follows that the normals on the plane are parallel.

$$[0001]_S \parallel [111]_{Nb}. \qquad (5.15)$$

Thus, the three-fold axes of the two crystals, Nb and Al_2O_3 (sapphire), are parallel.

The OR described by Equations 5.12 through 5.14 is independently identified at epitaxially grown Nb/Al_2O_3 interfaces for substrate surfaces parallel to: $(0001)_S$, $(1\bar{1}00)_S$, and $(1\bar{2}10)_S$. In this chapter, only the results for $(0001)_S$ substrates are reported. Observations for

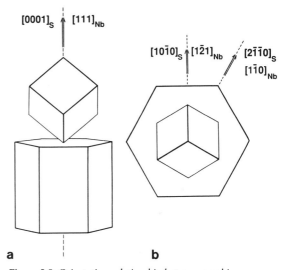

Figure 5.8. Orientation relationship between sapphire substrates and niobium overlayers. (a) The three-dimensional orientation relationship $[0001]_S \parallel [111]_{Nb}$ (S = sapphire) holds for all substrate orientations. (b) The orientation relationship for different directions within the $(0001)_S \parallel (111)_{Nb}$ plane.

---

*The orientation relationship (OR) between two crystals of different lattice structure is uniquely described by one coinciding plane (in both lattices) and one set of coinciding directions in that plane.

the other substrate orientations are reported elsewhere (Mayer et al. 1990b).

Direct lattice imaging of near interface regions allows for the determination of the atomistic structure of the interface as well as for the analysis of defects associated with the interface, such as misfit dislocations. To obtain interpretable HREM images, the electron beam should be incident along high symmetry directions in both crystals, and should be parallel to the plane of the interface. A three-dimensional analysis of the structure requires HREM images taken under different directions of the incident electron beam with respect to the interface orientation. These conditions are fulfilled if the electron beam is parallel to direction A and B, respectively (see Equations 5.13 and 5.14 and Figure 5.8). High-resolution electron micrographs were taken from the same interface in both directions by simply tilting the specimen inside the ARM.

Figure 5.9 shows an overview of a large area of a near interface region. The defocus of the objective lens is slightly more negative ($\Delta f = -70$ nm) than the Scherzer defocus ($\Delta f = -55$ nm). At this defocus, the atomic distance corresponding to the (200) planes with $d = 0.165$ nm becomes clearly visible (Rühle et al. 1990). Lattice planes can readily be identified in Al_2O_3 and Nb. The foil thicknesses of Nb and Al_2O_3 are identical. In Nb, regions of good matching (M) and poor matching (D) alternate at the interface. Steps can also be identified (S). The region of good matching (see Figure 5.9 M) is imaged at a higher magnification as shown in Figure 5.10. Figure 5.10(a) shows the interface with the electron beam parallel to direction A. Nb and Al_2O_3 possess the same thickness, and lattice planes transfer continuously from Nb to Al_2O_3. Figure 5.10(b) is a micrograph of the same interface viewed along orientation B. Only $(10\bar{1})$ lattice planes with a spacing of 0.233 nm are visible in the Nb crystal in regions of good matching M. The (222) lattice planes (perpendicular to $(10\bar{1})$) possess a spacing of 0.095 nm, which is beyond the resolution and information limits of the ARM. Therefore only $(10\bar{1})$ lattice fringes are visible. In both orientations seen in Figure 5.8(a) and (b), a perfect match of the Nb and Al_2O_3 lattice at the interface is visible.

5.2.4 Misfit Dislocations

The mismatch of the $(0\bar{1}10)_S$ and $(11\bar{2})_{Nb}$ planes that are perpendicular to the interface is only ~1.9%, and this misfit is accommodated by localized defects (misfit dislocations) in the regions of poor matching (Figure 5.11) (Mayer et al. 1992). This allows the Nb lattice that is between these defects to expand slightly along the interface (the Nb lattice possesses the smaller lattice plane spacing), resulting in extended regions of perfect matching. The expansion of the lattice plane spacings parallel to the interface is limited to regions close to the interface. The lattice planes of the Nb, especially near the misfit dislocations, are bent, resulting in a continuous transition to the undistorted Nb lattice further away from the interface (this can be seen by viewing Figure 5.11 under grazing incidence). No dislocations or lattice distortions can be seen in the Al_2O_3 lattice. The misfit dislocations in the Nb do not show any "stand-off" distance from the interface.

In HREM images, only projections of the foil on its exit surface can be analyzed. Dislocations can thus only be identified if viewed edge on, that is, if the dislocation line lies parallel to the beam. The dislocation shown in

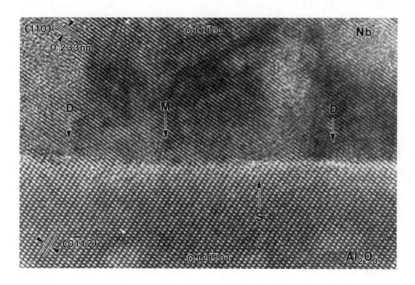

Figure 5.9. High-resolution images of a Nb/Al_2O_3 interface. The direction of incoming electrons is parallel to $[1\bar{1}0]_{Nb}$ and defocus value is $\Delta f = -70$ nm (Scherzer value -55 nm). Lattice planes can be clearly identified in both sapphire and Nb. Foil thickness ~10 nm. At the interface, regions of good matching (M) and poor matching (D) alternate. S steps in the substrate.

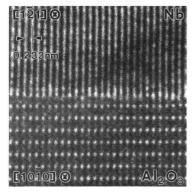

Figure 5.10. High-resolution image of Nb/Al$_2$O$_3$ interface. Region of good matching (*M*). (a) Direction A with $[2\bar{1}\bar{1}0]_S \parallel [1\bar{1}0]_{Nb}$. (b) Direction B with $[10\bar{1}0]_S \parallel [1\bar{2}0]_{Nb}$.

Figure 5.11 fulfills this requirement and, by making use of the continuous transition of lattice planes between the niobium and the sapphire, a Burgers circuit can be performed. A projected Burgers vector of $\underline{b} = 1/2\,[11\bar{1}]$Nb was determined. Because only the projection can be seen in the HREM image, this does not preclude the existence of a possible screw component parallel to the viewing direction.

While in HREM, individual dislocations can be viewed edge-on. Imaging of a two-dimensional array of dislocations lying parallel to an interface requires that this interface be inspected in plan-view. The dislocations can then be imaged with conventional TEM techniques.

In the case of our system, two difficulties arise: (1) All the low-indexed reflections of the Nb that can be used to image the dislocations are very close to corresponding reflections of the sapphire (the distances between the reflections are given by the 1.9 % misfit). In bright-field or dark-field images, this causes strong moiré contrast with the same periodicity as the misfit dislocations. (2) In order to obtain strong diffraction contrast of the dislocations (utilizing dark-field imaging techniques), bent lattice planes have to be present in the vicinity of the dislocation core. In our system, the core of the misfit dislocations is located directly at the interface. The bent lattice planes around the dislocation core terminate abruptly at the interface and additional relaxations have to be expected in the niobium at the interface to the rigid sapphire lattice. It is thus unclear whether the simple $g \cdot \underline{b}$ criterion that is used to select possible imaging conditions for bulk dislocations is also valid in the case of misfit dislocations without "stand-off."

A schematic drawing of the atomistic structure of a 1/2 $[11\bar{1}]_{Nb}$ dislocation in bulk Nb is shown in Figure 5.12. The model is purely geometric and relaxations in the vicinity of the dislocation core have not been taken into account. The dislocation core consists of three inserted $(22\bar{2})$ planes. The position of the (111) plane forming the terminating plane at the interface is indicated in Figure 5.12. In the experimental image (see Figure 5.11), it is virtually impossible to geometrically identify the three additional $(22\bar{2})$ planes forming the dislocation core. However, if viewed along the arrowed directions, an additional (110) plane (corresponding to a Burgers vector of 1/2 $[110]_{Nb}$) and an additional (002) plane (corresponding to a Burgers vector of 1/2 $[001]_{Nb}$) can clearly be identified in both the experimental image of Figure 5.11 and in the schematical drawing Figure 5.12. This description is equivalent to decomposing the Burgers vector into two components according to: 1/2 $[11\bar{1}]_{Nb}$ = 1/2 $[110]_{Nb}$ + 1/2 $[00\bar{1}]_{Nb}$. Especially for the 1/2 $[110]_{Nb}$ component, strong bending of the lattice planes and a strong localization of the dislocation core can be recognized in the HREM image (see Figure 5.11).

Bright-field images showing strong moiré contrast are depicted in Figure 5.13(a–c). The images were taken under dynamic scattering conditions that were obtained

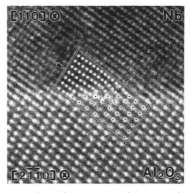

Figure 5.11. High-resolution image of a region of poor matching. A misfit dislocation forms with no stand-off distance from the interface. The core of the misfit dislocation can be readily identified, as outlined in the picture. The arrow indicates inserted planes in the Nb; additional (110) plane.

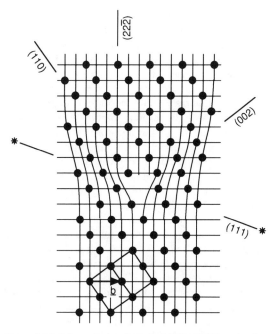

Figure 5.12. Schematic drawing of a 1/2 [111] dislocation in bulk Nb. Relaxations of the three inserted (222) planes have not been taken into account. One unit-cell, including the Burgers vector, is outlined. The position of the terminating (111) plane at the Nb/sapphire interface is marked (*--*).

by tilting only slightly away from the [111] zone axis so that a 110 systematic row of reflections was excited. The tilt angle was about 10 degrees and the same tilt was applied in three different directions related by symmetry. In these projections, the dislocation lines should be inclined by 30 degrees with respect to the moiré fringes. However, under the given imaging conditions, the contrast of the dislocations is almost extinct and the dislocations cannot be seen because of the strong moiré contrast.

The diffraction contrast caused by the dislocations and the moiré fringes can only be distinguished by weak-beam imaging (Pond 1984). In weak-beam images, diffraction contrast is caused by a specific strain component (Cockayne 1972). The image obtained from a dislocation is narrower (1 nm to 2 nm) because the imaging conditions are such that only areas close to the dislocation core, where lattice planes are strongly bent, contribute to the image. As stated earlier, the strong bending required to produce weak-beam images might not be present in the case of misfit dislocations. We have thus tried a variety of g-vectors that could be used to image bulk dislocations with $b = 1/2 <111>_{Nb}$. We failed to detect the dislocations with g-vectors of types $1\bar{1}0$, $11\bar{2}$ and $22\bar{2}$ in orientations close to the [111] zone axis. Only images that were obtained by tilting to a 110

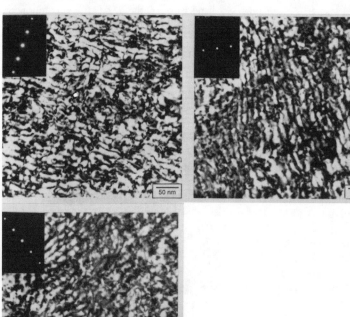

Figure 5.13. Bright-field images of a plan-view specimen of the interface. The images were taken along systematic orientations close to the [111] zone axis of Nb. Moiré fringes parallel to the {110} planes indicate that the mismatch is distributed evenly in all three directions. In this projection, dislocation lines should be inclined with respect to moiré fringes. However, under the given imaging conditions, their contrast is almost extinct and thus cannot be seen because of the strong contrast of the moiré.

systematic orientation close to the [001] zone axis revealed the arrangement of the dislocation lines. Figure 5.14 shows a bright-field image obtained under kinematic (equivalent to weak beam) conditions. The specimen area is bent, leading to a variation of the excitation error across the image. In the dark bend contour, strong moiré fringes can be seen whereas sharp dislocation lines become visible in the areas adjacent to the bend contour. The visibility in this orientation is in agreement with our observation that strong bending of lattice planes can only be seen in the HREM images if the dislocation is viewed in grazing incidence along a [001] direction. In Figure 5.14, only one set of the misfit dislocations forming the complete array can be seen. It consists of a parallel arrangement of continuous dislocation lines. The other two sets are related to the set shown by the three-fold symmetry, that is, two 120-degree rotations. From this we can conclude that the misfit dislocations are forming a triangular array at the interface between the Nb and the sapphire. We were not able to image the other two sets of parallel dislocations in one sample area because of the limited tilt-angle capacity of the microscope.

5.2.5 Atomistic Structure of Commensurate Regions at Interfaces

Quantitative HREM requires computer simulation. The atomistic configuration is obtained if experimental images are identical to images simulated for specific atomistic models. This analysis requires knowledge of the exact focusing value and the foil thickness. These values have to be determined in a first step (Spence 1987). The determination of the foil thickness is most accurate if observed images of the Nb and Al_2O_3 are compared to calculated lattice images for varying thicknesses and focus values (Stadelmann 1987). Good agreement between the simulated images and the experimental image is obtained for a defocus of $\Delta f = -40$ nm and a foil thickness of $t = 7$ nm in Nb and Al_2O_3. From the simulated images, it is also possible to determine the positions of the atoms in both crystals with respect to the intensity distribution in the experimental image. The next step is to identify the translational state $\underline{T}$ of the two crystals with respect to each other. Such a translational state only exists in the areas of good matching. In these areas, the two crystals become commensurate at the interface by expanding the lattice plane spacing of the Nb (Figure 5.15). The translational vector $\underline{T}$ ($T1$, $T2$, $T3$) can be constructed so that the components $T1$ and $T2$ lie within the interface plane and $T3$ is perpendicular to it.

An inspection of the HREM images of Figures 5.11 and 5.18 reveals that certain lattice fringes in Nb and Al_2O_3 transfer continuously into each other across the interface. Furthermore, the positions of the atoms with respect to these lattice fringes in the experimental images are known. However, due to the noise present in the experimental image the position of the atomic columns cannot be located accurately (see Figure 5.16(a)). In order to reduce the noise of our experimental images, we have applied a Fourier filtering technique, which has recently been developed for interfaces (Möbus et al. 1992). The result is shown in Figure 5.16(b). From the known positions of the atoms in both the Nb and the Al_2O_3, we have determined the translational state of both lattices across the interface. From this, a model of the atomic structure at the interface can be derived. If

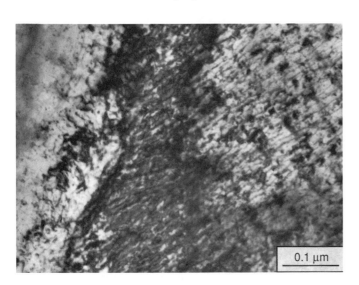

Figure 5.14. Bright-field image in a systematic orientation close to the [001] zone axis (tilt angle 55 degrees from [111] foil normal) with $g = 110$. In the dark bend contours (Bragg condition), only moiré fringes can be seen. Sharp dislocation lines (arrowed) become visible in the areas adjacent to the bent contour (kinematic diffraction condition).

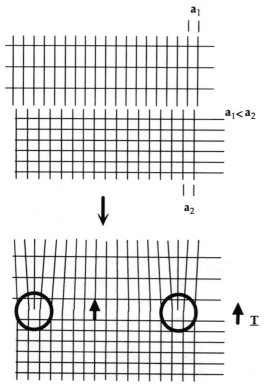

Figure 5.15. The lattice plane spacings of two different materials are incommensurate. If two lattice planes of both materials are brought together in one point, the same match will never occur again along the interface. Because of relaxations along the interface, an epitaxial fit can be achieved at metal/ceramic interfaces. The mismatch is accommodated by localized misfit dislocations in the metal (in circles). In the regions of good matching between these defects, a unique translational state is obtained between both lattices, which can be described by a translational vector T.

we assume that the lattice continues undisturbed up to the interface, then the atomistic structure represented in Figure 5.17 is obtained. That is, Al_2O_3 is terminated by an oxygen layer and the Nb atoms of the first Nb layer fit accurately in the A sites (Figure 5.17(b)) of the oxygen layer under which no Al ions are positioned. An obvious choice for the origin of the translational vector T is given by one of the Al-sublattices as indicated in Figure 5.17(a) and (b). The first Nb layer possesses exactly the same three-fold symmetry and the same atomic distances as the individual layers of the Al-sublattice parallel to the interface.

The third and final step is to simulate HREM images for the structure model shown in Figure 5.17 and to vary the individual parameters until a best fit is obtained. Four examples of the simulated images for different parameters are shown in Figure 5.16(c–f). Figure 5.16(c) shows the image with the best fit, from which we determined the translational vector T. Figure 5.16 (d) and (e) show the result for a decrease or an increase of the distance between the two lattices, that is, the component $T3$, by 0.04 nm. The relative shift of the lattice planes can be seen by viewing the images under grazing incidence. Figure 5.16(f) shows an image that was obtained by removing the last oxygen layer of the sapphire lattice. This results in a contrast along the interface that is clearly different from the one in the experimental image and also leaves a gap at the interface where the atoms would not be in contact with each other.

From Figure 5.16(c) the components $T1$, $T2$, and $T3$ of the translational vector result in:

$$T1 = a_0\sqrt{3}/6 = 0.137 \text{ nm,}$$
$$T2 = a_0/2 = 0.238 \text{ nm,}$$
$$T3 = c_0/6 = 0.216 \text{ nm} \qquad (5.16)$$

where a_0 and c_0 are the lattice constants of sapphire. The experimental error in determining each individual component is ± 0.02 nm. The vector $(T1, T2, T3)$ given above is a lattice vector of the Al sublattice indicated in Figure 5.17(a). Therefore, within the experimental error, the Nb atoms of the first layer are positioned exactly in the sites where the Al ions of the next layer would be placed if the sapphire lattice would be continued.

5.2.6 Limitations of the HREM

No twins could be identified in the (nearly) perfect Nb film. It was demonstrated that a unique atomistic relationship exists between the sapphire surface and the monoatomic Nb layer. There exists only one set of Nb atom positions on the unreconstructed sapphire surface, which leads to a twin-free film (Figure 5.17(b)). Nb atoms must be located on the terminating O^{2-} layer of sapphire on top of "empty sites." Comparison of experimental HREM images with the corresponding simulated images verified this hypothesis. All three components of the translational vector between the two lattices joining at a metal/ceramic interface could be determined.

The location of the Nb atoms on top of the terminating O^{2-} layer is such that the Al sublattice that would form the next layer of the sapphire is continued. This implies that the distance between an Nb atom of the first layer and the three neighboring O^{2-} ions is shorter than the distance that would be expected for neutral Nb atoms. However, a good match is obtained for the radius of the Nb^{3+} ion. A layer of Nb^{3+} would also account for the charge balance across the interface, assuming that a complete O^{2-} layer is terminating the sapphire at the interface.

Figure 5.16. Comparison between (a) the experimental image taken along direction A, and (b) the same image after Fourier-filtering, and (c–f) simulated images for various conditions. (c) Simulated image showing the best fit with the experimental image, (d) 0.04 nm shorter distance between the two lattices, (e) 0.04 nm longer distance, and (f) after removing the terminating oxygen layer of the sapphire.

In conclusion, the epitaxial growth of Nb on basal plane sapphire seems to be dictated by a continuation of the cation sublattice of the sapphire in both location and ionicity of the Nb atoms of the first layer. Nb grows well on sapphire because the resulting first layer has the same symmetry and atomic arrangement as the (111) plane of bulk Nb, with only a 1.9% mismatch. The model presented here was derived mainly from a geometric evaluation of the HREM images. It therefore does not take into account the effects of possible interdiffusion leading to a partial substitution of Nb by Al near the interface. HREM as a method is fairly insensitive to such a change in chemistry that leaves the structure unaffected. The chemistry of the interface will thus have to be studied by high-resolution chemical analysis.

5.3 Structure Determination by Grazing Incidence X-Ray Scattering

5.3.1 GIXS—The Technique

High-resolution electron microscopy (HREM) allows for the determination of the structure of interfaces of small areas. However, it is often interesting to also obtain information averaged over large areas. The most promising technique that recently became of importance is the scattering of X-rays, the so-called Grazing Incidence X-Ray Scattering, GIXS (Marra et al. 1979; Eisenberger and Marra 1981). In this technique, a grazing incidence X-ray beam of high intensity is reflected not only by the bulk material, but also by the reflectivity of interface buried slightly beneath the surfaces. Following the recent publication of the first GIXS experiments on structural relaxations at the Al-Gas interfaces in 1979 (Marra et al. 1979), were subsequent demonstrations of the monolayer surface sensitivity in 1981 (Eisenberger and Marra 1981). The potential of this technique has now been demonstrated in more studies. Recently, it was applied to the study of the atomistic structure of the Nb/Al interface, actually the same films studied by HREM (Lee et al. 1991, Liang et al. 1992).

As revealed by TEM, there exists a lattice mismatch of ~ 1.8%. This mismatch is relaxed by ordered arrays of misfit dislocations. Also, incommensurate structures with misfit dislocations have been extensively studied for adsorbed monolayer systems in the context of commensurate-incommensurate transitions. Such studies have not yet been performed specifically for buried interfaces.

5.3.2 Experimental

The GIXS measurements were carried out in a Z-axis surface scattering spectrometer (Fuoss et al. 1992) at Exxon beamline X10A of National Synchrotron Light

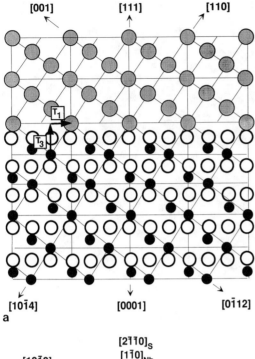

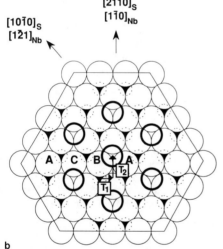

Figure 5.17. Schematic drawings of the atomic positions at the Nb/Al$_2$O$_3$ interface. (a) View parallel to the interface. O^{2-} ions: light large circles, Al^{3+}: black circles, Nb atoms: grey circles. Note the continuous transition of (110)$_{Nb}$ to (10$\bar{1}$4)$_S$ planes and (001)$_{Nb}$ to (0$\bar{1}$12)$_S$ planes. (b) View perpendicular to the interface. O^{2-} ions: light large circles, Al^{3+}: dark sections of small circles, Nb atoms: bold, medium-sized circles. It is assumed that the O^{2-} ions form the outermost layer. The Nb atoms of the first Nb layer are positioned above the empty sites of the first Al^{3+} layer.

Source, Brookhaven National Laboratory, Massachusetts. The X-rays from the electron storage ring were focused by a bent cylindrical mirror and monochromatized by a double crystal monochromator of Ge(111). Pairs of slits are employed as an analyzer in the diffracted beam with an in-plane resolution of about 0.0005 nm^{-1}.

The Nb thin films were grown on the sapphire substrates by an MBE method (Du and Flynn 1990). X-ray measurements were carried out in both surface-parallel (in-plane) and surface-normal directions at different grazing incidence angles. The critical angle of X-rays of Nb is 0.294 degrees for the 0.1129 nm X-rays employed in this study. The thickness of the films is about 12 nm as determined from specular X-ray reflectivity measurements. The crystal mosaic of the sapphire substrates is 0.006 degrees. Nb films were found to be plane despite a miscut of ~1.4 degrees. The in-depth structure of the films was probed with GIXS by varying the incidence angle of X-rays.

In our studies, three groups of films were investigated that were grown on different surfaces of sapphire crystals. These films have the following orientation relations in the growth directions: (111)$_{Nb}\|$(0001)$_S$, (112)$_{Nb}\|$(01$\bar{1}$0)$_S$ and (110)$_{Nb}\|$(01$\bar{2}$0)$_S$. In this chapter, only the results from the first case will be discussed. The results on samples of two other orientations will be published elsewhere.

5.3.3 Experimental Results

For Nb(111) films grown on the (0001) surface of the sapphire substrate, the in-plane orientation was found to be (1$\bar{1}$0)$_{Nb}\|$($\bar{2}$110)$_S$ with a six-fold symmetry. The miscut angles of the substrate and lattice misfits of the films studied are given in Table 5.1. From the lattice constants measured, we find that the Nb lattice is expanded in the in-plane direction by ~0.4% and contracted in the surface-normal direction by ~1.6%.

Figure 5.19 shows the in-plane radial scans near the Nb($\bar{2}$20) peak at incidence angles varying from 0.15

Table 5.1. Summary of Crystallographic Measurements on Nb(111) Film on Sapphire (0001) Substrate

In-plane orientation	Miscut (degrees)	Misfit		Strain	
		Cal.	Exp.	e$\|$	e+
[$\bar{1}$10]/[11$\bar{2}$0]	1.5	−1.8%	−0.6%	0.4%	−1.6%
[1$\bar{1}$2]/[1$\bar{1}$00]	0.0	−1.8%	−0.7%		

e$\|$ and e+ denote the strain within in-plane and surface-normal directions, respectively.

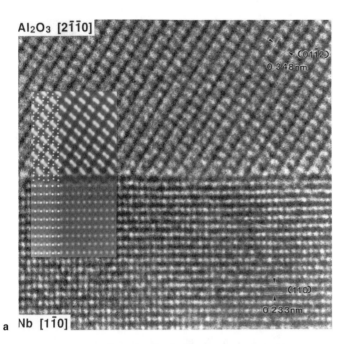

Figure 5.18. Lattice images in orientation II. (a) Focus value $\Delta f = 40$ nm; (b) focus value $\Delta f = 50$ nm. The insets represent the simulated images for the interface model C, in which the terminating plane of the oxide is an Al layer with additional Al atoms at the boundary. (•) Nb, (♦) Al, (▶) O.

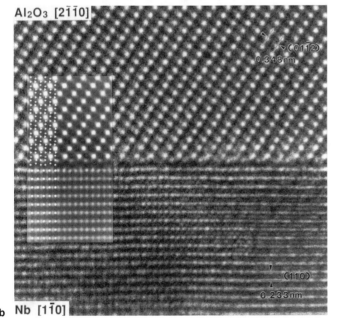

degrees to 0.6 degrees. In Figure 5.19, three features are quite noticeable: (1) a very sharp sapphire $(22\bar{4}0)$-peak, which is instrument-resolution limited, (2) a broad Nb$(\bar{2}20)$ peak for which the peak position is shifted from that for bulk Nb and the lineshape changes with incidence angle of x-rays, and (3) a clear satellite peak that appears at lower Q as the incidence angle is increased.

We note in Figure 5.19 that the diffraction profile observed at 0.15-degree incidence angle has a simple Gaussian lineshape (coherence length ~20 nm) with its peak position shifted from the bulk Nb lattice position (label C) toward a coherent lattice match position. At this incidence angle, the X-rays can only penetrate a few nanometers below the surface. This observation indicates that the surface of the 12 nm Nb film is still strained by the substrate. This result is supported by recent HREM studies (Mayer et al. 1992). As the incidence angle increases (the X-rays probe deeper toward

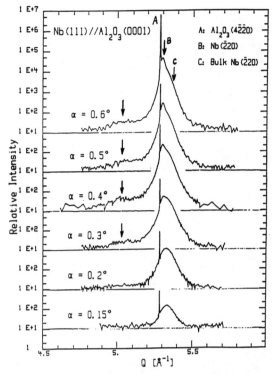

Figure 5.19. In-plane radial scans around the Nb($\bar{2}20$) peak with incidence angles of X-rays varying from 0.15 degrees to 0.6 degrees.

the interface), the Nb peak becomes asymmetric. The observed lineshape suggests that a new peak (label B) is emerging at a position close to that for bulk sapphire (label A). This new structural feature is believed to be associated with the interface.

As the incidence angle is increased, we notice that the satellite peak at the low-Q side can only be seen at the grazing incidence angle higher than 0.3 degrees, indicating that the satellite originates from the interface. The results are similarly seen on the $(112)_{Nb} \parallel (01\bar{1}0)_S$ sample with scans taken along the in-plane $[\bar{1}10]_{Nb} \parallel [2\bar{1}\bar{1}0]_S$ direction. We suggest that the satellite peak comes from more-or-less regularly spaced misfit dislocations at the interface. From the observed separation between the satellite and the principal Bragg peak, the distance between dislocations is estimated to be about 6 nm.

We now discuss the surface-normal rod profile measurements that help to explore the possible origin of the satellite peak. The results of these measurements, for scans taken at fixed incidence angle of 0.4 degrees, are shown in Figure 5.20. The rod profile of the satellite is shown to be relatively flat when compared with that of the main Nb peak. This indicates that the thickness of the interface layer associated with the satellite is smaller than that associated with the Nb peak. Noting that an ideal two-dimensional structure should have a constant rod profile, the observed drop in intensity of the satellite with increased Q suggests a finite thickness of this structure, presumably located at the interface.

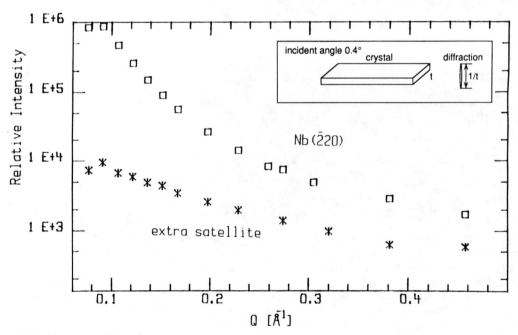

Figure 5.20. Comparison of the rod profiles of the principal Nb($\bar{2}20$) peak and its satellite; the incidence angle of X-rays was fixed at 0.4 degrees.

It is interesting to see that X-ray reflectivity measurements as shown in Figure 5.21 reveal an amplitude-modulated fringe pattern. By assuming a three-layer model, we are able to obtain a reasonable fit of the data. The model includes an oxide layer at the top surface, a normal Nb layer, and an interface layer between the Nb and the sapphire. The reflectivity data are fitted by following the procedure of Tidswell et al. (1990). In the fitting, an electron density of the oxide 1ayer of $1.8 \cdot 10^{24}$ e/cm^3 is assumed. We find that the fitting is relatively insensitive to the oxide layer within a reasonable thickness range (< 1 nm). The most striking result to come from this analysis is evidence for the existence of the interface layer with a thickness about one-third that of the Nb film thickness and an electron density ($1.66 \cdot 10^{24}$ e/cm^3) between those of Nb ($2.16 \cdot 10^{24}$ e/cm^3) and sapphire ($1.17 \cdot 10^{24}$ e/cm^3). This interface layer could be an Nb layer with a high density of dislocations.

Finally, we would like to briefly compare our results with those on internally oxidized samples. We note that the orientations of the epitaxial films are different from those of the Nb-Al$_2$O$_3$ interfaces observed in the internally-oxidized NbAl alloys (Burger et al. 1987; Mader 1989), which had the surface plane orientation of $(011)_{Nb} \| (0001)_S$ and the crystallographic directions in the plane: $[1\bar{1}0]_{Nb} \| [\bar{2}110]_S$ and $[100]_{Nb} \| [01\bar{1}0]_S$. Comparing these results, we note that the in-plane crystallographic orientation, $[1\bar{1}0]_{Nb} \| [\bar{2}110]_S$, is shared by both cases. Interestingly, it was in this orientation that misfit dislocations were observed in a previous TEM imaging study at the interface of the oxide precipitates (Mader 1989).

5.4 Chemical Processes at Metal/Ceramic Interfaces

5.4.1 Theoretic Considerations

In multicomponent two-phase systems, nonplanar interfaces or two-phase product regions can evolve from initially planar interfaces (Backhaus-Ricoult and Schmalzried 1985), even at constant temperature and pressure. Under the same conditions, the interface stays planar in binary systems. This difference originates with the thermodynamic degrees of freedom, f. For a binary system $f = 0$, whereas for ternary or higher-order systems, $f > 0$ (Gibbs' phase rule). Consequently, in the latter, interface compositions are partially controlled by kinetics.

Not all (higher-order) interfaces necessarily develop an unstable morphology during a high-temperature treatment. Analysis of the phenomenon is needed to assess susceptibility. Similar problems exist for solidification and for the oxidation of alloys (Mullins and Sekerka 1963, 1964). Mathematical treatments predict the time evolution of small interface perturbations. The perturbations may occur either as a result of initial roughnesses or upon small transport fluctuations caused

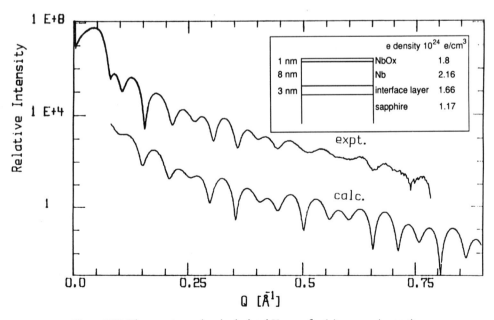

Figure 5.21. The experimental and calculated X-ray reflectivity curves (see text).

by changes in temperature or by the presence of defects. If perturbations increase in amplitude with time, initially planar interfaces become morphologically unstable. The critical conditions for instability depend primarily on the mobility of the constituents and the thermodynamic properties of the system.

The formalism previously developed for ternary systems (Backhaus-Ricoult and Schmalzried 1985) can be adapted to metal/ceramic couples, with the three independent components being the two cations and the anion. A schematic ternary phase diagram and the expected concentration profiles are shown in Figure 5.22 (Backhaus-Ricoult 1987). In general, the problem is complicated by having several phase fields present that may form intermediate phases, usually being intermetallics with noble metals and spinel (or other oxides) with less noble metals. The actual phases depend on the geometry of the tie lines, as well as on the diffusion paths in the ternary phase field, and cannot be predicted a priori.

The diffusion problem has thus far been examined (Backhaus-Ricoult 1987) for the simple case wherein no product phases had formed, the interfacial stresses were negligible, mass transport occurred by bulk diffusion, and local equilibrium was imposed everywhere. Even then, a general analytical solution was not possible. However, for several metal/ceramic systems, some further simplifications are appropriate. The oxygen and the metal atoms diffuse on different sublattices, allowing the interaction term in the diffusion coefficients to be neglected. (In niobium, oxygen diffuses on interstitial sites while the aluminum diffuses by vacancies.) Negligible solubility of the metal in the oxide (grad μ_{BO} = 0) allows point-defect relaxation only in the ceramic. Therefore, for a stoichiometric ceramic, the remaining defect fluxes are small compared with the fluxes in the metal and can be ignored. Subject to the above simplifications, solutions have been obtained for Nb/Al$_2$O$_3$ (Backhaus-Ricoult 1987; Backhaus-Ricoult 1992; Burger and Rühle 1989).

Rather high and probably unrealistic values of solute concentrations are calculated at the interface. In particular, the aluminum concentration profile in the niobium is steep and reveals an enrichment very near the interface, whereas the oxygen concentration profile is more extended, but with small absolute values. Such results do not agree with experimental observations, as elaborated below, suggesting that several assumptions are invalid. More sophisticated numerical models will thus require development as additional experimental results become available.

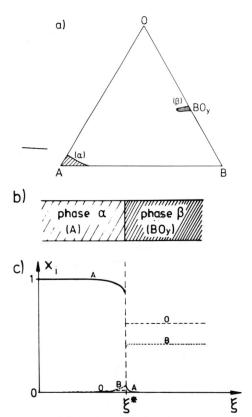

Figure 5.22. (a) Schematic ternary phase diagram. Extended phase fields exist near metal A and oxide BO$_y$. (b) Schematic drawing of diffusion couple. Equilibrium concentration profiles during diffusion bonding between metal phase A and oxide BO$_y$.

5.4.2 Experimental Observations of Systems Without Reaction Products

Detailed scanning electron microscopy (SEM) and TEM studies performed for Nb/Al$_2$O$_3$ (Burger and Rühle 1989; Burger et al. 1987; Rühle et al. 1986; Rühle et al. 1987) have shown that no reaction layer forms (see Figure 5.18) (Knauss and Mader 1991). Concentration profiles determined on cross-sections of rapidly cooled specimens revealed that, close to the interface, the concentration of aluminum is below the limit of detectability. However, with increasing distance from the interface, the concentration of aluminum, c_{Al} increases to a saturation value, $c_{Al}^* = 0.75$ wt %, at a distance of about 2.5 μm (Figure 5.23). The c_{Al} remains at that level up to a distance of $d^* \approx 16$ μm after bonding for 2 hours, and then decays exponentially. The magnitude of d^* depends on the bonding time. The corresponding oxygen content is below the limit of detectability. These measurements suggest that at the bonding temperature,

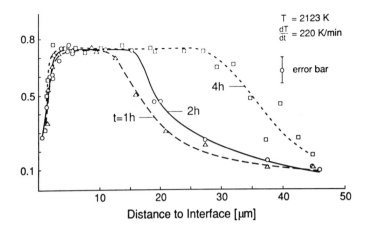

Figure 5.23. Measured concentration profiles of aluminum in niobium as a function of the distance of bonding time and of the interface. Diffusion bonding conditions: 2123 K, dynamic vacuum (10^{-4} torr), and "fast" cooling 220 K min^{-1}. Diffusion bonding times are marked in the figure.

$c_{Al}{}^*$ at the interface possesses a value governed by the solubility limit. This limiting concentration would then extend into the niobium to a characteristic distance that depends on the bonding temperature and time. Upon cooling, the solubility of aluminum in niobium decreases, causing some of the dissolved aluminum (as well as oxygen) to diffuse back to the interface and condense as Al_2O_3.

"Slow" cooling after bonding results in completely different observations. Instead, small precipitates of Θ-Al_2O_3 form in the niobium at distances between 8 μm and 14 μm from the interface (Burger et al. 1987; Rühle et al. 1986). Furthermore, close to the zone wherein precipitation occurs, c_{Al} is very small. The precipitation presumably occurs during slow cooling, because the time requirements for precipitate nucleation are satisfied.

Bonding between platinum and Al_2O_3 subject to an inert atmosphere also occurs without chemical reaction. Specifically, no aluminum can be detected by Auger spectroscopy on the platinum side of an interfacial fracture surface (Klomp 1987a; Klomp 1987b). However, for bonds formed subject to a hydrogen atmosphere containing about 100 ppm H_2O, aluminum is detected in the platinum, indicative of Al_2O_3 being dissolved by platinum. More detailed studies concerning local chemical compositions are clearly required for a better understanding of the bonding processes involved.

5.4.3 Experimental Observations of Systems Forming a Reaction Product

For systems that form interphases, it is important to be able to predict those product phases created during diffusion bonding (given the possible phases present in the phase diagram). However, even if all the thermodynamic data are known, so that the different phase fields and the connecting tie lines can be calculated, the preferred product phase still cannot be unambiguously determined. The problem involves kinetic considerations. Specifically, the diffusion paths in phase space are controlled by different diffusion coefficients and, consequently, interface compositions also depend on the diffusivity ratios. Sometimes, small changes in the initial conditions can influence the reaction path dramatically, as exemplified by the Ni-Al-O systems (Wasyncuk and Rühle 1987). Under high-vacuum conditions (activity of oxygen $a_O < 10^{-12}$), the diffusion path in the extended nickel phase field follows the side of the miscibility gap that is rich in aluminum and low in oxygen (Figure 5.24, path I). This is due to the more rapid diffusion of oxygen in nickel-aluminum than in nickel. This interface composition is directly connected by a tie line to the Al_2O_3 phase field, such that no product phase forms. However, whenever nickel contains sufficient oxygen (about 500 ppm solubility), the Ni(O)/Al_2O_3 diffusion couple yields a spinel product layer. Under these conditions, spinel forms, because the new diffusion path in the nickel phase field requires that the tie line connects the metal and spinel field (Figure 5.24, path II). The associated thermodynamic and atomistic consideration pertinent to spinel layer formation have been addressed (Rühle et al. 1987; Wasyncuk and Rühle 1987; Trumble and Rühle 1991). However, available observations do not unequivocally identify the operative mechanism. It is also noted that the interface between spinel and nickel seems to be unstable, with morphological instabilities becoming more apparent with increasing spinel layer thickness.

Bonding of copper to Al_2O_3 seems to require a thin layer of oxygen on the surface of copper prior to bonding and $CuAl_2O_3$ or $CuAl_2O_4$ form (Wittmer 1985). The spinel thickness can be reduced by annealing

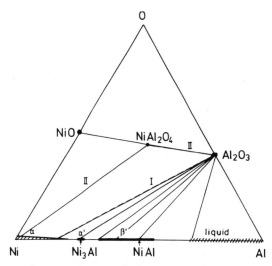

Figure 5.24. Ni-Al-O phase diagrams (schematically for $T = 1600$ K). Two reaction paths are possible when nickel is bonded to Al_2O_3: (I) Low oxygen activity: no reaction product forms; (II) high oxygen activity: spinel forms.

under extremely low oxygen activities leading first to a "non-wetting" layer of Cu_2O and then to a direct Cu/Al_2O_3 bond. The mechanical stability of spinel-free Cu/Al_2O_3 specimens has not yet been investigated.

Bonding of titanium to Al_2O_3 results in the formation of the intermetallic phases TiAl or Ti_3Al, which probably also include oxygen. The thickness of the reactive layer increases with increasing bonding time, and morphologically unstable interfaces develop. A detailed study is again required for the identification of the different stable phases.

Similar studies have been performed for other ceramic partners such as simple cubic oxides (MgO, NiO, ...), sesquioxides (Cr_2O_3, Mn_2O_3, ...), Si_3N_4, and SiC. The situation is much more complicated for Si_3N_4 and SiC because impurities or sintering additives frequently diffuse to the interface and form a glassy film. The bonding is then governed by the interfaces between the glass film and both the ceramic and the metal.

5.5 Case Study: Structure and Bonding of Ag/MgO Interfaces

5.5.1 Introduction and Results of ab-Initio Calculations

To gain the best possible understanding of the physics fundamental to bonding, it is important to compare the experimentally determined structure with models derived from physical principles. Both experimental structure determination and numeric modeling of interface structures are presently limited to particular model-like situations and, therefore, the variety of interfaces that can simultaneously be studied experimentally and theoretically is rather small.

Ag/{100}MgO interfaces are promising systems in this respect. Sphere-on-plate experiments of Au and Cu on {100}MgO substrates suggest that the energy of an Ag/{100}MgO interface reaches its absolute minimum when the Ag and the MgO lattices are in parallel orientation (Fecht and Gleiter 1985).

The lattice mismatch δ of Ag and MgO, defined as

$$\delta = \frac{|a_{MgO} - a_{Ag}|}{(a_{MgO} + a_{Ag})^{1/2}} \qquad (5.17)$$

amounts to 3%. This is sufficiently small for ab-initio modeling of *commensurate* regions of the parallel epitactic {100}Ag/MgO interface, as carried out by the group of Andersen (Blöchl et al. 1990; Schönberger et al. 1992). In these calculations, the lattice mismatch was balanced by a tetragonal distortion of the Ag lattice, while the high common symmetry of the Ag and MgO crystals in parallel orientation permitted a tractably small supercell.

The calculated electronic structure of the {100} Ag/MgO interface indicates that the bonding between Ag and MgO is mainly *electrostatic* with negligible covalent contributions. Another important result of the ab-initio calculations is the energetically most favorable translation state of Ag and MgO. The atoms in the terminating {100} layer of the Ag crystal tend to sit on top of the O ions in the first {100} MgO layer. A value of 0.25 nm was calculated for the spacing between the atomic layers terminating the adjacent crystals. This value corresponds to an excess volume of approximately 20% in the interfacial planes.

With the intention of comparing the results of the ab-initio calculations to experimental observations, we studied the {100} Ag/MgO interface by HREM. As the {111} and {200} spacings of both Ag and MgO are larger than the point resolution of the present generation HRTEM instruments, the structure of the {100} interface between Ag and MgO in parallel orientation can be imaged in a common <100> as well as in a common <110> direction of Ag and MgO. Imaging in crystallographically different directions is important in order to determine the translation state of the two crystals at a given section of the interface.

Besides the translation state, our interest is also focused on the structure of mismatched regions in the Ag/MgO interface, which were not included in the ab-initio calculations. Considering the nondirectional nature of electrostatic bonding, it is interesting to question the extent that the contact of Ag and MgO introduces coherency strains in the elastically softer Ag crystal.

Large coherency strains would be observed as "physical" misfit dislocations in the terminology of Bollmann (Bollmann 1970), which means that atomic positions at the interface *relax* to maintain coherence of corresponding planes across the interface. In the closely related {100} interface between Au and MgO in parallel orientation (δ = 3%), Hoel et al. (1989) could not detect the strain fields of such dislocations; they were also undetected by HREM in noble metal/oxide interfaces with larger lattice mismatch, such as Cu/Al$_2$O$_3$ (δ = 10%) or Ag/CdO (δ = 14%) (Ernst et al. 1991; Necker and Mader 1988). The following section reports some HREM investigation of the {100} Ag/MgO interface (Trampert et al. 1992).

5.5.2 Experimental Details and Specimen Preparation for HREM

Ag layers were deposited onto {100} MgO substrates in an MBE growth chamber (Flynn 1988) with a typical growth rate of one monolayer per second (Flynn 1990). The MgO substrate surfaces were prepared by cleaving single crystals along {100} planes and subsequent annealing between 650°C and 1100°C, in ultra-high vacuum. During Ag deposition in a dynamic vacuum of $\sim 10^{-8}$ Pa, the substrates were heated to different temperatures ranging from 50°C to 200°C.

5.5.3 Specimen Preparation and Electron Microscopy

The preparation of cross-sectional TEM specimens turned out to be extremely difficult. The weak bonding between Ag and MgO requires very careful handling, and the significantly different sputtering rates during ion beam milling render standard preparation techniques unsuccessful. Therefore, a novel method for the preparation of cross-sectional TEM samples had to be developed (Strecker et al. 1992). With this method, the specimens are first embedded in a ceramic tube filled with epoxy.

Slices with a thickness of 500 μm are cut from this tube and mechanically thinned to a thickness of 150 μm. After dimpling, the slices are ion beam milled with 6 keV Ar ions using a suitable shutter to protect a direct thinning of the Ag film. A subsequent milling with 3 keV Ar ions reduced the amorphous layer of the foil.

5.5.4 Structural Studies by HREM

Defect structure of the interface. Figure 5.25 presents an overview of the Ag/MgO interfaces in specimens *a* and *b*, respectively. These micrographs were recorded at optimum underfocus $z = 1.2 \cdot \sqrt{C_S \lambda} \approx 50$ nm for the JEM 4000EX. (C_S and λ denote the spherical aberration constant and the wavelength of the electrons, respectively). At this focus, the lattice image exhibits {200} planes in Ag and MgO with singes parallel to the Ag/MgO interface. Figure 5.25 indicates that the Ag crystal is not oriented exactly parallel to the MgO crystal but is slightly tilted around the <100> direction parallel to the Ag/MgO interface. As expected, the lattice images of both specimens do not show any trace of a chemical reaction between Ag and MgO.

Along the Ag/MgO interface, regions of good (G) and poor (P) lattice match alternate. In regions of poor lattice match, the Ag lattice planes are bent in such a way that this bending restores the continuity of Ag and MgO {200} planes across the interface.

In order to exclude the bending of the {200} planes as an artefact of HREM imaging, we simulated the lattice image of the Ag/MgO interface for the *hypothetical* case of an *un*strained Ag lattice. This "rigid-lattice model" is shown in Figure 5.26(a) and the image simulation for the appropriate underfocus and foil thickness is shown in Figure 5.26(b).

The rigid-lattice image simulation correctly reproduces the regions of good lattice match. In the regions of poor lattice match, however, the simulation does not exhibit the smooth bending of {200} planes observed in the experimental image. Therefore, we conclude that the observed distortions of the Ag lattice represent real coherency strains associated with "physical" misfit dislocations. The dislocation cores are imaged end-on and characterized by an Ag {200} plane that has no counterpart in the MgO crystal. In contrast to the Ag lattice, the MgO lattice appears unstrained up to the interface.

The average spacing of the misfit dislocation cores in the experimental image amounts to 32 ± 5 {200} lattice planes, which corresponds to a distance $D = (6.53 \pm 1.02)$ nm. In order to image the lattice plane bending more clearly, the experimental image of specimen A was digitally Fourier filtered. The details of the procedure used are given elsewhere (Möbus et al. 1992). In the filtered image seen in Figure 5.27(a), the position of the misfit dislocation core can be determined unambigiously. Parallel to the Ag/MgO interface, the misfit dislocation has a long-range strain field; the bending of lattice planes is still observable up to ten {200} spacings away from the core position.

The Burgers circuit around a misfit dislocation core in Figure 5.27(b) yields a Burgers vector component of $b_\perp = 1/2 \, a_{Ag}$ <100> normal to the electron beam and parallel to the Ag/MgO interface. The spacing of the misfit dislocations agrees well with the geometrically expected spacing $D = b_\perp / \delta$.

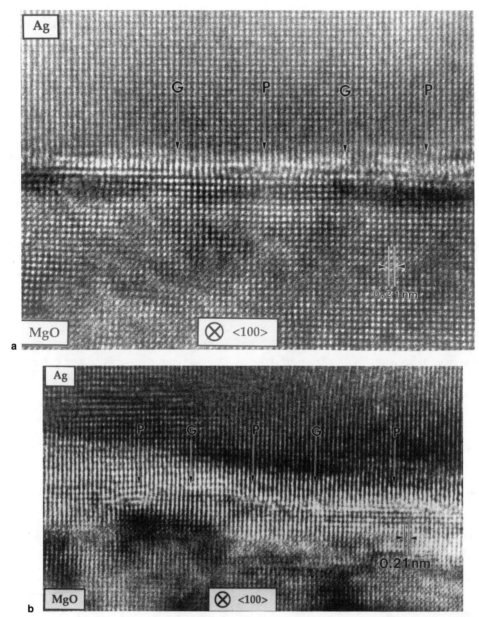

Figure 5.25. <100> HREM image of the {100}Ag/MgO interface in (a) specimen A and (b) specimen B. These micrographs were recorded in a JEM 4000EX (JEOL). This instrument operates at 400kV and has a point resolution of 0.175 nm. The objective aperture included four 200 and four 220 beams. {200} lattice planes are resolved in both crystals.

Translation state. In the commensurate regions between the misfit dislocations, where the coherency strains are minimal, two components of the relative translation between the Ag and the MgO crystal can be determined. In Figure 5.28, the bright spots on every side of the interface represent the atomic columns for the following reasons: assuming that the core structure of an edge dislocation (parallel to the interface in Figure 5.28) has mirror symmetry with respect to the additional half-plane, we conclude that the bright spots coincide with the Ag positions. Image simulations of the MgO crystal at a defocus of 50 nm and at different thicknesses also show that the atomic columns in MgO correspond to the bright spots in the image. The appearance of additional half-period fringes is helpful to determine the MgO foil thickness.

Structure and Chemistry of Metal/Ceramic Interfaces   105

1990). According to their model, the energetically most favorable translation state would be a "lock-in" configuration of the Ag and the MgO crystal, because the interfacial energy decreases with decreasing distance between the ions and their image charges. Our experimental observations show, however, that the most favorable translation state of Ag and MgO corresponds to a "ball-on-ball" rather than a lock-in configuration.

The translation component normal to the Ag/MgO interface can be sensitively determined from the course of {220} fringes crossing the interface at 45 degrees. The fringes show no kink in the interfacial region (Figure 5.29). This is only possible if the spacing between the terminating Ag {200} layer and the first MgO {200} layer is (20 ± 5)% larger than the MgO {200} spacing.

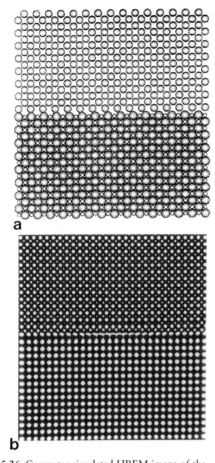

Figure 5.26. Computer-simulated HREM image of the unrelaxed Ag/MgO interface ("rigid lattice model").
(a) Atom position in the basis plane of the three-dimensional super cell with dimensions $x = yy = 1/2b$.
(b) Corresponding HREM image simulating for a foil thickness of $T \approx 8$ nm at optimum underfocus $z_{opt} = 50$ nm.

Thus we conclude that in *both* Figure 5.25 as well as in Figure 5.28, the bright spots of the lattice image of the crystals represent the atomic columns. In unstrained regions, the bright spots form straight lines without kinks across the interface. This means that the ions in the final Ag layer sit on top of either the Mg or the O ions of the first MgO layer. As in the <100> projection, the atomic columns contain both Mg and O ions, and we cannot decide whether the Ag ions sit on top of Mg or O ions. Also HREM imaging of a <110> projection of the interface cannot clarify this problem. We will return to this point later.

Stoneham and Tasker proposed that the adhesion between a noble metal and an oxide originates from Coulomb forces between the ions of the oxide and the image charges in the metal (Stoneham and Tasker

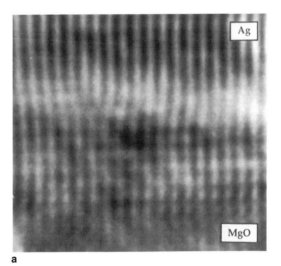

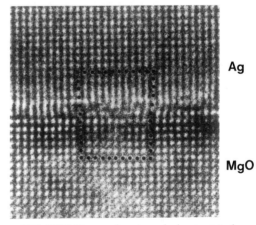

Figure 5.27. HREM image of a mismatched resion (p) of Figure 5.25. (a) After digital filtering (specimen A); (b) Burgers circuit marked (specimen B).

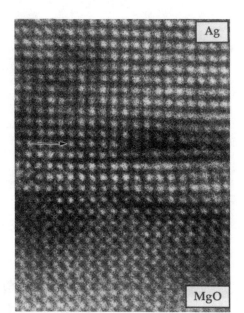

Figure 5.28. Commensurate region of specimen B identifying the atom positions at every side of the interface.

Thus, the excess volume predicted by the ab-initio calculations is experimentally confirmed within the limits of error.

Summary

High-resolution electron microscopy and grazing incident X-ray spectroscopy allow the determination of the atomistic structure of heterointerfaces with great accuracy. However, this technique is, as it was shown in the paper, restricted to certain well-defined grain boundaries, which possess a certain misorientation that allows the imaging in the HREM as well as in the GIXS. These results are essential for an understanding of the structure and chemistry of interfaces, and a continuation of this work will lead to a better understanding of the properties (such as fracture resistance) of these interfaces.

Acknowledgements

The author acknowledges helpful discussions with Dr. F. Ernst, Dr. W. Mader and Dr. J. Mayer. Mrs. G. Poech processed patiently and efficiently many versions of the manuscript. The work on high-resolution electron microscopy was supported by a grant from Volkswagen-Stiftung, under contract numbers I 62802 and I 63572. (Contract manager Dr. H. Steinhardt.)

References

Arbab, M., Chottimer, G.G., and R.W. Hoffmann. 1989. *MRS Symp. Proc.* 153:63–69.

Backhaus-Ricoult, M. 1987. *Ber. Bunsenges. Phys. Chem.* 90:684–691.

Backhaus-Ricoult, M. 1992. *Acta Metall. Mater.* 40:S95–S103.

Backhaus-Ricoult, M., and H. Schmalzried. 1985. *Ber. Bunsenges. Phys. Chem.* 89:1323–1330.

Balluffi, R.W., Brokman, A., and A.H. King. 1982. *Acta Metall.* 30:1453–1470.

Barbier, J., Hiraga, K., Otero-Diaz L.C., et al. 1985. *Ultramicroscopy.* 18:211–234.

Blöchl, P., Das, G.P., Fischmeister, H.F., and U. Schönberger. 1990. *In:* Metal/Ceramic Interfaces (M. Rühle, A.G. Evans, M.F. Ashby, J. Hirth, eds.), 9–14, Oxford: Pergamon Press.

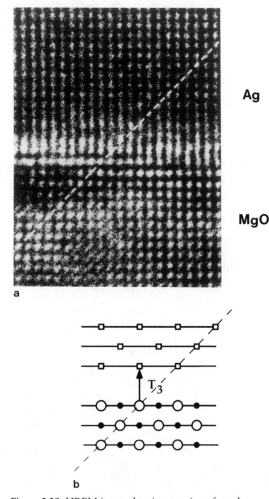

Figure 5.29. HREM image showing a region of good matching (G) of Figure 5.25. The course of {220} fringes across the interface depends sensitively on the translation component normal to the interface.

Bollmann, W. 1970. Crystal Defects and Crystalline Interfaces. Berlin: Springer-Verlag.

Bourret, A. 1985. *J. de Physique.* 46:C4-27–C4-38.

Burger, K., Mader, W., and M. Rühle. 1987. *Ultramicroscopy.* 22:1–14.

Burger, K., and M. Rühle. 1989. *Ceram. Eng. Sci. Proc.* 10:1549–1566.

Busek, P.R. (ed.) 1985. *Ultramicroscopy.* 18:1–4.

Busek P., Cowley J., and L. Eyring. 1988. High-Resolution Transmission Electron Microscopy. Oxford University Press, New York and Oxford.

Cockayne, D.J.H. 1972. *Z. Naturf.* 27a:452–461.

Dhingra, A.K., and S.G. Fishman. (eds.) 1986. Interfaces in Metal-Matrix Composites. The Metallurgical Society, Inc.

Doyema, M., Somiya, S., and R.P.H. Chang. (eds.) 1989. Metal-Ceramic Joints, Proc. MRS Intern. Meeting 8.

Du, R., and C.P. Flynn. 1990. *J. Phys.: Cond. Mat.* 2:1355.

Egerton R.F. 1986. Electron Energy-Loss Spectroscopy. New York: Plenum Press.

Eisenberger, P., and W.C. Marra. 1981. *Phys. Rev. Lett.* 46:1081–1084.

Ernst, F., Pirouz, P., and A.H. Heuer. 1991. *Phil. Mag.* A63:259–277.

Fecht, H.J., and H. Gleiter. 1985. *Acta Metall.* 33:557–562.

Fischmeister, H., Mader, W., Gibbesch, B., and G. Elssner. 1988. *MRS Symp. Proc.* 122:529–540.

Florjancic, M., Mader, W., Rühle, M., and M. Turwitt. 1985. *J. de Physique* 46:C4-133.

Flynn, C.P. 1988. *J. Phys. F: Met. Phys.* 18:L195.

Flynn, C.P. 1990. *In:* Metal/Ceramic Interfaces (M. Rühle, A.G. Evans, M.F. Ashby, J. Hirth, eds.), 168–177, Oxford: Pergamon Press.

Friedel, G. 1926. Leçon de Cristallographia. Paris: Blanchan.

Fuoss, P.H., Liang, K.S., and P. Eisenberger 1992. *In:* Synchrotron Radiation Research: Advances in Surface and Interface Science (R. Z. Bachrach, ed.), 385–419, New York: Plenum Press.

Gibson, J.M., and L.R. Dawson. (eds.) 1985. Layered Structures, Epitaxy and Interfaces. *MRS Symp. Proc.* 37.

Giess, E.A., Tu, K.N., and D.R. Uhlmann. (eds.) 1985. Electronic Packaging Materials Science. *MRS Symp. Proc.* 40.

Grimmer, H., Bullman, W., and D.H. Warrington. 1970. *Acta Cryst.* A30:197.

Hetherington, C.J.D., Nelson, E.C., Westmacott, K.H., et al. 1989. *Mat. Res. Soc. Symp. Proc.* 139:277–282.

Hoel, R.H., Penisson, J.M., and H.U. Habermeier. 1989. *J. Physique.* 51:C1-842.

Hren, J.J., Goldstein, J.I., and D.C. Joy. (eds.) 1979. Introduction to Analytical Electron Microscopy. New York: Plenum Press.

Huang, T.C., Cohen, P.J., and D.J. Eaglesham. (eds.) 1991. Advances in Surface and Thin Film Diffraction. *MRS Symp. Proc.* 208.

Ishida, Y. (ed.) 1987. Fundamentals of Diffusion Bonding. Amsterdam: Elsevier.

Jackson, K.A., Pohanka, R.C., Uhlmann, D.R., and D.R. Ulrich. (eds.) 1986. Electronic Packaging Materials Science. *MRS Symp. Proc.* 72.

Joy, D.C., Romig, A.D., Jr., and J.I. Goldstein (eds.) 1979. Principles of Analytical Electron Microscopy. New York: Plenum Press.

Kasper, E., and E.C.H. Parker (eds.) 1989. Silicon Molecular Beam Epitaxy. In: Thin Solid Film. 183:1–466.

Kinloch, A.J. 1981. *J. Mat. Sci.* 15:2141–2166.

Klomp, J.T. 1985. *MRS Symp. Proc.* 40:381.

Klomp, J.T. 1987a. *In:* Ceramic Microstructure 486: Role of Interfaces. (J.A. Pask and A.G. Evans, eds.), 307–317, New York: Plenum.

Klomp, J.T. 1987b. *In:* Fundamentals of Diffusion Bonding. (Y. Ishida, ed.), 3, Amsterdam: Elsevier.

Knauss, D., and W. Mader. 1991. *Ultramicroscopy.* 37:247–262.

Knowles, K.M., Alexander, K.B., Somekh, R.E., and W.M. Stobbs. 1987. *Inst. Phys. Conf. Ser. (EMAG 87).* 90:245–248.

Kronberg, M.L., and F.H. Wilson. 1949. *Trans. Amer. Inst. Min. Engrs.* 185:501–519.

Kuwabara, M., Spence, J.C.H., and M. Rühle. 1989. *J. Mat. Res.* 5:972–977.

Lee, C.H., Liang, K.S., Shieu, F.S., et al. 1991. *MRS Symp. Proc.* 209:679–683.

Lemkey, F.D., Fishman, S.G., Evans, A.G., and J.R. Strife. (eds.) 1988. High Temperature/High Performance Composites. *MRS Symp. Proc.* 120.

Liang, K.S. 1992. *In:* Metal-Ceramic Interfaces. Special Issue. (M. Rühle, W. Mader, A.H. Heuer, M.F. Ashby, eds.) *Acta Metall. Mater.* 40:5368.

Mader, W. 1989. *Z. Metallkunde.* 80:139–151.

Mader, W. 1987. *MRS Symp. Proc.* 82:403–408.

Mader, W., and M. Rühle. 1989. *Acta Metall.* 37:853–866.

Marra, W.C., Eisenberger, P., and A.Y. Cho. 1979. *J. Appl. Phys.* 50:6927–6933.

Mattox, D.M., Baglin, J.E.E., Gottschall, R.J., and C.D. Batich. (eds.) 1988. *MRS Symp. Proc.* 119.

Mayer, J., Mader, W., Phillipp, F.O. et al. 1989. *Inst. Phys. Conf. Ser. (EMAG 89).* 98:349–355.

Mayer, J., Mader, W., Knauss, D. et al. 1990a. *MRS Symp. Proc.* 183:55–58.

Mayer, J., Flynn, C.P., and M. Rühle. 1990b. *Ultramicroscopy.* 33:51–61.

Mayer, J., Gutekunst, G., Möbus, G. et al. 1992. *In:* Metal-Ceramic Interfaces. Special Issue. (M. Rühle, W. Mader, A.H. Heuer, M.F. Ashby, eds.) *Acta Metall. Mater.* 40:S217–S225.

Mehrabian, R. (ed.) 1983. Rapid Solidification Processing, Principles and Technologies III. Gaithersburg, MD: National Bureau of Standards.

Merkle, K.L. 1987. *MRS Symp. Proc.* 82:383–402.

Mitchell, T.E., and P.K. Davies. (eds.) 1988. *J. Electron. Micr. Tech.* 8:247–341.

Möbus, G., Necker, G., and M. Rühle. 1992. *Ultramicroscopy.* 43:46–65.

Mullins, W.W., and R.F. Sekerka. 1963. *J. Appl. Phys.* 34:323–329.

Necker, G., and W. Mader. 1988. *Phil. Mag. Lett.* 58:205–212.

Nicholas, M.G. (ed.) 1990. Joining of Ceramics. London: Chapman and Hall.

O'Keefe, M.A. 1985. Electron Image Simulation: A Complimentary Processing Technique, Electron Optical Systems, 209-220, O'Hare, Chicago: SEM, Inc., AFM Proc.

Pond, R.C. 1984. *J. Microscopy.* 135:213-240.

Rühle, M., Backhaus-Ricoult, M., Burger, K., and W. Mader. 1987. *In:* Ceramic Microstructure '86: Role of Interfaces. (J.A. Pask, A.G. Evans, eds.), 295-305, New York: Plenum Press.

Rühle, M., Balluffi, R.W., Fischmeister, H., and S.L. Sass. (eds.) 1985. *J. Phys. Colloq.* 46:C4.

Rühle, M., Burger, K., and W. Mader. 1986. *J. Microsc. Spectrosc. Electron.* 11:163-177.

Rühle, M., Evans, A.G., Ashby, M.F., Hirth, J. (eds.) 1989. Metal/Ceramic Interfaces. Acta-Scripta Metall. Proc. Series, Vol. 4, Pergamon Press, Oxford.

Rühle, M., Evans, A.G., Ashby, M.F., and J.P. Hirth (eds.) 1990. Metal/Ceramic Interfaces. Oxford: Pergamon Press.

Rühle, M., Heuer, A.H., and M.F. Ashby. 1992. Metal-Ceramic Interfaces. *Suppl. Acta Metall. Mater.* 40:S1-S368.

Schönberger, U., Anderson, O.K., and M. Methfessel. 1992. *Acta Metal Mater* 40:S1-S10.

Shimida, J. (ed.) 1990. *ISIJ International.* 30:1011-1150.

Spence, J.C.H. 1988. Experimental High-Resolution Electron Microscopy. 2nd Ed., New York: Oxford University Press.

Stadelmann, P.A. 1987a. *Ultramicroscopy.* 21:131-146.

Stoneham, A.M., and P.W. Tasker. 1990. *In:* Ceramic Microstructure '86: Role of Interfaces. (J.A. Pask, A.G. Evans, eds.), 155-165, New York: Plenum Press.

Strecker, A., Salzberger, U., and J. Mayer. 1993. *Praktische Metallographic.* In press.

Suresh, S., and A. Needleman. (eds.) 1989. Interfacial Phenomena in Composites: Processing, Characterization and Mechanical Properties. *Mat. Sci. Eng. A.* 107:3-280.

Sutton, A.P., and R.W. Balluffi. 1987. *Acta Mater.* 35:2177-2201.

Tidswell, I.M., Ocko, B.M., Pershan, P.S., et al. 1990. *Phys. Rev. B.* 41:1111-1127.

Trampert, A., Ernst, F., Flynn, C.P., Fischmeister, H.F., and M. Rühle. 1992. *Acta Mater* 40:5227-5236.

Trumble, K.P., and M. Rühle. 1991. *Acta Mater* 39:1915-1924.

Wasyncuk, J.A., and M. Rühle. 1987. *In:* Ceramic Microstructure '86: Role of Interfaces. (J.A. Pask, A.G. Evans, eds.), 341-348, New York: Plenum Press.

Wittmer, M. 1985. *MRS Symp. Proc.* 40:393-398.

Yoo, M.H., Clark, W.A.T., and C.L. Briant. (eds.) 1988. Interfacial Structure, Properties and Design. *MRS Symp. Proc.* 122.

Chapter 6
Microstructural Evolution in Whisker- and Particle-Containing Materials

NIELS HANSEN
CLAIRE Y. BARLOW

The microstructure and texture of a metal-matrix composite are determined by the manner in which it is processed: the microstructure and texture of the composite determine its mechanical properties. As a result, studies have concentrated on the relationship between the primary and secondary processing parameters, as well as on the parameters that characterize the reinforcement (volume-fraction, size, shape, and distribution) and the matrix (grain size, subgrain size, dislocation density and distribution, and crystallographic texture). Numerical and micromechanical models have been developed in parallel with microstructural analysis that relate structure and texture parameters to mechanical properties. This approach has, in recent years, led to the development of metal-matrix composites with significantly improved properties. Such composites are strengthened either by continuous fibers or whiskers, or by discontinuous fibers, whiskers, or particles. Apart from the case of metal/long metal fiber composites, the ductility of composites containing continuous fibers is not substantial enough to allow significant deformation during secondary processing. The composites that are considered in this chapter are therefore restricted to those containing discontinuous fibers, whiskers, and particles. The experimental work used to illustrate the main points has been performed on aluminum-based composites produced by powder routes.

6.1 Effect of Processing Route on Microstructure

The distribution of phases in metal-matrix composites is determined by the processing route followed, which will also have an effect on the microstructure of the metal-matrix. The significance of the distribution of reinforcement is that mechanical properties will be affected by an anisotropic distribution (Christman et al. 1989). Phase alignment leads to anisotropy of the elastic modulus, yield strength, and toughness. These characteristics can be used to advantage by designing components that make use of the improved properties along the most highly stressed directions. An inhomogeneous dispersion of the reinforcement, however, leads to undesirable local variations in these properties (Arsenault et al. 1991a). At worst, clusters of particles or whiskers may result in premature failure (Lloyd 1991).

The principal routes of processing utilize liquid-state mixing followed by casting, co-spray deposition, infiltration of a preform by liquid metal, extrusion of powder-blended material, and in-situ phase transformations (including eutectic alloys). The first two processes are commonly followed by a secondary process such as extrusion or forging. This increases the homogeneity of a dispersion and causes some alignment parallel to the streamlines. The matrix shows a deformed structure, with elongated grains and a well-developed subgrain structure.

Cast products normally have an isotropic distribution of reinforcement, thus, grain sizes tend to be small as the large number of reinforcement particles facilitates nucleation of matrix grains from the melt. Co-spray deposition leads to homogeneous material with a fine matrix grain size and an excellent dispersion of reinforcing whiskers or particles. The distribution of reinforcement in infiltrated products is predetermined, and the matrix microstructure can be controlled by adjusting the processing parameters. The phase distribution in composites formed by in-situ phase transformation is determined by standard metallurgical processing methodology, which can lead to discontinuous or continuous reinforcement.

Extrusion causes alignment of the phases, so that whiskers will be aligned with the extrusion direction. An additional characteristic of materials produced using a powder route arises from the fact that the metal powder is generally coated with oxide (Hansen 1969). This oxide is retained and dispersed in the final extruded product in the form of stringers of very fine particles aligned with the extrusion direction. The extrusion results in a deformed microstructure with large elongated grains and well-developed subgrains. The subgrain structure tends to be reasonably uniform throughout the material, although there is some reduction in subgrain size close to the reinforcing particles or whiskers (Juul Jensen et al. 1991). The overall microstructure is similar to that of composites manufactured by liquid-state mixing followed by thermomechanical processing.

6.2 Effects of Differential Thermal Contraction Coefficients

The microstructure of metal-matrix composites is influenced not only by the processing route but also by the inevitable cooling from the final stage in the processing. This is because a ceramic reinforcement has a coefficient of thermal contraction that is always lower that that of the metal matrix by a factor of at least five. The coefficients of thermal contraction of aluminum and silicon carbide, for example, are 32×10^{-6} and 4.7×10^{-6} K^{-1} respectively at 773K. The matrix is therefore impeded from contracting by the reinforcement. The consequences on the material are that stress fields build up, that may be partially relaxed by diffusion or dislocation nucleation. The magnitudes of these effects will depend on the size, shape, and volume fraction of the reinforcement, as well as on the processing temperature and the cooling rate (Lilholt and Juul Jensen 1987; Withers et al. 1989). The different manufacturing and secondary processing routes are associated with different cooling rates. Spray-formed composites, for example, are subjected to a very high cooling rate, while the cooling rate for large castings may be very slow.

Dislocations have been shown to accommodate the matrix strain that results from the mismatch in thermal contraction coefficients by nucleating on the whiskers or particles (Arsenault and Fisher 1983; Arsenault and Shi 1986). Single dislocations and loose tangles will normally be seen (Barlow and Hansen 1991), as illustrated in Figure 6.1. Dislocation nucleation is particularly observed around the ends of whiskers, where the local stresses are greatest (Povirk et al. 1990) and punching of prismatic loops has been observed (Vogelsang et al. 1986; Dunand and Mortensen 1991). The stress fields around particles are less intense than those around whisker ends, leading to a comparatively small amount of dislocation nucleation occurring on equiaxed particles. Where the particles are angular, as is generally the case in practical composites, dislocation nucleation tends to occur on asperities, where stress concentrations build up.

The distribution of dislocations after cooling depends on the distance the dislocations have moved. This punching distance has been the subject of some debate. There is broad agreement that the punching distances are of the order of the particle or whisker diameters and, for a matrix such as aluminum, are greater by a factor of two or three around the ends of whiskers (Christman and Suresh 1988; Taya and Mori 1987). The distances to which dislocations move are influenced not only by the stress that builds up at the reinforcement, but also on the deformation characteristics and flow stress of the matrix. In a cell-forming, high

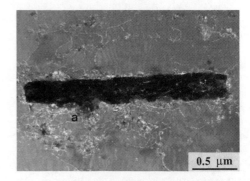

Figure 6.1. Dark-field micrograph showing dislocation nucleation around an SiC whisker in aluminum as a result of differential thermal contraction. Dislocations have been trapped close to the whisker by small alumina particles at a. *(Reprinted by permission of Risø National Laboratory, Denmark, from C.Y. Barlow 1991.)*

stacking-fault energy material such as aluminum, the dislocations emitted tend to form tangles, particularly in regions of high dislocation density such as around the whisker ends. These tangles reduce the mobility of following dislocations and thus, the punching distances are reduced. When a neat row of loops is punched from a whisker end, as seen in experiments on silver chloride (Dunand and Mortensen 1991), punching distances can be much enhanced. A high matrix flow stress naturally hinders the movement of the dislocations, and this may, in some cases, be locally higher in the vicinity of the particles or whiskers. For example, in materials where powder routes have been followed there is a higher density of oxide particles in these regions, which results in an enhanced flow stress. Consequently, the dislocations will be trapped close to the reinforcing phase, as seen in Figure 6.1. As the early stages of deformation are heavily influenced by the microstructure close to the reinforcing particles or whiskers (Barlow et al. 1991), this may produce a greater increase in the yield stress of the material than would be predicted simply on the basis of the overall increase in dislocation density.

The dislocation density resulting from the mismatch in thermal contraction coefficients has been calculated using various theoretic models (e.g. Arsenault and Shi 1986; Taya et al. 1991). To give an example, a simple way of calculating the dislocation density around a whisker is to assume that the strain is relieved by nucleating dislocations (Barlow and Hansen 1991). The strain along the length of the whiskers is then accommodated by nucleating prismatic dislocation loops at the ends of the whisker. The number of loops is calculated by dividing the mismatch in length between the whisker and the matrix by the Burgers vector of the matrix. For a whisker 3 mm in length, about one hundred dislocations could be nucleated from the whisker ends. Predicted densities have been compared with measured densities from analysis using HVEM (High Voltage Electron-Microscopy). The dislocation density predicted using the above model for aluminum containing 2% silicon carbide whiskers, 2.5 mm long and 0.5 mm in diameter, is 4.7×10^{12} m^{-2}. This is in reasonable agreement with measured densities in a related material (Arsenault et al. 1991b), assuming that the dislocation density increases in proportion to the volume fraction of reinforcement.

Not all of the stress resulting from the differential contractions is relaxed by dislocation nucleation. Some relaxation may also occur by diffusion, particularly around small reinforcing particles (Hirsch and Humphreys 1969). There will, however, be a residual elastic stress that will locally reach the stress for dislocation nucleation. The average stress in the matrix will be tensile, although where there is significant whisker alignment, the stresses will be anisotropic, and may in places be compressive (Withers et al. 1989). The tensile stresses show a significantly greater component in the direction of the long axes of the whiskers. This reduces the yield stress for tensile deformation, and thereby tends to counterbalance the effect of the increase in dislocation density.

6.3 Secondary Processing

Secondary processing is an integral part of the manufacturing process of metal-matrix composites. Such processing can improve the properties of the composites by consolidation of a porous matrix, by homogenization of the distribution of the reinforcement, and by optimization of some parameter of the matrix, such as grain size.

6.3.1 Cold Deformation Microstructures

Pure metals and largely single-phase alloys have certain deformation features that are not present in more complex materials, and vice versa. We will outline the salient points, and indicate the ways in which metal-matrix composites are distinctive.

In single-phase cell-forming metals and alloys, the deformation structure develops with increasing strain from tangled dislocations to well-defined cells and subgrains, the size of which decreases with increasing strain to a limiting value. From a very early stage in deformation, such materials show relatively large-scale inhomogeneities of flow, resulting from the need to accommodate neighboring volumes that are deforming on different combinations of slip systems. Microbands, transition bands, and segments of high-angle grain boundaries perform this function. The addition of a moderate volume fraction of small second-phase particles (such as 4% of alumina particles with diameters of the order of 10nm) homogenizes the slip, and may prevent the formation of these structural inhomogeneities (Barlow and Hansen 1989). The particles also increase the dislocation density, as geometrically necessary loops and tangles form around them, and the onset of subgrain formation is consequently accelerated.

The addition of even a small volume fraction of large particles, with diameters of the order of a micron or greater, also has the effect of reducing the incidence of formation of the macroscopic slip inhomogeneities, although plastic instabilities such as shear bands have been observed. As with the submicron particles, the large particles result in an increase in the dislocation density, although the mechanisms are not entirely comparable

(Humphreys 1980). With increasing strain, the formation of localized deformation zones at whiskers and particles becomes a dominant feature. These zones consist of small subgrains of approximately 0.1μm in diameter, with relatively large misorientations across the subgrain walls. The sizes and shapes of typical deformation zones around whiskers and particles are shown schematically in Figure 6.2(a) and (b).

6.3.2 Hot Deformation

The evolution of the deformation structure in cell-forming materials such as pure aluminum, aluminum alloys with a low solute content (Sheppard et al. 1986; McQueen and Jonas 1990), and alloys containing low concentrations of small particles (Hansen 1969) at elevated temperature (above about $0.5 T_M$), follows a similar pattern to the evolution of deformation structure at room temperature. The subgrains that form at elevated temperatures are more equiaxed and perfect than those that form at room temperature because of the increase in dislocation mobility and the decrease in friction stress. The frequency of microstructural inhomogeneities, such as microbands, transition bands, and shear bands is lower than that for cold deformation, so that the deformation microstructure becomes more homogeneous with increasing deformation temperature.

The macroscopic effect of reinforcing particles and whiskers is to increase the dislocation density, both during cooling (as discussed in section 6.2) and during hot deformation. Extruded structures show dislocations arranged in subgrain structures, although with a relatively high density of dislocations within the subgrain interiors. The subgrain size increases with increasing extrusion temperature and decreases with increasing volume fraction of reinforcement (Juul Jensen et al. 1991).

The effect of hard reinforcing particles on hot deformation structures is, as in the cold-deformation structures, seen as the generation of deformation zones around the particles (Brown 1982; Humphreys and Kalu 1987). As the temperature increases, so does the incidence of cross-slip and climb. The dislocation density at the particles decreases in consequence, and results in a reduction in the size of the deformation zone and in the associated lattice rotation. At 500°C and strain rates up to 0.1 s^{-1}, the decrease in dislocation accumulation at large particles is such that deformation zones no longer appear (Humphreys and Kalu 1987). At higher strain rates (such as those encountered during rolling and extrusion) deformation zones are observed even at these high temperatures (Juul Jensen et al. 1991).

6.3.3 Recrystallization

Recrystallization nuclei may form in material that has been cold or hot deformed. Nucleation sites tend to be regions of high dislocation density, such as original grain boundaries, intragranular high-angle boundaries formed during deformation, and shear bands (Doherty 1980). In addition, large particles may act as sites for nucleation, an effect that is enhanced if the particles are clustered (as a result of particle pushing during solidification, or by agglomeration of the reinforcement during powder forming). The relative importance of different sites may change as the degree of deformation and deformation temperature are altered. For example, it has been found in pure metals that the original grain boundaries are favored as nucleation sites when the degree of deformation is low or when the deformation temperature is high (Bay and Hansen 1984; Humphreys 1990). The same may be true under these conditions for

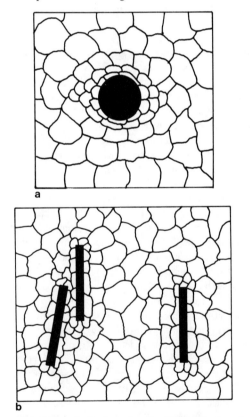

Figure 6.2. Schematic diagrams showing the distribution of subgrains following cold deformation around (a) an SiC particle, and (b) SiC whiskers, singly and grouped. Small subgrains form particularly at whisker ends (a) and between grouped whiskers (b). *(Reprinted by permission from Liu et al. 1991.)*

metal-matrix composites containing whiskers and particles. At larger strains, however, the reinforcement may form the dominant nucleation sites, resulting in the frequently reported particle-stimulated nucleation (Humphreys et al. 1990; Liu et al. 1989). At very large strains, there may be competition between nucleation sites, with perhaps high-angle grain boundaries and shear bands in combination with the reinforcement (Hansen and Juul Jensen 1986).

In composites where the nucleation sites are limited, normal recrystallization will take place by nucleation and grain growth, and the recrystallized grain size will consequently tend to be large. This will occur when the volume fraction of reinforcement is low, or when the size of the reinforcement particles is so small that nucleation at particles is rare. This type of behavior can also be observed if particle-stimulated nucleation is hindered by the presence of small, second-phase particles, such as alumina particles in aluminum composites manufactured by a powder route (Liu et al. 1989, 1992b). The tendency toward particle-stimulated nucleation in-

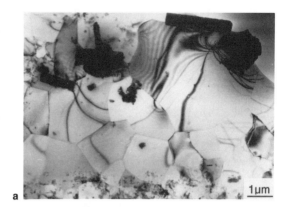

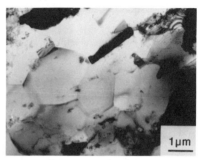

Figure 6.4. 0.1% alumina particles, diameter 80nm. Material cold-rolled 90% and annealed at 300°C for 1 hour. (a) Continuous subgrain growth occurring both in the matrix and by the whiskers. Subgrains impinge on each other before developing into nuclei. (b) Material cold rolled 90% and annealed at 400°C for 1 hour. Fully recrystallized by the mechanism of continuous subgrain growth, resulting in a grain size of the order of microns. *(Courtesy Liu et al. 1992b.)*

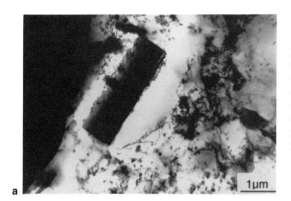

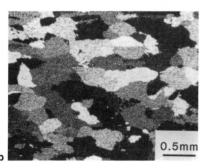

Figure 6.3. Aluminum containing 2% SiC whiskers and 0.8% alumina particles, diameter 30nm. (a) Material cold rolled 90% and annealed at 300°C for 1 hour. A nucleus is seen by a whisker, elsewhere subgrain growth is very limited. (b) Material cold rolled 90% and annealed at 400°C for 1 hour. Fully recrystallized by the mechanism of high-angle grain boundary migration, with a resulting grain size of hundreds of microns. *(Courtesy Liu et al. 1992b.)*

creases with reinforcement particle size, and the number of recrystallization nuclei increases as the volume fraction of reinforcement goes up, and as the volume fraction of small second-phase particles goes down (Liu et al. 1989). As a result of abundant nucleation in combination with enhanced subgrain growth at the reinforcement, normal recrystallization can be replaced by a process of continuous subgrain growth or extended recovery. The resulting grain size will be very small (microns), and the misorientation across the grain boundaries will be smaller than that which is characteristic of ordinary grain boundaries (Liu et al. 1991, 1992a). Typical microstructures resulting from these two mechanisms are illustrated in Figures 6.3 and 6.4.

The number of recrystallization nuclei is affected not only by the size of the reinforcements but also by their shape. The deformation zones formed at particles do not surround the whole particle and have smaller subgrain wall misorientations than those at whiskers. A higher nucleation rate, lower recrystallization temperature, and smaller grain size are therefore found in whisker-containing materials (Jiang and Liu 1992).

6.4 Texture

6.4.1 Cold-Rolling Texture

The cold-rolling texture of high stacking-fault materials, including pure aluminum, is classified as a copper-type rolling texture. This texture can be described as an orientation tube running through the Euler space close to $\{110\}<112>$, $\{123\}<634>$ and $\{112\}<111>$ orientations (the so-called Brass, S, and Copper orientations) (Hirsch and Lucke 1988).

The addition of small second-phase particles may strengthen the texture (Juul Jensen et al. 1988; Hansen and Juul Jensen 1990). Conversely, large particles and whiskers cause some weakening in the texture both in materials with and without small particles (Liu et al. 1991; Bowen et al. 1991). This weakening can be ascribed to the generation of deformation zones around the large particles and whiskers. The texture in these zones is not random but can be described as a spread deformation texture. The formation of the deformation zones thus causes an overall weakening of the cold-rolling texture. This weakening increases as the volume fraction of reinforcement rises, but is not very sensitive to the size of the reinforcement (Bowen et al. 1991). In addition to the normal texture components, a $\{100\}<uvw>$ component has been found in cold-rolled aluminum containing whiskers. Whisker ends tend to be associated with regions of $\{100\}<uvw>$ orientation, and the incidence is still stronger for regions between pairs or groups of aligned whiskers (Liu et al. 1991).

6.4.2 Recrystallization Texture

The recrystallization texture can be described as a mixture of cube, retained-rolling and random texture components. These components are related to the orientation of the nucleation sites for recrystallization, which may be original grain boundaries or intragranular sites. Additional sites are present in composites containing whiskers and particles. As such sites are mostly associated with spread rolling or random orientation, this particle-stimulated nucleation will lead to weakening of the recrystallization texture (Juul Jensen et al. 1988; Bowen et al. 1991). An additional component is a rotated cube orientation close to $\{100\}<013>$ (Juul Jensen et al. 1989; Bowen et al. 1991). It has been suggested (Juul Jensen et al. 1989) that this component may be formed by a rotation in the deformation zone of the $\{100\}<uvw>$ component observed in the cold-rolled state, and that the growth rate is fastest for this component. In recrystallized composites containing whiskers, the typical texture components are therefore spread rolling, rotated cube, and random (Liu et al. 1991). Similar texture components have been found in particle-containing composites (Bowen et al. 1991).

In a whisker-containing material, the concentrations of the texture components have been related to the grain-size distribution after recrystallization (Juul Jensen et al. 1989). It has been found that most of the large grains have a $\{100\}<013>$ orientation. The apparently high growth rate of grains of this orientation leads to the significant concentration of this component in the recrystallization texture. It has also been found that the strength of this component is strongly affected by annealing conditions, such as temperature and heating rate (Juul Jensen and Hansen 1991).

6.5 Stress-Strain Behavior

Figures 6.5 and 6.6 show typical tensile stress-strain curves for a series of aluminum-based composites, together with the matrix material and matrix with fine alumina. The fine alumina increases the yield stress and the initial work-hardening rate. Silicon carbide whiskers and particles produce an increase in yield stress only when present at the higher volume-fractions, and cause a steady increase in flow stress with increasing volume-fraction. They increase the initial work-hardening rate significantly, but in contrast to the fine alumina, they also provide some increase in work-hardening rate over the whole strain range. Particles are seen to be less effective at increasing flow stresses than the same volume fraction of whiskers. Clearly, composites that have undergone some sort of deformation as the final stage of their fabrication will contain a well-developed subgrain structure and will, consequently, show a higher flow stress than those that have not. The extrusion microstructure provides a noticeable improvement in mechanical properties only for perhaps the first 3% to 5% of subsequent deformation, and thereafter the stress-strain curves merge.

On loading recrystallized material, the deformation initiates at the particles or whiskers, where there is a rapid build-up in internal stress when the plastic deformation begins. In whisker-containing material, this build-up is particularly great around the ends of the whiskers (Christman et al. 1989). Figures 6.7, 6.8, and 6.9 demonstrate the pattern of dislocation nucleation

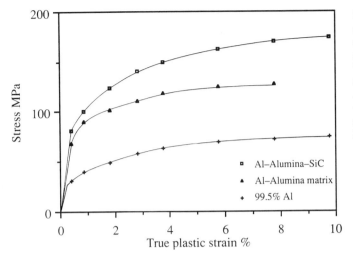

Figure 6.5. Stress-strain curves for recrystallized materials: aluminum-0.8% alumina-2% silicon carbide whiskers; aluminum-0.8% alumina; 99.5% aluminum. The alumina increases the yield stress and flow stress by a set increment of about 50MPa, but the silicon carbide increases the work-hardening rate over the whole strain range. *(Reprinted by permission of Risø National Laboratory from Barlow 1991.)*

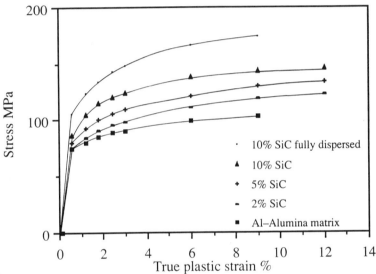

Figure 6.6. Stress-strain curves for extruded aluminum composites containing different volume fractions of silicon carbide whiskers. The two curves for 10% addition of whiskers indicate the benefit of gaining a homogeneous dispersion of whiskers without any clustering. *(Reprinted by permission of Risø National Laboratory from Barlow 1991.)*

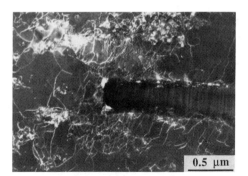

Figure 6.7. Dark-field micrograph showing the formation of dislocation nets (and the beginning of subgrain formation) in regions of high dislocation density at the end of a whisker in material strained 3%. *(Reprinted by permission of Risø National Laboratory from Barlow 1991.)*

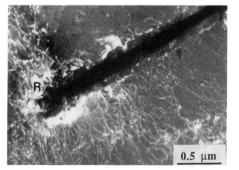

Figure 6.8. Dark-field micrograph showing extensive dislocation nucleation along the length of a whisker in material strained 5%. A rotated region (labeled R) has formed at the whisker end. *(Reprinted by permission of Risø National Laboratory for Barlow 1991.)*

Figure 6.9. Dark-field micrograph showing extensive subgrain formation in material strained 10%. A rotated region (R) is indicated. The whisker has cracked (C), a common occurrence at this strain. *(Reprinted by permission of Risø National Laboratory.)*

around whiskers. In the initial 1% of strain, dislocations are seen only at the ends of the whiskers. As strain increases, nucleation spreads along the sides of the whiskers until by 8% to 10% strain they are seen along the whole length of all whiskers. The deformation pattern at this strain is homogeneous, with little variation close to the whiskers. Subgrain walls develop progressively with strain, forming initially close to whisker ends and in the vicinity of clusters of whiskers. Rotated zones, designated R, form at the regions of highest strain at the whisker ends. Comparison with a similar alloy without whiskers shows the essential difference attributable to the whiskers to be that the dislocation density is significantly higher in the whisker-containing material at all stages.

On loading extruded material, the deformation patterns are much less clearly seen. The nucleation of dislocations around the ends of whiskers causes a local decrease in the subgrain size. In addition, the interiors of subgrains, particularly those in the vicinity of whiskers, contain significantly increased numbers of dislocations that are only loosely tangled and have not formed sharp subgrain walls.

6.5.1 Flow Stress and Microstructure

The flow stress is affected principally by the parameters related to the deformed matrix microstructure and to the reinforcement. The contribution from any changes in texture is small, and may be quantified through changes in the Taylor M-factor. The structure contains a number of barriers to dislocation motion, and their contribution to the flow stress may be estimated on the basis of structure-related strengthening parameters and micromechanical models relating these parameters to the flow stress. The different contributions to the flow stress are additive in some sense. As an approximation, the summation may be done for example by linear addition (Lilholt 1983; Humphreys et al. 1991; Barlow 1991; Barlow and Hansen 1991). The barriers are as follows:

1. The dislocation density generated during cooling and during plastic deformation. At relatively small strains, the dislocations arrange into a cell or subgrain structure. The dislocation density may be used as a strengthening parameter, but only in the early stages may the density be measured with any confidence (by analysis of HVEM micrographs). However, the cell or subgrain size is readily measured from analysis of conventional TEM micrographs, and therefore provides a useful measure of the strengthening parameter. The flow stress may be estimated from the dislocation density, from the cell size, or from the subgrain size in the appropriate deformation regime (Barlow and Hansen 1991). In hot deformed composites, the subgrain size may usefully be taken as the strengthening parameter.
2. The original grain boundaries are formed by recrystallization, and the grain size is a strengthening parameter. The contribution from grain size to the flow stress is normally calculated according to the Petch-Hall relationship, $\tau = \tau_0 + kd^{-1/2}$. The constant k in this equation is expected to depend on the characteristics of the grain boundaries, so that boundaries containing whiskers and particles may have different values from those formed in pure metals (Hansen 1985).
3. In composites containing second-phase particles, the relevant strengthening parameter is the particle spacing. The particle contribution to the flow stress may be calculated according to the Orowan equation (Barlow and Hansen 1991).

In addition to the flow stress contribution arising directly from the microstructure and texture, internal stresses also contribute to the flow stress of metal-matrix composites. These stresses are generated during both cooling and elastic and plastic deformation, and their magnitude can be calculated (Pedersen and Brown 1983; Chapter 10, this volume). However, an unknown proportion of the internal stresses may be relaxed by a number of processes such as diffusion, generation of dislocations, dislocation glide, interfacial decohesion and cracking of the reinforcement (Ashby 1970; Withers et al. 1989). The residual stress cannot, at the time of this writing, be determined microscopically, although the technique of electron back-scattering looks promising as an analytical technique for this purpose (Harris and Morgan 1991). The macroscopic residual stress can, however, be measured with good accuracy using neutron diffraction (Lorentzen and Sørensen 1991).

Summary

The results presented in this chapter indicate areas in which further research is needed, such as those regarding the mechanical and chemical interaction between the reinforcement and the matrix, the mechanisms of stress relaxation, the interaction between reinforcement particles themselves, particularly when present in high concentrations, and the combined effects of small and large particles. New understanding should lead to metal-matrix composites with improved properties, and to better theoretic models for the mechanical behavior of composites. The results presented also show that the qualitative relationships are fairly well understood between processing parameters, microstructure and texture, and the mechanical properties of metal-matrix composites. Such relationships demonstrate that further process optimization is possible to produce composites with improved properties, particularly strength, ductility, and fracture toughness.

References

Arsenault, R.J., and R.M. Fisher. 1983. *Scripta Metall.* 17:67–71.
Arsenault, R.J., and N. Shi. 1986. *Mater. Sci. Eng.* 81:175–187.
Arsenault, R.J., Shi, N., Feng, C.R., and L. Wang. 1991a. *Mater. Sci. Eng.* A131:55–68.
Arsenault, R.J., Wang, L., and C.R. Feng. 1991b. *Acta Metall. Mater.* 39:47–57.
Ashby, M.F. 1970. *Phil. Mag.* 21:399–424.
Barlow, C.Y. 1991. *In:* Metal Matrix Composites. Proceedings of the 12th Risø International Symposium (N. Hansen, et al., eds.), 1–15, Roskilde, Denmark: Risø National Laboratory.
Barlow, C.Y., and N. Hansen. 1989. *Acta Metall.* 37:1313–1320.
Barlow, C.Y., and N. Hansen. 1991. *Acta Metall. Mater.* 39:1791–1979.
Barlow, C.Y., Liu, Y.L., and N. Hansen. 1991. Interfacial Phenomena in Composite Materials, Leuven, Belgium (I. Verpoest, F. Jones, eds.), 171–174, Stoneham, MA: Butterworth-Heinemann.
Bay, B., and N. Hansen. 1984. *Metall. Trans.* 15A:287–297.
Bowen A.W., Ardakani M., and F.J. Humphreys. 1991. *In:* Metal Matrix Composites. Proceedings of the 12th Risø International Symposium (N. Hansen, et al., eds.), 241–246, Roskilde, Denmark: Risø National Laboratory.
Brown, L.M. 1982. Fatigue and creep of composite materials. Proceedings of the 3rd Risø International Symposium (H. Lilholt, R. Talreja, eds.), 1–18, Roskilde, Denmark: Risø National Laboratory.
Christman, T., Needleman, A., and S. Suresh. 1989. *Acta Metall.* 37:3029–3050.
Christman, T., and S. Suresh. 1988. *Acta Metall.* 36:1691–1704.
Doherty, R.D. 1980. Recrystallisation and Grain Growth. Proceedings of the 1st Risø International Symposium (N. Hansen, A.R. Jones, T. Leffers, eds), 57–70, Roskilde, Denmark: Risø National Laboratory.
Dunand, D.C., and A. Mortensen. 1991. *Acta Metall. Mater.* 39:127–139.
Hansen, N. 1969. *Trans. Metall. Soc. AIME.* 245:2061–2068.
Hansen, N. 1985. *Metall. Trans.* 16A:2167–2190.
Hansen, N., and D. Juul Jensen. 1986. *Metall. Trans.* A17:253; 1186.
Hansen, N., and D. Juul Jensen. 1990. Recrystallisation '90. (T. Chandra ed.), 79–88, Warrendale, PA: The Minerals, Metals and Materials Society.
Harris, S.J., and P.C. Morgan. 1991. Interfacial Phenomena in Composite Materials. (I. Verpoest, F. Jones, eds.), 223–226, Stoneham, MA: Butterworth-Heinemann.
Hirsch, P.B.. and F.J. Humphreys. 1969. Physics of strength and plasticity, ed. A. Argon, MIT Press, Cambridge, Mass., p189.
Hirsch, P.B., and K. Lucke. 1988. *Acta Metall.* 36:2863–2882.
Humphreys, F.J. 1980. Recrystallisation. Proceedings of the 1st Risø International Symposium (N. Hansen, A.R. Jones, T. Leffers, eds.), 35–44, Roskilde, Denmark: Risø National Laboratory.
Humphreys, F.J. 1990. *In:* Recrystallisation 90. (T. Chandra, ed.), 113–122, Warrendale, PA: The Metallurgical Society.
Humphreys, F.J., Basu, A., and M.R. Djazeb. 1991. *In:* Metal Matrix Composites. Proceedings of the 12th Risø International Symposium (N. Hansen, et al., eds.), 51–66.
Humphreys, F.J., and P.N. Kalu. 1987. *Acta Metall.* 35:2815–2829.
Humphreys, F.J., Miller, W.S., and M.R. Djazeb. 1990. *Mater. Sci. Tech.* 6:1157–1166.
Jiang, Z.J., and Y.L. Liu. 1992. Proceedings of the 3rd International Conference on Al alloys, *Effect of SiC parameters on recrystallization behavior of Al-SiC composites* (L. Arnberg, D. Lohne, E. Nes, N. Ryum, eds.), 507–512. University of Trondheim, Norwegian Institute of Technology, Norway.
Juul Jensen, D., and N. Hansen. 1991. Proceedings of the 9th International Conference on the Texture of Metals. 14–18, 853–858.
Juul Jensen, D., Hansen, N., and F.J. Humphreys. 1988. Proceedings of the 8th International Conference on Textures of Metals. (J.S. Kallend, G. Gottstein, eds.), 431–444, Warrendale, PA: The Metallurgical Society.
Juul Jensen, D., Hansen, N., and Y.L. Liu. 1989. Materials Architecture. Proceedings of the 10th Risø International Symposium (J.B. Bilde-Sørensen et al., eds.), 409–414, Roskilde, Denmark: Risø National Laboratory.
Juul Jensen, D., Liu, Y.L. and N. Hansen. 1991. Metal-Matrix Composites. Proceedings of the 12th Risø International Symposium (N. Hansen et al., eds.), 417–422, Roskilde, Denmark: Risø National Laboratory.
Lilholt, H. 1983. Multi-Phase Materials. Proceedings of the 4th Risø International Symposium (J.B. Bilde-Sørensen et al., eds.), 381–392, Roskilde, Denmark: Risø National Laboratory.
Lilholt, H., and D. Juul Jensen. 1987. Proceedings of the Second International Conference on Testing, Evaluation

and Quality Control of Composites, TEQC-87. (J. Herriot, ed.), 156–160, Seven Oaks, U.K.: Butterworths.

Liu, Y.L., Hansen, N., and D. Juul Jensen. 1989. *Metall. Trans.* A20:1743–1753.

Liu, Y.L., Hansen, N., and D. Juul Jensen. 1991. Metal-Matrix Composites. Proceedings of the 12th Risø International Symposium (N. Hansen et al., eds), 67–80, Roskilde, Denmark: Risø National Laboratory.

Liu, Y.L., Hansen, N., and D. Juul Jensen. 1992a. *Metall. Trans.* A23:807–819.

Liu, Y.L., Juul Jensen, D., and N. Hansen. 1992b. Recrystallization '92: International Conference on Recrystallization and Related Phenomena. (M. Fuentes and J.G. Sevillano, eds.), 539–544, Brookfield, VT: Trans Tech.

Lloyd, D.J. 1991. *Acta Metall. Mater.* 39:59–71.

Lorentzen, T., and N. Sørensen. 1991. Metal-Matrix Composites. Proceedings of the 12th Risø International Symposium (N. Hansen et al., eds.), 489–496, Roskilde, Denmark: Risø National Laboratory.

McQueen, H.J., and J.J. Jonas. 1990. Proceedings of the 2nd International Conference on Aluminum Alloys. Beijing.

Pedersen, O.B., and L.M. Brown. 1983. Deformation of Multi-Phase Materials. Proceedings of the 4th Risø International Symposium. (J.B. Bilde-Sørensen et al., eds.), 83–101, Roskilde, Denmark: Risø National Laboratory.

Povirk, G.L., Needleman, A., and N.R. Nutt. 1990. *Mater. Sci. Eng.* A125:129–140.

Sheppard, T., Zaidi, M.A., Hollinshead, P.A., and N. Raghunathan. 1986. Microstructural Control in Aluminium Alloys. (E.H. Chia and H.J. McQueen, eds.), 19–43, Warrendale, PA: The Metallurgical Society.

Taya, M., Lulay, K.E., and D.J. Lloyd. 1991. *Acta Metall. Mater.* 39:73–87.

Taya, M., and T. Mori. 1987. *Acta Metall.* 35:155–162.

Vogelsang, M., Arsenault, R.J., and R.M. Fisher. 1986. *Metall. Trans.* 17A:379–389.

Withers, P.J., Stobbs, W.M., and O.B. Pedersen. 1989. *Acta Metall.* 37:3061–3084.

Chapter 7
Aging Characteristics of Reinforced Metals

SUBRA SURESH
KRISHAN K. CHAWLA

Precipitation-hardenable or age-hardenable alloys, such as aluminum alloys, are attractive matrix materials in metal-matrix composites. These alloys offer the flexibility to drastically alter the microstructural constitution and mechanical properties of the matrix material in response to systematic and controlled aging or precipitation treatments. They also make feasible the ability to tailor the overall composite properties by a judicious selection of the size, shape, and distribution of the reinforcement phase. Age-hardenable ductile alloys can be synthesized by conventional casting or by powder metallurgy routes, which enhance their appeal for use as matrices in metal-matrix composites.

The addition of a brittle reinforcement to a precipitation-hardenable alloy can significantly alter the nucleation and growth kinetics of precipitation in the matrix, as compared to those in an unreinforced alloy. Because the overall composite properties can be influenced by such changes in the precipitation characteristics of the matrix in a composite as a result of the introduction of the reinforcement phase, the study of aging characteristics in reinforced metals is of considerable scientific and practical interest.

The purpose of this chapter is to provide an overview of the mechanisms and kinetics of precipitation in reinforced metals and alloys. Available results for a variety of metallic systems are reviewed for both continuous and discontinuous reinforcements, and the implications of such precipitation to the overall properties of the composites are highlighted, wherever possible.

7.1 Background

7.1.1 Basics of Aging Characteristics in Unreinforced Metals

Precipitation hardening (or age hardening) is a versatile method to strengthen certain metallic alloys. It is an especially popular technique for aluminum alloys. The aging treatment involves precipitation of a series of metastable and stable precipitates out of a homogeneous, supersaturated solid solution. Various metastable structures offer different levels of resistance to dislocation motion. Figure 7.1 shows a schematic of a typical aging curve, that is, the variation of hardness (or strength) with aging time, at a fixed aging temperature, for an Al–Cu binary alloy. Also shown in this figure are the different precipitate types that occur during the aging treatment. Peak hardness or strength corresponds to a critical dispersion of coherent or semi-coherent precipitates.*

The shape of the aging curve in Figure 7.1 can be rationalized as follows. Primarily solid solution hardening is active immediately following quenching from the solutionizing temperature. As Guinier-Preston (GP) zones form during the initial stages of aging, the

---

*Full details of the micromechanisms and kinetics of precipitation hardening in unreinforced alloys can be found in any standard textbook on mechanical metallurgy or physical metallurgy. For example, see Haasen 1978; Meyers and Chawla 1984; Courtney 1990.

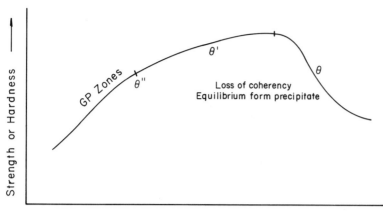

Figure 7.1. Schematic showing the variation of strength (or hardness) with aging time (at a fixed aging temperature) or precipitate size in an Al–Cu alloy. See text for details.

strength or hardness of the alloy increases because additional stresses are necessary for dislocations to cut through (shear) the coherent zones during plastic deformation. The hardness increases as the size of the GP zones increases with time, making it even more difficult for the dislocations to shear them. The peak hardness or strength is associated with a critical dispersion of coherent or semicoherent (θ') precipitates. With still further increase in aging time, stable, equilibrium (θ) precipitates with incoherent interfaces with the matrix begin to form and the dislocations become increasingly more capable of bypassing or looping around the precipitate particle as they move through the alloy during plastic deformation. This circumventing of particles by dislocations results in the formation of so-called Orowan dislocation loops around particles as the particles are left behind by the advancing dislocations. The stress required for the bowing of dislocations between particles (as a consequence of the Orowan mechanism) varies inversely with interparticle spacing. Continued aging results in an increase in precipitate size and interparticle spacing. Consequently, a progressively lower strength results as the bowing of dislocations around particles becomes easier with increasing aging time.

The strengthening precipitates can nucleate homogeneously in the alloy or heterogeneously along dislocations, grain boundaries, interfaces or other inhomogeneities; the latter process requires less time for particle nucleation. The growth of the precipitates is controlled by diffusion of the solute in the alloy, which is dictated by the precipitation temperature and aging time. Many Cu-containing aluminum alloys develop strengthening precipitates even at room temperature. This process is commonly referred to as "natural aging." When precipitation is effected in the alloy at (controlled) elevated temperatures, the resulting changes in the microstructure are referred to as "artificial aging."

7.1.2 Accelerated Aging in Reinforced Metals

In a metallic matrix containing a brittle fiber, whisker or particle, the difference in the thermal expansion coefficient, $\Delta \alpha$, between the matrix and the reinforcement can be so large (as much as a factor of 4.7 between aluminum and SiC), that even a small change in temperature will generate thermal residual stresses in the matrix. (See Chapters 4 and 10 for quantitative discussions of this effect.) The matrix can undergo plastic yielding as the magnitude of the residual stresses locally exceeds the yield strength. The consequent development of dislocations in the matrix gives rise to a greater dislocation density in the matrix of the composite than in the unreinforced alloy. These dislocations can serve as heterogeneous sites for the nucleation of strengthening precipitates and can provide short circuit diffusion paths for solute atoms. As a result, both the nucleation and growth of precipitates in the matrix can be drastically altered by the presence of the reinforcements. Experimental results, to be discussed in the following sections of this chapter, clearly show that, in a wide variety of precipitation-hardenable aluminum alloys, the alloy with the brittle reinforcements exhibits a significantly shorter aging time to achieve peak strength than does the unreinforced matrix alloy at the same aging temperature. This faster development of strengthening precipitates in the composite is commonly referred to as "accelerated aging."* Figure 7.2 shows an

---

*There have been some preliminary reports of the possibility of a lower rate of precipitation in the composite than in the control alloy (Appendino et al. 1991). This behavior apparently occurs in response to natural aging in some rare circumstances in which complete solutionizing of the material is not accomplished prior to aging. A similar trend has also been reported by Kumai et al. (1992) for a cast A356 aluminum alloy with SiC particles. In addition, Lee et al. (1991) report that when oxide inclusions are also added to a

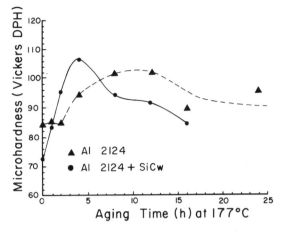

Figure 7.2. Variation of matrix microhardness as a function of aging time at 177°C for powder metallurgy processed 2124 aluminum alloy with and without 13.2 vol % SiC whiskers. *(From Christman and Suresh 1988a.)*

example of this phenomenon in a 2124 aluminum alloy (powder metallurgy processed) with 13.2 vol % SiC whiskers (aspect ratio = 5), where the variation of Vickers (diamond pyramid indentor under 5 g load) microhardness in the matrix is plotted as a function of aging time at a fixed aging temperature of 177°C (after Christman and Suresh 1988a). Also shown in Figure 7.2 is the aging curve for the unreinforced 2124 aluminum alloy with the same processing history as that of the composite. Although the peak aging time for the unreinforced alloy is 10 to 12 hours, the peak hardness is reached in the composite after only 3–4 hours of aging because of the precipitation of metastable S' Al_2CuMg precipitate, which in the advanced stages of aging, is transformed to the stable S precipitate. In this particular example, note that the peak hardness for the composite is essentially the same as that for the unreinforced alloy despite a three-fold reduction in peak aging time, which resulted from accelerated aging. Mechanisms responsible for this trend are discussed in later sections of this chapter.

7.1.3 Fundamental and Practical Significance

Because the precipitation characteristics of an alloy can be significantly altered by reinforcements, optimizing the mechanical response of metal-matrix composites requires a thorough understanding of the aging characteristics. A large number of investigations conducted to date on reinforced metals have made use of heat treatment procedures that are predicated primarily upon the precipitation characteristics of the unreinforced matrix alloys. Such approaches may lead to erroneous interpretations about the mechanical, thermal, and environmental response of the composite materials. For example, the overall yield strength and strain hardening exponent of some metal-matrix composites show little variation in response to matrix aging treatments because strengthening derived from reinforcements is substantially more than that arising from precipitation (Christman et al. 1989). However, the (accelerated) aging of the composite matrix can have an important effect on ductility (LLorca et al. 1991), ultimate tensile strength (Mahon et al. 1990), fracture resistance (Lewandowski et al. 1989; Christman 1990; Derby and Mummery 1993), fatigue-crack growth resistance (Christman and Suresh 1988b; Davidson 1989; Suresh 1991), and wear resistance (Pan et al. 1990). Furthermore, the resistance to stress corrosion cracking and corrosion fatigue, which is known to be influenced by aging condition in aluminum alloys, would be expected to be affected by accelerated aging (LLorca 1992).

A knowledge of the accelerating aging phenomenon is also essential for the development of micromechanically based strengthening theories for metal-matrix composites. As an example, consider the widely used analytical and numerical theories for composite strengthening, in which one commonly derives the overall strength on the basis of the isolated properties of the phases of the composite (see Chapter 10). If accelerated aging is not accounted for during the estimation of matrix properties, an incorrect estimate for the effective properties for the composite may be obtained.

7.2 General Observations

Experimental work by a number of researchers has provided clear evidence for the occurrence of accelerated aging and for the precipitation characteristics of a wide range of metal-matrix composites. A summary of works on the aging characteristics of reinforced metals is provided in Table 7.1. These studies have employed diverse experimental and analytic tools that include transmission electron microscopy (TEM), differential scanning calorimetry (DSC), microhardness measurements in the matrix of the composite, macrohardness and strength measurements of composites, and electrical conductivity measurements. In this section, we review some salient results with a view toward developing an understanding of the micromechanisms of precipitation hardening in metallic composites.

---

powder metallurgy aluminum alloy containing SiC particles, no accelerated aging occurs.

Table 7.1. Summary of Work on Aging Kinetics of Particle-Reinforced Metal-Matrix Composites

Matrix (Al)	Reinforcement Size and Shape	Reinforcement Vol. Fraction (%)	Processing Technique	Techniques Used	Key Results	References
6061	B_4C_p (5 μm)	23	PM	Macrohardness TEM	Accelerated aging response in composite	Nieh and Karlak (1984)
6061	SiC_w	20	PM	Macrohardness TEM	Accelerated matrix aging kinetics at 175°C, slower kinetics at 125°C	Rack (1987)
2124	SiC_w (l/d=6–10)	13.2	PM	TEM Vickers microhardness	Increase in dislocation density due to SiC reinforcement accelerated aging response in the composite	Christman and Suresh (1988a)
2124	SiC_w (l/d=6–10)	13.2	PM	TEM SEM Vickers microhardness	The fatigue threshold stress intensity factor ΔK_0 for composite about twice that of controlled alloy; ΔK_0 insensitive to variation in aging treatments	Christman and Suresh (1988b)
2124 2219 6061 7475	SiC_w (l/d=3) SiC_p(10 μm)	8, 20 20 20 10, 15	PM PM IM	DSC	Age hardening sequence in composites is similar to that in unreinforced alloys; precipitation kinetics affected by SiC	Papazian (1988)
6061	SiC_w (l/d=3–25)	10	PM	TEM DSC	Two dominant mechanisms of accelerated aging in MMC; the dislocation density dominant for MMC with large reinforcement size and high dislocation density; and the residual stress dominant for MMC with small reinforcement size and low dislocation density	Dutta and Bourell (1989)
7xxx	SiC_p (16 μm)	20	PM	TEM SEM quantitative metallography	Fracture affected by matrix microstructure; fracture of SiC predominant in the matrix and near the interface in the overaged materials	Lewandowski and Liu (1989)
2024	SiC_p (10 μm)	6, 13, 20	IM	TEM Vickers microhardness	Accelerated aging in composites with different vol. fraction of SiC_p due to enhanced dislocation density; extent of accelerated aging insensitive to vol. fraction of SiC_p	Suresh et al. (1989)
6061	SiC_w (l/d=3–25)	10	PM	SEM DSC	Theoretic model predicting the aging response of composite	Dutta and Bourell (1990)
8090	SiC_p (3 μm)	20 wt %	PM	TEM DSC	Rapid formation of GP zone or increased dislocation reaction led to more rapid aging hardening of the composite	Hunt et al. (1990)
2124	SiC_w (l/d=6–10)	13.5	PM	TEM EDX	Overaged composites had a lower ultimate tensile strength and ductility than those of underaged composites because of precipitation of S phase at SiC_w/Al interface	Mahon et al. (1990)
6061	SiC_p	13.9	PM	Brinell hardness	Effect of solution treatment, comparison of natural aging, and precipitation hardening	Appendino et al. (1991)

Alloy	Reinforcement	vol%	Processing	Techniques	Observations	Reference
2014	SiC_p	17	PM	SEM, TEM, DSC, Vickers microhardness	Amount of acceleration in matrix aging is a function of the aging temperature	Chawla et al. (1991); Esmaeili et al. (1991)
6061	$(Al_2O_3)_p$ (0.5–2.5 μm)	10, 15	IM	Vickers microhardness, TEM, DSC, Electrical resistivity	Accelerated nucleation and growth of precipitates in matrix; the reinforcement amount affected the type and amount of precipitates in the matrix	Dutta et al. (1991)
8090	SiC_p (3 μm)	17	Cold rolling to various degree of reduction	TEM, SEM, Vickers microhardness	Rapid aging of composite not due to S'; aging at/or above 175°C had a deleterious effect on fatigue crack growth	Knowles et al. (1991)
6061	$(Al_2O_3)_p$	15	PM	DSC	Cold rolling changed precipitation kinetics and precipitation sequence in both composite and unreinforced alloy	Lee et al. (1991)

Note: Subscripts p and w indicate particle and whisker forms of reinforcement, respectively. IM and PM indicate ingot metallurgy and powder metallurgy, respectively. *l/d* indicates the aspect ratio (length/diameter) of a whisker.

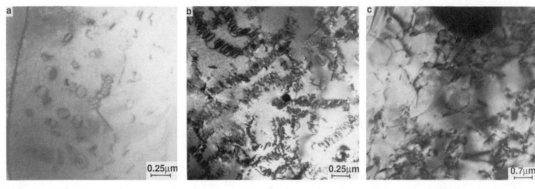

Figure 7.3. Dislocations in the matrix of an underaged 2124 aluminum alloy. (a) Unreinforced. (b) and (c) Reinforced with 13.2 vol % SiC whiskers. *(From Christman and Suresh 1988a.)*

7.2.1 Enhanced Dislocation Density in the Composite Matrix

The generation of a greater density of dislocations in reinforced metals (as compared to the unreinforced condition) has now been documented for matrices that include pure Cu (Chawla and Metzger 1972), Mg–Zn alloy (Choudhry et al. 1991), annealed pure aluminum (Vogelsang et al. 1986), Al–Cu binary alloys (Suresh et al. 1989), 2xxx series aluminum alloys (McDanels 1985; Arsenault and Shi 1986; Christman and Suresh 1988a; Papazian 1988; Dutta et al. 1988; Dutta and Bourell 1989, 1990; Mahon et al. 1990; Chawla et al. 1991), 6xxx series aluminum alloys (Nieh and Karlak 1984; Vogelsang et al. 1986; Papazian 1988; Dutta and Bourell 1989; Rack 1989; Appendino et al. 1991; Dutta et al. 1991; Lee et al. 1991), 7xxx series aluminum alloys (Papazian 1988), and 8xxx series aluminum alloys (Hunt et al. 1990; Oh et al. 1990; Chawla 1992), with W, SiC, B_4C or Al_2O_3 as continuous or discontinuous reinforcements.

Because dislocations can serve as preferential nucleation sites for precipitates, and because dislocations can provide a short-circuit path for the diffusion of solute atoms, the enhanced dislocation density in the matrix of the composite would be expected to significantly alter the precipitation kinetics of the matrix (see also Chapter 6). This line of reasoning has been amply discussed in the literature. For example, Arsenault and Fisher (1983), employing transmission electron microscopy, observed a high density of dislocations (4×10^{12} m^{-2}) clustered around the ends of SiC whiskers and forming low angle cell boundaries in a 20 vol % SiC whisker reinforced 6061 aluminum alloy. They also observed very fine strengthening precipitates decorating the dislocations.* Subsequently, in-situ observations of dislocations generated by thermal mismatch were made using a high-temperature stage in a high-voltage transmission electron microscope (Vogelsang et al. 1986). Here, the sample viewed with the whiskers lying in the plane of the TEM foil showed punching of dislocations from whisker ends. However, the sample viewed end-on to the whiskers showed only a rearrangement of dislocations. Figures 7.3(a) and (b) show examples of the differences in the matrix dislocation density in an underaged 2124 aluminum alloy in both an unreinforced condition and with 13.2 vol % SiC whiskers, respectively. Note the substantially higher dislocation density in the latter case.

In addition to the prismatic loops and dislocation tangles, TEM micrographs often reveal the presence of helical dislocations in the matrix of the composite (Christman and Suresh 1988a; Mahon et al. 1990). Figure 7.3(b) shows an example of such helical dislocations in the composite matrix. These helices are formed as a result of condensation of excess vacancies onto pre-existing screw dislocations. The helices form during the early stages of quenching from the solutionizing temperature, when the temperature is still high enough for the vacancies to be mobile.

Evidence for plastic relaxation of thermal residual stresses in reinforced metals has also been provided by Dunand and Mortensen (1991a) using a model composite with silver chloride as matrix and alumina fibers or glass microspheres as brittle reinforcements. Silver chloride, a transparent salt, shows a ductility of up to 400% during tensile elongation as a result of dislocation mechanisms similar to those found in metals. Single dislocations in AgCl can be decorated with metallic silver at room temperature. This makes the dislocations

---

*These authors also found that the presence of SiC particles hindered subgrain growth in the aluminum-alloy matrix. It was noted that the dislocation density around the reinforcement particles depended on the size of the particle, with an increase in particle size causing an increase in dislocation density around the particle.

around reinforcing fibers or particles visible in high-resolution transmission optical microscopy. The results showed that rows of prismatic loops were punched out in the AgCl matrix by borosilicate glass spheres (1 mm to 5 mm in diameter), with the number of loops emitted being proportional to the sphere diameter. For the alumina-fiber reinforcement, the mechanism of prismatic loop punching at the fiber ends was observed. Along the circumference of the fibers, plastic zones containing tangled dislocations were noticed. The stress within the plastic zone, estimated from the curvature of the pinned dislocations, was three times the resolved shear yield stress of the matrix, which indicates that the matrix is locally strain hardened. Figure 7.4(a) shows decorated prismatic loops punched into the AgCl matrix by a glass sphere. Figure 7.4(b) provides an example of the development of a yield zone around an alumina short fiber embedded in the AgCl matrix because of the plastic relaxation of thermal residual stresses in the matrix.

For an isolated ellipsoidal SiC inclusion in a 2124 aluminum-alloy matrix, the punching distance for dislocation loops generated during cooling, as a result of the thermal contraction mismatch between the inclusion and the matrix, has been estimated (Christman and Suresh 1988a). The change in the total potential energy of the composite (estimated using Eshelby's equivalent inclusion method) was equated with the change in the work done by the movement of the prismatic dislocation loop (estimated using the analyses of Tanaka and Mori 1972 and Taya and Mori 1987). Figure 7.5(a) schematically illustrates an ellipsoidal SiC inclusion (of major axis and minor axis radii c and a, respectively) and the maximum distance (c', measured from the center of the inclusion) over which the dislocations punched out from the ends of the inclusion have moved. Figure 7.5(b) shows the predicted punching distance c'/c as a function of the aspect ratio of the SiC inclusion c/a for a volume fraction of 13.2% for the inclusion. It is seen that the normalized punching distance c'/c decreases with increasing aspect ratio of the inclusion. The dashed line in Figure 7.5(b) indicates the punching distance required to cover the entire matrix with dislocations, assuming uniform distribution of inclusions and a volume fraction of 0.132. The results of this idealized model reveal that for volume fractions of reinforcement that are typically smaller than 15% to 20%, a significant fraction of the matrix is likely to be covered with dislocations arising from the relief of residual stresses. Consequently, one would not expect variations in the density of dislocations, as a function of distance, from the interface to the regions in the matrix away from the interface. This expectation finds experimental backing in a number of TEM studies of SiC whisker-reinforced aluminum alloy matrices (Christman and Suresh 1988a; Suresh et al. 1989).* It is also interesting to note that experimental measurements in the model composite of an AgCl–glass sphere reveal a plastic zone radius in the silver halide matrix that is two to three times the radius of the glass sphere reinforcement (Dunand and Mortensen 1991a). This result is in reasonable agreement with the predictions illustrated in Figure 7.5(b).

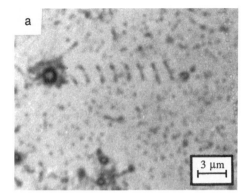
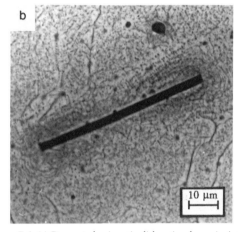

Figure 7.4. (a) Decorated prismatic dislocation loops in AgCl punched by a glass sphere. Note the presence of unresolved dislocation tangles in directions where no loops are created. (b) Plastic zone formed around an Al_2O_3 fiber in AgCl. *(From Durand D.C., and A. Mortensen, Acta Metallurgica et Materialia, vol. 39, On Plastic relaxation of thermal stresses in reinforced metals, 1988. Oxford, Pergamon Press Ltd.)*

---

*Some researchers have also reported a gradual reduction in dislocation density with distance, from the regions close to the whisker–matrix interface to regions of matrix away from the interface (Appendino et al. 1991). Whether or not dislocations are distributed uniformly in the matrix of the composite depends on such factors as the shape and volume fraction of the reinforcement.

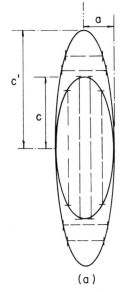

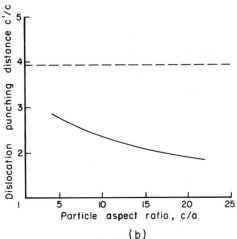

Figure 7.5. (a) Schematic showing the geometry of ellipsoidal inclusion and the associated nomenclature. (b) Variation of the normalized punching distance c'/c as a function of the inclusion aspect ratio, c/a. Predicted by Christman and Suresh (1988a) on the basis of the analyses by Tanaka and Mori (1972) and Taya and Mori (1987).

Although the results described in Figure 7.5 are predicated upon Eshelby's equivalent inclusion method, Dunand and Mortensen (1991b) have presented an alternate approach for analyzing dislocation emission at fiber ends. Their analysis uses the shear-lag model, which accounts for load transfer between the elastic fiber and the elastic–perfectly plastic matrix. The underlying premise of the analysis is that the central portion of the fiber is strained by elastic and plastic interfacial shear until it displays no mismatch with the surrounding matrix. It is found that the number of prismatic dislocation loops punched out at the fiber end reaches a constant value when the fiber reaches a certain minimum length.

7.2.2 Experimental Methods

A number of different experimental methods have been employed to study microstructural evolution and accelerated aging in metal-matrix composites. These involve direct observational methods, such as TEM, as well as indirect methods, such as differential scanning calorimetry, hardness measurements, and electrical conductivity measurements.

The most direct method of observing matrix precipitation in reinforced metals involves use of TEM, wherein the details of the matrix microstructure (involving the size, shape and distribution of precipitates) can be systematically examined as a function of aging time (Christman and Suresh 1988a; Mahon et al. 1990). TEM can also provide quantitative information on dislocation generation caused by thermal mismatch, heterogeneous precipitation along dislocations, and interfacial versus matrix precipitation. Furthermore, with the availability of high-temperature stages for electron microscopes, in-situ observations can be made on precipitation in response to controlled heat treatments. However, caution should be exercised to ensure that the artifacts of TEM foil preparation do not cause precipitation or lead to an apparent modification of the precipitation sequence observed during aging.

Differential scanning calorimetry, DSC (Papazian 1988; Dutta and Bourell 1989; Petty-Gallis and Goolsby 1989; Chawla et al. 1991; Lee et al. 1991) has become an important analytical tool for determining subtle microstructural transformations in metal-matrix composites. In basic terms, one determines the change in thermal energy or enthalpy resulting from an endothermic or exothermic transition or reaction in the sample as a function of temperature or time, during a DSC scan. In precipitation hardenable alloys, DSC allows one to study the various transformations such as GP zone formation, and metastable and stable precipitation. In principle, it is feasible with this technique to obtain the peak temperature, enthalpy, and the activation energy for a phase transition. In practice, it is more common to identify the presence or absence of a given transition. In order to detect phase transitions during the aging treatment of an alloy, specimens in the form of disks are solutionized and quenched, and then analyzed in a differential scanning calorimeter over a certain temperature range (for example, from 273 K to 833 K for aluminum alloys) at a fixed heating rate such as 10 K/min. The same procedure is repeated for an alloy aged at a given temperature. A comparison of the thermograms obtained for the unaged and aged alloys will enable the identification of the phase transitions during the aging process. The DSC technique is a very useful complement to the microscopy and hardness techniques.

Because precipitation results in changes in the electrical conductivity of an alloy, electrical conductivity measurements performed on the unreinforced and reinforced alloys over the entire range of aging conditions (spanning the very underaged to the severely overaged microstructures) can be used to gauge the differences in their precipitation response. For this purpose, the relative values of electrical conductivity, based on the International Annealed Copper Standard (IACS), are plotted against the aging time (at a fixed aging temperature) for both the reinforced and unreinforced alloy. Because the reinforced metal can show a lower conductivity than the control alloy (because SiC whiskers are very effective electron scatterers), and because the total change in conductivity is also lower for the reinforced alloy (because the fractional volume of material undergoing change in conductivity during aging is less than that in the unreinforced alloy), some normalizations are made to interpret the results. For example, one way of defining a normalized conductivity is to consider the ratio of the difference between the instantaneous conductivity and the initial conductivity (in the as-quenched condition) to the difference between the saturation conductivity (in the severely overaged condition) and the initial conductivity. An example of the use of electrical conductivity measurements for the analysis of accelerated aging can be found in Christman and Suresh 1988a.

One of the popular methods of examining the modification of the aging response of the matrix material due to the addition of the reinforcement phase is to examine the variation of hardness as a function of aging time at a fixed aging temperature. Some researchers have employed macroscopic hardness measurements (such as the Rockwell B or the Brinnell hardness scales) to document the differences between the precipitation characteristics of the reinforced and unreinforced metals (Nieh and Karlak 1984; Appendino et al. 1991). Although this method is straightforward, it can lead to ambiguities in interpretation in the sense that the hardness values for the composite include both the matrix hardness (including that from precipitation) and the hardness contribution from the reinforcements. An overall estimate of differences in precipitation kinetics between the reinforced and unreinforced alloys is obtained by comparing the shapes of the hardness versus aging-time curves (especially the shift of the peak hardness point in the aging curve). Such estimates can be prone to considerable error in alloys where the extent of strengthening from precipitation is substantially smaller than that caused by reinforcement. In an attempt to overcome these limitations, microhardness measurements are often performed in the matrix of the composite material and in the unreinforced alloy to document the variation of matrix strength with aging (Christman and Suresh 1988a; Chawla et al. 1991). In this method (involving Vickers pyramid indentor under a small load), the hardness measurements are made in the regions of the matrix of the composite that are relatively free of the reinforcement particles. Because whisker-rich and whisker-poor regions exist even in metal-matrix composites with large reinforcement content (approximately 20 vol %), such microhardness measurements can be made in most discontinuously reinforced metal-matrix composites. The drawback of all hardness techniques for estimating aging response is that they are prone to considerable experimental scatter, especially because of the clustering of the reinforcement phase (see the data shown in Figure 7.9, for example).

7.3 Role of Thermomechanical and Processing Variables

7.3.1 Effects of Aging Temperature

Because precipitation is a diffusion-aided process, both aging temperature and time have a strong influence on the precipitation kinetics of an age-hardenable alloy. In addition, an interesting aspect of the effect of a brittle reinforcement on the aging kinetics of a ductile matrix is that the extent of accelerated aging is a very sensitive function of the aging temperature. Chawla et al. (1991) examined the aging behavior at four different aging temperatures of the matrix in two cast composites with SiC or Al_2O_3 particle-reinforced 2014 aluminum alloys and compared the results with those obtained for unreinforced 2014 aluminum alloy. The aging curves at the different aging temperatures for the composites with SiC or Al_2O_3 particulates in the same 2014 aluminum matrix were almost identical. However, the aging temperature and the presence of the ceramic reinforcement had a significant effect on the precipitation kinetics of the 2024 aluminum-alloy matrix. For the sake of brevity, only the results for the SiC particulate reinforced 2024 aluminum alloy are presented here. The aging response represented by the Vickers diamond pyramid indentor microhardness (under 0.25 N load) as a function of aging time is presented in Figures 7.6 and 7.7 for the unreinforced and reinforced 2014 aluminum alloy, respectively. It is seen that the

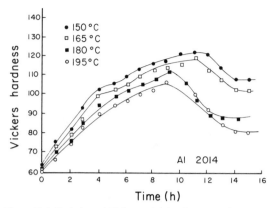

Figure 7.6. Variation of Vickers microhardness as a function of aging time (at different aging temperatures) in an unreinforced 2014 aluminum alloy *(After Chawla et al. 1991.)*

time to peak age in the Al 2014 matrix of the composite decreased much more with increasing aging temperature than it did in the unreinforced alloy. This point becomes more clearly evident if we compare the aging characteristics of the composite matrix and the unreinforced alloy at 195°C and 150°C, two extreme aging temperatures. At 195°C, the reinforced metal reached peak hardness in about 2.5 hours, while the unreinforced alloy took about 9.5 hours. At 150°C, both materials peak-aged around the same time, with no accelerated aging resulting from the presence of the brittle reinforcement.

The time to reach peak strength in the 2014 aluminum alloy and in the composite is shown in Figure 7.8 as a function of aging temperature. As the aging temperature increased, the peak aging time for the matrix decreased rather drastically. Furthermore, as the aging temperature decreased, the difference in peak aging

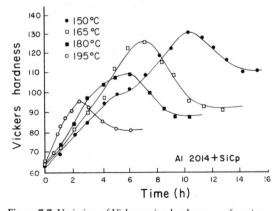

Figure 7.7. Variation of Vickers microhardness as a function of aging time (at different aging temperatures) in a 2014 aluminum alloy reinforced with xxx vol % SiC particulates. *(After Chawla et al. 1991.)*

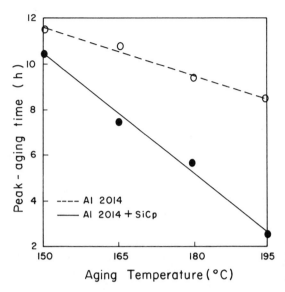

Figure 7.8. Peak aging time as a function of the aging temperature for unreinforced and SiC particle reinforced 2014 aluminum alloy. *(After Chawla et al. 1991.)*

time between the matrix of the composite and the unreinforced 2014 alloy became smaller; at 150°C, the difference in peak aging time is essentially negligible, while it is 7 hours at 195°C. Similar results were also observed for the 2014 alloy reinforced with Al_2O_3 particulates. Studies have also revealed similar trends for SiC-reinforced lithium-containing 8090 aluminum-alloy matrices (Chawla 1992).

It is well known that the formation of GP zones and their transition to θ'' and θ' in the Al–Cu system is possible only when the alloy is aged at a temperature below the GP zone solvus (Lorimer 1978; Porter and Easterling 1981). If, for example, the aging temperature is above the θ'' solvus but below the θ' solvus, there will be no GP zones and the first precipitate to come out of the supersaturated solid solution will be θ', which is heterogeneously nucleated along dislocations. One would expect a similar phenomenon to occur during heat treatment of the Al–Cu–Mg (2xxx series) alloy. At low aging temperatures, the first precipitates to nucleate are the coherent Cu-rich GP zones. The nucleation of GP zones at the lower aging temperatures is mostly homogeneous and their formation is significantly aided by the introduction of excess vacancies upon quenching from the solutionizing temperature. Therefore, the ability to generate a higher density of heterogeneous nucleation sites (such as dislocations) in the composite material is not a major factor in determining the precipitation kinetics. At high aging temperatures, the (S') precipitates nucleate heterogeneously on dislocations and grow as laths on (210) planes along [001] directions

(Wilson and Partridge 1965). Because the composite material has a greater number of heterogeneous sites (a higher dislocation density due to the thermal contraction mismatch between the matrix and the reinforcement), faster precipitation would be expected in the composite than in the unreinforced alloy. This line of reasoning provides a rationale for the experimental trends seen in Figures 7.6 through 7.8, in which the extent of accelerated aging increases significantly with increasing aging temperature. Such an argument also finds experimental backing from differential scanning calorimetry analyses. DSC scans showed the presence of GP zones and S' precipitates in both the composite and the control alloy aged at 150°C (Chawla et al. 1991). However, at 195°C, there was no evidence of GP zones in both materials; only S' precipitation was detected.

The effect of aging temperature on precipitation in metal-matrix composites was also studied in a 6061 aluminum alloy reinforced with SiC and Al_2O_3 particles (Salvo et al. 1990). They also observed that the matrix in the composite materials aged faster than the unreinforced alloy above a certain critical aging temperature, which they determined to be 190°C for their material.

7.3.2 Effects of Reinforcement Content

Suresh et al. (1989) reported a study of the effects of reinforcement volume fraction on the evolution of matrix microstructure in response to controlled artificial aging in a cast Al–3.5% Cu alloy. The changes in matrix microhardness (using Vickers diamond pyramid indentor) of the composite matrix were determined from the as-quenched to the severely overaged conditions for SiC particulate reinforcements with volume fractions of 6, 13, and 20%. The aging characteristics of the composites were compared with those of the unreinforced Al–3.5% Cu alloy, which was processed and heat treated identically to the composite.

Figure 7.9 is a plot of Vickers diamond pyramid microhardness of the matrix as a function of the aging time, at a fixed aging temperature of 190°C for the unreinforced Al–3.5% Cu alloy, and for the composites reinforced with 6%, 13%, and 20% SiC particles. Also indicated for the peak aging conditions are the average standard deviations in the microhardness values for the four different materials. (The standard deviations for the other aging conditions had comparable values.) It is seen that the unreinforced alloy and the matrix of the composites exhibit essentially the same microhardness in both the solutionized and the as-quenched conditions. However, the peak aging time at 190°C for the Al-Cu alloy decreases from about 60 hours for the unreinforced condition to about 16 to 24 hours for the three composites. (Given the inherently large scatter in the microhardness data, any differences in the peak aging time for the three composites could not be pinpointed.) Quantitative measurements of dislocation densities, ρ, for the four materials were also reported (Suresh et al. 1989). For the unreinforced alloy, $\rho \approx 6 \times 10^{13}$ m^{-2}, while $\rho \approx 2 \times 10^{14}$ m^{-2} for all three composite materials. These results appear to show that a certain critical value of dislocation density is generated even at small reinforcement content, which serves to provide adequate heterogeneous nucleation sites for the strengthening precipitates, thereby causing accelerated aging in the composite materials. TEM observations of heterogeneous precipitate nucleation at earlier aging times in the composite materials also corroborated the microhardness measurements in this study.

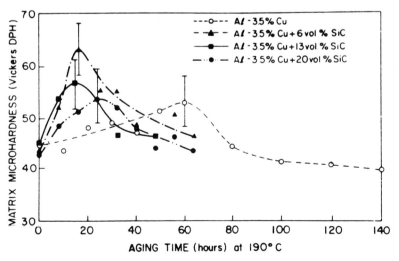

Figure 7.9. Matrix microhardness as a function of aging time at 190°C for an Al–3.5%Cu binary alloy reinforced with different amounts of SiC particulates. *(After Suresh et al. 1989.)*

7.3.3 Effects of Cold Work

As noted earlier, a number of investigators have proposed that dislocations generated in the matrix of the composite materials (because of the plastic relaxation of thermal residual stresses generated upon cooling from the solutionizing temperature) serve as heterogeneous nucleation sites for the precipitates during aging treatments. Such early nucleation, accompanied by the possibility of enhanced pipe diffusion along dislocations promoting precipitate growth, provides one of the principal causes for accelerated aging in the metal-matrix composite. If a higher density of dislocations can be introduced in the matrix by methods other than reinforcements (such as by cold work), accelerated aging in the cold worked material compared to the unworked alloy would be expected (Christman and Suresh 1988a; Dutta and Bourell 1990). Such trends have been clearly documented for metal-matrix composites and corresponding control alloys, wherein the role of enhanced dislocation density on the precipitation kinetics of the matrix of the composite becomes abundantly clear.

Dutta and Bourell (1990) have presented a quantitative comparison of the effects of reinforcement and cold work on the precipitation kinetics of a 6061 aluminum alloy, using differential scanning calorimetry. Figure 7.10 shows the variation of the percent β' precipitation as a function of aging time, based on the data obtained from DSC scans at three different heating rates. The results in these figures pertain to an unstrained 6061 aluminum alloy, 6061 Al cold worked to a plastic strain of 0.36% after the solutionizing quench, and 6061 aluminum reinforced with 10 vol % SiC particulates (unstrained). In all three plots shown in Figure 7.10, β' precipitation is considerably faster in the composite than in the unstrained control alloy. A plastic strain of 0.36% causes a sufficient shift in both the nucleation and the growth rate of the precipitates such that the strained control alloy and the unstrained composite exhibit essentially the same precipitation kinetics at the heating rate of 40°C/min in the DSC analysis. At lower heating rates, it is clear that the composite exhibits a higher nucleation rate and a lower growth rate for the β' precipitates than for the strained control alloy. (Possible reasons for this trend are discussed in Section 7.4.)

Using DSC analysis, the effect of prior cold work (imposed by rolling) on the precipitation kinetics in a 6061 aluminum alloy with and without 15 vol % Al_2O_3 particulate reinforcement was examined (Lee et al. 1991). Figures 7.11(a) and (b) show the heat flow of the precipitation reactions as a function of temperature for the 6061 aluminum control alloy and the composite subjected to several different levels of cold work. These figures show that both reinforcement and cold work

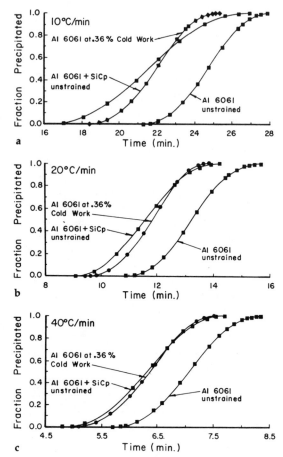

Figure 7.10. Fraction of β' precipitation as a function of aging time for unreinforced and unstrained 6061 Al, 6061 Al cold worked to a plastic strain of 0.36% following the quench from the solutionizing temperature, and unstrained 6061 Al reinforced with 10 vol % SiC particles. These values are based on DSC scans on the three materials, each of which was solutionized and quenched before the DSC scan. (After Dutta and Bourell 1990.)

have a significant effect in aiding the precipitation. In both the composite and the control alloy, precipitation is accelerated by cold rolling; the initial temperature for the precipitation reaction also decreases in both cases as a result of cold rolling, suggesting that preferential nucleation at defects is the dominant cause for this trend. However, increasing the amount of cold work beyond 25% does not significantly alter the initial or peak reaction temperatures. This result is fully consistent with the trends shown in Figure 7.9, in which an increase in reinforcement content beyond 6% does not have a significant effect on accelerated aging in the Al–Cu alloy. The results of Figure 7.9 and Figure 7.11

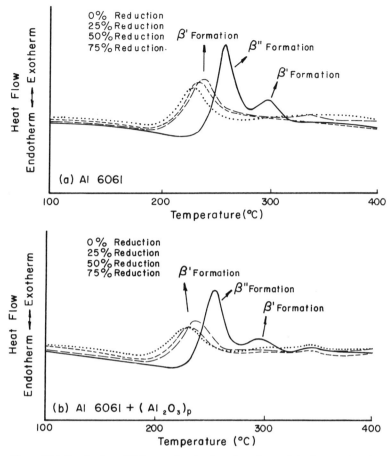

Figure 7.11. Results from DSC scans showing heat flow for precipitation reactions in (a) 6061 Al and (b) 6061 Al reinforced with 15 vol % Al_2O_3. Both materials were subjected to the indicated amounts of cold reduction. *(After Lee et al. 1991.)*

thus appear to reveal that precipitation kinetics are not significantly influenced by reinforcement content after a certain critical density of defects are introduced either by the reinforcement or by cold work. The relative amount of metastable precipitates was also significantly reduced in the cold-rolled materials as compared to the unstrained control alloy or the composite.

7.4 Mechanisms and Models

On the basis of experimental data gathered for a variety of composites with age-hardening matrices, many conceptual and quantitative models have been developed. A feature common to most of these models is the underlying notion that accelerated aging primarily results from the enhanced dislocation density in the composite matrix, which enhances the rate of precipitate nucleation, growth, or both. As reviewed earlier, TEM, microhardness, and DSC analyses revealed that in the 2xxx and 6xxx series aluminum matrices, the addition of a ceramic reinforcement does not change the precipitation sequence of the matrix. However, both dissolution kinetics and precipitation kinetics are modified by the reinforcement. Furthermore, some usually quench-insensitive alloys such as 6061 aluminum become quench-sensitive as a consequence of reinforcement (see Figure 7.10).

By employing transmission electron microscopy, the precipitation kinetics of a powder metallurgy processed 2124 aluminum alloy with and without 13.2 vol % SiC whiskers (aspect ratio of 5) was measured (Christman and Suresh 1988a). Figure 7.12 shows the variation of the average size of the S' precipitates as a function of aging time (at a constant aging temperature of 190°C) in both the composite and the control alloy. While it takes up to 4 hours of aging for the S' precipitation to nucleate in the control alloy, the matrix in the composite exhibits S' precipitation heterogeneously along

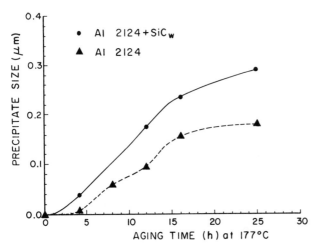

Figure 7.12. TEM observations of S' precipitate size (the largest dimension) as a function of aging time at 177°C for 2124 aluminum alloy reinforced with 13.2 vol % SiC whiskers. *(After Christman and Suresh 1988a.)*

dislocations immediately following the commencement of artificial aging. By the time S' precipitates begin to form in the control alloy, the matrix attains peak hardness.

7.4.1 Precipitate Nucleation as a Function of Dislocation Density

Dutta and Bourell (1990) have presented a model, based on the analysis of Avrami (1939, 1940), for precipitation in metal-matrix composites. They assume that (1) the nucleation of precipitates depends on dislocation density only or (2) both nucleation and (self-similar) growth of precipitates depend on dislocation density only. They consider a one-dimensional problem by modeling the matrix as a slab with a plate reinforcement at one end, Figure 7.13(a). It is assumed that the plastic zones around the ends of the reinforcements extend a distance L from the interface. The dislocation density ρ (i.e., the number of dislocation lines intersecting a unit cross-sectional area) outside the plastic zone is considered negligible to that within the plastic zone. As shown in Figure 7.13(b), the dislocation density inside the plastic zone is assumed either to be uniform ($\rho = M\bar{\rho}$, where M is a numerical factor equal to 1 to 1.5) or to decrease linearly with distance from the interface so that it reaches a zero density at the end of the plastic zone ($\rho = 2M\bar{\rho}x/L$).

For the case in which only precipitation nucleation is affected by dislocation density, the nucleation rate per unit volume is

$$N = \rho N\perp \qquad (7.1)$$

where $N\perp$ is the nucleation rate per unit dislocation line. The fraction precipitated at time t for dislocation-aided heterogeneous nucleation with a uniform density of $M\bar{\rho}$ is then found to be

$$X_u(t) = 1 - \exp\left[-M\left(\frac{t}{\Lambda_n}\right)^n\right] \qquad (7.2)$$

where

$$\Lambda_n = \left(\tfrac{2}{5}JD^{3/2}N\perp\bar{\rho}\right)^{-1/n} = \text{time constant} \qquad (7.3)$$

In this equation, J is a material constant, D is the diffusion coefficient of the solute in the matrix (assumed independent of dislocation density), and n is an exponent with a value between 1.5 and 2.5. For a linear variation in dislocation density with an average dislocation density of $M\bar{\rho}$,

$$X(t) = 1 - \left(\tfrac{1}{2M}\right)\left(\tfrac{\Lambda_n}{t}\right)^n [1 - \{1 - X_u(t)\}^2] \qquad (7.4)$$

The plot of fraction precipitated for uniform and linearly varying dislocation densities, $X_u(t)$ and $X(t)$, respectively, against aging time t (normalized by Λ_n) is shown in Figure 7.14 (from Equations 7.2 and 7.4) for the case of 6061 Al with SiC whiskers where only the nucleation rate of precipitates depends on the dislocation density. Here, as expected, an increase in average dislocation density causes an increase in the rate of precipitation. However, a gradient in dislocation density results in a lower precipitation rate; this slowing effect being greater during the later stages of precipitation. Note the similarity of the shapes of the predicted curves in Figure 7.14 to the TEM observations of precipitation depicted in Figure 7.12.

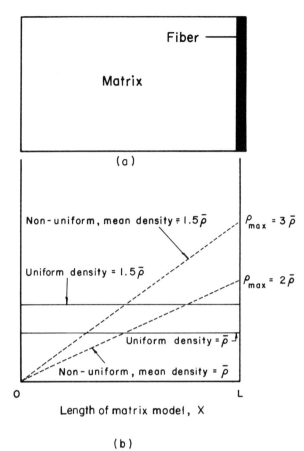

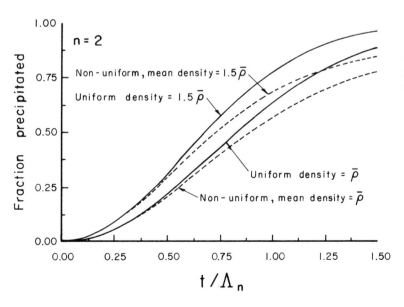

Figure 7.13. (a) An idealized model of a matrix of length L adjoining a fiber. (b) Schematic of different dislocation distributions in the matrix. *(After Dutta and Bourell 1990.)*

Figure 7.14. Fraction of β' precipitation as a function of normalized aging time when only the precipitate nucleation rate is influenced by the dislocation density. *(After Dutta and Bourell 1990.)* See text for details.

7.4.2 Precipitate Nucleation and Growth as a Function of Dislocation Density

Dutta and Bourell (1990) also presented a simple model for the combined nucleation and growth of strengthening precipitates as a function of the dislocation density. They invoked the assumption that both the nucleation rate N and the apparent diffusivity D are proportional to the dislocation density ρ, whereas the growth rate $\mathcal{G}$ of the precipitate (that is, the increase in the average radius of the precipitate $r(t)$ with time) is proportional to the dislocation density on the basis of the relationship:

$$\mathcal{G} = \frac{dr(t)}{dt} \propto t^{-1/2} \sqrt{D_p \rho} \quad (7.5)$$

when there is no incubation time for precipitate nucleation.

For both nucleation and growth of precipitates affected by the dislocation density, the fraction precipitated $X_u(t)$ at time t with uniform dislocation density $M\bar{\rho}$ can be obtained by combining Equations 7.2 through 7.5 such that

$$X_u(t) = 1 - \exp\left[-M^{5/2}\left(\frac{t}{\Lambda_{n+\mathcal{G}}}\right)^n\right] \quad (7.6)$$

where

$$\Lambda_{n+\mathcal{G}} = \left(\frac{2J}{5 \times 10^{-27} D_p^{3/2} N_\perp \bar{\rho}^{5/2}}\right)^{-1/n}$$
$$= \text{time constant.} \quad (7.7)$$

Similarly, when the dislocation density varies with distance according to the expression $\rho = 2M\bar{\rho} \cdot x/L$ (with an average dislocation density of $M\bar{\rho}$), the fraction precipitated is found to be

$$X(t) = \frac{2}{5} \sum_{j=2}^{\infty} \frac{(-1)^j (t/\Lambda_{n+\mathcal{G}})^{n(j-1)} (2M)^{(j-1)5/2}}{(j-\{3/5\})(j-1)!} \quad (7.8)$$

The predictions of Equations 7.7 and 7.8 for precipitation, whose nucleation and growth are both affected by the dislocation density, are plotted in Figure 7.15 for two different values of M. First, it is evident that the qualitative trend in this figure is different from that found in Figure 7.14 for dislocation-aided nucleation only. When both nucleation and growth are affected by dislocation density, a gradient in dislocation density leads to faster precipitation in early stages of aging than does the case involving uniform dislocation density. The situation is reversed at later aging times.

The predictions shown in Figure 7.15 can now be used to rationalize the experimental trends plotted in Figure 7.10. When a uniform dislocation density is created in an age-hardenable alloy by cold work, the rate of precipitation is faster in the early stages of aging than it is in the case of a metal-matrix composite where some gradients in dislocation density are expected. However, such trends in aging kinetics are reversed for the two cases in the advanced stages of aging. Thus the Avrami-type theoretic model of Dutta and Bourell (1990), despite its simplicity, provides an appealing rationale for the experimentally observed trends in the aging response of cold worked and reinforced alloys. It should be noted that gradients in dislocation density in the composite are likely to disappear at larger volume fractions of the reinforcement. Therefore, systematic

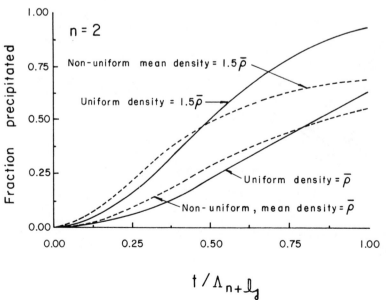

Figure 7.15. Fraction of β' precipitation as a function of normalized aging time when both the nucleation and the growth rates of the precipitate are influenced by the dislocation density. *(After Dutta and Bourell 1990.)* See text for details.

experiments are needed to verify if with increasing reinforcement content the cross-over between curves for the cold worked control alloy and the unstrained composite disappears.

Summary

The results presented in this chapter demonstrate the significant effect of reinforcement, matrix composition, processing variables, aging treatments, and prior mechanical working on the evolution of microstructure in precipitation-hardenable alloys. These descriptions point to the flexibility available to the materials scientist or the engineer when tailoring the microstructure of the matrix material in a composite, which can have a significant effect on the overall mechanical, thermal, and environmental performance. This flexibility also poses the inevitable problem of having to develop a thorough understanding of the mechanisms and kinetics of microstructure evolution in response to the metallurgical and thermomechanical processing of the composite material. Although the primary focus of this chapter has been on the mechanisms associated with the aging behavior of reinforced metals, the effects of aging response on the overall properties cannot be overlooked. Such effects are discussed in several publications (Christman and Suresh 1988a; Davidson 1989; Christman et al. 1989; LLorca et al. 1991) and in Chapters 8, 11, 13, 14, and 15 of this volume.

The majority of available experimental evidence points to the possible role of enhanced dislocation density in the matrix of the composite in influencing the precipitation response. However, several researchers have advanced other points of view that need further exploration in future studies. For example, Hunt et al. (1990) hypothesize that the precipitation response of conventionally cast 2124 aluminum alloy with SiC particles is influenced by the increased efficiency of hardening by the copper solute in the strained matrix. Furthermore, as mentioned in Section 7.1.2, Appendino et al. (1991) note that in some reinforced aluminum alloys, the precipitation sequence may even be altered by the brittle reinforcing phase which, in some rare cases, might result in decelerated aging of the composite matrix. Other factors such as interfacial reactions (Friend et al. 1988), internal stress fields (Prangnell and Stobbs 1991), and particle oxidation (Ribes and Suéry 1989; Salvo et al. 1991; Towle and Friend 1992), are also known to play an important role in influencing the aging response of metal-matrix composites.

Micromechanisms of aging response as well as theoretic and conceptual models for the nucleation and growth of strengthening precipitates in reinforced metals have been the focus of a number of research studies in recent years, particularly for discontinuously reinforced aluminum-alloy matrices. These investigations have provided the groundwork for a fundamental understanding of microstructural evolution by employing such diverse experimental techniques as optical and electron microscopy, indentation hardness techniques, differential scanning calorimetry, and electrical conductivity analyses, as well by drawing on the wealth of information available from decades of research on the aging characteristics of unreinforced metals and alloys. Despite such advances, further work is necessary, for both scientific and practical purposes, to elucidate and clarify several basic issues pertaining to this topic. Such issues include: (1) the role of vacancy formation and of its influence on precipitation in reinforced alloys, and a comparison of the effects of vacancies with those seen in unreinforced alloys; (2) experimental studies of precipitation in a broad variety of metal-matrix composites, including different classes of continuously reinforced alloys where the higher concentration of the reinforcement (than in the discontinuously reinforced metals) would be expected to cause a stronger change in the aging response of the matrix; (3) high-resolution transmission electron microscopy of precipitation at the interface between the matrix and the reinforcement, and in the near-interface regions for different sizes, shapes, distributions, and concentrations of the reinforcement phase; (4) in-situ precipitation studies where the effects of aging temperature, reinforcement shape, reinforcement concentration, and prior cold work are quantitatively examined in high-purity model alloys; (5) thermodynamic studies of the possible occurrence of Ostwald ripening and spinodal decomposition in reinforced metals, including the effects of reinforcement content on the such precipitation effects, and (6) the effects of an aggressive environment on microstructural modification through electrochemical reactions with the precipitates, grain boundaries, and interfaces. Such experimental studies are necessary before available theories for aging response in metal-matrix composites can be further refined.

Acknowledgements

Subra Suresh acknowledges the support of the Materials Research Group on Micromechanics of Failure-Resistant Materials in the Division of Engineering at Brown University, which is funded by a grant from the National Science Foundation (Grant No. DMR-9002994). Additional support for this work was provided by the Richard P. Simmons Chair at M.I.T. Krishan K. Chawla acknowledges the support from the NSF Center for Micro-Engineered Ceramics and the Naval Surface Weapons Center.

References

Appendino, P., Badini, C., Marino, F., and A. Tomasi. 1991. *Mater. Sci. Eng.* A135:275–279.
Arsenault, R.J., and N. Shi. 1986. *Mater. Sci. Eng.* 81:175–187.
Arsenault, R.J., and R.M. Fisher. 1983. *Scripta Metall.* 17:67–71.
Avrami, M. 1939. *J. Chem. Phys.* 7:1103–1112.
Avrami, M. 1940. *J. Chem. Phys.* 8:212–224.
Chawla, K.K., and M. Metzger. 1972. *J. Mater. Sci.* 7:34–39.
Chawla, K.K., Esmaeili, A.H., Datye, A.K., and A.K. Vasudevan. 1991. *Scripta Metall. Mater.* 25:1315–1319.
Chawla, K.K. 1992. Unpublished results. New Mexico Institute of Technology, Socorro, NM.
Choudhry, P.K., Rack, H.J., and B.A. Mikuchki. 1991. *J. Mater. Sci.* 26:2343–2347.
Christman, T. 1990. Ph.D. Thesis, Division of Engineering, Brown University, Providence, RI.
Christman, T., Needleman, A., and S. Suresh. 1989. *Acta Metall.* 37:3029–3050.
Christman, T., and S. Suresh. 1988a. *Acta Metall.* 36:1691–1704.
Christman, T., and S. Suresh. 1988b. *Mater. Sci. Engin.* A102:211–216.
Courtney, T. 1990. Mechanical Behavior of Materials. New York: McGraw-Hill.
Davidson, D. 1989. *Engin. Fract. Mech.* 33:965–977.
Derby, B., and P. Mummery. 1993. Chapter 14, this volume.
Dunand, D.C., and A. Mortensen. 1991a. *Acta Metall. Mater.* 39:127–139.
Dunand, D.C., and A. Mortensen. 1991b. *Acta Metall. Mater.* 1405–1416.
Dutta, I., Bourell, D.L., and D. Latimer. 1988. *J. Comp. Mater.* 22:829–849.
Dutta, I., and D.L. Bourell. 1989. *Mater. Sci. Eng.* A112:67–77.
Dutta, I., and D.L. Bourell. 1990. *Acta Metall.* 38:2041–2049.
Dutta, I., Allen, S.M., and J.L. Hafley. 1991. *Metall. Trans. A.* 22A:2553–2563.
Esmaeili, A.H., Chawla, K.K., Datye, A.K., et al. 1991. *In:* Proceedings of the International Conference on Composite Materials. ICCM/VII. Honolulu, HI, p. 17F1–17F10.
Friend, C.M., Horsfall, I., Lexton, S.D., and R.J. Young. 1988. *In:* Cast Reinforced Metal Composites. (S.G. Fishman and A.K. Dhingra, eds.), 309–315, Metals Park, OH: ASM International.
Haasen, P. 1978. Physical Metallurgy. Cambridge, England: Cambridge University Press.
Hansen, N., and C. Barlow. Chapter 6, this volume.
Hunt, E., Pitcher, P.D., and P.J. Gregson. 1990. *Scripta Metall. Mater.* 24:937–941.
Kumai, S., Hu, J., Higo, Y., and S. Nunomura. 1992. *Scripta Metall. Mater.* 27:107–110.
Lee, H.-L., Lu, W.H., and S.L. Chan. 1991. *Scripta Metall. Mater.* 25:2165–2170.
Lewandowski, J.J., Liu, C., and W. Hunt. 1989. *Mater. Sci. Engin.* A107:241–255.
LLorca, J. 1992. Unpublished results. Polytechnic University of Madrid, Spain.
LLorca, J., Needleman, A., and S. Suresh. 1991. *Acta Metall. Mater.* 39:2317–2335.
Lorimer, G.W. 1978. Precipitation Process in Solids. Warrendale, PA: The Metallurgical Society, p. 88.
Mahon, G.J., Howe, J.M., and A.K. Vasudevan. 1990. *Acta Metall.* 38:1503–1512.
McDanels, D. 1985. *Metall. Trans. A* 16A:1105–1115.
Meyers, M.A., and K.K. Chawla. 1984. Mechanical Metallurgy. New York: Wiley, pp. 402–437.
Nieh, T.G., and R. Karlak. 1984. *Scripta Metall.* 18:25–28.
Oh, K.H., Lee, H.I., Kim, T.S., and T.H. Kim. 1990. *In:* Fundamental Relationships Between Microstructure and Mechanical Properties of Metal-Matrix Composites. (P.K. Liaw and M.N. Gungor, eds.), 179–184, Warrendale, PA: The Minerals, Metals and Materials Society.
Pan, Y.M., Fine, M.E., and H.S. Cheng. 1990. *Scripta Metall. Mater.* 24:1341–1345.
Papazian, J. 1988. *Metall. Trans. A* 19A:2945–2953.
Petty-Galis, J.L., and R.D. Goolsby. 1989. *J. Mater. Sci.* 24:1439–1446.
Porter, D.A., and K.E. Easterling. 1981. Phase Transformations in Metals and Alloys. New York: Van Nostrand Reinhold Company, p. 308.
Prangnell, P.B., and W.M. Stobbs. 1991. *In:* Proceedings of the 12th Risø Conference on Metallurgy and Materials Science: Metal-Matrix Composites—Processing, Microstructure and Properties. (N. Hansen, D. Juul-Jensen, T. Leffers, et al., eds.), 603–610, Roskilde, Denmark: Risø National Laboratory.
Rack, H.J. 1989. *In:* Powder Metallurgy Composites. (P. Kumar, K. Vedula, and A. Ritter, eds.), 155–168, Warrendale, PA: Metallurgical Society of AIME.
Ribes, H., and M. Suéry. 1989. *Scripta Metall.* 23:705–709.
Salvo, L., Suéry, N., and F. Decomps. 1990. *In:* Fabrication of Particulate Reinforced Metal Composites. Metals Park, OH: ASM International, pp. 139–144.
Salvo, L., Suéry, N., Legoux, J.G., and G. l'Ésperance. 1991. *Mater. Sci. Eng.* A126:129–133.
Suresh, S. 1991. Fatigue of Materials. Cambridge, England: Cambridge University Press.
Suresh, S., Christman, T., and Y. Sugimura. 1989. *Scripta Metall. Mater.* 23:1599–1602.
Tanaka, K., and T. Mori. 1972. *J. Elast.* 2:199–206.
Taya, M., and T. Mori. 1987. *Acta Metall.* 35:155–163.
Towle, D.J., and C.M. Friend. 1992. *Scripta Metall. Mater.* 26:437–441.
Vogelsang, M., Arsenault, R.J., and R.M. Fisher. 1986. *Metall. Trans. A.* 17A:379–389.
Wilson, R.N., and P.G. Partridge. 1965. *Acta Metall.* 13:1321–1327.

PART III
MICROMECHANICS AND MECHANICS OF DEFORMATION

Chapter 8
Crystal Plasticity Models

PETER E. McHUGH
ROBERT J. ASARO
C. FONG SHIH

Computational mechanics is applied extensively to the analysis of the complex deformation and failure processes in material microstructures. Detailed descriptions of microstructural deformation have been developed in terms of physically based continuum constitutive theories. In this chapter, we review the use of one of these theories, crystal plasticity, for modeling the behavior of discontinuously reinforced metal-matrix composite (MMC) materials during deformation. One of the features that make this a physical theory is that plastic deformation is crystallographic, meaning that plastic slip occurs only along certain well-defined planes, representing the motion of dislocations along lattice glide planes. On this kinematic basis, the plastic constitutive description is phrased in terms of slip-system-resolved shear stresses and shear strains, which allows actual slip system shear stress/shear strain data to be used to evaluate stain-hardening and rate-dependence parameters. Crystal plasticity theory is a continuum theory because individual dislocations are not represented; plastic deformation is not represented as being composed of discrete dislocation motion events. A significant feature of this theory, natural for continuum theories, is that size scales are not incorporated; one implication of this is that there is no representation of lattice plane spacing. In Chapters 4 through 7 of this volume, the heterogeneous and nonuniform nature of the microstructures of these materials is clearly evident. Such nonuniformity increases the complexity of microscale deformation and failure behavior in comparison to (for example) single-phase metals, and renders the modeling task more difficult. To make the modeling problem more tractable, certain assumptions and idealizations are made. In the work reviewed here, two-dimensional composite models with periodic microstructures are used. The volume fraction (V_f) and the distribution pattern of the reinforcement are controlled by the amount and configuration of reinforcement within a unit cell. A variety of morphologies are produced, the matrix being treated both as a single crystal and as a polycrystal. The result of this is then, essentially, *ideal microstructures*.

The initial goal of the work reviewed in this chapter is to assess material microstructural performance under variation in microstructural parameters such as reinforcement volume fraction, microstructural morphology and material properties of the matrix. Consider performance to mean deformation behavior, the development of strengthening mechanisms and the development of failure mechanisms. These mechanisms are revealed by relating microscopic behavior to macroscopic behavior. The effects of variation in overall loading state are also considered.

Processing is an important variable in the experimental analysis of MMCs and thus, for the microstructural modeling, to be realistic and predictive, must address the issue of thermomechanical processing. It is important to determine the extent to which the representation of processing alters the predictions of microscale and macroscale behavior for the cases where processing is ignored. This determination allows the assessment of processing significance in the modeling context and, therefore, the assessment of the extent to which physically based modeling must incorporate physical phenomena. In this chapter, initial attempts at the modeling of processing are described and their effects are assessed.

One of the main motivating factors for using the physical modeling approach for the analysis of MMCs is the nonuniformity of microstructure regarding geometric and material properties, particularly because this nonuniformity is at grain level; reinforcing fibers or particulates may be sized similarly to the matrix grains and may be separated by only a few grains. If the nonuniformity was on a larger scale, for instance, reinforcement size and spacing on the order of hundreds of grains, then a phenomenological constitutive description is arguably all that is required. This chapter reviews comparisons of physical and phenomenological approaches in terms of microscale and macroscale behavior predictions.

One of the significant features of the composite models is that slip system strain hardening properties are homogeneous throughout the matrix. This essentially assumes that, except for lattice orientation, the matrix has a uniform microstructure. It has been shown experimentally that the matrix of a composite, when aged, will have a different microstructure and dislocation density than will an unreinforced polycrystal (Nieh and Karlak 1984; Christman and Suresh 1988; Suresh et al. 1989; Arsenault 1984; Vogelsang et al. 1986) or a single crystal of the same alloy for the same aging treatment. However, precipitate density is quite uniform (Christman and Suresh 1988), and therefore justifies the assumption that this would be true. However, in some cases, a nonuniform distribution can result with, for example, preferential precipitation at grain boundaries and surrounding precipitate-free zones.

Section 8.1 reviews the use of crystal plasticity over recent years. It provides examples on modeling the deformation and failure of metals and metal-based materials. Section 8.2 presents the theoretical formulation and most common numerical implementation of crystal plasticity. Section 8.3 describes the composite models, of which there are two main types. The first, as used by Needleman and Tvergaard (1993) and Needleman et al. (1992), is a fiber-reinforced material with a single-crystal matrix. The second, as used by McHugh et al. (1989, 1991, 1993a, 1993b, 1993c and 1993d), is a particulate-reinforced material with a polycrystal matrix. Section 8.4 presents the results of the various simulations for microscale and macroscale behavior. Section 8.5 discusses the results for the single-crystal and polycrystal model types. These results are compared with those of models based on a phenomenological theory of plasticity, specifically, J_2 flow theory. Some general conclusions are presented the Summary.

8.1 Review of Crystal Plasticity Modeling

Crystal plasticity theory has been used successfully for modeling large deformations in metallic single crystals and polycrystals. This chapter assesses the significance of crystal kinematic and constitutive properties on deformation. The analyses presented have been both analytic and computational, much of the latter in the context of the finite element method. Examples are presented in the text that follows.

8.1.1 Modeling Using the Finite Element Method

Nonuniform deformations in single crystals that are subject to tensile loading have been analyzed by Pierce et al. (1982, 1983) using the finite element method. In Pierce et al. (1982) a rate-independent version of the theory is used and the simulations follow the deformation of the crystal through necking and shear band formation. Lattice misorientations within the shear bands occur, and these misorientations increase the resolved shear stress on the dominant slip system, that is, they induce a geometric softening. The issues of self and latent hardening of the crystal are discussed. Pierce et al. (1983) considers a rate-dependent formulation that eliminates nonuniqueness of solution, a characteristic feature of rate-independent theories. The development of shear bands depends greatly on the initial configuration of the crystal slip systems, with primary and secondary slip controlling the extent of the geometric softening. Dependence of strain localization on rate sensitivity is significant; the lower the rate sensitivity the greater the tendency for localization. The predictions of the rate-dependent theory were compared with experimental results for the tensile deformation of internally nitrided single crystals of Fe-Ti-Mn (Dève et al. 1988). Electron diffraction studies of the deformation structures near shear band interfaces showed lattice reorientations, as observed in the simulations. There was also good quantitative agreement between theory and experiment. Harren et al. (1988) investigated shear band formation in single crystals and polycrystals of an Al-3 wt % Cu alloy in plane-strain compression, both experimentally and computationally. For the single crystals, excellent agreement was obtained between both the experimentally observed and the computed forms of the shear bands that developed, both methods exhibiting lattice reorientation and geometric softening within the bands. Shear band development in polycrystals, as studied by Harren et al. (1988) and Dève and Asaro (1989), is a more complex process as it is influenced by grain boundaries and lattice mismatch between grains.

Once again the predictions of the theory regarding the phenomenology of localized deformation and shear band development were in close accord with experimental observations. Shear bands tended to initiate at grain boundary triple points and propagate across the microstructure with seemingly little change in trajectory. The process of shear band transmission across grain boundaries involved an alignment of the lattices, or more specifically, of the slip systems that dominated flow. This is another form of geometric softening. The same polycrystal models were used by Harren and Asaro (1989) to investigate tensile and shear deformations of polycrystals. Crystalline deformation in all of the mentioned finite element models was considered to be purely two dimensional, with out-of-plane slip not explicitly represented. The single crystal and the polycrystal models of Harren et al. (1988) have been extended to FCC crystals in generalized plane-strain deformation by Becker (1991) and Becker and Lalli (1991). This affords the representation of compression in a direction perpendicular to the plane. Nonuniform deformation patterns developed. In the polycrystal case, many grains exhibited sharp transitions in their deformation patterns.

Crystal plasticity has also been used in models that examine crack-tip fields and simulate fracture. For example, Mohan et al. (1992), as part of a series on various aspects of crack-tip fields in crystals, have examined the case of a crack at a bicrystal interface using fully three-dimensional FCC models. Varias et al. (1990) have modeled decohesion and crack propagation at bimaterial interfaces, typically between a ductile single crystal and an elastic substrate. The crystal was modeled using both crystal plasticity theory and J_2 flow theory. This work focused on the investigation of the dependence of deformation and failure modes on interface strength.

8.1.2 Averaging Models for Polycrystals

Crystal plasticity theory has been used extensively for the prediction of overall properties and textures of polycrystals using averaging models such as those based on that of Taylor (1938). An enhanced version of this model for the rate dependent case is developed in Asaro and Needleman (1985). The influence of texture development on overall strain hardening was also examined. Harren et al. (1989) applied this model to the study of large strain shear in FCC polycrystals. The results are compared with those of a variety of phenomenological constitutive theories. In Harren and Asaro (1989), finite element and aggregate model results are compared for an idealized two-dimensional polycrystal, allowing the validity of the aggregate model to be assessed.

8.2 Theory and Numerics

The constitutive theory that is used in the analysis is a finite deformation, elastic-plastic theory. The plasticity is rate dependent and crystallographic. Crystal plasticity theory originates from the work of Taylor (1938), and has undergone rigorous development in Hill (1966), Rice (1971), Hill and Rice (1972), and Asaro and Rice (1977). The theory is reviewed in Asaro (1979, 1983). For the single crystal, plasticity is described solely in terms of continuum shear flows that occur along the various slip systems of the crystal. Hyperelasticity and the representation of rigid motions complete the material description. A version of the constitutive theory that incorporates deformation caused by temperature change is presented later. The constitutive theory is developed in detail in McHugh et al. (1993a). For clarity of presentation in this and subsequent sections, a standardized notation is used that may differ from that used in the various articles reviewed; however, the relationships presented are *identical* in meaning to those of the original articles. In the text that follows, summation over repeated Latin and nonparenthetical Greek indices is implied, and repeated parenthetical Greek indices are not summed.

8.2.1 Single Crystal Theory

The deformation of the crystal from the undeformed stress-free state (considered to be the reference configuration), to the current configuration, is described by the deformation gradient, F, which is subject to the multiplicative decomposition

$$F = F^* \cdot F^\theta \cdot F^p \qquad (8.1)$$

F^p represents the shear flow of material through the undeformed lattice along the various slip systems of the crystal. The spatial gradients of velocity of this plastic shear flow are given by

$$\dot{F}^p \cdot F^{p-1} = \dot{\gamma}_\alpha s_\alpha m_{(\alpha)} \qquad (8.2)$$

where $\dot{\gamma}_\alpha$ is the shear rate on the αth slip system, the latter being defined by the orthogonal pair of unit vectors (s_α, m_α). In the reference configuration s_α is aligned with the αth slip direction and m_α is normal to the αth slip plane. F^θ represents the deformation of the crystal caused by change in temperature θ. The spatial gradients of velocity of this thermal deformation are given by

$$\dot{F}^\theta \cdot F^{\theta-1} = \dot{\theta} a \qquad (8.3)$$

where a is a tensor whose components are thermal expansion coefficients. F^* represents the elastic distortion and rigid body rotation of the crystal. For convenience, we define

$$\hat{F} = F^\theta \cdot F^p; \quad \bar{F} = F^* \cdot F^\theta \quad (8.4)$$

Slip on slip system α is driven by the resolved shear stress on that system, τ_α, where

$$\tau_\alpha = m^*_\alpha \cdot \tau \cdot s^*_\alpha; \quad s^*_\alpha = \bar{F} \cdot s_\alpha; \quad m^*_\alpha$$
$$= m_\alpha \cdot \bar{F}^{-1}; \quad \tau = J\sigma \quad (8.5)$$

$J = \det(F)$ is the Jacobian determinant of the deformation gradient, τ is the Kirchhoff stress, and σ is the Cauchy (or true) stress. The constitutive description of the plasticity on each slip system relates the slip rate (or shear rate) to the resolved shear stress on that system,

$$\dot{\gamma}_\alpha = \dot{a}\,\text{sgn}\{\tau_\alpha\}\left|\frac{\tau_{(\alpha)}}{g_{(\alpha)}}\right|^{\frac{1}{m}} \quad (8.6)$$

where $g_\alpha > 0$ is the current value of the slip system hardness. In Equation 8.6, m is the rate sensitivity exponent and $\dot{a} > 0$ is a reference shear rate, both of which are the same for each slip system. The term sgn{.} means the sign of {.}. Also, m and $\dot{a}$ are taken as constants. The slip system hardness g_α is treated as an internal variable whose current value is obtained by path-dependent integration of the evolution equation

$$\dot{g}_\alpha = h_{\alpha\beta}(\gamma_a,\theta)|\dot{\gamma}_\beta| + g^\theta_\alpha(\gamma_a,\theta)\dot{\theta};$$

$$\gamma_a \int_0^t \sum_\alpha |\dot{\gamma}_\alpha|\,dt \quad (8.7)$$

where $h_{\alpha\beta}$ is a matrix of (non-negative) hardening moduli, g^θ_α is the rate of change of slip system hardness with respect to temperature alone, and γ_a is the accumulated slip that is a measure of total plastic strain, similar to effective plastic strain used in phenomenological plasticity theories. The initial conditions are $g_\alpha(\gamma_a = 0, \theta = \theta_0) = g_0(\theta_0)$, where θ_0 is an initial temperature. g_0 is essentially then the initial slip-system strength. Regarding thermal deformation, the components of a are considered constant, being unaffected by deformation or temperature change. It is important to note that in this formulation, the material temperature distribution and the thermal history are considered to be externally prescribed and are not affected by the deformation; for example, the effects of plastic dissipation on temperature are not incorporated. But, these simplifications do not imply any fundamental limitations on the theory. The specification of the single crystal's elasticity and the derivation of the overall constitutive law have been given different treatments in the literature. In Pierce et al. 1983; Asaro and Needleman 1985; Needleman and Tvergaard 1993; and Needleman et al. 1992, elasticity is phrased in terms of the lattice Jaumann rate of Kirchhoff stress and the elastic part of the rate of deformation tensor. Here however, we follow Harren et al. (1988), Harren and Asaro (1989), and McHugh et al. (1989, 1991, 1993a) where elasticity is phrased in terms of the lattice-based second Piola-Kirchhoff stress, $S^* = F^{*-1} \cdot \tau \cdot F^{*-T}$, and the Lagrangian lattice strain, $E^* = 1/2(F^{*T} \cdot F^* - I)$, where I is the second-order identity tensor. Both forms are entirely equivalent. In rate form;

$$\dot{S}^* = K \cdot \dot{E}^*; \quad K_{ijkl} = \frac{\partial^2 \Phi}{\partial E^*_{ij}\partial E^*_{kl}} \quad (8.8)$$

where Φ is the Helmholtz potential energy of the lattice per unit reference volume, and K_{ijkl} and E^*_{ij} are the components of K and E^* respectively, in a time-independent, orthonormal Cartesian basis. We assume that the crystal is elastically isotropic, which is usually sufficient for metals. Therefore, K is the usual isotropic linear elastic fourth-order tensor. We also assume that the elastic constants are independent of temperature. The entire constitutive theory is phrased in terms of the second Piola-Kirchhoff stress, $S = F^{-1} \cdot \tau \cdot F^{-T}$, and the Lagrangian strain, $E = 1/2(F^T \cdot F - I)$. Straightforward tensor manipulations of the above relations, as detailed in McHugh et al. (1993a), yield the governing rate form

$$\dot{S} = L:\dot{E} - \dot{\gamma}_\alpha X_\alpha - \dot{\theta}Y \quad (8.9)$$

where

$$L_{ijrn} = \hat{F}^{-1}_{ik}\hat{F}^{-1}_{jl}K_{klpq}\hat{F}^{-1}_{rp}\hat{F}^{-1}_{nq} \quad (8.10)$$

(the components of the tensors being with respect to the Cartesian basis) and

$$X_\alpha = \hat{F}^{-1} \cdot \{K:A_\alpha + 2H_\alpha\} \cdot \hat{F}^{-T};$$
$$Y = \hat{F}^{-1} \cdot \{K:B + 2Q\} \cdot \hat{F}^{-T}$$
$$A_\alpha = \text{sym}\{F^{*T} \cdot F^* \cdot F^\theta \cdot \{s_\alpha m_{(\alpha)}\} \cdot F^{\theta-1}\};$$
$$B = \text{sym}\{F^{*T} \cdot F^* \cdot a\}$$
$$H_\alpha = \text{sym}\{F^\theta \cdot \{s_\alpha m_{(\alpha)}\} \cdot F^{\theta-1} \cdot S^*\};$$
$$Q = \text{sym}\{a \cdot S^*\} \quad (8.11)$$

8.2.2 Numerical Implementation

In the reviewed articles, the rate-constitutive equations of the theory are integrated using the one-step, explicit *rate tangent* algorithm, originally developed and fully described by Pierce et al. (1983). An enhanced version of that algorithm, incorporating temperature change, required for the version of the theory that integrates Equation 8.9, is presented in detail in McHugh et al. (1993a).

Boundary value problems can be solved using the finite element method. The formulation is based on the principle of virtual work and has been well documented in the literature (Needleman 1972; Pierce et al. 1983; Dève et al. 1988; Needleman and Tvergaard 1993; McHugh et al. 1993a). The details are not repeated here. In the reviewed articles, the type of finite element used is the "crossed triangle" quadrilateral, in which four constant strain triangles are arranged with edges along the diagonals of the quadrilateral and having the central node eliminated by static condensation. This results in an element that is well suited to handling incompressible plasticity (Nagtegaal et al. 1974).

8.3 Model Descriptions

8.3.1 Single Crystal with Embedded Fibers

In the work of Needleman and Tvergaard (1993) and Needleman et al. (1992), the composite's matrix is modeled as a single crystal embedded with a doubly periodic array, that is, periodic in both the horizontal and vertical directions of the reinforcing fibers/whiskers. This structure can be envisaged as being composed of a doubly periodic array of identical rectangular unit cells, of dimensions $2\omega_0$ in the horizontal direction and $2L_0$ in the vertical direction. During plane-strain tensile or compressive deformation in the vertical direction, the boundaries of the unit cell remain straight, do not rotate, and support no shear tractions. This also applies to horizontal and vertical lines through the center of the unit cell. These constraints result from periodicity and the added requirement of mirror symmetry of the array. One can view the unit cell as being constructed from its upper-right-hand quadrant in the following way. First, reflect the material in the quadrant about its bottom edge, generating the lower-right-hand quadrant and then reflecting the right half about its left edge, generating the left half. Under these conditions, the behavior of one quadrant of the unit cell characterizes the behavior of the complete unit cell and, hence, of the whole composite. The upper-right-hand quadrant is analyzed numerically with the following boundary conditions

$$\dot{u}_2 = 0, \quad \dot{T}_1 = 0 \quad \text{on} \quad x_2 = 0$$
$$\dot{u}_2 = \dot{U}_2 = \dot{\varepsilon}_{ave}(L_0 + U_2), \quad \dot{T}_1 = 0 \quad \text{on} \quad x_2 = L_0$$
$$\dot{u}_1 = \dot{U}_1, \quad \dot{T}_2 = 0 \quad \text{on} \quad x_1 = \omega_0 \quad (8.12)$$

Here, $\dot{u}_1$ and $\dot{u}_2$ are components of the material particle velocity, $\dot{T}_1$ and $\dot{T}_2$ are components of the nominal traction rate, and x_1 and x_2 are material particle coordinates in the reference configuration, all with respect to an orthonormal Cartesian basis and an origin at the center of the unit cell. Also, $\dot{\varepsilon}_{ave}$ is a prescribed constant tensile or compressive strain rate, with magnitude taken equal to $\dot{a}$ of Equation 8.6 in the simulations. $\dot{U}_1$, the normal velocity of the right edge of the quadrant, is determined from the condition of zero-total loading rate in the horizontal direction, that is,

$$\int_0^{L_0} \dot{T}_1 dx_2 = 0 \quad \text{on} \quad x_1 = \omega_0 \quad (8.13)$$

The crystal matrix has a two-dimensional, idealized lattice geometry. There are three slip systems with slip directions making angles of 60, 120, and 180 degrees with the positive x_1 axis. The slip systems can be viewed as forming an equilateral triangle. Perfect bonding at the matrix fiber interface is assumed. All deformations are considered isothermal. The boundary value problem is solved using a combined finite element-Galerkin method. The macroscopic behavior of the composites is quantified in terms of overall strain, $\varepsilon_{ave} = \ln(1 + U_2/L_0)$, and overall stress is defined as

$$\sigma_{ave} = \frac{1}{\omega_1 + U_1} \int_0^{\omega_0} T_2 dx_1 \quad (8.14)$$

Regarding the strain hardening description of the crystal, the hardening moduli, $h_{\alpha\beta}$ of Equation 8.7, which completely determine the evolution of the slip system hardness for the isothermal case, are given by

$$h_{\alpha\beta} = qh(\gamma_a) + (1-q)h(\gamma_a)\delta_{\alpha\beta} \quad (8.15)$$

q determines the *latent hardening* nature of the crystal. The value $q = 1$, as used in the simulations, corresponds to *isotropic* hardening (Taylor 1938). The function $h(\gamma_a)$ is given by the following power-law

$$h(\gamma_a) = h_0\left(1 + \frac{\gamma_a}{\gamma_0}\right)^{N-1} \quad (8.16)$$

where h_0 is the initial slip system hardening rate, γ_0 is a reference strain, and N is the strain hardening exponent.

In Needleman and Tvergaard (1993), each unit cell contains a single rigid fiber aligned with the vertical or loading direction. Only tensile deformations are considered. Reinforcement volume fractions (V_f) of 10%, 20%, and 30% are used, with a fixed fiber length to width ratio of 4 and a fixed unit cell length to width (L_0/ω_0) ratio of 4. Some simulations are performed for a matrix lattice geometry different from the one described above; two slip systems with slip directions oriented at 60 degrees and 120 degrees to the horizontal. The crystal's material properties are as follows: E (Young's modulus) = $533.3g_0$ and ν (Poisson's ratio) = 1/3, $h_0 = 10g_0$, $\gamma_0 = 0.01$, $N = 0.1$, and $m = 0.005$.

In Needleman et al. (1992), the emphasis is on investigating clustering effects and the cell models of Christman et al. (1989) are used. Each unit cell contains eight fibers, all aligned with the loading direction. In this case, the fibers are elastic. Three different morphologies are considered, as shown in Figure 8.1, and the configurations are such that in each case the unit cell can be considered to consist of eight subcells, two per quadrant. The V_f value is 13.2%, with a fixed fiber length to width ratio of 5 and a fixed subcell length to width ratio of 6. In Figure 8.1(a), the fiber ends are aligned. In Figure 8.1(b), the fibers have undergone a 50% horizontal shift, the amount of shifting being quantified by $[(d-d')/d] \times 100\%$. In Figure 8.1(c), the fibers are shifted vertically by 50%, the shifting quantified by $[(a-a')/a] \times 100\%$. The crystal's material properties are as follows: $E_{matrix} = 500g_0$ and $\nu = 1/3$, $h_0 = 68g_0$, $\gamma_0 = 0.01$, $N = 0.14$, and $m = 0.005$. For the elastic reinforcement, $E_{fiber} = 5.48E_{matrix}$ and $\nu = 1/3$. An 18×66 quadrilateral finite element mesh is used for the uniform and 50% horizontal clustering distributions, while a 36×33 mesh is used for the 50% vertical distribution. The models are used to perform both tensile and compressive simulations.

8.3.2 Polycrystalline Particulate Composites

In the work of McHugh et al. (1989, 1991, 1993a, 1993b, 1993c and 1993d) a particular type of composite material is modeled: Al-3 wt % Cu alloys reinforced with SiC particulates. Experimental stress-strain data is used to determine the strain hardening behavior of the matrix. The composite model is based on the two-dimensional polycrystal model of Harren et al. (1988) and Harren and Asaro (1989). In this model, all grains are the same size and have the same hexagonal shape in the reference configuration. As with the models described in the previous section, the polycrystal consists of a doubly periodic array of rectangular unit cells. The

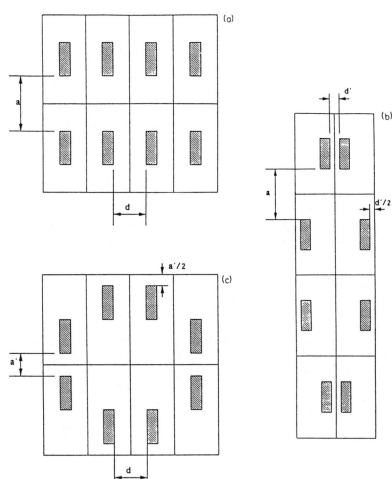

Figure 8.1. Unit cells used to analyze distribution effects in the single crystal composites. (a) Uniform distribution, (b) 50% horizontal clustering, and (c) 50% vertical clustering. (Reprinted by permission from Needleman, A., Suresh, S., and Tvergaard, V. 1992. Deformation of a Metal-Ceramic Composite with a Crystal Matrix: Reinforcement Distribution Effects. Berlin: Springer-Verlag.)

unit cell has horizontal and vertical dimensions of $2H_0$ and $2L_0$, respectively, in the reference configuration. The constraints on the deformation of the unit cell as a result of periodicity, and the preservation of mirror symmetry, are the same as described above. That is, the boundary of a quadrant of the unit cell must remain rectangular in shape during deformation and must support no shear tractions. Bear in mind that the complete unit cell is generated by reflection of the upper right hand quadrant about its bottom and left edges. There are 27 grains (or single crystals) per quadrant. The lattice geometry is the same as that described above for the single crystal composites, meaning that the slip systems form an equilateral triangle. In the polycrystal model, the triangle's initial angle of orientation with respect to the reference axes, ψ, is randomly chosen for 23 of the grains so that each of these grains has its own randomly assigned initial lattice orientation. As explained in Harren et al. (1988), Harren and Asaro (1989), and McHugh et al. (1993a) the remaining four *half grains* are assigned a ψ value of zero. Because the companion halves of these grains are generated by reflection, if ψ had some other value, these grains would then be split by a subgrain boundary. Two-dimensional particulate-reinforced composites are modeled by replacing selected grains within the unit cell with elastic reinforcing particulates. Each grain constitutes a volume fraction of 4%. To investigate the effects of variation in volume fraction and morphology, the replacement is done systematically, with the chosen configurations shown in Figure 8.2. V_f values of 4%, 8%, 16%, and 20% are considered with two distribution patterns, referred to as *a* and *b*, for the 8% and 16% cases. Two basic types of simulations are performed, deformation at a constant room temperature and thermomechanical processing involving temperature change and deformation.

In regard to the crystal strain hardening description, the hardening moduli are as given in (8.14), with $q = 1$. Because the model is used for thermomechanical processing simulations in addition to isothermal deformation simulations, the evolution of slip system hardness will generally depend on γ_a and θ, (Equation 8.7). To account for this, the following is assumed

$$h(\gamma_a) = \frac{\partial g(\gamma_a, \theta)}{\partial \gamma_a} \quad g_\alpha^\theta(\theta) = \frac{\partial g(\gamma_a, \theta)}{\partial \theta} \quad (8.17)$$

where the hardness function $g(\gamma_a, \theta)$ is taken as

$$g(\gamma_a, \theta) = g_0(\theta) + h_\infty \gamma_a + [g_\infty - g_0(\theta_R)] \tanh\left\{\gamma_a \left[\frac{h_0 - h_\infty}{g_\infty - g_0(\theta_R)}\right]\right\} \quad (8.18)$$

θ_R is a reference temperature, g_∞ represents the asymptotic slip-system strength when $h_\infty = 0$ and $\theta = \theta_R$; h_0

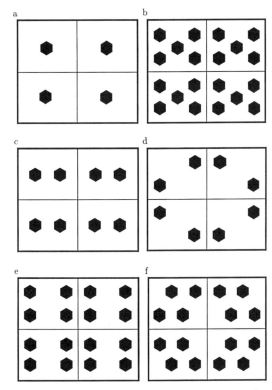

Figure 8.2. Unit cell morphologies used in the polycrystalline composite simulations. V_f values and morphologies are: (a) 4%, (b) 20%, (c) 8%a, (d) 8%b, (e) 16%a, and (f) 16%b. (Reprinted by permission from McHugh, P.E., Asaro, R.J., and Shih, C.F. 1993. *Acta Metall. Mater.* 41: 1461, Oxford: Pergamon Press.)

and h_∞ are the initial and asymptotic rates of hardening respectively, at a fixed temperature. Implicit in the above is the significant simplifying assumption that temperature dependence enters through the initial slip-system strength only.

The room temperature strain hardening properties of the matrix are obtained from the experimental data of Chang and Asaro (1981) for room temperature tensile deformation of Al-3 wt % Cu single crystals. The crystals were deformed in single slip so that tensile stress-strain data can be readily reduced to yield actual slip-system shear/stress-shear strain (τ–γ) data. For a single active slip system in monotonic loading, $\gamma_a = \gamma$ so that the function in Equation 8.18 is fit directly to the τ–γ data. In fact, obtaining an accurate fit to actual slip-system stress strain curves is the main motivation for using the functional form given in Equation 8.18. θ_R is taken as room temperature. In the experiments, crystals containing a nominal solid solution (*SS*) were used as were others that had been age hardened to produce microstructures containing GP-II zones and θ', each of the three types having different strain hardening characteristics. This selection allows assessment of the effects of matrix hardening properties on performance. The values obtained, normalized by the g_0 value for the θ' alloy, are as follows, θ': $g_0 = 1.0, g_\infty = 1.8, h_0 = 8.9$, and $h_\infty = 0.0$; GP-II: $g_0 = 0.948, g_\infty = 1.715, h_0 = 2.205$, and $h_\infty = 0.287$; SS: $g_0 = 0.924, g_\infty = 1.279, h_0 = 2.722$, and $h_\infty = 0.557$. It should be emphasized however, that using the single crystal properties does not necessarily allow direct assessment of the dependence of composite behavior on aging treatment because experiments show that accelerated aging occurs in the matrix (Nieh and Karlak 1984; Christman and Suresh 1988; Suresh et al. 1989). As has been fully documented in Chapter 7, the above mentioned authors as well as others have attributed the accelerated aging to the higher dislocation densities produced in the matrix during cooling, as a result of thermal expansion mismatch between matrix and reinforcement. It is interesting to note that the experiments of Suresh et al. (1989), for an Al-3.5 wt % Cu alloy reinforced with SiC particulates, indicate that V_f does not significantly affect aging acceleration above a critical value of approximately 6%. The data of l'Esperence et al. (1984) for the deformation of solutionized 2014 Al alloy is used to approximate the variation of g_0 with temperature for the SS case. This data shows an approximate twenty-fold reduction in proof strength for increasing the temperature from 22° C to 500° C. The matrix viscoplastic properties are $\dot{a} = 0.001 s^{-1}$ and $m = 0.005$ in all cases. The reinforcement is assumed to be elastically isotropic, with the same elastic constants as the matrix, which in turn are assumed to be the same for the three microstructure types. The values are, $E = 1000g_0$ (g_0 for the θ' alloy) and $\nu = 0.3$. Isotropic thermal expansion is assumed, with thermal expansion coefficients of $2.2 \times 10^{-5} K^{-1}$ for the matrix and $0.49 \times 10^{-5} K^{-1}$ for the reinforcement. A 40×56 quadrilateral finite element mesh is used, the initial dimensions chosen so that the grains have regular hexagonal shapes. All grain boundaries coincide with either an edge or a diagonal of a quadrilateral element. Particular composites are referred to by using a symbol that indicates matrix type, reinforcement volume fraction, and microstructural morphology. For example, SS-16%a signifies a composite with an SS matrix, and a V_f of 16% with morphology a. Also, all composites with the 16% reinforcement volume fraction and the a morphology will be referred to as 16%a composites and all composites with GP-II strengthened alloy matrices will be referred to as GP-II composites. The macroscopic behavior of the composites is quantified in terms of overall strain, $\varepsilon = \ln(1 + U_2/L_0)$, and the overall stress is defined as

$$\bar{\sigma} = \frac{1}{H_0 + U_1} \int_0^{H_0} T_2 dx_1 \qquad (8.19)$$

Room temperature tension simulations of up to 5% engineering strain, and compression simulations of up to 25% engineering strain are performed, where engineering strain is defined as $\varepsilon_{eng} = (1 + U_2/L_0)$. These are referred to as the isothermal deformation simulations. Two types of thermomechanical processing simulations are then performed. The first corresponds to a simple quench from 495° C to room temperature (22° C). The temperature is considered to be spatially homogeneous and to decrease exponentially with time (time constant = 1 sec). During the quench, the matrix alloy is considered to be in the solutionized state. The quench is followed by an aging to produce θ' in the matrix. During aging, it is assumed that no further deformation occurs and that the implementation of aging simply requires the replacement of the solutionized alloy strain-hardening properties with those of the θ' strengthened alloy. This process is designated by the symbol Q. In the second type of process, the material is isothermally compressed at 495° C up to an overall engineering strain of 15% in the vertical direction, after which it is isothermally unloaded at the same temperature. It is then quenched in the same way as in the Q process. The compressive level of 15% was chosen to be reflexive of the large strains encountered during extrusion and rolling, given that the actual strains encountered during such processing could not be represented with any great accuracy using the current modeling approach. This second type of process is designated by the symbol $C\&Q$. Composites that have been processed are identified by a symbol that indicates the process that they have undergone, the reinforcement volume fraction, and the microstructural morphology. For example, Q-16%a signifies a composite with a V_f of 16% and a morphology, a, having undergone the Q process.

8.4 Results

On the microscale, the presence of the reinforcement results in highly nonuniform deformation patterns. Stress distributions are also nonuniform. On the macroscale, all simulations show an increase in composite flow strength or overall hardness with increase in V_f. This increase depends on the microstructural morphology.

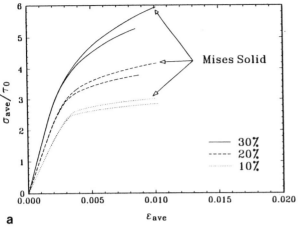

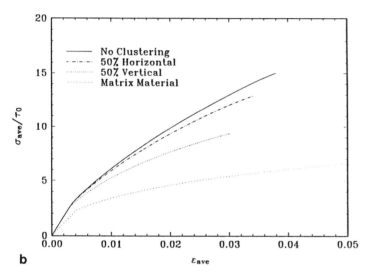

Figure 8.3. Plane strain tensile stress-strain curves for single crystal composites. (Note that τ_0 is equivalent to g_0). (a) Fiber ends are aligned and the crystal matrix has three slip systems. Curves for corresponding J_2 flow theory composites are also shown. (b) Curves for the unit cells shown in Figure 8.1. A curve for unreinforced matrix is also shown, (·) marks the maximum stress point.

8.4.1 Single Crystal with Embedded Fibers

On the macroscale, an increase in V_f has the effect of increasing the elastic modulus and making the elastic-plastic transition more gradual, as is evidenced by the stress-strain curves of Figure 8.3(a).[1] The effects of reinforcement clustering are shown in Figure 8.3(b). The effects are negligible in the elastic region although differences become increasingly pronounced as plastic straining proceeds. The shifting of the fibers leads to a softening, particularly for the 50% vertical shift material that is significantly softer than either of the other two materials. The compressive stress-strain curves essentially coincide with the tensile curves of Figure 8.3(b). Varying the number of slip systems (using only two slip systems), is shown by Needleman and Tvergaard (1993) to have the effect of increasing overall hardening rates.

On the microscale, the simulations of Needleman and Tvergaard (1993) for aligned arrays of fibers show that plastic strains are nonuniform and are concentrated in regions adjacent to the fiber corners. The dominant slipping occurs on the second slip system, the one making an angle of 120 degrees with the positive x_1 axis, and strain is localized to a narrow band essentially parallel to the initial orientation of that slip system. Strain rates reach very high values in these highly strained regions. The simulations of Needleman et al.

---

[1] In Needleman and Tvergaard (1993) for a V_f of 30% a stress maximum was reached beyond which the stress dropped. Similar stress drop were reported in Needleman et al. (1992). The authors later found these stress drops to be incorrect. They were caused by a programming error. The results prior to stress drop were unaffected by the error.

(1992) show the effects on the deformation patterns of change in microstructural morphology through fiber clustering for a fixed V_f. Deformation patterns are controlled by the positions and shapes of the fibers. γ_a contour maps, given in Needleman et al. (1992), show that for 50% vertical shift, bands of strain traverse the matrix connecting fiber corners. They also show that for 50% horizontal shift, case localized straining tends to be confined to the vicinity of a single fiber. Plastic strain levels are higher for the 50% vertical shift composite.

8.4.2 Polycrystalline Particulate Composites— Isothermal Deformations

On the macroscale for both tension and compression, an increase in V_f results in an increase in overall hardening rate, especially at lower plastic strains. Unlike the single-crystal composites, the introduction of the reinforcement into the polycrystal does not affect the elastic behavior because it is assumed that the elastic properties of the matrix and reinforcement are equal. In compression, the 16%(a) and 20% composites exhibit a significant increase in hardening rate at high strain. These observations are illustrated by the compressive stress-strain curves for the GP-II and θ' composites shown in Figures 8.4(a) and (b), respectively. The strengthening or hardening effect of the reinforcement can be quantified by a *hardness increment*, which we define as the fractional increase in hardness of a composite over that of its corresponding unreinforced polycrystal at a given strain level. (That is, a measure of composite strength relative to unreinforced matrix strength.) At a given overall strain, hardness increments increase but do not quite scale with V_f. Hardness increments also vary with strain and the variation is different from composite to composite. Strengthening depends on morphology and this dependence becomes greater with higher values of V_f. Matrix strain hardening properties influence the hardness increments. The average hardness increments for the SS, GP-II and θ' composites are approximately equal to $0.9V_f$, $1.0V_f$, and $0.5V_f$, respectively, at strain $\epsilon = -0.01$. Composites are slightly stronger in compression than in tension, with the exception of the 16%b materials.

It is useful to compare microscale deformation behavior of polycrystalline composites with that of the unreinforced polycrystals, because the work of Harren et al. (1988), as discussed above, has shown that nonuniform deformation patterns develop in single-phase polycrystals. This nonuniformity of strain is, of course, not caused by the presence of a second phase and, for the composites, it is of interest to determine the extent to which the second phase changes the single-phase microscale deformation behavior. It is observed that, in the composites, nonuniform deformation patterns develop from the onset of macroscopic yield and at lower overall strains than those for the unreinforced material. The composite deformation patterns are different than those of the unreinforced material in that strain tends to localize in different regions of the microstructure. These regions are usually in the vicinity of the reinforcing particles and are thus determined by the positions and shapes of the particles. But, strain localization does occur, albeit to a lesser extent, in grains remote from the particles. This is because of the polycrystalline nature of the matrix. At the same overall strain level, localization intensity and peak strain levels in the composites are higher than they are in the unreinforced polycrystal. Figure 8.5(a) and (b) shows two γ_a maps for the GP-II-16% composites at overall compressive strains of 5%. Note the tendency for plastic strain to localize into bands parallel to the oblique facets of the particle, showing the flow of matrix material along these facets. These bands are a characteristic feature of the deformation patterns of all the polycrystalline composites. The general trends are that increasing V_f results in (1) enhanced localization of strain; in a high V_f composite there are larger volumes of matrix at very low plastic strains than there are in a low V_f composite, and (2) the related increase in peak strain. The maps of Figure 8.5 illustrate the general effects of morphological change for a constant V_f. Localization intensity and peak strain are affected as both are higher for the 16%a material in comparison with the 16%b material. Deformation patterns are strongly affected with different modes of deformation becoming active. In 16%a during compression, particles approach in the vertical direction and cause intense straining of the separating matrix. In 16%b, stretching of grains separating particles in the horizontal direction is evident. Regarding the effects of changing matrix strain hardening properties one finds that deformation patterns are not significantly affected, as they are mostly determined by the microstructural morphology. Localization intensity and peak strain levels are, however, affected. The θ' composites exhibit higher localization intensities and peak strain levels than do the GP-II and SS composites (deformation in the latter two being more diffuse). Deformation patterns, at least at small strains, are not significantly affected by change in loading sense (tension versus compression). At larger strains, differences arise because of the finite displacement of particles, which essentially results in morphologic changes.

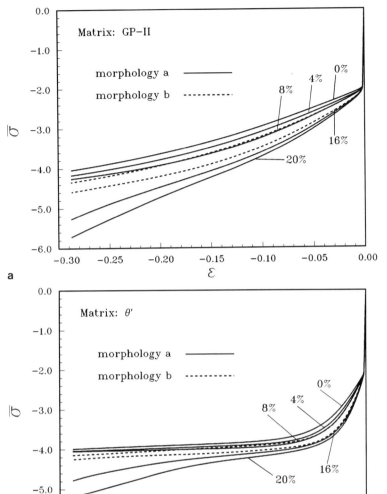

Figure 8.4. Plane strain compressive stress-strain curves for polycrystalline composites. (a) Matrix: GP-II and (b) matrix: θ'. The curves for the unreinforced polycrystals are also shown. Stresses are normalized by g_0 for the θ' alloy. (Reprinted with permission from McHugh, P.E., Asaro, R.J., and C.F. Shih. 1993. *Acta Metall. Mater.* 41:1477, Oxford: Pergamon Press.)

8.4.3 Polycrystalline Particulate Composites — Processing and Post-Processing Deformations

The Q process induces residual stresses and strains in the material. Plastic strains are induced in the matrix surrounding the particles, with highest intensity at the particle-matrix interface, especially at the vertices where they are on the order of 4%. At further distances into the matrix, deformation tends to be more intense in the direction of an adjacent particle, although this depends on the particle deformation pattern as a whole. The general trend, as illustrated in McHugh et al. (1993d), is that plastic strain in the matrix is most intense in directions that contain a succession of relatively closely spaced particles. Contour maps of hydrostatic stress show that particles are generally in compression and that peak tensile values are reached in the matrix halfway between two particles. Although peak γ_a values are not proportional to V_f, and actually increase up to 16% and dropping for higher V_f, volume average γ_a values (being approximately an order of magnitude smaller than peak values) are very nearly proportional to V_f. The $C\&Q$ process induces residual stresses and strains of considerably higher magnitude than does the Q process. The deformation patterns and peak stress and strain levels are similar to those produced during the isothermal deformation simulations. This is because the dominant part of the $C\&Q$ process is the high constant temperature compression. It is

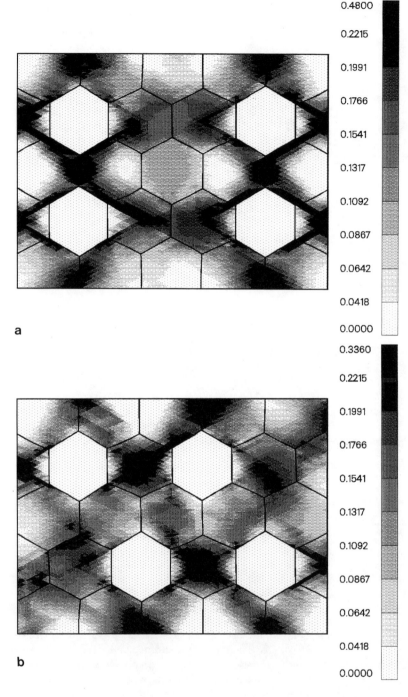

Figure 8.5. Contour maps of accumulated slip, γ_a, for the GP-II-16% (polycrystalline) composites at $\varepsilon_{eng} = -0.05$. (a) Morphology "a." (b) Morphology "b." (Reprinted with permission from McHugh, P.E., Asaro, R.J., and C.F. Shih. 1993. *Acta Metall Mater.* 41:1461, Oxford: Pergamon Press.)

interesting to note that peak stress values still remain high in the processed material even though it is unloaded before quenching. Isolating the strains produced during the quench step of the *C&Q* process shows that deformation is considerably more localized than it was for the *Q* process. Peak γ_a values are quite similar, but average γ_a values are only approximately half those of the *Q* process, illustrating the small contribution of the quench step to the development of the residual fields in the *C&Q* process.

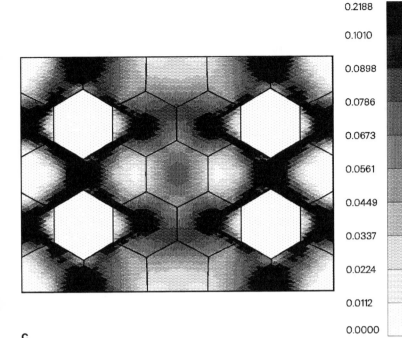

Figure 8.5. *continued* (c) Contour map of effective plastic strain, ε_{eff}, for the GP-II-16%a (J_2 flow theory) composite at $\varepsilon_{eng} = -0.05$. (Reprinted with permission from McHugh, P.E., Asaro, R.J., and C.F. Shih. 1993. *Acta Metall. Mater.* 41:1461 and 1489, Oxford: Pergamon Press.)

c

The post-processing behavior of the materials is compared with the behavior of the corresponding untreated materials, that is, the results of the isothermal deformation simulations for the θ' composites. Attention is restricted to the 16% and 20% composites. On both the macroscale and microscale, the behavior of quenched materials does not really differ from that of untreated materials having the same morphologies. Small differences between macroscopic tensile- and compressive-yield behavior are introduced. Yielding also becomes more gradual with identifiable yield strains being approximately doubled. This is illustrated by the stress-strain curves shown in Figure 8.6 for the 20% materials. On the microscale, for the same overall strain level, peak strain values are increased and peak hydrostatic stress values are reduced by amounts of about 5%. The effects of the *C&Q* process are more significant, as this is also apparent on the macroscale (Figure 8.6). Plastic strain patterns on the microscale are similar in tension and compression, implying that there are no major differences in plastic flow development

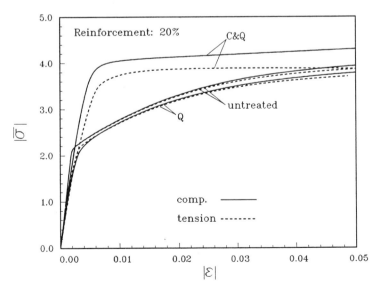

Figure 8.6. Plane strain tensile and compressive stress-strain curves for the 20% V_f untreated *Q* and *C&Q* (polycrystalline) composites. Stresses are normalized by g_0 for the θ' alloy. (Reprinted with permission from McHugh, P.E., Asaro, R.J., and C.F. Shih. 1993. *Acta Metall. Mater.* 41:1501, Oxford: Pergamon Press.)

between reverse loading and continued compressive loading from the initial highly compressed state.

8.5 Discussion

8.5.1 Composite Strengthening and Failure

The strengthening effect of the reinforcement can be understood by correlating macroscopic stress-strain behavior with microscopic deformation and stress patterns.

This part of our discussion will be confined to polycrystalline composites. Consider the effects of the presence of the reinforcement on the strength or hardness of the matrix, given that the slip systems harden with strain. As we have observed, plastic strain intensity, at any given strain, is higher in the matrix than it is in the unreinforced polycrystal. First, there is less material with a volume fraction of $(1 - V_f)$ that can deform plastically. Second, plastic strain is more localized. Both factors combine to yield much higher plastic strains in regions where the majority of the macroscopic deformation is accommodated. Therefore, the strain hardening of the slip systems in the composite's matrix is more advanced than it is in the unreinforced polycrystal. Based on this, the strength of matrix and hence of the composite is expected to be greater than that of the unreinforced polycrystal. This effect is enhanced with an increase in V_f and is affected by morphologic changes. This rationale was first proposed by Drucker (1966). This strengthening mechanism, based solely on the strain hardening of the matrix, is termed *matrix hardness advancement* in McHugh et al. (1993b). If slip system strain hardening rates are increased, the enhanced plastic strain levels in the composite's matrix will result in more significant increases in matrix strength. Therefore, composite hardness increments, for a fixed V_f and morphology (because these essentially set the deformation pattern), will be greater for composites with higher matrix strain hardening rates. The results for the polycrystalline composites illustrate this effect. Stress-strain curves for all but two of the θ' composites have essentially flattened out beyond 8% compressive strain (as shown in Figure 8.4b) with hardness increments on the order of $0.5V_f$. This value is lower than the average hardness increments for the GP-II and SS composites, as reported in the previous section, and is also lower than the average hardness increment value for the θ' composites themselves at compressive strain of less than 5%. This relates to the fact that the slip-system strain hardening rate of the θ' alloy decreases to a zero limiting value hardness saturation ($h_\infty = 0$) with increased plastic strain.

Matrix hardness advancement alone does not completely explain the stress-strain response of Figure 8.4(b). At overall strains in excess of approximately 10%, the localized strains in both the unreinforced and the reinforced materials are well beyond those that would cause saturation of slip system strain hardness (the saturation value is $g_\infty = 1.8g_0$). With no further slip-system hardening possible, the rationale for matrix hardness advancement would predict no strengthening and zero hardness increments. The behavior can be explained in terms of the geometric constraints on plastic flow in the matrix imposed by the reinforcement, as first discussed by Drucker (1966). Localization accompanies constraint because plastic strain is confined to geometrically favorable paths, indeed localization intensity is a fairly good indicator of constraint strength. When plastic flow is constrained, volumetric or compressible deformation becomes significant. Compressible deformation is completely elastic and determines the hydrostatic (or mean) stress state. It has been shown that the overall stress in the loading direction, as plotted in the stress-strain curves, is essentially proportional to the volume-average hydrostatic stress so that plastic constraint is a significant strengthening mechanism (McHugh et al. 1993b). Plastic constraint is essentially the only active mechanism in the θ' composites at high compressive strain; the fact that it explains the strong morphologic dependence of the 16% composites is illustrated by the hydrostatic stress distribution functions for these materials as plotted in Figure 8.7. The area under a distribution curve between any two hydrostatic stress levels equals the volume fraction of material supporting hydrostatic stresses between the two levels. The 16%b material is characterized by very little change in hydrostatic stress state with increase in overall strain; in contrast, the curves for the 16%a material flatten out with progressively larger volumes of material experiencing high hydrostatic compression. Contour maps of hydrostatic stress show this buildup to be in the particles and the matrix separating them in the vertical direction, with the local matrix constraint being enhanced by the approach of the particles (recall the deformation patterns of Figure 8.5). For composites with strain hardening matrices, both the matrix hardness advancement and plastic constraint are active.

The same is true for single-crystal composites. Contour maps of hydrostatic stress for the two shifted morphologies show significant differences in hydrostatic stress levels, indicating significant differences in constraint strength, and correlating with the macroscopic stress-strain behavior shown in Figure 8.3b (Needleman et al. 1992).

Thus far, no reference has been made to lattice reorientation. One might expect it to modify the above

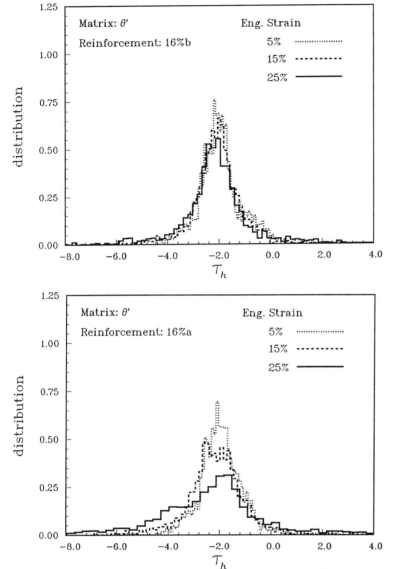

Figure 8.7. Kirchhoff hydrostatic stress, τ_h, distribution curves for the θ'-16% (polycrystalline) composites at various indicated engineering strain levels in compression. Stresses are normalized by g_0 for the θ' alloy. (Reprinted with permission from McHugh, P.E., Asaro, R.J., and C.F. Shih. 1993. *Acta Metall. Mater.* 41:1477, Oxford: Pergamon Press.)

conclusions on strengthening because the investigations discussed in Section 8.1.1 have shown it to be a geometric softening mechanism in both single crystals and polycrystals. In polycrystals, an additional factor affecting strength is constraint caused by grain interaction, a natural consequence of plastic strain tending to occur in different preferred directions in adjacent grains. The compressive stress-strain curve for the unreinforced θ' polycrystal (see Figure 8.4b) shows that, at least for the unreinforced materials, the combined effect of these two factors is small because, in the absence of slip system strain hardening, which is essentially complete at 10% strain, the curve deviates only slightly from the horizontal. It would seem reasonable then to assume that these factors do not significantly change the above interpretation of the results on strengthening for the polycrystalline composites. Lattice reorientation would be expected to be more significant for the single-crystal composites, with geometrical softening occurring alone in the absence of grain interaction constraint.

Depending on the processing method, thermomechanical processing influences the behavior of the polycrystalline composite model during subsequent deformation to different extents. The effects of the C&Q process are probably unrealistically large, because only initial strength and not strain hardening rates are assumed to vary with temperature (hardening rates are too high at high temperatures). Also, the constitutive theory does not include any microstructural recovery mechanism.

In addition to failure through localization, other modes of failure can be inferred by microstructural stress levels. High tensile hydrostatic stresses develop in the matrix at reinforcement corners and this has implications for void nucleation and growth as well as for interfacial decohesion. Also, the reinforcement is subjected to high hydrostatic tension, which could lead to particle fracture.

8.5.2 Comparison with Results of Phenomenological Models

Much analytic and numerical modeling has been done to investigate the behavior of discontinuously reinforced metal-matrix composites using phenomenological theories. These include J_2 flow theory and J_2 deformation theory, which describe matrix plasticity. It is beyond the scope of this chapter to review all such work, however more complete surveys appear in articles such as Christman et al. (1989), McHugh et al. (1993a), and in other chapters of this volume. However, this body of work is relevant in the context of this chapter because nonuniform matrix deformation, plastic constraint, and the importance of matrix strain hardening rate are revealed and explained using the phenomenological theories (e.g., Christman et al. 1989; Bao et al. 1991; Brockenbrough et al. 1991). It is important to assess the differences in the predictions of the crystal theory and the phenomenological theories. The articles reviewed report on finite element simulations performed using composite models with the geometries described in Section 8.3, where J_2 flow (Mises) theory with isotropic power-law strain hardening is used to characterize matrix plasticity. The finite deformation rate dependent formulation of Pierce et al. (1984) is used and the details are not repeated here (see also Needleman and Tvergaard 1993; McHugh et al. 1993c). The elastic and viscoplastic properties are the same as those used in the crystal plasticity simulations. The strain hardening behavior of the J_2 flow theory material is specified as follows. In Needleman and Tvergaard (1993), the strain hardening exponent is the same as that used in the crystal plasticity simulations and the reference stress (essentially the plane strain tensile strength) is taken as $2.039g_0$, in which 2.039 is the plane strain Taylor factor for the model lattice geometry with $\nu = 1/3$. In McHugh et al. (1991, 1993c), the strain hardening properties are determined by fitting the plane strain stress-strain curve for the unreinforced J_2 flow theory material to the computed curve for the unreinforced polycrystal for each of the three alloy types.

Regarding macroscopic behavior, the effects of increasing V_f and changing morphology are similar for both types of matrix characterizations. This is illustrated by the stress-strain curves of Figures 8.3(a) and 8.8(a) here, and other figures given in Needleman and Tvergaard (1993) and McHugh et al. (1991, 1993c). Needleman et al. (1992) report that the influence of morphology on the overall stress-strain behavior in Figure 8.3(b), is qualitatively the same as that obtained by Christman et al. (1989) for composites with the same geometries and J_2 flow theory matrices. The differences in the predictions of overall stress levels increase with an increase in V_f and are morphology dependent. Similarly, microscale deformation and stress patterns show good agreement although strain tends to be less uniform and localization intensity tends to be higher in the crystal plasticity composites. This trend mostly results from the fact that the crystalline plastic material is essentially a solid with a vertex on the yield surface (Asaro 1979, 1983), as noted in Needleman and Tvergaard (1993) and Needleman et al. (1992), and this promotes localization of plastic strain (Rice 1977). McHugh et al. (1993c) report that, for the polycrystalline composites, strain localization close to the particle is more intense for highly constrained microstructures (16%a), but for less constrained cases (16%b), localization intensity is similar close to the particles but more strain nonuniformity occurs remote from the particles. An ε_{eff} map for the GP-II-16%a material with the J_2 flow theory matrix is shown in Figure 8.5(c). ε_{eff} is the effective plastic strain defined as

$$\varepsilon_{\text{eff}} = \int_0^{t_1} \sqrt{\tfrac{2}{3} D^p : D^p}\, dt \quad (8.20)$$

where D^p is the plastic part of the rate of deformation tensor. A comparison of this figure with the γ_a map in Figure 8.5(a) shows that the ε_{eff} distribution is less localized (the contour levels in Figure 8.5(c) are proportional to those in Figure 8.5(a)). The most significant differences in macroscopic behavior occur in the comparisons of single crystal composites. In these cases, the J_2 flow theory composites show higher overall strain hardening (except for the case with a two-slip-system crystal). This is relative to the differences in localization intensity, as is illustrated by the γ_a and ε_{eff} contour maps in Needleman and Tvergaard (1993). Thus, localization causes geometric softening, a phenomenon that is not accounted for by J_2 flow theory.

In McHugh et al. (1991 and 1993c), comparisons are made between the stress-strain behavior of the polycrystalline composites and the behavior predicted by an analytic/numerical model (Duva 1984; Duva and Storm 1989; Bao et al. 1991) of a rigidly reinforced power-law

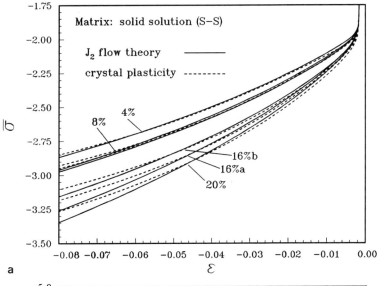

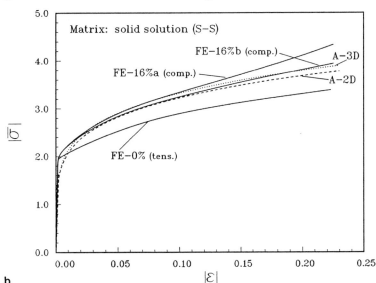

Figure 8.8. Plane strain stress-strain curves. (a) GP-II (polycrystalline) composites and the corresponding J_2 flow theory composites in compression. (b) SS-16% (polycrystalline) composites and the corresponding unreinforced polycrystal (denoted by "FE"), and the predicted curves of the two-dimensional and three-dimensional analytic models (denoted by "A-2D" and "A-3D," respectively). Stresses are normalized by g_0 for the θ' alloy. (Reprinted with permission from McHugh, P.E., Asaro, R.J., and C.F. Shih. 1993. *Acta Metall. Mater.* 41:1489, Oxford: Pergamon Press.)

hardening J_2 deformation theory material. The model is based on the tenet that the macroscopic power-law hardening exponent of pure matrix and composite are the same and that the introduction of the reinforcement simply results in a scaling of the overall flow strength. The scaling factor is determined in an approximate fashion, as a function of V_f and the hardening exponent by the differential self-consistent scheme. The reinforcement is assumed to be randomly distributed so that distribution effects cannot be assessed. As was done for the J_2 flow theory models of McHugh et al. (1991 and 1993c), the matrix hardening properties are determined from fits to the unreinforced polycrystal stress-strain curves. The predictions of both two-dimensional and three-dimensional versions of the model, assuming circular and spherical reinforcing particles, respectively, compare well with the finite element results illustrated by the stress-strain curves of Figure 8.8(b). However, Bao et al. (1991), in comparisons of the predictions with the results of finite element simulations for composites with J_2 flow theory matrices, found that the model significantly underestimates strengthening for V_f values above 20%.

The quench simulations show the development of localized residual strains in the matrix adjacent to the particles. Similar localized strains are observed in the heat-treating simulations of Povirk et al. (1990) and Povirk et al. (1991) for whisker/fiber reinforced composites with J_2 flow theory matrices. The fact that the residual fields alter peak stress and strain values during subsequent deformation could have implications for the modeling of microstructural failure such as interfacial decohesion. That is, processing effects should be incorporated into the modeling of failure. However, Povirk et al. (1991) have shown that even for weak interface strengths, the residual fields have only a small effect on the initiation of decohesion.

Summary

Some of the main issues in composite strengthening are certainly well explained by the phenomenological theories but, less intense localization is predicted by these theories as compared with crystal plasticity theory, and this can lead to significant differences in overall stress-strain behavior. The differences in localization levels have important implications for the modeling of microscopic failure mechanisms. Xu and Needleman (1993) show that intense localization in a crystal matrix close to the reinforcement, which does not occur for a J_2 flow theory matrix, retards the process of decohesion at the reinforcement-matrix interface.

The processing simulations using the crystal theory, like those using J_2 flow theory, yield strain patterns that are localized. This does not conform to experimentally observed dislocation distributions produced during heat treatment (Nieh and Karlak 1984; Christman and Suresh 1988; Suresh et al. 1989; Arsenault 1984; Vogelsang et al. 1986), which tend to be more uniform throughout the matrix. This is an example of one of the shortcomings of crystal plasticity theory. Although physically based it is still, like J_2 flow theory, a continuum theory, the phenomenon of dislocations being punched out from the particle-matrix interface well into the matrix during cooling is not adequately represented. Another shortcoming is the absence of size scales in continuum theories. In the context of composites, this means that reinforcement and matrix grain size effects cannot be examined. The method of eliminating these shortfalls may lie in incorporating discrete dislocation effects, or may be, as suggested by Povirk et al. (1992), based on solutions to many dislocation problems such as those considered by Lubarda et al. (1993).

Acknowledgements

This work was supported by the Ford Motor Company. P.E. McHugh also received support from Grant MSM-8957816 and Grant DMR-9002994 (Materials Research Group at Brown University), both funded by the National Science Foundation. The computations were carried out at the San Diego Supercomputer Center, the John von Neumann National Supercomputer Center and the Solid Mechanics Computational Mechanics Facility at Brown University.

References

Arsenault, R.J. 1984. *Mater. Sci. Eng.* 64:171.
Asaro, R.J. 1979. *Acta Metall.* 27:445.
Asaro, R.J. 1983. *J. Appl. Mech.* 50:921.
Asaro, R.J., and J.R. Rice. 1977. *J. Mech. Phys. Solids.* 25:309.
Asaro, R.J., and A. Needleman. 1985. *Acta Metall.* 33:923.
Bao, G., Hutchinson, J.W., and R.M. McMeeking. 1991. *Acta Metall. Mater.* 39:1871.
Becker, R. 1991. *Acta Metall. Mater.* 39:1211.
Becker, R., and L.A. Lalli. 1991. *Textures and Microstructures.* 14:145.
Brockenbrough, J.R., Suresh, S., and H.A. Wienecke. 1991. *Acta Metall. Mater.* 39:735.
Chang, Y.W., and R.J. Asaro. 1981. *Acta Metall.* 29:241.
Christman, T., Needleman, A., and S. Suresh. 1989. *Acta Metall.* 37:3029.
Christman, T., and S. Suresh. 1988. *Acta Metall.* 36:1691.
Dève, H.E., and R.J. Asaro. 1989. *Metall. Trans. A.* 20A:579.
Dève, H.E., Harren, S.V., McCullough, C., and R.J. Asaro. 1988. *Acta Metall.* 36:341.
Drucker, D.C. 1966. *J. Mater.* 1:873.
Duva, J.M., 1984. *J. Eng. Mat. Tech.* 106:317.
Duva, J.M., and D. Storm. 1989. *J. Eng. Mat. Tech.* 111:368.
Harren, S.V., and R.J. Asaro. 1989. *J. Mech. Phys. Solids.* 37:191.
Harren, S.V., Dève, H.E., and R.J. Asaro. 1988. *Acta Metall.* 36:2435.
Hill, R. 1966. *Proc. R. Soc. Lond.* A326:131.
Hill, R., and J.R. Rice. 1972. *J. Mech. Phys. Solids.* 20:401.
l'Esperence, G.L., Loretto, M.H., Roberts, W.T., et al. 1984. *Metall. Trans. A.* 15A:913.
Lubarda, V.A., Blume, J., and A. Needleman. 1993. *Acta Metall. Mater.* 41:625.
McHugh, P.E., Asaro, R.J., and C.F. Shih. 1991. In: Modelling the Deformation of Crystalline Solids (T.C. Lowe, et al. eds.), 369, Warrendale, PA: TMS.
McHugh, P.E., Asaro, R.J., C.F. and Shih. 1993a. *Acta Metall. Mater.* 41:1461.
McHugh, P.E., Varias, A.G., Asaro, R.J., and C.F. Shih. 1989. *Future Generation Computer Systems* 5:295.
McHugh, P.E., Asaro, R.J., and C.F. Shih. 1993b. *Acta Metall. Mater.* 41:1477.
McHugh, P.E., Asaro, R.J., and C.F. Shih. 1993c. *Acta Metall. Mater.* 41:1489.
McHugh, P.E., Asaro, R.J., and C.F. Shih. 1993d. *Acta Metall. Mater.* 41:1501.
Mohan, R., Ortiz, M., and C.F. Shih. 1992. *J. App. Mech.* 59:84.
Nagtegaal, J.C., Parks, D.M., and J.R. Rice. 1974. *Comp. Meth. Appl. Mech. Engng.* 4:153.
Needleman, A. 1972. *J. Mech. Phys. Solids.* 20:111.
Needleman, A., Suresh, S., and V. Tvergaard. 1992. In: Local Mechanic's Concepts for Composite Material Systems (J.N. Reddy, and K.L. Reifsnider, eds.), Berlin: Springer-Verlag.

Needleman, A., and V. Tvergaard. 1993. *J. Appl. Mech.* 60:70.
Nieh, T.G., and R.F. Karlak. 1984. *Scripta Metall.* 18:25.
Pierce, D., Asaro, R.J., and A. Needleman. 1982. *Acta Metall.* 30:1087.
Pierce, D., Asaro, R.J., and A. Needleman. 1983. *Acta Metall.* 31:1951.
Pierce, D., Shih, C.F., and A. Needleman. 1984. *Comp. Struct.* 18:875.
Povirk, G.L. 1992. Ph.D. Thesis. Division of Engineering, Brown University, Providence, RI.
Povirk, G.L., Needleman, A., and S.R. Nutt. 1990. *Mater. Sci. Eng.* A125:129.
Povirk, G.L., Needleman, A., and S.R. Nutt. 1991. *Mater. Sci. Eng.* A132:31.
Rice, J.R. 1971. *J. Mech. Phys. Solids.* 19:433–455.
Rice, J.R. 1977. *In:* Theoretical and Applied Mechanics (W.T. Koiter, ed.), 207, Amsterdam: North-Holland.
Suresh, S., Christman, T., and Y. Sugimura. 1989. *Scripta Metall.* 23:1599.
Taylor, G.I. 1938. *J. Inst. Metals.* 62:307.
Tvergaard, V. 1990. *Acta Metall. Mater.* 38:185.
Varias, A.G., O'Dowd, N.P., Asaro, R.J., and C.F. Shih. 1990. *Mater. Sci. Eng.* A126:65.
Vogelsang, M., Fisher, R., and R.J. Arsenault. 1986. *Metall. Trans. A.* 17A:379.
Xu, X.-P., and A. Needleman. 1993. *Modelling and Simulation in Materials Science and Engineering* 1:111.

Chapter 9
Continuum Models for Deformation: Discontinuous Reinforcements

JOHN W. HUTCHINSON
ROBERT M. MCMEEKING

Underlying the mechanics of particle reinforcement of ductile matrix is the assumption that the size of the particles and the spacing between the particles are sufficiently large such that continuum plasticity can be used to characterize the deformation of the matrix material. The continuum results to be discussed are very different in character from dislocation-based models of precipitation hardening. For example, there is a strong size effect in precipitation hardening, independent of particle volume fraction, while there is no size effect predicted by the conventional continuum models because the constitutive model for the matrix has no length scale associated with it. The length scale characterizing the transition between the two approaches has not been established and is undoubtedly material dependent and may even depend on whether the description is for rate-independent plasticity or for creep. For continuum plasticity to be valid, it is necessary for the particle size and spacing to be large as compared with the dominant scale of the dislocation motion (such as cell size). Generally, for typical metal-matrix materials, it is felt that particle sizes and spacings of several microns or more should ensure that conditions are met for validity of the continuum description.

An independent issue is the assignment of in situ plastic properties of the matrix material, such as the flow strength and strain hardening index. It is not uncommon for the in situ properties of the matrix to be altered from the bulk properties as a result of processing of the reinforced composite system. In the approach described later, the properties of the matrix must be regarded as in situ properties. Further discussion of this matter will be made in a later section, in which we will make comparisons of theory and experiment.

Section 9.1 of this chapter considers the flow strength of perfectly plastic matrix materials. In particular, the effects of particle shape, volume fraction, alignment, and distribution will be discussed. The perfectly plastic idealization has a number of advantages for presenting and discussing the role of particle reinforcement. It isolates the primary strengthening effect and it provides a setting for discussing the effect of matrix hardening in rate-independent materials, which is the topic of Section 9.2. This chapter does not address the important question of damage development. It will be assumed that the reinforcing particles are well bonded to the matrix and that the interface does not separate and the particles do not crack. Thus, the macroscopic stress-strain responses of the composites predicted here must be regarded as limiting responses for systems that are free of damage. Section 9.3 deals with residual stress development and with the effect of these stresses on overall stress-strain behavior. Section 9.4 discusses the application of the approach to creep reinforcement, including the effect of diffusional relaxation of the reinforcement.

Much of the material presented in this chapter is drawn from two recent papers by two of the authors (Bao et al. 1991a, 1991b). Primary emphasis is placed on the role of particulate reinforcements, although results on the *transverse* behavior of continuous aligned fiber reinforced composites are also to be included because this behavior falls naturally within the framework of discontinuous reinforcement. There are a number of relatively recent papers in the literature based on the continuum approach, and other chapters in this book draw on this work. Information especially relevant to the present chapter is contained in the papers by Christman et al. 1989; Levy and Papazian 1990; Tvergaard 1990; and Dragone and Nix 1990, and in Chapter 10.

9.1 Flow Strength of Composites with Nonhardening Matrices

With the exception of creep behavior, discussed in Section 9.4, the matrix material will be taken to be rate independent. In this section, the matrix will be further idealized to be elastic-perfectly plastic with a tensile flow stress σ_0. A Mises yield condition is assumed such that $\sigma_e = \sigma_0$, where the effective stress depends on the deviator stress components, s_{ij}, according to $\sigma_e = \sqrt{(3s_{ij}s_{ij}/2)}$. Plastic strain increments are directed along the outward normal to the yield surface, proportional to s_{ij}. The primary emphasis in this section is on the strengthening of the composite as a result of the discontinuous reinforcements, which is best reflected by the increase in the overall limit stress of the composite. Let $\bar{\sigma}$ denote the overall tensile stress applied to the composite in some direction. For aligned reinforcements, we will be concerned with overall tension (or compression) applied parallel to the direction of alignment, except for some examples of continuous fiber reinforcement in which the loading will be transverse to the direction of alignment. For an elastic-perfectly plastic matrix reinforced by discontinuous reinforcements that do not deform plastically, there is a limit value of $\bar{\sigma}$, which is denoted here by $\bar{\sigma}_0$. This *overall limit yield stress* does not depend on the elastic moduli of either the matrix or the reinforcement nor does it depend on the absolute size of the reinforcing particles. It does depend, however, on the volume fraction of the reinforcement phase and on other details of the reinforcement such as shape and alignment. It is these dependencies that will be presented here.

9.1.1 Aligned Discontinuous Reinforcements

An axisymmetric cell model is used to calculate the overall limit stress. This method has been widely employed and is used by some of the authors of other chapters in this text. The population of reinforcing particles is assumed to have identical size, shape, and alignment and, in addition, is assumed to be uniformly distributed such that each particle and its surrounding matrix deforms in the same way as every other particle-matrix neighborhood. In this way, a single particle-matrix cell with special periodic boundary conditions can be used to compute the response of the entire composite. Usually, to reduce the computations, a cell with full three-dimensional geometry will be replaced, as an approximation, by an axisymmetric cell under axisymmetric boundary conditions. This final step in the modeling process, which is illustrated in the insert in Figure 9.1 for a three-dimensional array of spherical particles, has been shown to introduce little error at moderate volume fractions (Hom 1992). The volume fraction of the reinforcing phase in the cell is identified with the volume fraction of that phase in the composite. The height to diameter of the cell can be chosen to model the axial to transverse spacing of the particulates in the composite. Some further discussion of the role of the cell aspect ratio can be found in Bao et al. (1991a). In this chapter, all results for aligned reinforcements will have been calculated using a cell whose height to diameter ratio equals that of the reinforcing particles.

The results of the cell model computation for the strengthening ratio, $\bar{\sigma}_0/\sigma_0$, as a function of particle volume fraction f for spherical particle reinforcements in an elastic-perfectly plastic matrix, is shown in Figure 9.1. The limit-yield stress of the composite becomes unbounded as the volume fraction approaches the value where the particles make contact, which for the cell model is $f = 2/3$. The most notable feature of these predictions is how little strengthening effect spherical particles have. A 10% increase in limit-yield stress requires almost a 20% volume fraction of particles. This, of course, is in marked contrast with the relatively large strengthening possible from volume fractions of less than even 1% of equiaxed precipitates whose sizes are sufficiently small that discrete particle/dislocation interactions govern strengthening. Larger equiaxed particles, in the range in which continuum plasticity governs, are remarkably inefficient strengthening agents.

Elongated or disc-shaped particles can be effective strengthening agents, as seen in Figures 9.2 and 9.3. The effect of aligned ellipsoidal particles on the strengthening ratio is displayed in Figure 9.2, in which the prolate

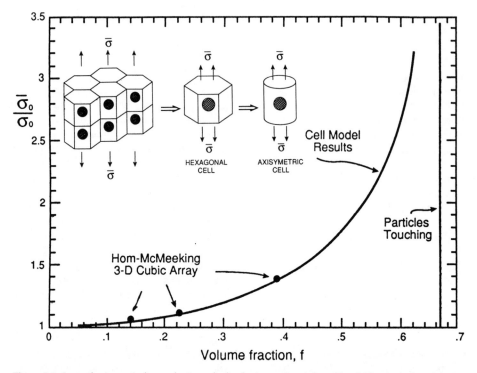

Figure 9.1. Strengthening ratio for an elastic-perfectly plastic matrix reinforced by rigid, spherical particles. The insert illustrates the steps leading to an axisymmetric cylindrical cell for a uniform distribution of particles.

axisymmetric particles have an aspect ratio of $a/b < 1$ and the oblate particles have an aspect ratio of $a/b > 1$. The effect of either pronounced elongation or flattening of the particles is striking in comparison with the effect of the spherical particles, which is labeled in Figure 9.2 by $a/b = 1$. Particles in the shape of right circular cylinders (rods $a/b < 1$, unit cylinders $a/b = 1$, or discs $a/b > 1$) are somewhat more effective than ellipsoids at the same

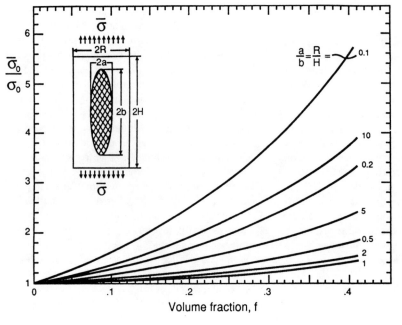

Figure 9.2. Strengthening ratio for an elastic-perfectly plastic matrix reinforced by aligned ellipsoidal particles. *(Data from Bao et al. 1991a.)*

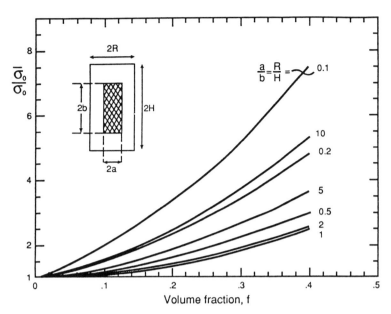

Figure 9.3. Strengthening ratio for an elastic-perfectly plastic matrix reinforced by aligned cylindrical particles. *(Data from Bao et al. 1991a.)*

volume fraction (Figure 9.3). The unit cylinders are still not very efficient, although they are almost twice as effective as spheres (Christman et al. 1989). Thus, details of particle shape in addition to aspect ratio can be important. Arrangement of the reinforcing phase can also be important, especially when the particles are highly elongated with significant strengthening capacity. The results for the elongated particles shown in Figures 9.2 and 9.3 represent an arrangement where the particles are end-to-end with no overlap, by virtue of geometry of the axisymmetric cell model. Alternative arrangements of highly elongated particles, allowing for varying degrees of particle overlap, have shown that results such as those discussed above based on the simplest cell model usually predict the largest possible strengthening (Levy and Papazian 1990; Tvergaard 1990; and Dragone and Nix 1990).

9.1.2 Transverse Strengthening of Continuous Fiber-Reinforced Composites

A composite with aligned continuous fibers that do not deform plastically (nor fracture) will not have a limit-yield stress for stressing parallel to the fiber direction, and nonlinear response of the composite for such loadings will be addressed elsewhere in this book. The fibers have a strengthening effect for stressing in tension or compression in directions perpendicular to the fibers, which is analogous to that described for the discontinuous particulate reinforcements with a well-defined limit-yield stress. For this reason, the results of Jansson and Leckie (1992) and the unpublished work of Schmauder and McMeeking for the transverse strengthening of fiber-reinforced composites will be included in this chapter. As in the case of the discontinuous reinforcements, the fibers are assumed to be perfectly bonded to the matrix, which is elastic-perfectly plastic with tensile yield stress σ_0. The strengthening ratio, $\bar{\sigma}_0/\sigma_0$, is plotted as a function of the fiber volume fraction f in Figure 9.4 for round fibers in square and hexagonal arrangements. It should be noted that results for biaxial stress states can be obtained from Figure 9.4 by superposition of a hydrostatic stress.

At small volume fractions there is very little strengthening except for the imposition by the fibers of plane strain flow in the matrix. Thus, the limit at zero volume fraction is a strengthening ratio of $2/\sqrt{3}$, consistent with plane strain. The minimal effect of fibers on the transverse strength is analogous to the situation with spherical reinforcements discussed previously. The results plotted in Figure 9.4 make it clear that there is a strong effect of fiber arrangement, which has been observed previously in calculations by Brockenbrough et al. 1991 and others. When the stress is applied parallel to the diagonal of the square packing, the strengthening is $2\sqrt{3}$ for all of the volume fractions studied by Schmauder and McMeeking, which were as high as 75%. The lack of strengthening beyond the plane strain level occurs in this case because plastic shear strain can occur without constraint on planes parallel to the diagonal of the packing. This deformation is sustained by the yield stress in shear parallel to the diagonals and therefore by a transverse tension of $\bar{\sigma} = 2\sigma_0/\sqrt{3}$. This particular mechanism of deformation occurs when the square-packed composite is loaded parallel to the

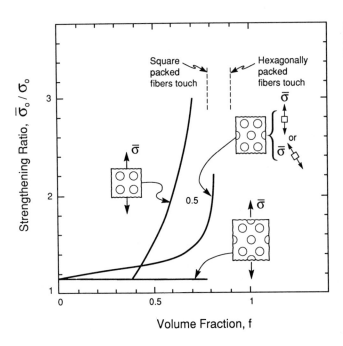

Figure 9.4. Strengthening ratio for the transverse loading of a perfectly plastic material reinforced by long rigid fibers.

diagonal. It is only precluded when the fibers touch each other, which takes place when $f = \pi/4$. Thus, the strengthening ratio of $2/\sqrt{3}$ should prevail up to just below this volume fraction of fibers. When $f = \pi/4 = 0.79$ and the fibers touch, the square-packed composite has an unbounded limit load analogous to the behavior of the material reinforced with spheres.

When the material with fibers packed in a square arrangement is loaded parallel to the fiber rows, there is no strengthening beyond the plane strain level up to a volume fraction of $\pi/8 = 0.39$, as can be seen in Figure 9.4. Below this volume fraction, shear strain can occur on uninterrupted planes at 45 degrees to the tensile axis. Thus, a shear stress, at 45 degrees to the fiber rows, equal to the shear yield strength, is sufficient to ensure yielding of the square-packed material at volume fractions below 39%. Consequently, the transverse limit strength in these circumstances is $2\sigma_0/\sqrt{3}$. When the volume fraction exceeds 39%, the planes at 45 degrees to the square-packed fiber rows are interrupted by the fibers. As a result, shear strain can no longer occur freely on those planes and the constraint leads to an elevation of the transverse strength. Figure 9.4 shows that this constraint rises rapidly as the volume fraction is increased. Indeed, the strength would become unbounded at a volume fraction of $\pi/4 = 79\%$ when the fibers are touching. Therefore, the strengthening ratio must rise rapidly between the volume fractions $\pi/8 = 39\%$, where it is $2/\sqrt{3}$, and $\pi/4 = 79\%$, where it is infinity. Comparison of the results in Figure 9.4 for the square-packed fiber composite loaded in the two orientations reveals a significant anisotropy at volume fractions above 39%.

In contrast, the material with the fibers arranged in a hexagonal array exhibits minimal anisotropy. This was noted by Brockenbrough et al. (1991) and Jansson and Leckie (1992) who cite unpublished work by Jansson. A single line has been used in Figure 9.4 to represent all results for the hexagonal packed composites, although with a perfectly plastic matrix there is a small difference in the transverse strength if the stress is applied parallel to the fiber rows or at 30 degrees to the fiber rows. At a given volume fraction, other results lie in-between. It can be seen in Figure 9.4 that there is little strengthening when the volume fraction of fibers is low. In this regard, the hexagonally packed fibers in terms of transverse strength are similar to spherical particulates in their ineffectiveness as strengthening agents. It requires a 50% volume fraction of fibers to increase the transverse strength to 30% above the matrix uniaxial strength, and about half of that effect comes from plane strain. In contrast to the square-packed fiber case, there is a small effect at low volume fractions of hexagonally packed fibers. This arises because there are no uninterrupted shear planes in the case of hexagonally packed fibers, so that even a small number of fibers causes some constraint, although the net effect is modest. The strengthening ratio rises more strongly above volume fractions of 70%. Presumably, the more rapid rise occurs because the volume fraction is approaching the level of $\pi/2\sqrt{3} = 91\%$, at which the hexagonally packed fibers touch and the strength becomes unbounded.

9.1.3 Randomly Oriented Versus Aligned Discontinuous Reinforcements

Very few results are available to illustrate strengthening effects for reinforcements other than those that are aligned. The computational cell models are not readily extended to other arrangements than those that are aligned. One exception is the set of results obtained by Bao et al. (1991a) for randomly oriented elongated ellipsoidal and disc-shaped oblate ellipsoidal particles embedded in an elastic-perfectly plastic matrix. The composite was assumed to have a packet morphology with grain-like regions containing a number of aligned particles. The grain-like packets are randomly oriented such that the overall behavior of the composite is isotropic. The analysis of Bao et al. involved two steps: a three-dimensional cell model analysis was performed to obtain the multiaxial limit yield surface of the grain-like packet; this result was then used in conjunction with a Bishop-Hill procedure, averaging over all orientations of the "grains" relative to the tensile stressing axis, to obtain an upper bound to the limit yield stress, $\bar{\sigma}_0$, of the composite. The strengthening ratios for such randomly oriented prolate ellipsoids ($a/b = 0.1$) and oblate ellipsoids ($a/b = 10$) are shown in Figure 9.5. Included in this figure are the results from Figure 9.2 for the same particles when they are aligned and stressed parallel to their direction of alignment. The particles are obviously not nearly as effective when they are randomly oriented as they are when they are aligned. Of course, the strengthening effect for the randomly oriented reinforcement holds for any orientation of the tensile axis, whereas the strengthening for the composite with aligned particles applies only for stressing that occurs parallel to the alignment. Its transverse yield strength is lower than the yield strength of the composite with the randomly oriented particles.

Relatively high aspect ratio disc-shaped particles packed to volume fractions of about 20% or more must assume a packet-like morphology if they are randomly oriented, but needle-shaped particles (for example, chopped fibers) need not assume the packet morphology. A more common morphology involves little orientation correlation between neighboring particles. It is an open question as to whether morphologies giving isotropic behavior other than the packet morphology would give more significant strengthening than that observed in Figure 9.5.

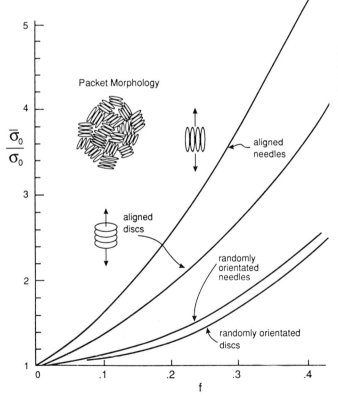

Figure 9.5. Strengthening ratio for an elastic-perfectly plastic matrix reinforced by randomly oriented ellipsoidal needles ($a/b = 1/10$) and randomly oriented ellipsoidal platelets ($a/b = 10$), in which the particles possess a packet-like morphology. The strengthening ratios for the corresponding aligned reinforcements from Figure 9.2 are included for reference. *(Data from Bao et al. 1991a.)*

9.1.4 Spatial Distribution of the Reinforcement: Nonuniform Versus Uniform

The particle/matrix computational cells model a particle distribution that is necessarily uniform, both in space and with respect to the size and shape of the particles. Even when the cell is embellished to reflect details such as local arrangement of particles, the overall spatial distribution is uniform. Some methods for estimating the overall elastic moduli of composite materials lend themselves to the study of the effect of nonuniformity (Torquato 1991). Nevertheless, there is surprisingly little guidance available from the literature for material designers to go by in the form of simple "rules of thumb" on the role of nonuniformity, even for elastic properties. A recent study of the effect of a special form of nonuniform spatial distribution of reinforcement on limit flow strength (Bao et al. 1991b) leads to the clear-cut conclusion that nonuniformity increases the strength of the composite relative to its uniform counterpart, at least for the class of nonuniformities envisioned. The procedures leading to this result will now be described for the case of reinforcement of an elastic-perfectly plastic matrix with isotropic distributions of rigid spherical particles.

Self-consistent calculations were carried out for the overall limit flow stress $\bar{\sigma}_0$ for a two-phase elastic-plastic composite where each phase is isotropic and elastic-perfectly plastic with flow stress $\bar{\sigma}_0^{(i)}$ and volume fraction $f^{(i)}$, $i = 1,2$. Each of these two phases is assumed to be isotropically distributed so that the overall behavior of the composite is isotropic. The self-consistent calculations of $\bar{\sigma}_0$ of Bao et al. (1991b), which will not be reported here, employed a three-shell model with an inner sphere representing the 'particulate' phase #1 (with $\sigma_0^{(1)}$ and $f^{(1)}$), an intermediate shell representing the 'matrix' phase #2 (with $\sigma_0^{(2)}$ and $f^{(2)}$), and an outer region extending to infinity endowed with the unknown properties of the composite. For ratios of $\sigma_0^{(1)}/\sigma_0^{(2)}$, differing from unity by less than a factor of two, the uniform strain rate upper bound (the rule of mixtures) to the overall limit tensile stress gives an excellent approximation.

$$\bar{\sigma}_0 = f^{(1)}\sigma_0^{(1)} + f^{(2)}\sigma_0^{(2)} \tag{9.1}$$

The nonuniformity in the distribution of the spherical particles is also depicted in Figure 9.6(a). The average volume fraction of the spherical particles taken over the whole composite is $\bar{c}$. It is assumed that there are particle-rich subregions and particle-poor subregions. Moreover, it is assumed that the spacing between the particles in each of these regions is small compared to the size of the subregions. Thus, within each of the respective subregions, the results of Figure 9.1 for a uniform distribution of spherical particles (sketched also in Figure 9.6(a) and denoted by $\Sigma(c)$) can be used to specify the flow stress. Specifically, let c_1 be the volume fraction of the particles in the isolated subregions comprising volume fraction $f^{(1)}$ of the composite and whose flow stress $\sigma_0^{(1)}$ is read off the curve for the uniformly distributed particles at the value c_1. The volume fraction of the particles in the contiguous subregions is c_2 with associated values $f^{(2)}$ and $\sigma_0^{(2)}$, which is also read off the same curve. The relation between the local particle volume fractions, the average particle volume fraction, and the volume fractions of the two subregions ($f^{(1)}$ and $f^{(2)} = 1 - f^{(1)}$) is

$$\bar{c} = f^{(1)}c_1 + f^{(2)}c_2 \tag{9.2}$$

In this way, the self-consistent results for the two-phase composite can be used to estimate the effect of the nonuniformity.

An example of the outcome of the calculation just described is shown in Figure 9.6(b) for three levels of average particle volume fraction $\bar{c}$. In this example, the volume fraction of the isolated (particulate phase) and contiguous regions (matrix phase) are taken to be the same ($f^{(1)} = f^{(2)} = 1/2$). The measure of the nonuniformity is taken as $c_1 - \bar{c}$, so that $c_1 = c_2 = \bar{c}$ gives the uniform distribution. Each curve in Figure 9.6(b) corresponds to a fixed value of $\bar{c}$, and thus it is noted that the uniform distribution gives the *minimum* estimate of $\bar{\sigma}_0$. Any nonuniformity, whether corresponding to particle-rich contiguous subregions or particle-rich isolated subregions, leads to an increase in limit flow stress relative to the uniform distribution. An analytic expression for the limit flow stress, valid for sufficiently small nonuniformities, brings out this feature very clearly:

$$\bar{\sigma}_0 = \Sigma(\bar{c}) + \frac{1}{2}\left(\frac{\partial^2 \Sigma}{\partial c^2}\right)_{c=\bar{c}}\left(\frac{f^{(1)}}{f^{(2)}}\right)(c_1 - \bar{c})^2 \tag{9.3}$$

Thus, as long as the curvature of the relation of flow stress to particle volume fraction is positive for the uniformly distributed case, any nonuniformity of the class discussed here will enhance the flow strength. This conclusion seems to be borne out by the numerical results for the transverse stress-strain behavior of continuous fiber reinforced metal-matrix composites discussed by Suresh and Brockenbrough in Chapter 10. These authors have compared transverse behavior for various nonuniform distributions of circular fibers (random arrays) with the corresponding behavior for uniform triangular arrays, which give rise to nominally isotropic transverse behavior. The transverse stress-strain curves of the composites with the nonuniform distributions of fibers lie well above that for the triangular array at the same volume fraction of reinforcement.

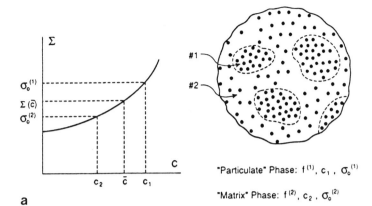

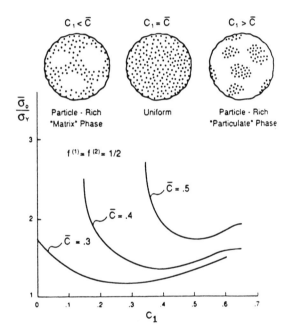

Figure 9.6. (a) Scheme for applying the results for a dual-phase composite to a matrix reinforced by equisized rigid particles that are nonuniformly distributed. (b) Strengthening ratio as a function of nonuniformity for three levels of average particle volume fraction $\bar{c}$. In this example, the volume fraction of each of the nonuniform "phases" is taken to be 1/2, and the tensile yield stress of the elastic-perfectly plastic matrix is taken to be σ_Y.

9.2 Aligned Reinforcement of Elastic-Strain Hardening Matrices

Now consider isotropic matrix materials whose tensile stress-strain curve is specified by the Ramberg-Osgood relation

$$\varepsilon = \frac{\sigma}{E} + \alpha \frac{\sigma_0}{E}\left(\frac{\sigma}{\sigma_0}\right)^n \quad (9.4)$$

where E is the Young's modulus, σ_0 is now a reference yield stress, and n is the stress hardening exponent. Let $\bar{\sigma}$ and $\bar{\varepsilon}$ be the overall tensile stress and strain of the composite in the direction of the aligned reinforcement. An approximation to the tensile stress-strain curve of the composite developed by Bao et al. (1991a) is

$$\bar{\varepsilon} = \frac{\bar{\sigma}}{\bar{E}} + \alpha \varepsilon_0 \left(\frac{\bar{\sigma}}{\bar{\sigma}_N}\right)^n \quad (9.5)$$

where $\varepsilon_0 = \sigma_0/E$ is the reference yield strain of the matrix and $\bar{\sigma}_N$ is the reference stress of the composite elaborated on below.

The elastic modulus of the composite in the direction of alignment $\bar{E}$ can be computed using a cell model or it can be estimated in a number of ways. The prediction of elastic properties of composites as dependent on the constituent properties is a well-developed subject, and thus the estimation of $\bar{E}$ will not be discussed in depth in this chapter.

To understand the origin of the second term in Equation 9.5, consider a pure-power matrix material reinforced by rigid particles (or, in the case of the transverse behavior of a continuous fiber composite, a pure power law matrix surrounding by rigid fibers). The incompressible matrix has the following tensile and multiaxial behavior:

$$\frac{\varepsilon}{\varepsilon_0} = \alpha\left(\frac{\sigma}{\sigma_0}\right)^n \quad \text{and} \quad \frac{\varepsilon_{ij}}{\sigma_0} = \frac{3}{2}\alpha\left(\frac{\sigma_e}{\sigma_0}\right)^{n-1}\frac{s_{ij}}{\sigma_0} \quad (9.6)$$

As discussed in detail by Bao et al. (1991a), the composite with perfectly bonded rigid particles also has pure power law behavior. The uniaxial stress-strain relation for the composite for stressing in the direction of particle alignment is

$$\frac{\bar{\varepsilon}}{\varepsilon_0} = \alpha\left(\frac{\bar{\sigma}}{\bar{\sigma}_N}\right)^n \quad (9.7)$$

The composite reference stress $\bar{\sigma}_N$ depends on n as well as on the same parameters influencing $\bar{\sigma}_0$ for the perfectly plastic solid f, particle shape, etc. In the limit for $n \to \infty$, $\bar{\sigma}_N \to \bar{\sigma}_0$. The dependence of the reference stress on n must be computed. A large number of such computations reported in Bao et al. (1991a) indicate that the dependence on $N \equiv 1/n$ is accurately approximated by

$$\bar{\sigma}_N \cong \bar{\sigma}_0 + cN(\bar{\sigma}_0 - \sigma_0) \quad (9.8)$$

The coefficient c has a weak dependence on f and particle shape (Figure 9.7), but is in the range from 2 to 2.5 for most systems of interest. Strain hardening of the matrix enhances the flow strength of the composite as reflected by Equation 9.7 in two ways: through the stress exponent n and through the increase of the reference stress above $\bar{\sigma}_0$.

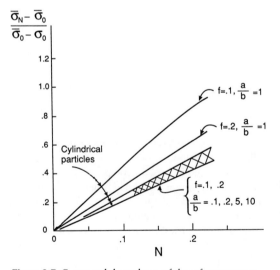

Figure 9.7. Computed dependence of the reference stress, $\bar{\sigma}_N$, on the hardening index, $N = 1/n$, for a variety of particle shapes and volume fractions.

Now consider again the composite with an elastically deforming reinforcement phase well bonded in the Ramberg-Osgood matrix (Equation 9.4). Neither the elastic properties of the reinforcement nor the elasticity of the matrix influences the asymptotic "large strain" behavior of the composite for overall strains that become large compared to ε_0. Thus, the asymptotic behavior of the composite is given precisely by the pure-power relation (Equation 9.7). The approximate Ramberg-Osgood relation (Equation 9.5) for the elastic-plastic behavior of the composite was proposed as a formula to interpolate between the elastic limit and the asymptotic limit for "large strains." The formula reasonably accurately captures the response of the composite. This can be seen in Figure 9.8, in which results for an example for a composite with aligned disc-shaped cylindrical reinforcements ($f = 0.2$ and $a/b = 5$) are displayed for three levels of hardening. The matrix curves of Equation 9.4 are shown for reference. The dashed curves represent the approximation (Equation 9.5) with the relevant values of $\bar{E}$ and $\bar{\sigma}_N$, and the solid line curves are the results of numerical calculations using a cell model with the complete Ramberg-Osgood relation (Equation 9.4) for the matrix. Other examples are shown by Bao et al. (1991a). The example in Figure 9.8 illustrates the point that the flow stress enhancement derives from both the exponentation and the dependence of the reference stress $\bar{\sigma}_N$ on n. It should also be noted that the approximate formula (Equation 9.5) tends to overestimate the stress in the knee of the composite stress-strain curve. This may be of some consequence because, for many metal-matrix composites, the strain range of interest may not significantly exceed the region of the knee. More accurate predictions will require more detailed computations such as those shown as solid line curves in Figure 9.8 and as reported elsewhere in this volume.

A series of squeeze-cast composites with a matrix of aluminum/magnesium and reinforced by silicon carbide particles were prepared and tested to obtain uniaxial stress-strain data by Yang et al. (1990). This study was notable for the range of volume fractions, particle sizes, and shapes considered, and for the careful attempt to establish the in situ matrix stress-strain behavior. Particle sizes ranged from several microns to more than one hundred microns. Over this size range, it was established that there was very little dependence on particle size once the volume fraction and particle shape were fixed. Two sets of stress-strain data are shown in Figure 9.9, in which comparison with the Ramberg-Osgood estimation procedure just described is made. These figures were taken from Yang et al. (1991), which contains a fuller discussion of the composites and their preparation. Figure 9.9(a) shows tensile stress-strain data for the

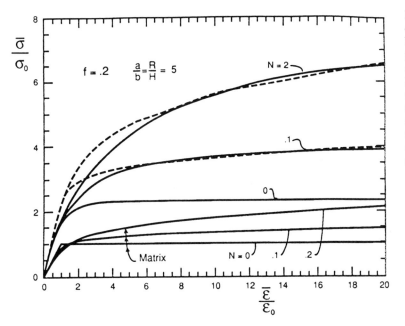

Figure 9.8. Tensile stress-strain curves for composites reinforced by aligned disc-shaped cylindrical particles ($a/b = 5$ and $f = 0.2$). The matrix material has the Ramberg-Osgood stress-strain curves shown. The solid line curves for the composite were computed using a cell model, while the dashed line curves were obtained using the estimation scheme described in the text.

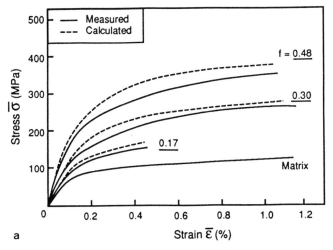

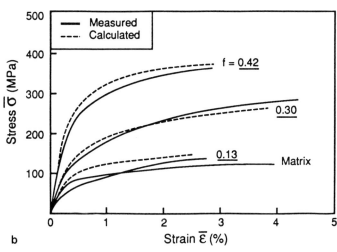

Figure 9.9. Comparison of experimental data for composites of an Al/Mg matrix material reinforced by SiC particles with stress-strain curves predicted by the estimation scheme described in the text. (From Yang et al. 1991.) (a) Uniaxial tensile curves for matrices reinforced by equiaxed particles. (b) Uniaxial compression curves for matrices reinforced by randomly oriented platelets.

matrix as well as for composites reinforced by three volume fractions of equiaxed particles whose average size was 9 microns. The estimation scheme described above was applied by fitting the Ramberg-Osgood curve (Equation 9.4) to the matrix curve to obtain n and σ_0 (α was taken to be 3/7). The value of $\bar{E}$ in (Equation 9.5) was taken from the experimental curve (which in turn was shown to agree well with self-consistent predictions), and $\bar{\sigma}_N$ was determined from Equation 9.8, using the results for $\bar{\sigma}_0$ from Figure 9.3 for the unit cylindrical particles. The dashed-line curves in Figure 9.9(a) are the result of the estimation procedure. The same procedure was applied to the uniaxial compression data in Figure 9.9(b) for composites reinforced by randomly oriented platelets whose average maximum diameter was 25 microns. In this case, the platelets are taken to have a 10:1 aspect ratio and the results for $\bar{\sigma}_0$ for the randomly oriented ellipsoidal platelets in Figure 9.5 were employed to estimate $\bar{\sigma}_N$.

9.3 The Influence of Residual Stress on Composite Yielding

The results presented so far in this chapter are for materials initially free of residual stress in the matrix and in the reinforcements. Most metal-matrix composites are processed at high temperatures and, upon cooling, develop residual stresses as a result of thermal expansion mismatch between the matrix and the reinforcements. Although the residual stresses have no effect on the purely elastic response of the composite, it is of interest to determine the effect on the yielding of reinforced materials. Such effects can occur because the residual stress can have a deviatoric component and can thus influence the process of yielding in the matrix. The effect has been considered by several investigators including Povirk et al. (1991). However, Zahl and McMeeking (1991), have provided a series of results for strongly bonded elastic reinforcements in perfectly plastic matrices showing the influence of the thermal strain mismatch relative to the volume fraction of reinforcements and the yield strain of the matrix.

The results of Zahl and McMeeking (1991) were obtained by the unit cell method with finite elements used for the analysis. The residual stresses were first generated by cooling the material while the matrix was permitted to respond elastoplastically. Thereafter, loads were applied to cause macroscopic deformation. The magnitude of the residual stresses generated were controlled by the parameter $\Delta\alpha\Delta T/\varepsilon_0$, where $\Delta\alpha$ is the thermal expansion coefficient of the reinforcement minus the thermal expansion coefficient of the matrix, and ΔT is the current temperature minus the temperature at which the composite material is free of residual stress in both the matrix and the reinforcement. The parameter ε_0 is, as before, the yield strain in tension of the matrix material. The calculations were carried out with an elastic modulus for the reinforcement that is 6.62 times the elastic modulus of the matrix.

Figure 9.10 shows the stress-strain curves for spherical reinforcements in a perfectly plastic matrix with $\Delta\alpha\Delta T/\varepsilon_0 = 1$. This case corresponds to SiC particles in an Al alloy matrix 256C below the stress-free temperature. A softening of the composite response results at strains comparable to the matrix yield strain for both tension and compression with the effect much more pronounced in the compressive cases. The compressive stress-strain curve is up to 30% below the tensile curve in terms of strength at the same strain magnitude. However, as the strain increases beyond $\varepsilon_0 = \Delta\alpha\Delta T$, the compressive and the tensile stress-strain curves converge toward the curve for the material without initial residual stress. The limit strength is thus the same whether or not there are initial residual stresses. Because the compressive stress-strain curve also represents tension applied to a material with $\Delta\alpha\Delta T/\varepsilon_0 = -1$ (the sign of the residual stresses reversed), the limit strength is unaffected by whether the residual stresses in the matrix are tensile or compressive. Because the limit strength of reinforced materials is independent of the initial residual stresses, the behavior of the composite material when the strain greatly exceeds $\varepsilon_0 = \Delta\alpha\Delta T$ is correspondingly independent of them too. However, the limited ductility of particulate composites means that such large strains are rarely achieved in tension and are unusual in compression unless accompanied by internal damage. As a consequence, tension-compression asymmetries in the yielding of particulate composites are to be expected, and will generally persist until fracture of the material occurs.

The degree of yielding caused by thermal expansion mismatch between the matrix and the reinforcements depends on the magnitude of $\Delta\alpha\Delta T/\varepsilon_0$. When this parameter equals 1, as in the case discussed above, approximately 50% of the matrix around spherical particles is yielded. When $\Delta\alpha\Delta T/\varepsilon_0 = 2$, 60% of the matrix has yielded. When $\Delta\alpha\Delta T/\varepsilon_0 = 5$, the entire matrix has deformed plastically upon cooling. The effect of these different degrees of yielding on a material with 20% of elastic spheres on the compression and tension stress-strain curves is shown in Figures 9.11 and 9.12. It can be seen that in both tension and compression the greater thermal expansion misfit causes a softer response. Because the matrix is fully yielded when $\Delta\alpha\Delta T/\varepsilon_0 = 5$, any magnitude of $\Delta\alpha\Delta T/\varepsilon_0$ larger than 5 will give rise to the same tension and compression stress-strain curves as occur for $\Delta\alpha\Delta T/\varepsilon_0 = 5$. It is of

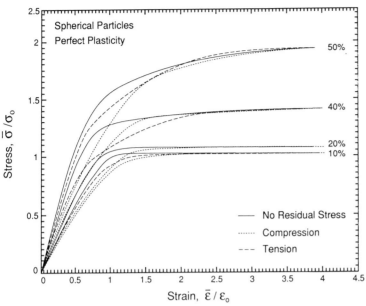

Figure 9.10. Stress-strain curves for 10, 20, 40, and 50 vol % of spherical reinforcements in a perfectly plastic matrix with $\Delta\alpha\Delta T/\varepsilon_0 = 1$.

interest that the characteristic strain, during which the thermally induced transient occurs, is ε_0 rather than $\Delta\alpha\Delta T$. At a macroscopic strain of $2\varepsilon_0$, the transient effect in each case has largely disappeared.

Zahl and McMeeking (1991) have also given results for 20% by volume of aligned, well-bonded elastic short fibers in a perfectly plastic matrix. The aspect ratio of the fibers is 10 and they are circular cylinders. The stress-strain curves are shown in Figure 9.13 for compression and 9.14 for tension. As before, the thermal residual stresses cause a transient softening of the response that is more marked in compression than in tension. However, in both cases, the effect is not great and, in the tensile case, it is almost negligible. In compression, the maximum softening is only about 15%. Additionally, the matrix is almost fully yielded by thermal stresses when $\Delta\alpha\Delta T/\varepsilon_0 = 1$ and when only a region beyond the fiber ends is still elastic. Increasing values of $\Delta\alpha\Delta T/\varepsilon_0$ do not change this situation much. It is likely that this effect is caused by the fact that, in the calculations, the fibers are modeled as being fairly close together side-by-side but far apart end-to-end. In this case, when there are no initial residual strains, the limit strength is reached at a strain of about $5\varepsilon_0$. When there are initial thermal stresses, the transient they cause is noticeable up to this strain. Thus it can be concluded that, in general, the transient persists up

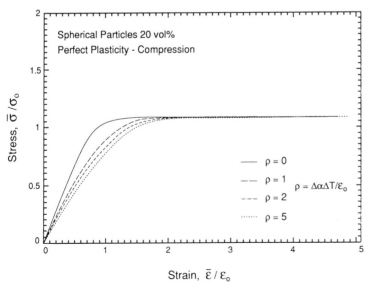

Figure 9.11. Compressive stress-strain curves for 20 vol % of spherical particles with different amounts of thermal expansion misfit and a perfectly plastic matrix.

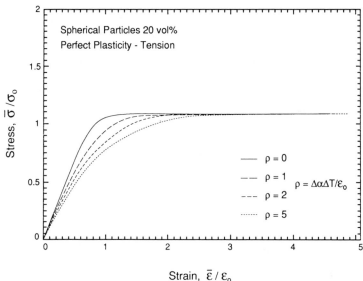

Figure 9.12. Tensile stress-strain curves for 20 vol % of spherical particles with different amounts of thermal expansion misfit and a perfectly plastic matrix.

to the strain at which the limiting behavior sets in when there are no thermal residual strains.

9.4 Reinforcement Against Creep

9.4.1 Steady-State Power Law Creep

A simple correspondence between behavior for pure power law plasticity and steady-state power law creep permits the results of Equation 9.7 to be translated immediately to give insight into reinforcement against creep. It is again assumed that the particle/matrix bond is perfect and that, additionally, the sole mechanism of inelastic deformation is power law creep. The consequences of interface sliding and diffusional relaxation of the reinforcement are taken up in the next subsection. In power law creep, the tensile and multiaxial stress-strain rate relations are the exact counterparts to Equation 9.6:

$$\frac{\dot{\varepsilon}}{\dot{\varepsilon}_0} = \alpha\left(\frac{\sigma}{\sigma_0}\right)^n \quad \text{and} \quad \frac{\dot{\varepsilon}_{ij}}{\dot{\varepsilon}_0} = \frac{3}{2}\alpha\left(\frac{\sigma_e}{\sigma_0}\right)^{n-1}\frac{s_{ij}}{\sigma_0} \quad (9.9)$$

where $\dot{\varepsilon}_0$ is now a reference strain rate and α is an adjustable constant that may be taken to be unity. The steady-state creep behavior of the composite also has a pure power law form, and the tensile behavior in the direction of the aligned particles converts from Equation 9.7 to

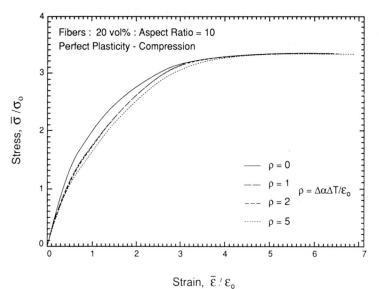

Figure 9.13. Compressive stress-strain curves for 20 vol % cylindrical fibers with a 10 to 1 aspect ratio with different amounts of thermal expansion misfit and a perfectly plastic matrix.

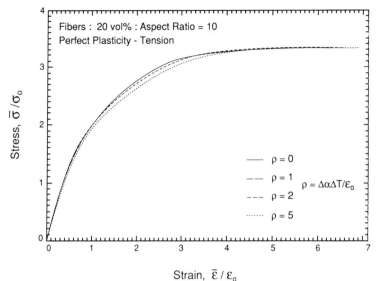

Figure 9.14. Tensile stress-strain curves for 20 vol % cylindrical fibers with a 10 to 1 aspect ratio with different amounts of thermal expansion misfit and a perfectly plastic matrix.

$$\frac{\dot{\bar{\varepsilon}}}{\dot{\varepsilon}_0} = \alpha \left(\frac{\bar{\sigma}}{\bar{\sigma}_N}\right)^n \quad (9.10)$$

where $\bar{\sigma}_N$ is precisely the same composite reference stress as it is for the rate-independent matrix. The strengthening in creep is reflected by the ratio $\bar{\sigma}_N/\sigma_0$, which in turn is given by Equation 9.8, together with the plots of $\bar{\sigma}_0/\sigma_0$ for the various forms of reinforcement presented earlier.

9.4.2 Diffusional Relaxation of the Reinforcement

At high temperature, in addition to matrix power law creep, mass transport by diffusion can occur within grains, on grain boundaries, and on interfaces. Matter diffuses from regions of low stress to regions of high stress and the result is a macroscopic deformation that is proportional to the applied stress, usually in a linear manner (Frost and Ashby 1982). This is known as Nabarro-Herring or Coble creep, depending on the diffusion path. Macroscopically, it can be modeled as a continuum linear viscosity of the matrix or, in cases where it is nonlinear, as a continuum power law creep with a low creep index. However, the presence of the interface between the reinforcement and the matrix in a composite material provides an additional path for mass transport that is not accounted for by either matrix properties or by reinforcement properties alone. Furthermore, the interface is typically a rapid path for mass transport and as such, diffusion on the interface can lead to a relatively fast creep of the composite material. As a consequence, it is desirable to model this mode of creep. Rosler et al. (1991) have used this mechanism to explain trends in the creep strength of composite materials. There is reason to believe that when temperature is sufficiently high, an otherwise strongly bonded interface will be capable of sliding. Therefore, the sequence that is envisioned is that which occurs as the temperature is increased; the resistance to sliding on the interface diminishes and eventually disappears. As the temperature continues to increase, diffusion is activated in the interface and occurs in an environment of zero resistance to sliding.

Little comprehensive modeling has been done on this problem. However, Sofronis and McMeeking (1992) have provided theoretical results for the creep strength of a material containing 20% by volume rigid spheres of radius R. The results were obtained using the cell model method as described previously in this chapter, but with interfaces that are capable of sliding and are simultaneously subject to mass transport within the interface. In the results of Sofronis and McMeeking (1992), the resistance to sliding in the form of a shear strength τ_s was considered to be proportional to the relative velocity v_s of sliding across the interface and thus

$$\tau_s = \mu v_s \quad (9.11)$$

where μ is an effective interface viscosity. As the temperature rises, μ will diminish and eventually disappear. Mass transport is controlled by an effective diffusion parameter $\mathcal{D}$ such that

$$v_n = -\mathcal{D}\nabla^2 \sigma_n \quad (9.12)$$

where v_n is the relative velocity normal to the interface of the matrix relative to the reinforcement, $\underline{\nabla}$ is the gradient operator in the surface of the interface, and σ_n is the normal component of the stress across the interface.

A summary of the results of Sofronis and McMeeking (1992) is shown in Figure 9.15. The creep strength $\bar{\sigma}/\sigma$ is plotted against the inverse interface sliding viscosity and against the effective interface diffusion coefficient. The creep strength is the stress required to cause a given strain rate in the composite, divided by the stress to cause the same strain rate in the matrix alone. In the left-hand side of the figure $\mathcal{D} = 0$ so that no diffusion takes place on the interface. In the right-hand side of the figure $\mu = 0$ so that shear-free sliding occurs at the interface. In terms of the whole plot, the temperature is relatively low (but in the creep range) at the far left of the diagram and continually increases from left to right. However, the scale would almost certainly not be linear or continuous in terms of temperature from left to right as drawn in Figure 9.15. It is evident that the creep strength falls as the resistance to sliding diminishes and mass transport increases. When the matrix is linear ($n = 1$), the creep strength is independent of strain rate. However, the power law creeping matrix leads to an effect, in conjunction with interface sliding or mass transport, that is sensitive to strain rate. A slower strain rate leads to a lower creep strength, indicating that the slow straining permits the sliding behavior or the diffusion process to become active relative to matrix creep. A fast strain rate seems to preclude those interface processes to some extent.

Sliding with no mass transport (see left-hand side of Figure 9.15) leads to a reduction of the creep strength of the composite material. The creep strength in each case reaches an asymptote as the sliding resistance diminishes, and these asymptotic values represent the creep strength when there is no shear strength at the interface. In the linear case, the strength falls from over twice the matrix strength for a well bonded nonsliding interface down to only 30% above the matrix strength when the interface slides freely.

When mass transport by diffusion becomes active, creep strength is further diminished. As before, an asymptotic value of the creep strength is approached as the diffusion becomes rapid. In each case, the asymptotic creep strength is below that of the matrix strength, indicating that because of the presence of the interface,

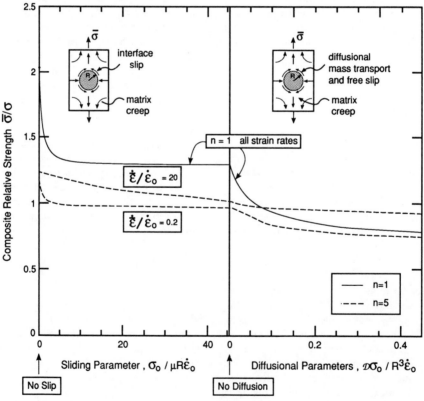

Figure 9.15. Creep strength of a material containing 20% vol of rigid spheres when sliding occurs on the interface and when shear free sliding and mass transport occur on the interface. The radius of the spherical reinforcements is R.

the reinforcements actually weaken the composite material if the diffusion rate is high enough.

Another important feature of the results of Figure 9.15 is effect of particle size. The material parameters μ and $\mathcal{D}$ inherently contain length scales. As a result, the continuum analysis predicts a behavior which, in turn, depends on the particle size. For a given interface and volume fraction of reinforcements, a larger particle size improves the creep strength. In the case of sliding, the effect can be understood in terms of the reduction in the total area of particle surface as the size is increased, thereby diminishing the effective strain produced by the same velocity of sliding. In the case of mass transport, the larger particle imposes greater diffusion distances, hence reducing the macroscopic strain rate. This effect in relation to grain size is well known in diffusion controlled creep of homogeneous materials (Frost and Ashby 1982).

Acknowledgments

The work of J.W.H. was supported in part by a grant from the National Science Foundation (Grant No. NSF-MSS-92-02141), by the DARPA URI (Subagreement P.O.#VB38639-0 with the University of California, Santa Barbara, ONR Prime Contract N00014-86-K-0753), and by the Division of Applied Sciences, Harvard University. The work of R.M.M. was supported by the DARPA URI at the University of California, Santa Barbara (ONR Contract N00014-86-K-0753).

References

Bao, G., Hutchinson, J.W., and R.M. McMeeking. 1991a. *Acta Metall. Mater.* 39:1871–1882.

Bao, G., Hutchinson, J.W., and R.M. McMeeking. 1991b. *Mech. Mater.* 12:85–94.

Brockenbrough, J., Suresh, S., and H.A. Wienecke. 1991. *Acta Metall. Mater.* 39:735–752.

Christman, T., Needleman, A., and S. Suresh. 1989. *Acta Metall. Mater.* 37:3029–3047.

Dragone, T.L., and W.D. Nix. 1990. *In:* Proc. TMS International Conference on Advanced Metal and Ceramic Composites, Anaheim, CA.

Frost, H.J., and M.F. Ashby. 1982. Deformation Mechanism Maps. Oxford: Pergamon Press.

Hom, C. L. 1992. *J. Mech. Phys. Solids* 40:991–1008.

Jansson, S., and F.A. Leckie. 1992. *J. Mech. Phys. Solids* 40:593–612.

Levy, A., and J.M. Papazian. 1990. *Metall. Trans.* 21A:411–420.

Povirk, G.L., Needleman, A., and R.S. Nutt. 1991. *Mater. Sci. Eng.* A132:31–38.

Rosler, J., Bao, G., and A.G. Evans. 1991. *Acta Metall. Mater.* 39:2733–2738.

Sofronis, P., and R.M. McMeeking. 1992. *Mech. Mater.* accepted for publication.

Torquato, S. 1991. *Appl. Mech. Rev.* 44:37–76.

Tvergaard, V. 1990. *Acta Metall. Mater.* 38:185–194.

Yang, Y., Pickard, S., Cady, C., et al. 1991. *Acta Metall. Mater.* 39:1863–1870.

Zahl, D.B., and R.M. McMeeking. 1991. *Acta Metall. Mater.* 39:1117–1122.

Chapter 10
Continuum Models for Deformation: Metals Reinforced with Continuous Fibers

SUBRA SURESH
JOHN R. BROCKENBROUGH

The deformation characteristics of fiber-reinforced metals have been the subject of analytic, experimental, and numerical investigations over the last several decades. Continuous fiber reinforced metal-matrix composites exhibit deformation responses that are distinctly different from those of conventional metals in many respects. First, the elastoplastic stress-strain behavior of these composites is highly anisotropic. The compressive strength is also generally much lower than the ultimate tensile strength because of the propensity for failure by kink band formation or fiber buckling in compression. Because the thermal expansion of coefficients of the constituents of the composite generally differ substantially, and because the volume fraction of the fibers is usually high (between 40% and 60%), thermal residual stresses that develop during cooling from the processing temperature can have a decisive effect on the deformation of the composite.

When the brittle fibers and the ductile matrix of a metal-matrix composite deform primarily elastically, the size, shape, or distribution of the fiber do not have any significant effect on the average constitutive response of the composite for loading parallel to the fiber direction, or for modest concentrations of the uniformly distributed reinforcement with the loading axis normal to the fiber direction. In such cases, available methods provide a fairly accurate prediction of the overall elastic deformation response (see, for example,

the reviews by Christensen 1979; Hashin 1983; and Bahei-El-Din and Dvorak 1989). However, for high concentrations of fibers and for transverse tension or shear loading, the elastic response can be sensitive to such factors as fiber distribution in the matrix. This situation is further compounded when the ductile matrix begins to deform plastically; the local properties become stress-dependent, the deformation becomes highly inhomogeneous, and the overall deformation begins to be affected by such factors as the loading path and the distribution and shape of the fibers. Consequently, analytic methods to describe the elastoplastic response often become intractable and the problem needs to be addressed by recourse to detailed numerical models. Such computational modeling of deformation and failure in metal-matrix composites has, in fact, been gaining increasing popularity in recent years.

10.1 Background

The objective of this chapter is to provide an overview of continuum models and analyses for the deformation of fiber-reinforced metals. Section 10.2 provides a brief summary of available methods with which to characterize the elastic deformation of fibrous composites. The approaches described here include the rule of mixtures, self-consistent analyses, Hashin–Shtrikman bounds, and periodic cell models based on analytical and computational formulations. A quantitative comparison of these models is made using the example of an aluminum alloy reinforced with boron fibers. Section 10.3 addresses plastic deformation of the composite material, with particular attention devoted to detailed numerical analyses of the effects of fiber distribution, concentration, orientation, and shape on the stress–strain response. It is shown that nonuniform deformation in the matrix cannot be uniquely described by recourse to analytical approaches, which neglect the effects of fiber shape and spatial distribution, and that the manner in which fibers are packed has a significant effect on the overall flow stress and strain hardening. Note that the approach adopted here focuses on modeling the elastic–plastic response of the matrix of the composite by using macroscopic continuum descriptions. The microscopic details that influence deformation, such as the grain size and mean dislocation spacing, are not taken into consideration. As a result, the approaches outlined here are effective as long as the characteristic microstructural dimensions of the composite constitutent, including substructure, are small compared to the mean separation distance of the fibers and the fiber diameter. Section 10.4 highlights the effects of thermal residual stresses on the elastic-plastic response. While the bulk of this chapter focuses on fiber-reinforced metallic composites in which perfect mechanical bonding exists between the fibers and the matrix, Section 10.5 briefly examines some simple, limiting cases, in which the fibers are debonded from the matrix. The chapter concludes with a discussion of critical issues and directions for future research in this area.

Throughout this chapter, results of the effects of geometric variables (such as fiber orientation and distribution) on the constitutive response of the composite are discussed with the aid of analytic, computational, and experimental results obtained for a specific composite with an aluminum-alloy matrix and boron unidirectional fibers. This choice was made because of the vast amount of published information available for this composite material. It should be noted, however, that the results presented for this system also apply for any other metal-matrix composite with similar matrix and interfacial deformation characteristics. Experimental results on deformation and failure mechanisms in other unidirectionally reinforced metal-matrix composite systems can be found in Chapters 9 and 13 of this volume.

Because there is a large body of literature on the deformation of fiber-reinforced composites, and because the overview presented in this chapter is necessarily brief, it is not feasible to provide all the derivations and original references. However, a sufficient number of citations to original work and key review articles is presented so that the interested reader can easily gain access to important sources of information on this topic.

10.2 Elastic Deformation

For a start, consider the overall elastic moduli of a two-phase, unidirectional, fiber-reinforced composite within the context of small linear elastic deformations. No particular geometry of the reinforcing phase needs to be assumed at this stage. Elastic constants are defined for a piece of the composite subjected to uniform conditions along the boundary. Under uniform stress (or strain), the overall moduli depend on average stresses (or strains) within each phase. In general, the mean fields in each phase are influenced by the volume fraction, shape, and spatial distribution of the fibers. To illustrate this point, consider the case of prescribed far-field surface tractions that are manifested in a uniform stress field. The far-field uniform stress in the composite $\hat{\sigma}$ is related to the average stress in the matrix and the fiber, $\hat{\sigma}_m$ and $\hat{\sigma}_f$, respectively, by

$$\hat{\sigma} = c_f \hat{\sigma}_f + c_m \hat{\sigma}_m \qquad (10.1)$$

where c_m and c_f are the volume fractions of the matrix and fiber, respectively, with $c_f + c_m = 1$, and the hat denoting a volumetric average that is defined as

$$\hat{f} = \tfrac{1}{V}\int f dV \qquad (10.2)$$

V is the volume and f is any field quantity (Hill 1963a, 1964). When both the fiber and the matrix undergo only elastic deformation,

$$\hat{\sigma}_f = L_f \hat{\epsilon}_f, \qquad \hat{\sigma}_m = L_m \hat{\epsilon}_m \qquad (10.3)$$

where L_f and L_m are the tensors representing elastic moduli of the fiber and the matrix, respectively. Combining Equations 10.1 and 10.3,

$$\hat{\sigma} = c_f L_f \hat{\epsilon}_f + c_m L_m \hat{\epsilon}_m \qquad (10.4)$$

where $\hat{\epsilon}_f$ and $\hat{\epsilon}_m$ are the average strains in the fiber and the matrix, respectively. When the far-field loading is uniform, the average elastic strain in each phase is related to the average far-field strain through the tensors A_f and A_m, so that

$$\hat{\epsilon}_f = A_f \hat{\epsilon}, \qquad \hat{\epsilon}_m = A_m \hat{\epsilon} \qquad (10.5)$$

where

$$c_f A_f + c_m A_m = I \qquad (10.6)$$

I is the identity tensor. Thus, the equation, $\hat{\sigma} = \bar{L}\hat{\varepsilon}$, determines the overall stiffness of the composite with

$$\bar{L} = c_f L_f A_f + c_m L_m A_m \quad (10.7)$$

The tensors A_f and A_m are usually referred to as concentration factor tensors and their components depend on the geometry and spatial distribution of each phase.* Thus, the overall moduli of the composite are not determined solely by the volume fraction of the constituent phases.

Various models have been proposed to determine the overall elastic moduli of the composite from a knowledge of the elastic properties of the constituent phases. In the subsections that follow, we briefly summarize the salient features of some prominent models and finite element analyses, and provide a quantitative comparison of their predictions for the particular example of an aluminum–boron composite.

10.2.1 Rule of Mixtures

In order to derive the longitudinal properties of continuously reinforced composites, it is reasonable to assume coupling of the averaged strains along the direction of fiber alignment such that

$$\varepsilon_A^{(f)} = \varepsilon_A^{(m)} = \hat{\varepsilon}_A \quad (10.8)$$

Here the subscript A denotes axial property and the superscripts (f) and (m) refer to the fiber and the matrix, respectively. Within the context of small-strain theory, exact volume averaging shows the average stress tensor in the composite to be that given in Equation 10.1. In the classic "rule of mixtures" formula, the effective moduli of the composite is expressed as

$$L_{RM} = c_f L_f + c_m L_m \quad (10.9)$$

wherein the components of L_{RM} do not depend on either the fiber distribution within the matrix, or the fiber shape. From the principle of minimum potential energy (Hill 1963b),

---

*All of the methods discussed in Sections 10.2.1 through 10.2.3 for estimating the effective properties of metal-matrix composites invoke the argument that the stress and strain fields in each phase of the composite can be represented by averaged fixed values. Theories predicated on this line of reasoning are widely referred to as *mean field theories*. The mean field theories can be expressed by relationships of the form of Equation 10.5, in which the mean strain fields in each phase are related to the overall strain field in the composite via strain concentration tensors A. Similar expressions can also be written for the stress fields.

$$\bar{L} \leq c_f L_f + c_m L_m \quad (10.10)$$

L_{RM} is an upper bound for the effective moduli of the composite.

For loading parallel to the fiber axis, the axial stresses in the fibers and the matrix under an average applied axial stress $\hat{\sigma}_A$ are given by

$$\hat{\sigma}_A^{(f)} = \frac{\hat{\sigma}_A E_A^{(f)}}{\bar{E}_A}, \quad \hat{\sigma}_A^{(m)} = \frac{\hat{\sigma}_A E_A^{(m)}}{\bar{E}_A} \quad (10.11)$$

where the superscripts (f) and (m) refer to the fiber and the matrix, respectively. E_A is the axial Young's modulus. Equation 10.11 can be modified (Sutcu and Hillig, 1990) to incorporate the average axial thermal residual stresses $\hat{\sigma}_{A,r}$ such that

$$\hat{\sigma}_A^{(f)} = \frac{\hat{\sigma}_A E_A^{(f)}}{\bar{E}_A} + \hat{\sigma}_{A,r}^{(f)}$$

$$\hat{\sigma}_A^{(m)} = \frac{\hat{\sigma}_A E_A^{(m)}}{\bar{E}_A} + \hat{\sigma}_{A,r}^{(m)} \quad (10.12)$$

Because residual stresses must be self-equilibrating,

$$c_f \hat{\sigma}_{A,r}^{(f)} + c_m \hat{\sigma}_{A,r}^{(m)} = 0 \quad (10.13)$$

Although Equation 10.9, when applied to the axial deformation of fibrous composites, invokes a Voigt-type approximation in which the fibers and the matrix are treated as parallel springs, a Reuss-type approximation is often useful to provide a crude estimate of the effective transverse properties of unidirectionally reinforced composites. In this case, the average stresses in the fiber and the matrix and the average composite stress are assumed equal. The overall Young's modulus in the transverse direction can then be written as

$$\bar{E}_T = \frac{E_T^{(f)} E^{(m)}}{c_f E^{(m)} + c_m E_T^{(f)}} \quad (10.14)$$

where the subcript T refers to the transverse direction. Note that this expression is valid for transversely isotropic fibers and an isotropic matrix. However, as discussed later in this chapter, the transverse elastic properties are sensitive to the shape and distribution of fibers in the matrix. Therefore, Equation 10.14 can deviate significantly from the actual transverse elastic response of unidirectional composites.

10.2.2 Hashin–Shtrikman Bounds

Improved bounds for the elastic moduli can be constructed for unidirectionally reinforced continuous-fiber composites by considering materials that are statistically transversely isotropic (Hashin 1979, 1983; Hashin and Shtrikman 1962, 1963). This implies that the effective stress–strain relation is independent of any rotation of coordinates about the fiber axis. With x_3 as the direction parallel to the fiber axis (in a Cartesian coordinate system with x_1, x_2 and x_3 are reference axes), the stress–strain relation for transverse isotropy can be written as

$$\sigma_{11} = (\bar{k} + \bar{m})\epsilon_{11} + (\bar{k} - \bar{m})\epsilon_{22} + \bar{l}\epsilon_{33}$$
$$\sigma_{22} = (\bar{k} - \bar{m})\epsilon_{11} + (\bar{k} + \bar{m})\epsilon_{22} + \bar{l}\epsilon_{33}$$
$$\sigma_{33} = \bar{l}\epsilon_{11} + \bar{l}\epsilon_{22} + \bar{n}\epsilon_{33}$$
$$\sigma_{12} = 2\bar{m}\epsilon_{12}$$
$$\sigma_{23} = 2\bar{p}\epsilon_{23}$$
$$\sigma_{31} = 2\bar{p}\epsilon_{31}. \quad (10.15)$$

Here, $\bar{k}$ is the plane strain bulk modulus for lateral dilation without longitudinal extension, $\bar{m}$ is the transverse shear modulus, $\bar{n}$ is the modulus for longitudinal axial straining, $\bar{l}$ is the associated cross modulus, and p is the longitudinal shear modulus. The overbar denotes the effective moduli for the composite. By application of the extremum principles of elasticity, bounds can be obtained for the various elastic moduli solely in terms of the fiber volume fraction (Hashin 1979, 1983). These bounds are the best possible in terms of volume fraction alone. The bounds for $\bar{k}$ are

$$k_{lb} \le \bar{k} \le k_{ub} \quad (10.16)$$

where

$$k_{lb} = k_m + \frac{c_f(k_f - k_m)(k_m + m_s)}{(k_f + m_s) - c_f(k_f - k_m)}$$
$$m_s = \min\{m_f, m_m\} \quad (10.17a)$$

$$k_{ub} = k_m + \frac{c_f(k_f - k_m)(k_m + m_g)}{(k_f + m_g) - c_f(k_f - k_m)}$$
$$m_g = \max\{m_f, m_m\} \quad (10.17b)$$

and where the subscripts f and m refer to the fiber and the matrix, respectively (Hashin and Shtrikman 1963). The lower and upper bound for the transverse shear moduli are

$$m_{lb} = m_m + \frac{c_f(m_f - m_m)(m_m + \gamma_s)}{(m_f + \gamma_s) - c_f(m_f - m_m)}$$
$$\gamma_s = \frac{m_s k_s}{(k_s + 2m_s)} \quad k_s = \min\{k_f, k_m\} \quad (10.18a)$$

$$m_{ub} = m_m + \frac{c_f(m_f - m_m)(m_m + \gamma_g)}{(m_f + \gamma_g) - c_f(m_f - m_m)}$$
$$\gamma_g = \frac{m_g k_g}{(k_g + 2m_g)} \quad k_g = \max\{k_f, k_m\} \quad (10.18b)$$

The bounds for the longitudinal shear moduli are

$$p_{lb} = p_m + \frac{c_f(p_f - p_m)(p_m + p_s)}{(p_f + p_s) - c_f(p_f - p_m)}$$
$$p_s = \min\{p_f, p_m\} \quad (10.19a)$$

$$p_{ub} = p_m + \frac{c_f(p_f - p_m)(p_m + p_g)}{(p_f + p_g) - c_f(p_f - p_m)}$$
$$p_g = \max\{p_f, p_m\} \quad (10.19b)$$

Finally, the two remaining moduli, $\bar{l}$ and $\bar{n}$, are solved from the Hill connections (Hill 1948, 1963a)

$$\bar{l} - c_f \nu_f + c_m \nu_m$$
$$= \frac{(l_f - l_m)}{(k_f - k_m)} (\bar{k} - c_f k_f - c_m k_m) \quad (10.20a)$$

$$\bar{l} - c_f \nu_f + c_m \nu_m$$
$$= \frac{(k_f - k_m)}{(l_f - l_m)} (\bar{n} - c_f n_f - c_m n_m) \quad (10.20b)$$

Here ν is the Poisson ratio. The five constants $\bar{m}$, $\bar{p}$, $\bar{k}$, $\bar{l}$, and $\bar{n}$ characterize the stiffness of a transversely isotropic composite. In particular, Young's modulus in the transverse direction is given by

$$\bar{E}_T = 4\left(\frac{\bar{n}}{\bar{n}\bar{k} - \bar{l}^2} + \frac{1}{\bar{m}}\right)^{-1} \quad (10.21)$$

and the axial Young's modulus is

$$\bar{E}_A = \bar{n} - \frac{\bar{l}^2}{\bar{k}} \quad (10.22)$$

The Hashin-Shtrikman bounds apply regardless of the cross-sectional geometry of the reinforcing fibers. Therefore, when the spread between the bounds is small, the reinforcement shape has little effect on the average elastic response.

10.2.3 Self-Consistent Methods

For the continuous fiber composite, the basic self-consistent method employs a single reinforcing fiber embedded in an infinite matrix of the overall composite. There are many modifications of these approaches (Hill 1965; Willis 1977; Christensen 1990), such as the generalized self-consistent method (Christensen and Lo 1979), which considers a single fiber embedded in a shell of matrix material, which is then embedded in a matrix of the overall composite.

Self-consistent methods provide approximate estimates of the overall elastic response of the composite by explicity accounting for the phase geometry. These methods utilize the classical Eshelby approach of a single ellipsoidal inclusion embedded in an infinite matrix for the overall solution (Eshelby 1957, 1959). As part of this inclusion analysis, the phase geometry enters through the volume fraction, shape, and orientation of the reinforcing inclusion.* The compound comprising the fiber and the matrix is embedded in a continuum that has the same effective properties as those of the composite. The assumption of transverse isotropy for the effective medium implies that the spatial distribution of fibers renders the composite isotropic in the transverse plane. This restricts the representation of the composite to have either a random distribution of fibers or hexagonal (triangular) symmetry. Thus, the results extracted from this approach compensate for interaction effects in an averaged sense. An example of the role of fiber cross-sectional geometry and fiber volume fraction in influencing the effective elastic moduli of ribbon-shaped fiber reinforced composites can be found in Zhao and Weng (1990).

The analytic solution based on the self-consistent method can be used to estimate the concentration factor tensors appearing in the definition of overall moduli in Equation 10.7 (Walpole 1966). That is, a single inclusion is embedded in an infinite matrix material consisting of the overall composite material. The concentration tensor A_f is obtained from the solution to the inclusion problem. The self-consistent method for a continuous fiber composite leads to the following two connected relations from which the composite moduli $\bar{m}$ and $\bar{p}$ are extracted:

$$\frac{c_m}{\bar{p}-p_f} + \frac{c_f}{\bar{p}-p_m} = \frac{1}{2\bar{p}} \quad (10.23)$$

and

$$2\left(\frac{c_f m_m}{\bar{m}-m_m} + \frac{c_m m_f}{\bar{m}-m_f}\right) + \frac{c_f k_f}{k_f-\bar{m}} + \frac{c_m k_m}{k_m-\bar{m}} = 0 \quad (10.24)$$

With $\bar{m}$ and $\bar{p}$ obtained from Equations 10.23 and 10.24, $\bar{k}$ is determined from

$$\frac{c_f}{k_f+\bar{m}} + \frac{c_m}{k_m+\bar{m}} = \frac{1}{\bar{k}+\bar{m}} \quad (10.25)$$

Subsequently, $\bar{n}$ and $\bar{l}$ are obtained from the Hill connections of Equations 10.20a and 10.20b.

---

*Modifications of the Eshelby inclusion method that account for finite concentration effects, such as the Mori-Tanaka approach (Mori and Tanaka 1973), are commonly used in connection with discontinuously reinforced metal-matrix composites (where the volume fraction of the particulate or whisker reinforcements is typically 20% or less) or for modest concentrations of continuous fiber reinforcements (Lilholt 1988; Pedersen 1983; Taya and Arsenault 1989; Taya and Mori 1987).

10.2.4 Other Analytic/Semi-Analytic Methods

Aboudi (1989, 1990) has proposed a cell model based on a periodic square array of square fibers for metal-matrix composites. This model consists of a rectangular subcell for the fiber and three similar subcells for the matrix. By using a single material model for the matrix and linear variation of displacements within the subcells, the combined interactions among subcells and cells are incorporated in this analytic approach for thermoelastic composites (which can be extended into a semi-analytical form for inelastic deformation).

Nemat-Nasser et al. (1982) have suggested another method for modeling the deformation of unidirectional composites. In this method, the fibers are placed in periodic square arrays of the matrix material. Integral equations are developed for the square subvolumes with the aid of series expansions of the inhomogeneous local stress and strain fields. In this approach, and its modifications for both the elastic and inelastic deformation regimes, the accuracy and scope of the estimates can be enhanced by increasing the number of subvolumes. Furthermore, as shown in the next section, numerical models employing square unit cells with periodic boundary conditions have also been used to determine the elastoplastic response of metal-matrix composites.

In addition to the unit cell models involving packing of fibers in square cells, various periodic cells of hexagonal or triangular packing, employing both analytic and numerical techniques, have been utilized to generate both elastic and elastoplastic responses. Although the case of square packing inherently involves anisotropy in the transverse plane, hexagonal packing (commonly referred to as the periodic hexagonal array or PHA model) produces a composite response that is transversely isotropic for elastic deformation (provided that thermal residual stresses are either negligible or appropriately symmetrical). In Section 10.2.5, we discuss in some detail the modeling of fiber packing using hexagonal symmetry in the context of finite element formulations.

Different types of micromechanical models for continuously reinforced metal-matrix composites have also been proposed by Brown and Clarke (1977), Lilholt (1988), and Pedersen (1983, 1990), who classify the overall strengthening seen in the composite into three components: (1) Strengthening of the matrix phase, as in monolithic metals, due to the evolution of precipitation and dislocation substructures. (Additional contributions to matrix strengthening may also arise as a consequence of enhanced dislocation density in the matrix as a result of accelerated aging of the matrix; see Chapter 7.); (2) constraint on the local deformation in the ductile matrix as a result of the presence of brittle fibers; and (3) Strengthening arising from the interaction of the reinforcements with dislocations, which are described by such terms as the *source shortening stress* and *friction stress*. The first two factors can be described, to a reasonable degree of accuracy, by purely macroscopic, continuum calculations without regard to the characteristic microstructural size scale of the composite, while the last term is strongly influenced by the fiber size. These approaches provide fairly accurate predictions for axial deformation of fiber-reinforced composites, although for transverse loading, considerable differences may exist between predictions and real material response, especially in the plastic regime.

10.2.5 Numerical Models

Figure 10.1(a) is a cross section normal to the fiber axis in a unidirectional metal-matrix composite where the periodic hexagonal array (PHA) is depicted. A number of different unit cells can be extracted from this array for the purpose of finite element analysis. The rectangular unit cell in Figure 10.1(a) and the triangular unit cell in Figure 10.1(b) show two examples for finite element discretization of the unit cells conforming to the PHA packing arrangement (Teply and Dvorak 1988; Brockenbrough and Suresh 1990; Brockenbrough et al. 1991; Böhm 1991; Böhm and Rammerstorfer 1991). The unit cell in Figure 10.1(a) is a rectangle situated in a plane normal to the fiber direction. With generalized plane strain and the requirement that the rectangle remain a rectangle during deformation, this model involves two transverse planes that are free to move relative to each other in the longitudinal direction. The boundary conditions include strains and stresses that are constant in the axial direction and axial strains that are identical. The unit cell in Figure 10.1(b) is an equilateral prism whose vertices coincide with adjacent fiber centers; the unit cell contains one-sixth of three fibers. Boundary conditions are prescribed so that during deformation the unit cell will continually tile the space, with no gap existing between adjacent cells as viewed on the transverse plane. Thus, the edges of the unit cell need not necessarily remain straight. The boundary conditions so imposed allow any far-field stress to be modeled. Whereas the unit cell in Figure 10.1(b) is capable of incorporating transverse shear loads in the analysis of deformation (for details see Teply and Dvorak 1988; Brockenbrough et al. 1991), the PHA model in Figure 10.1(a) enables simpler boundary conditions for normal loading in the transverse plane.

Another common periodic arrangement for numerically modeling the deformation of metal-matrix com-

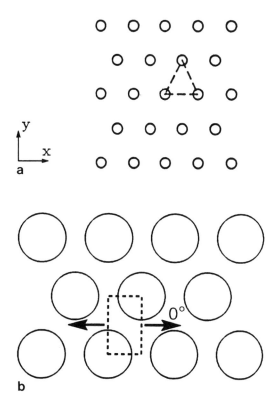

Figure 10.1. (a) Triangular unit cell and (b) rectangular unit cell for the periodic hexagonal array (PHA) model. See text for details.

posites involves square unit cells (Brockenbrough and Suresh 1990). Figure 10.2(a) shows the transverse cross section of the so-called square edge-packing of fibers where the dashed line represents the unit cell. This corresponds to the configuration where the fibers are located at the edge of the square or equivalently, at the center of the square (this can be termed zero-degree square packing). Here, the smallest spacing between adjacent fibers is along the edge of the square unit cell. If the lattice in Figure 10.2(a) is rotated 45 degrees about the fiber axis, the unit cell shown in Figure 10.2(b) is obtained. This is often referred to as square diagonal packing. Here, the smallest spacing between adjacent fibers is along the diagonal of the square unit cell. Periodic boundary conditions are chosen for these square unit cells.

Within the framework of analytic or computational modeling, unit cell formulations involving perfectly periodic and uniform distributions of fibers in the ductile matrix are commonly used to model the composite. In practice, however, such periodic arrangements are seldom observed in commercially synthesized composites. Some examples of nonuniform fiber distributions in unidirectionally reinforced metal-matrix composites are

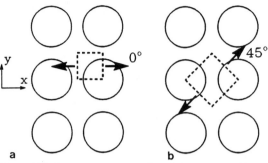

Figure 10.2. A transverse cross-section of the square packing of fibers. (a) Square edge packing and (b) square diagonal packing, which is obtained by rotating the arrangement in (a) by 45 degrees about the fiber axis.

shown in Figure 10.3. Figure 10.3(a) is a transverse sectional view of a liquid-infiltrated α-alumina fiber composite with an aluminum-lithium alloy matrix (Champion et al. 1978). Figure 10.3(b) is a transverse view of a 6061 aluminum alloy that is reinforced with 43 vol % (of pitch-55) boron fibers. This composite was synthesized by diffusion bonding of the constituent phases (DWA Composites, Chatsworth, CA).

In addition to the PHA and square packing arrangements, Brockenbrough and Suresh (1990), Brockenbrough et al. (1991), and Nakamura and Suresh (1993) have recently performed finite element simulations employing rectangular unit cells in which large numbers of *randomly placed fibers* are incorporated. Figure 10.4(a) and 10.4(b) shows examples of two such random fiber unit cells incorporating 30 and 60 fibers, respectively. Nakamura and Suresh (1993) have computationally modeled the elastic and plastic deformation along the transverse directions by employing a large number of random fiber unit cells (such as the ones shown in Figure 10.4), the average properties of which provide a reliable measure of the effective properties for a random distribution of fibers in the composite. Periodic boundary conditions within the context of generalized plane strain formulations are employed for the unit cells with random fiber arrangements.

10.2.6 A Comparison of Different Models

In this section, we present a quantitative comparison of the predictions of overall elastic moduli of fiber-reinforced metal-matrix composites using the various models previously discussed. The material chosen for this comparison is a 6061-o aluminum alloy reinforced with 46 vol % boron unidirectional fibers. (The effects of fiber packing on the transverse elastic properties have also been numerically examined by Wisnom (1990),

who conducted a finite element analysis of transverse tension deformation in a 6061 aluminum alloy reinforced with 48 vol % SiC unidirectional fibers.) The assumption of perfect mechanical bonding between the matrix and the reinforcement is known to be reasonable for this material (Böhm 1991). It is assumed that both the aluminum matrix and the boron fibers have isotropic elastic properties. The values of Young's modulus, shear modulus, and Poisson's ratio for the matrix are $E_m = 68.9$ GPa, $G_m = 25.9$ GPa, and $\nu_m = 0.33$, respectively. The corresponding values for the fiber are $E_f = 414$ GPa, $G_f = 172$ GPa, and $\nu_f = 0.2$. From a knowledge of Young's modulus and shear modulus for the matrix and reinforcement, the overall Young's modulus in the axial direction $\overline{E}_A$, the overall Young's modulus in the transverse direction $\overline{E}_T$, and the overall shear modulus in the transverse direction $\overline{G}_T$, were calculated assuming perfect mechanical bonding between the two phases. The models considered here for calculating the transversely isotropic elastic properties are the rule of

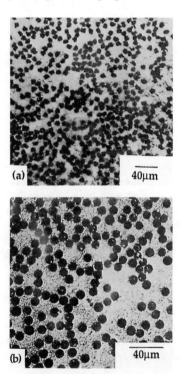

Figure 10.3. Transverse cross-sections of some fiber-reinforced metals indicating nonuniform spatial distribution of fibers. (a) Liquid-infiltrated aluminum-lithium alloy matrix that is reinforced with α-alumina fibers. *(Champion et al., 1978, Proc. Second International Conference on Composite Materials, pp. 882-904, The Metallurgical Society, Warrendale, PA. Reprinted with permission from The Metallurgical Society.)* (b) 6061 aluminum alloy reinforced with 43 vol % pitch-55 boron fibers.

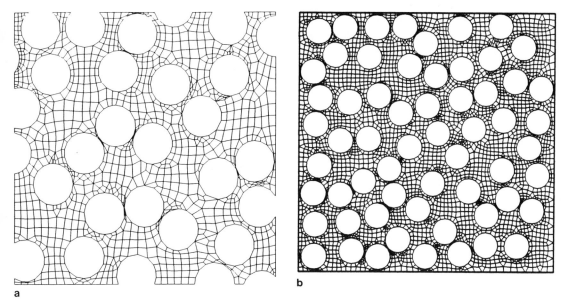

Figure 10.4. Unit cells consisting of (a) 30 and (b) 60 randomly placed unidirectional fibers and the finite element discretization.

mixtures, the self-consistent method, the upper bound and lower bound estimates of Hashin and Shtrikman, and the numerically based periodic hexagonal array model (with unit cell similar to that shown in Figure 10.1(b). Among elastically anisotropic arrangements, two periodic numerical models with square packing arrangements of fibers with the edge (or 0-degree rotation, Figure 10.2(a)) and diagonal (or 45-degree rotation, Figure 10.2(b)) were also considered. In addition, numerical results of elastic properties are shown for the two random-fiber finite-element models incorporating 30 and 60 fibers. These results are shown in Figure 10.5(a) and (b), where the transverse properties were computed along the x and y directions (z being the axial direction of fiber alignment).

Table 10.1 provides the numerical predictions of the elastic moduli for these various models, which are listed in the order of increasing transverse Young's modulus. Also shown in this table are the experimental results obtained for this composite material. The following conclusions can be derived from the results shown in Table 10.1.

1. The particular manner in which the fibers are packed in the unit cell has essentially no effect on Young's modulus in the axial direction, $\bar{E}_A$.
2. Among the models that have some validity for use in determining the transverse properties (setting aside the simple rule of mixtures), it is found that Young's modulus in the transverse direction $\bar{E}_T$ increases by 33.07%, from 127 GPa for the most compliant case of 45-degree loading (diagonal packing) of the square array numerical unit cell, to 169 GPa for the stiffest case predicted by the Hashin-Shtrikman upper bound.
3. Similarly, the transverse shear modulus $\bar{G}_T$ varies from 45 GPa to 64 GPa between these two extremes, representing an increase of as much as 42.2%. *These results clearly demonstrate that even the elastic properties of fiber reinforced metal-matrix composites (with fixed fiber concentration and cross-sectional geometry) are sensitive to the manner in which the fibers are arranged in the matrix.*
4. The models that possess transverse symmetry, specifically, the PHA method, the self-consistent method, and the random-fiber numerical methods, provide stiffness values that fall within the Hashin-Shtrikman bounds. The cell models based on square packing do not possess transverse symmetry and hence, they need not lie within these bounds. (It is also seen that the transverse moduli based on square diagonal packing fall below the Hashin-Shtrikman lower bound.) The difference between the elastic moduli predicted by the square edge packed and diagonal packed arrangements is a measure of the transverse anisotropy.
5. The experimental results are very close to those predicted by the PHA model and the two random-fiber numerical models.
6. The self-consistent method provides results that fall within the Hashin-Shtrikman bounds. However, they are stiffer than the transverse elastic moduli

Table 10.1. Comparison of Predictions of Elastic Moduli from Different Models for 6061-0 Aluminum Alloy, Reinforced with 46 vol % Boron Unidirectional Fibers.

Model	Geometry	Axial Young's modulus $\bar{E}_A$(GPa)	Transverse Young's modulus $\bar{E}_T$(GPa)	Transverse shear modulus $\bar{G}_T$(GPa)
Square diagonal packed (FEM)	Periodic	227	127	45
Hashin-Shtrickman lower bound	Transversely isotropic	227	134	49
Periodic hexagonal array (FEM)	Periodic	227	137	50
Unit cell with 30 random fibers (FEM)	Random periodic	228	138	51
Experiments: Becker et al. (1987)	Nonuniform distribution	228	138	57
Hill's self-consistent method	Transversely isotropic	228	144	53
Square edge packed (FEM)	Periodic	228	152	57
Hashin-Shtrickman upper bound	Transversely isotropic	228	169	64
Rule of Mixtures	Axial	227	—	—
	Transverse	—	112	—

derived from the PHA and the random-fiber models. The results of the self-consistent method are 5.1% and 6% higher than those of the PHA model for the transverse Young's modulus and shear modulus, respectively. This difference must be regarded as a consequence of the approximations inherent in the self-consistent method.

10.3 Plastic Deformation

As noted at the beginning of the chapter, the onset of plastic deformation in the metal matrix renders the local deformation highly inhomogeneous, with the result that the overall stress response becomes a strong function of the applied loading path as well as a function of the geometry of fiber shape and distribution within the ductile matrix. When the unidirectionally reinforced composite is loaded in tension along the axis of the fibers, the load is carried primarily by the brittle elastic fibers and there is very little plastic deformation in the intervening ductile matrix. In this case, the fiber packing arrangement has essentially no effect on the longitudinal stress-strain response of the metal-matrix composite. (Recent work by Nakamura and Suresh (1993) has shown that matrix yielding induced by the thermal residual stresses during cooling from the processing temperature can significantly lower the *apparent* Young's modulus of the composite during subsequent axial tension loading.)

When the loading axis is perpendicular to the direction in which the fibers are aligned, considerable plastic deformation can occur in the ductile matrix as the load is transmitted between the fibers. Consequently, fiber distribution has a strong effect on the deformation response in the transverse orientation. Such loading-path-dependent, highly inhomogeneous *local* deformation fields involving nonproportional strains are not amenable to simple analytic formulations. This situation is further exacerbated by fiber geometries that have sharp corners leading to intense local plastic deformation. Thus, it is not surprising that plasticity theories based on single-fiber-unit cell models are incapable of providing unique and accurate predictions of the average transverse constitutive response of the composite that are compatible with experimental observations. This difficulty can be somewhat overcome by means of detailed finite element models of inelastic deformation in the composite, where several different packing arrangements, including random distributions of large numbers of fibers in a ductile matrix, are examined.

The effect of fiber distribution on plastic response in metal-matrix composites is illustrated in Figure 10.5 (after Brockenbrough et al. 1991) with the aid of results obtained for axial and transverse deformation in 6061-o aluminum alloy reinforced with 46 vol % boron unidirectional fibers. In this work, the finite element simulations were conducted using the periodic and random unit cells similar to those described in Section 10.2. The matrix is modeled as an isotropically hardening elasto-plastic solid amenable to a Mises formulation (with a stress-strain response known from experiments), the fibers are modeled as linear elastic with isotropic properties, and the interface between the fibers and the matrix is taken to be perfectly bonded in a mechanical sense. Full details can be found in Brockenbrough et al. (1991).

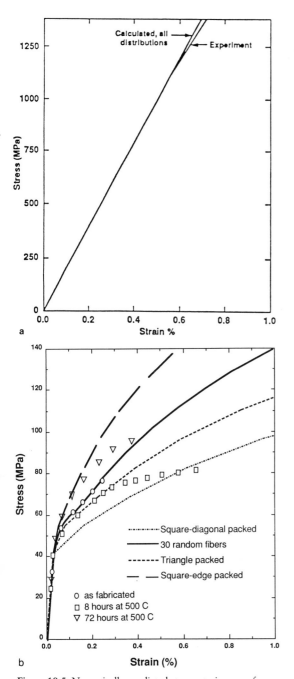

In Figure 10.5(a), the calculated curve shows the finite element predictions of axial stress-strain variation based on square-edge-packed, square-diagonal-packed, PHA, and 30-fiber random distributions. The numerical predictions closely match the experimental observations. (Both the predicted and experimental lines are terminated at the same far-field tensile stress, which corresponds to the stress at which final fracture occurred in the experiments.) Although there is no effect of fiber arrangement on the constitutive response in axial tension in Figure 10.5(b), a very different picture emerges for transverse tension. Here, transverse tensile loading at 0 degrees and 45 degrees with respect to the unit cell for square packing (see Figure 10.2) provides the strongest and weakest overall response, respectively, and the highest and lowest strain hardening exponent, respectively. Triangle packing involving the PHA model shown in Figure 10.1(b), and a 30-fiber random arrangement similar to the one shown in Figure 10.4(a) show an intermediate plastic response.* The same ranking of strengthening was found for the four packing arrangements for pure shear loading on the transverse plane (involving equal tension and compression along x and y axes, respectively). The effects of longitudinal shear loading in fiber reinforced metal-matrix composites have also been examined computationally by Sorensen (1991).

Figure 10.5(b) also shows the experimental data on transverse tensile stress-strain response measured by Kyono et al. (1986) on an aluminum alloy reinforced with 46 vol % boron unidirectional fibers.[†] The experimental data pertain to three heat-treatment conditions: as-fabricated, heat-treated for 8 hours at 500°C, and heat-treated for 72 hours at 500°C. On the basis of fractographic observations, Kyono et al. (1986) note that the as-fabricated composite exhibits the lowest ductility because of poor interfacial bonding. The heat treatment involving 8-hour exposure at 500°C improves interfacial bonding and ductility. Continued exposure up to 72 hours at 500°C leads to a deterioration of B fibers by the formation of a brittle AlB_2 phase that results in fiber splitting during transverse deformation.

Figure 10.5. Numerically predicted stress–strain curve for tensile deformation (a) along the fiber axis and (b) normal to the fiber axis for a 6061-o aluminum alloy with 46 vol % boron fibers. *(After Brockenbrough et al. 1991)* The experimental data in (a) *are from Becher et al. 1987. The experimental data in* (b) *are from Kyono et al. (1986).)*

*All the numerical results shown in Figures 10.5(a) and (b) are probably *upper bound* values because they have been obtained using the finite element code ABAQUS (1991), where a kinematically admissible *displacement field* is prescribed in each element. The displacement field is chosen to be the best possible for equilibrium.

[†]Although the composition of the matrix material in the experiment of Kyono et al. was different from that of the 6061-o aluminum alloy used in the numerical simulations by Brockenbrough et al. (1991), the two matrix materials have had essentially the same stress-strain response up to strain values that are of interest here.

It is interesting to note that the experimental data lie closest to the PHA and the random fiber numerical model predictions during the early to mid stages of plastic deformation.

Brockenbrough et al. (1991) found that the effect of fiber arrangement on the plastic response directly correlates with the extent of constraint developed in the matrix. For the transverse tension loading of the four fiber arrangements in Figure 10.5(b) at the fixed far-field tensile strain of 0.0023, Figure 10.6(a) shows the volume fraction of the matrix that has a hydrostatic stress σ_h greater than the value indicated on the abscissa. The magnitude of the matrix hydrostatic stresses for the different fiber arrangements fall within the same ranking scale as do the flow stress and the strain hardening exponent in transverse tension. A similar plot of the equivalent plastic strains for the four fiber arrangements is depicted in Figure 10.6(b). This plot shows that the ranking of plastic strains is the inverse of the ranking of overall strength shown in Figure 10.5(b), indicating that the fiber arrangement with the softest overall response to transverse tensile deformation exhibits the highest magnitude of matrix plastic strain.

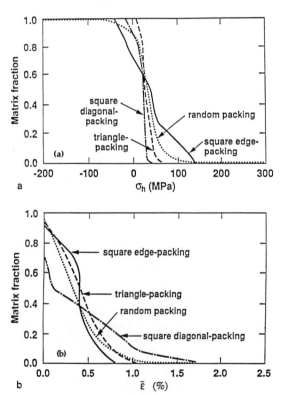

Figure 10.6. Numerically predicted volume fraction of the matrix that has (a) hydrostatic stress and (b) effective plastic strain greater than that indicated on the abscissa for a 6061 Al-boron composite subjected to far-field transverse tension at a strain of 0.0023. (After Brockenbrough et al. 1991.)

The role of constraint in influencing the overall stress-strain response of fiber-reinforced composites is evident from the above results. This role of constrained plastic flow in determining the apparent flow strength and strain hardening exponent can be further documented with the aid of results obtained for fibers with sharp corners. When the calculations for the various packing arrangements used earlier for cylindrical fibers with circular cross-sections are repeated with cylindrical fibers with square cross sections (with the same fiber concentration), a considerable elevation in apparent flow stress and strain hardening results. This is a consequence of elevated constraint caused by the sharp corners in the fibers. Figure 10.7 shows the numerically predicted constitutive response of the Al-boron composite in transverse tension for different fiber shapes and packing arrangements. The higher strengthening effect of the square cross section (for a fiber fixed packing arrangement) is evident.

Similar effects of strengthening from constrained plastic flow have been reported for whisker-reinforced metal-matrix composites (Christman et al. 1989; Dragone and Nix 1990; Tvergaard 1990; Levy and Papazian 1991; Bao et al. 1991). For example, both axisymmetric and plane strain models for discontinuously reinforced metal-matrix composites show that whiskers oriented parallel to the tensile axis and aligned end-to-end show the highest level of hydrostatic stress in the matrix and the highest overall flow strength. When adjacent whiskers are staggered in the axial direction, the average level of triaxiality in the matrix is diminished; this also results in a drop in the overall yield stress and a drop in the strain hardening exponent.

Because constraint on the ductile matrix increases with increasing concentration of the brittle reinforcement, and because constrained flow is a major contributor to the effects of reinforcement distribution on the overall flow strength, lower concentrations of reinforcement render the composite material less sensitive to reinforcement distribution. This conclusion is borne out by the results of numerical simulations by Brockenbrough et al. (1991).

10.4 Role of Thermal Residual Stresses

10.4.1 Elastic Deformation

Unidirectionally reinforced metal-matrix composites contain a large concentration of the reinforcement phase (typically between 40% and 60%). There exists a large thermal mismatch between the matrix and reinforcement phases (which can differ by as much as a factor of 4.7 between an aluminum alloy and a SiC fiber). Consequently, the thermal residual stresses that

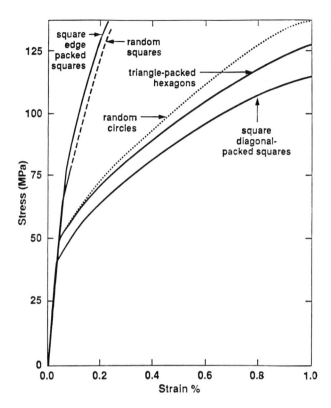

Figure 10.7. Numerically predicted constitutive response of the Al-boron composite in transverse tension for different fiber shapes and packing arrangements. *(After Brockenbrough et al. 1991.)*

are generated in the composite because of the differences in contraction of the two phases during cooling from the processing temperature, can have a noticeable impact on subsequent mechanical response. Detailed numerical analyses are necessary to determine thermal stresses by incorporating the temperature dependence of material deformation during cooling. However, simple expressions for the overall thermal properties of the unidirectional composite can be derived by assuming that the material data are independent of temperature between the processing and room temperatures.

Concentrating on axial deformation and assuming that (1) the deformation is linear elastic, and (2) the average strains in the matrix, the reinforcement, and the composite are all equal (noting that the residual thermal stresses in the matrix and fiber should be self-equilibrating), the following expression for the overall thermal expansion coefficient in the axial direction for uniform temperature conditions is obtained for the composite (Böhm 1991):

$$\bar{\alpha}_A = \frac{c_f E_A^{(f)} \alpha_A^{(f)} + c_m E^{(m)} \alpha^{(m)}}{\bar{E}_A} \qquad (10.26)$$

where α is the thermal expansion coefficient, A denotes that the subscripted quantity pertains to the axial direction, and the superscripts (f) and (m) denote that the superscripted quantities pertain to the fiber and the matrix, respectively. Using variational energy principles, Schapery (1968) has shown that the overall thermal expansion coefficient in the transverse direction can be written as

$$\bar{\alpha}_T = c_m(1+\nu^{(m)})\alpha^{(m)} + c_f(1+\nu_A^{(f)}\alpha_A^{(f)}) \\ - \bar{\alpha}_A(c_f \nu_A^{(f)} + c_m \nu^{(m)}), \qquad (10.27)$$

where ν is Poisson's ratio and the subscript A denotes its reference to the longitudinal direction. (The last term in this equation includes a rule-of-mixture-type formula for the overall Poisson ratio.) Further refinements to Equations 10.26 and 10.27 have been presented by Hashin and Rosen (1964).

In addition to the simple analytic models described above, Müller et al. (1991) have investigated both analytically and numerically the radial and circumferential residual stresses caused by thermal and elastic mismatch at the interfaces of circular elastic fibers embedded in an elastic matrix. They consider plane stress, plane strain, and generalized plane strain. Their numerical analysis involving the square and hexagonal packing arrangements also includes the effects of fiber volume fraction and the relative elastic properties of the constituent phases.

10.4.2 Plastic Deformation

Because the effective axial coefficient of thermal expansion is nearly independent of the fiber concentration for a fully plastic matrix (Böhm 1991),

$$\bar{\alpha}_A^{pl} \approx \alpha_A^{(f)} \qquad (10.28)$$

for a unidirectional composite with an elastic perfectly-plastic metallic matrix. For this composite with an ideally plastic matrix, the overall transverse coefficient of thermal expansion can be approximated by replacing $E^{(m)}$ in Equations 10.12 and 10.13 with the tangent (hardening) modulus $E_t^{(m)}$, and by replacing $\nu^{(m)}$ with 0.5 (an incompressible matrix). This approximation provides the following value of $\bar{\alpha}_T$ for the composite with a fully yielded matrix (Böhm 1991):

$$\begin{aligned}\bar{\alpha}_T^{pl} \approx\ & 1.5c_m\alpha^{(m)} + c_f(1 + \nu_A^{(f)})\alpha_A^{(f)} \\ & - \bar{\alpha}_A^{pl}(0.5c_m + c_f\nu_A^{(f)}).\end{aligned} \qquad (10.29)$$

More detailed analyses of the effects of thermal residual stresses on the constitutive response of fiber-reinforced metals are undertaken using the finite element method. With such analyses, the dependence of matrix yield strength on temperature (during cooling from the processing temperature), and the variation of the thermal expansion coefficient with temperature, can be incorporated into the analyses. Furthermore, these computational models are also capable of accounting

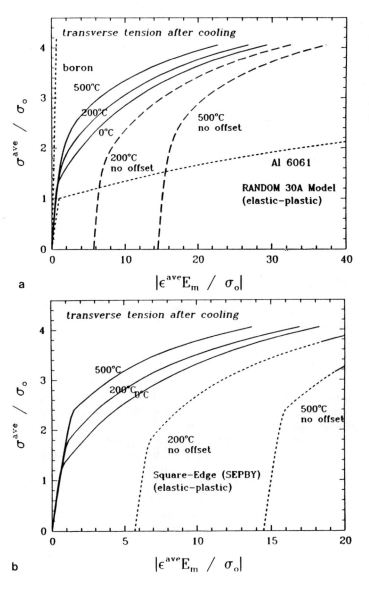

Figure 10.8. Effect of thermal residual stress (developed during different temperature drops) and transverse tension loading on the stress-strain response of a 6061 Al-boron composite. (a) 30 random fiber unit cell and (b) single fiber square edge packed unit cell. *(After Nakamura and Suresh 1993.)*

for the effects of fiber distribution on the thermal residual stress field and the attendant effect of this field on subsequent plastic deformation.

Using a generalized plane strain model within the context of unit cell finite element formulations for the deformation of a 6061 aluminum alloy reinforced with 48 vol % SiC fibers, Wisnom (1990) found that during cooling from the synthesis temperature, the whole of the composite matrix yielded plastically. Similar conclusions were also reached by Nakamura and Suresh (1993) in their simulations of the influence of thermal residual stresses on subsequent axial tension loading of aluminum-boron composites. The matrix yielding sets up a hydrostatic stress state with a nonuniform stress distribution. The analysis showed that the presence of residual stresses is beneficial to strengthening in the transverse direction because radial compressive stresses are induced at the interface during cooling from processing.

Nakamura and Suresh (1993) have carried out finite element simulations of the combined role of residual stresses and fiber distribution on the transverse tensile deformation of 6061-o aluminum alloy reinforced with 46 vol.% boron fibers. Figure 10.8(a) shows the variation of the average tensile stress $\bar{\sigma}$ in the transverse (y) direction (normalized by the matrix reference stress σ_0) as a function of the average strain in the composite $\bar{\epsilon}$ (divided by σ_0/E_m). The material illustrated is fiber-reinforced aluminum-boron composite (30 random fiber unit cell, similar to Figure 10.4(a)), which is subjected to no thermal residual stress (0°C cooling), and is cooled by a temperature drop of 200°C as well as by a temperature drop of 500°C. (The analyses incorporate the effects of changes in yield strength, thermal expansion coefficient, and elastic modulus with temperature during cooling.) The solid lines denote transverse tension deformation in which the initial matrix strain is offset to provide the same initial average strain for the three cases; the large dashed curves show no strain offset for the 200°C and 500°C temperature drops. Also shown for reference are the stress-strain response of the matrix material and the boron fiber. Figure 10.8(a) illustrates that thermal residual stresses can have a significant effect on the overall constitutive response of the composite in the transverse direction. Moreover, the thermal residual stresses apparently enhance the overall transverse yield stress well into the plastic region of deformation (as compared to the case where there are no residual stresses). This trend is at variance with that predicted for discontinuously reinforced metal-matrix composites (with typically less than 25 vol % of reinforcement), in which the effects of thermal residual stresses on the constitutive response are eliminated by the development of appreciable plasticity in the matrix.

The effect of fiber packing on the transverse stress-strain response (incorporating residual stresses) is evident from a comparison of parts (a) and (b) of Figure 10.8: Part (b) shows the effects of 0°C, 200°C, and 500°C temperature drops on subsequent transverse tensile deformation for a square edge packed fiber arrangement similar to the one shown in Figure 10.2(a). All three cooling conditions result in a stronger resistance to transverse deformation in the case of square edge packing than that found in the random case. Furthermore, the differences between the three different temperature drops is also slightly altered with the packing arrangement.

10.5 Interfacial Debonding

Interfacial debonding is a common damage mechanism in most fiber-reinforced composite materials. The transverse stress-strain characteristics of fiber reinforced composites are particularly sensitive to the extent of interfacial debonding. Detailed numerical studies that incorporate realistic trends of interfacial debonding in continuous fiber-reinforced metal-matrix composites are now unavailable.* Although in principle this problem can be handled with varying degrees of sophistication (from atomistic to macroscopic levels of modeling) within the framework of finite element formulations, experimentally verifiable micromechanical models for the onset and progression of debonding are not yet available. However, some potentially useful, idealized cases are examined here in an attempt to develop a perspective on the role of interfacial debonding on the constitutive response.

In their finite element analysis aimed at modeling the axial deformation of whisker-reinforced composites, Owen and Lyness (1972) employed a special element that allowed shear failure at a predetermined stress, followed by frictional sliding. A similar method was used in a three-dimensional computational analysis by Curiskis and Valliappan (1984). Recently, Wisnom (1990) investigated the role of interfacial strength in influencing the transverse deformation of 6061 aluminum alloy reinforced with 48% SiC fibers. In his finite element model, pairs of nodes on each side of the interface were coupled with stiff springs. When the normal stress reached the assumed tensile strength of the interface, the nodes were released. For an interface strength of 100 MPa in tension, the transverse strength of the composite was found to be 87 MPa. Failure was predicted at overall strain values that were significantly lower than the ductility of the matrix.

Using the periodic hexagonal array model (identical to that in Figure 10.1(a)), Böhm (1991) numerically

---

*Modeling of interfacial debonding in discontinuously reinforced metal-matrix composites is reviewed by Needleman et al. in Chapter 13 of this volume.

investigated the transverse tension constitutive response of 6061 aluminum-boron fiber composites using contact interfaces in the ABAQUS finite element code. Because the matrix is subject to strong clamping stresses as a result of differential contraction between the two phases, the composite is expected to display some tensile strength in the transverse direction even in the extreme case of zero bonding between the matrix and the fiber. Because no interfacial bond exists, the residual clamping stresses keep the interface from failing. Figure 10.9 shows the overall stress-strain response in a 6061-0 aluminum-boron composite (with a contact interface) that was cooled down from the processing temperature and subjected intermittently to tensile loads in the transverse direction (solid lines). The resistance of this case to plastic flow is significantly less than that of the same composite with a perfectly bonded interface (Figure 10.9, dashed lines).

From earlier discussions in this chapter, it is expected that different fiber distributions will produce different responses to transverse deformation for composites with fully or partially debonded interfaces. To explore some idealized cases of this problem, consider an aluminum-boron composite in which all the interfaces between the fiber and the matrix are fully debonded,

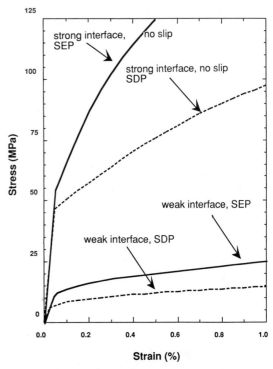

Figure 10.10. The effect of fiber packing and interfacial bonding on transverse tension stress-strain response in a 6061 Al-boron fiber composite.

with no frictional contact along the interfaces. (This, in effect, models the matrix material with 46 vol % unidirectional cylindrical holes.) Figure 10.10 shows the constitutive response for the bounding cases of square edge packing and square diagonal packing of 46 vol % boron continuous fibers in a 6061-o aluminum matrix, subjected to transverse tension. The top two curves represent the edge-packed and diagonal-packed fiber results for perfect interfacial bonding (taken from Figure 10.5(b)). The bottom curves represent the corresponding response for the same two fiber distributions when the interface is completely debonded. Here, the effect of fiber debonding on the transverse response is more severe than is the effect of fiber distribution. However, because debonding can be a local phenomenon in a real composite, it is expected that the local spatial arrangement of fibers will play a strong role in influencing the progression of debonding. Such trends are currently under investigation.

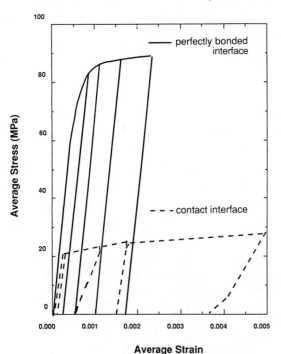

Figure 10.9. Stress-strain response for transverse tension in a 6061 Al-boron composite that is cooled down from the processing temperature. The solid lines and the dashed lines denote the debonded and perfectly bonded interface conditions, respectively. *(After Böhm 1991.)*

Summary

This chapter has provided an overview of the models and analyses for elastic and plastic deformation in unidirectional fiber reinforced metals. A comparison of the various approaches was presented, as were discussions of their significance, limitations, and of their relevant experimental data. These discussions clearly show that although analytic formulations can successfully model the deformation characteristics in axial deformation for fiber-reinforced metals, particular attention should be given to the effects of geometric factors such as fiber shape and spatial distribution on the overall deformation response. Because geometric factors strongly influence the *local* micromechanical response of the composite material and render the *local* deformation highly inhomogeneous, a complete understanding of the mechanical response of composites inevitably calls for detailed numerical simulations. With the availability of supercomputers, such computational studies are indeed becoming increasingly popular. These methods also offer the potential for providing realistic simulations of transverse and shear responses of actual composites by considering large numbers of fibers in the unit cell formulation. Furthermore, the numerical approaches offer a direct link with the geometric design and processing of composites in that they can potentially provide predictions of optimum fiber size, shape, and distribution for superior mechanical response; such information cannot often be directly obtained from experiments alone.

Although considerable progress has been made over the past several decades in developing a detailed quantitative understanding of the deformation characteristics of fiber-reinforced metals, there also remain many areas in which further research is needed. First, reliable and quantitative methods for characterizing the spatial distribution of fibers in the metallic matrix and for synthesizing such distributions in a controlled fashion are necessary in order to develop optimum reinforcement conditions for improved mechanical performance. Second, constitutive formulations that compensate for the limitations inherent in the continuum models (such as the J_2 flow theory with isotropic hardening) are necessary and third, corresponding analyses/computational simulations are needed for the matrix of the metallic composite. It may be recalled from Section 10.1 that the continuum models (such as those discussed) lose validity when the diameter and mean spacing of the reinforcing fibers becomes comparable to the characteristic microstructural dimensions of the matrix material (such as grain size and dislocation cell size). In this context, it is interesting to note that continuum formulations employing crystal plasticity theories (involving single- or multiple-slip systems) have recently been employed in conjunction with finite element formulations to simulate the deformation characteristics of discontinuously reinforced metal-matrix composites (McHugh et al. 1991, 1993; Needleman and Tvergaard 1993; Needleman et al. 1992). (Full details of such crystal plasticity theories can be found in Chapter 8 of this volume.) These models indicate that geometric effects of reinforcement distribution, shape, and volume fraction on the deformation of the discontinuously reinforced metal with a crystal plasticity matrix are qualitatively similar to those predicted by continuum elastic-plastic models (such as the fully dense von Mises theory). However, the crystal plasticity theories for the matrix predict a stronger propensity for shear localization than do the continuum models. Similar analyses need to be carried out for continuously reinforced metals. It is known that the introduction of a brittle fiber alters the dislocation density and precipitation characteristics of the ductile matrix (Chawla and Metzger 1974; Christman and Suresh 1988; Suresh et al. 1989). (Discussions of reinforcement-induced enhancements in dislocation density and accelerated aging in reinforced metals can be found in Chapter 7 of this volume.) The attendant changes in the mechanical properties of the matrix materials should be accounted for in the analyses of the overall deformation of the composite.

Although the majority of available theories and computational analyses of unidirectionally reinforced fibrous composites are applicable for situations where the fiber-matrix interface is fully bonded, many processing and service conditions in fiber-reinforced metals involve fibers that are partially or fully debonded from the surrounding matrix. Further work is necessary in which reliable constitutive formulations and micromechanical analyses are developed to predict the deformation response of fibrous composites with interfacial debonding. Inevitably, such studies also call for an understanding of the role of processing methods and environmental interactions in influencing interfacial decohesion.

Acknowledgments

S.S. acknowledges support of this work by the Office of Naval Research through Grant No. N00014-92-J-1360. The authors are also grateful to the Aluminum Company of America for the support received during the preparation of this manuscript. Thanks are due to Professor T. Nakamura for many useful discussions.

References

ABAQUS. 1991. Finite Element Code, Version 4.8, Hibbitt Karlsson and Sorensen Inc., Providence, RI.
Aboudi, J. 1989. *Appl. Mech. Rev.* 42:193–221.

Aboudi, J. 1990. *Int. J. of Plas.* 6:471–484.

Bahei-El-Din, Y.A., and G. Dvorak. 1989. *Metal-Matrix Composites: Testing, Analysis and Failure Modes*, ASTM STP 1032. (W.S. Johnson, ed.), 103–129, Philadelphia: American Society for Testing and Materials.

Becher, W., Pindera, M., and C. Herakovich. 1987. Engineering Report No. VPI-E-87-17. Department of Engineering Science and Mechanics, Virginia Polytechnic Institute and State University, Blacksburg, VA.

Bao, G., McMeeking, M., and J.W. Hutchinson. 1991. *Acta Metall. Mater.* 39:1871–1882.

Böhm, H.J. 1991. Ph.D. Thesis, Department of Aerospace Engineering, University of Vienna.

Böhm, H.J., and F.G. Rammerstorfer. 1991. *Mater. Sci. Eng.* A135:185–190.

Brockenbrough, J.R., and S. Suresh. 1990. *Scripta. Metall. Mater.* 24:325–330.

Brockenbrough, J.R., Suresh, S., and H.A. Wienecke. 1991. *Acta. Metall. Mater.* 39:735–752.

Brown, L.M., and D.R. Clarke. 1977. *Acta Metall. Mater.* 25:563–570.

Champion, A.R., Krueger, W.H., Hartmann, H.S., and A.K. Dhingera. 1978. *In:* Proceedings of the Second International Conference on Composite Materials. (B. Norton, R. Signorelli, K. Street, and L. Phillips, eds.), 882, Warrendale, PA: The Metallurgical Society.

Chawla, K.K., and M. Metzger. 1974. *J. Mater. Sci.* 7:34–39.

Christensen, R.M. 1979. Mechanics of Composite Materials, New York: John Wiley.

Christensen, R.M., and K.H. Lo. 1979. *J. Mech. Phys. Solids.* 27:315–330.

Christensen, R.M. 1990. *J. Mech. Phys. Solids.* 38:379–404.

Christman, T., Needleman, A., and S. Suresh. 1989. *Acta Metall.* 37:3029–3047.

Christman, T., and S. Suresh. 1988. *Acta Metall.* 36:1691–1704.

Curiskis, J.I., and S. Valliappan. 1984. *In:* 9th Australasian Conference on Mechanics of Structures and Materials, 349–354, Sydney: Elsevier Science.

Dragone, T., and W.D. Nix. 1990. *Acta Metall. Mater.* 38:1941–1953.

Eshelby, J.D. 1957. *Proc. Roy. Soc. London.* A241:376–391.

Eshelby, J.D. 1959. *Proc. Roy. Soc. London.* A252:561–569.

Hashin, Z. 1979. *J. Appl. Mech.* 46:543–550.

Hashin, Z. 1983. *J. Appl. Mech.* 50:481–505.

Hashin, Z., and B.W. Rosen. 1964. *J. Appl. Mech.* 31:223–228.

Hashin, Z., and S. Shtrikman. 1962. *J. Mech. Phys. Solids.* 10:335–342.

Hashin, Z., and S. Shtrikman. 1963. *J. Mech. Phys. Solids.* 11:127–140.

Hill, R. 1948. *Proc. Roy. Soc. London.* A193:189–297.

Hill, R. 1963a. *J. Mech. Phys. Solids.* 11:357–372.

Hill, R. 1963b. *In:* Progress in Applied Mechanics, Prager Anniversary Volume, 91–106, New York: Macmillan.

Hill, R. 1964. *J. Mech. Phys. Solids.* 12:199–212.

Hill, R. 1965. *J. Mech. Phys. Solids.* 13:189–213.

Kyono, T., Hall, I.W., and M. Taya. 1986. *ASTM STP* 964:409–431.

Levy, A., and J.M. Papazian. 1991. *Metall. Trans.* 21A:411–420.

Lilholt, H. 1988. *In:* Mechanical and Physical Behavior of Metallic and Ceramic Composites. (S.I. Andersen, H. Lilholt, and O.B. Pedersen, eds.), 89–107, Roskilde, Denmark: Risø National Laboratory.

LLorca, J., Needleman, A., and S. Suresh. 1991. *Acta Metall. Mater.* 39:2317–2335.

McHugh, P., Asaro, R.J., and C.F. Shih. 1991. *In:* Modeling the Deformation of Crystalline Solids. (T.C. Lowe, A.D. Rollet, P.S. Follansbee, and G.S. Daehn, eds.), 369–385, Warrendale, PA: The Minerals, Metals and Materials Society.

McHugh, P., Asaro, R.J., and C.F. Shih. 1993. *Acta Metall. Mater.* 41:1461–1510.

Mori, T., and K. Tanaka. 1973. *Acta Metall.* 21:571–574.

Müller, W.H., Schmauder, S., and R.M. McMeeking. 1991. Interface Stresses in Fiber Reinforced Materials with Regular Fiber Arrangements, Department of Mechanical Engineering, University of California, Santa Barbara, CA.

Nakamura, T., and S. Suresh. 1993. *Acta Metall. Mater.* 41:1665–1681.

Needleman, A., and V. Tvergaard. 1993. *J. Appl. Mech.* In press.

Needleman, A., Suresh, S., and V. Tvergaard. 1992. *In:* Local Mechanics Concepts for Composite Material Systems (J.N. Reddy, and K.L. Reifsneider, eds.), 199–213, Berlin: Springer-Verlag.

Nemat-Nasser, S., Iwakuma, T., and M. Hejazi. 1982. *Mech. Mater.* 1:239–267.

Owen, D.R.J., and J.F. Lyness. 1972. *Fiber Sci. Tech.* 5:129–141.

Pedersen, O.B. 1983. *Acta Metall.* 31:1795–1808.

Pedersen, O.B. 1990. *Acta Metall. Mater.* 38:1201–1219.

Schapery, R.A. 1968. *J. Compos. Mater.* 2:380–404.

Sorensen, N. 1991. A Planar-Type Analysis of the Elastic-Plastic Behaviour of Continuous Fibre-Reinforced Metal-Matrix Composites Under Longitudinal Shearing and Combined Loading. Report No. 418. The Technical University of Denmark, Lyngby, Denmark.

Suresh, S., Christman, T., and Y. Sugimura. 1989. *Scripta Metall.* 23:1602–1605.

Sutcu, M., and W.B. Hillig. 1990. *Acta Metall. Mater.* 38:2653–2662.

Taya, M., and R.J. Arsenault. 1989. Metal Matrix Composites: Thermomechanical Behavior. Oxford: Permagon Press.

Taya, M., and T. Mori. 1987. *Acta Metall.* 35:155–162.

Teply, J.L., and G.J. Dvorak. 1988. *J. Mech. Phys. Solids.* 36:29–58.

Tvergaard, V. 1990. *Acta Metall. Mater.* 38:185–194.

Walpole, L.J. 1966. *J. Mech. Phys. Solids.* 14:151–174.

Willis, J.R. 1977. *J. Mech. Phys. Solids.* 25:185–202.

Wisnom, M.R. 1990. *J. Comp. Mater.* 24:707–726.

Zhao, Y.H., and G.J. Weng. 1990. *J. Appl. Mech.* 112:158–167.

Chapter 11
Creep and Thermal Cycling

DAVID C. DUNAND
BRIAN DERBY

Reinforcements for metal matrix composites (MMCs) feature strong interatomic bonding, usually resulting in higher specific strength and stiffness, as well as higher melting temperatures and lower thermal expansion than those found in metallic materials (Table 11.1). The first two properties of the reinforcement result in composite strengthening at low temperature. The higher melting temperature ensures that, at elevated service temperatures, the reinforcement creeps at a lower rate than does the metallic matrix because it is at a lower homologous temperature. The resulting strengthening of the metal by a more creep-resistant reinforcement is the topic of the first part of this chapter.

Although the higher melting temperature and subsequently better high-temperature strength of the reinforcement are desirable properties, the concomitant lower coefficient of thermal expansion is often undesirable. The mismatch of thermal expansion between matrix and reinforcement upon temperature change results in internal stresses that affect both the microstructure and the mechanical properties of the composite. The effect of thermal cycling on MMCs with and without superimposed external load is reviewed in the second part of this chapter, with particular emphasis on the micromechanics and microstructural evolution of the composite.

11.1 Isothermal Creep

A common goal of MMC modeling has been to predict the creep properties of the composite from those of its constituents (Lynch and Kershaw 1972; Brown 1982; Lilholt 1982, 1984, 1985, 1988, 1991; McLean 1982a, 1983, 1985, 1988; Wolff 1989; Taya and Arsenault 1989; Taya 1991; Taya et al. 1991). Early experimental studies focused primarily on in situ composites with continuous fibers or lamellae formed by the directional solidification of a eutectic melt. The majority of the recent experimental data concerns synthetic composites (mostly aluminum MMCs) produced by incorporating a ceramic reinforcing phase into the metal.

11.1.1 Continuously Reinforced Composites

11.1.1.1 Uniaxial Creep

As is the case at lower temperature, maximum strengthening of a composite is achieved when the reinforcing phase is continuous. In the case of in situ composites, the melting temperature of the fibers and matrix are usually in the same range and both constituents are modeled as creeping. McDanels et al. (1967) assumed that the steady-state creep rate of the composite was equal to that of the individual constituents, matrix, and fibers. Partitioning the stresses according to the rule of mixture, and assuming that both fibers and matrix creep according to a simple power law

$$\dot{\varepsilon} = \dot{\varepsilon}_0 \left(\frac{\sigma}{\sigma_0}\right)^n \quad (11.1)$$

where $\dot{\varepsilon}$ is the creep rate, σ is the creep stress, n is the

Table 11.1. Mechanical and Physical Properties of a Number of Reinforcements and Matrix Alloys

Material	CTE (K^{-1})	Young's Modulus (GPa)	Yield Stress (MPa)	Fiber Diameter (μm)
Al_2O_3—FP	8.1×10^{-6}	380	—	20
Al_2O_3—Saffil	$6-8 \times 10^{-6}$	103	—	3
C—PAN HM longitudinal	$-0.5-0.1 \times 10^{-6}$	390	—	7–10
radial	$7-12 \times 10^{-6}$	12	—	
SiC—Nicalon	$1-2 \times 10^{-6}$	175–200	—	12–15
SiC—AVCO SCS6	1.5×10^{-6}	400–415	—	150
SiC—α particles	4.3×10^{-6}	440	—	—
B—fibers (W core)	8.3×10^{-6}	385	—	100–200
W—fibers	4.5×10^{-6}	411	1900	8–10
Ni (annealed)	13.3×10^{-6}	200	60	—
Cu (annealed)	17.7×10^{-6}	130	48	—
Ti-6Al-4V (annealed)	8.0×10^{-6}	106	860	—
Al 1100 (annealed)	23.5×10^{-6}	71	35	—
Al 2024 (annealed, aged)	22.5×10^{-6}	73	75–325	—

stress exponent, and $\dot{\varepsilon}_0$ and σ_0 are materials constants. The composite stress σ_c is given by

$$\sigma_c = V_f \sigma_{0f} \left(\frac{\dot{\varepsilon}_c}{\dot{\varepsilon}_{0f}} \right)^{1/n_f} + V_m \sigma_{0m} \left(\frac{\dot{\varepsilon}_c}{\dot{\varepsilon}_{0m}} \right)^{1/n_m} \quad (11.2)$$

where V is the volume fraction and c, f, and m are subscripts for the composite, fiber, and matrix, respectively. Equation (11.2) is insensitive to the microstructural dimension of the reinforcing fibers; however, the composite creep rate has been observed to be inversely proportional to the fiber radius for a large number of in situ composites (Bullock et al. 1977). To take this effect into account, McLean and co-authors (Bullock et al. 1977; McLean 1983; Goto and McLean 1991a) modified the above model by introducing, at the matrix/fiber interface, a third, strain-hardened phase deforming by power law.

In an approach based on the fiber backstress resulting from the elastic strain difference between the matrix and the fibers, Blank (1988) developed a model describing primary creep of the composite. Steady-state creep was then described by a linear viscoelastic model where fibers and matrix were assumed to be Maxwell solids with different mechanical characteristics. This model can also treat creep-fatigue, where the external load varies cyclically with time.

In the case of synthetic composites, the reinforcement usually exhibits a much higher melting temperature than that found in the matrix. As a result, the creep rate of the fibers is negligible and the composite can be modeled as a creeping matrix containing elastic fibers. This case was treated by McLean (1982b, 1983). As deformation proceeds, the load is transferred from the matrix to the fibers until the matrix is completely relieved and the fibers are deformed to a maximum elastic strain ε_∞. The matrix creep properties completely determine the composite strain rate that decreases monotonically with time as the strain tends asymptotically toward ε_∞. McLean (1982a) found a good match between theory and data for $Ni-Ni_3Al/Cr_3C_2$ composites.

Endo et al. (1991) used a similar derivation for the case of a matrix that deforms according to an exponential law, exhibited typically by metals subjected to high stresses:

$$\dot{\varepsilon} = \dot{\varepsilon}_0 \exp\left(\frac{\sigma_m}{\sigma_0} \right). \quad (11.3)$$

The composite creep strain was then derived as a function of time and found to be in agreement with experimental data for aluminum reinforced with SiC fibers.

Because fiber deformation is purely elastic, recovery of the composite strain is possible by annealing the deformed composite after unloading. The fibers recover their original shape elastically, inducing reverse creep in the matrix. This so-called "rejuvenation" was experimentally observed by Khan et al. (1980) and McLean (1982a).

The interfacial matrix region close to the fiber often exhibits properties different from those of the bulk matrix. This was taken into account by Goto and McLean (1989, 1991a), who included in their model a matrix interfacial zone with distinct creep characteristics. They considered two cases: (1) an interfacial zone weaker than the bulk matrix, leading to slippage at the interface and (2) a work-hardened interfacial zone stronger than the bulk matrix. The resulting three-component model is based on the rule of mixtures. The results show that a weak interfacial zone has little effect on the creep performance of the composite; conversely, a strong interfacial zone can temporarily carry some of the load before transferring it to the fibers, thus decreasing the average strain rate of the composite.

Tertiary creep—the rapid increase of strain rate preceding specimen fracture—is not predicted by Equation 11.2. This phenomenon is observed however in composites containing continuous fibers, and is the result of fiber fracture (McLean 1983; Evans et al. 1990) after which models for short fibers must be used.

11.1.1.2 Off-Axis Creep

Johnson (1977) gave a general treatment of the triaxial loading of aligned composites based on a generalized power law derived from a scalar dissipation potential. The model predictions for the special case of uniaxial off-axis loading were in close agreement with the creep rates of an in situ composite measured by Miles and McLean (1977) as a function of the angle θ between the fiber axis and the load direction. These authors also developed a model in which the multiaxial stress state is separated into three components: (1) a tensile stress parallel to the fibers (θ = 0 degrees), (2) a tensile stress perpendicular to the fibers (θ = 90 degrees), and (3) a shear stress in a plane parallel to the fibers. The first contribution leads to equal deformation of both fibers and matrix, as in the models described above, while the two other contributions can be controlled by the creep of the matrix alone. A power-law relationship is assumed for the three contributions, the sum of which yields the overall creep rate, resulting in close agreement with experimental data. Unlike the case of composites without misalignment, off-axis loading can lead to a steady-state creep rate as a result of the nonaxial stress components. Figure 11.1 shows the measured stress values necessary to induce a constant strain rate and a constant strain of 1% in 100 hours as a function of the angle θ: the composite strength decreases rapidly for angles between 0 degrees and 40 degrees and becomes nearly constant for higher values of θ. McLean (1985) also proposed to adapt the above model to composites with randomly oriented fibers. These can be idealized as an assembly of randomly oriented layers, each containing aligned fibers. Integrating the contribution of each layer given by the above model yields an expression for the overall behavior of the composite.

Another micromechanical model based on a matrix deforming by power law was developed by Yancey and Pindera (1990). They made use of the *Correspondence Principle*, which links by Laplace transformation the elastic boundary value problems to the corresponding viscoelastic problems. The authors found a good correlation between their predictions and the data collected

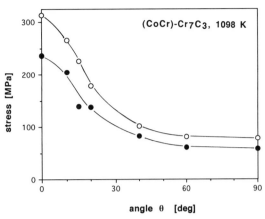

Figure 11.1. Measured stresses necessary to induce in (CoCr)-Cr$_7$C$_3$ at 1098 K a constant strain rate of 2.5 10^{-4}h^{-1} (open symbols) and a constant strain of 1% in 100 hours (full symbols) as a function of the angle θ between the stress and the fibers *(Data from Miles and McLean 1977)*.

under different loading angles θ on polymer matrix composites deforming by creep.

Lee et al. (1991) modeled the special case of transverse creep (θ = 90 degrees) using a solution for the stress fields in a matrix deforming by power law and containing a rigid fiber (Lee and Gong 1987). They predicted that voids nucleate at the location of maximum stress, found at some distance away from the interface; this result is in qualitative agreement with microstructural observations of deformed samples. An upper bound for the transverse creep rate of a composite was also derived by Binienda and Robinson (1991), who considered the case of strong elastic fibers preventing any plastic deformation in the longitudinal direction.

Equation 11.1 for power law creep can be rewritten to explicitly take into account the temperature dependence of the composite strain rate $\dot{\varepsilon}_c$:

$$\dot{\varepsilon}_c = A\sigma_c^{n_c} \exp\left(-\frac{Q_c}{RT}\right) \quad (11.4)$$

where A is a constant, n_c is the composite stress exponent, Q_c is the composite activation energy, R is the universal gas constant, and T is the absolute temperature. Measurements of transverse creep by Balis et al. (1988) on Al-2Li/Al$_2$O$_3$ yielded a composite stress exponent $n_c = 10$, larger than for the bulk matrix, and a value for the activation energy close to that for self diffusion in the matrix.

Finite element methods have also been used by Crossman and co-authors (Crossman et al. 1974; Crossman and Karlak 1976) to predict the matrix stresses and the composite creep rate under transverse loading. Parameters such as debonding, packing geometry, and residual thermal stresses for the Al/B system were investigated and correlated to data.

Finally, the effect of crossply configurations in Al/25%B was investigated by Shimmin and Toth (1972). Crossplied composites (±5 degrees and ±45 degrees) exhibited better creep properties than did unidirectional composites tested at the same angle (5 degrees and 45 degrees, respectively). The matrix shear that is responsible for the creep of the unidirectional composites was strongly reduced by the second set of fibers in the crossplied specimens. Tested crossplied specimens exhibited a rotation of the fibers toward the stress axis.

11.1.1.3 In Situ Composites

A wealth of data exists on in situ composites, and this data has been summarized in a series of review articles (Bibring 1973; Thompson and Lemkey 1974; Lawley 1976; Stoloff 1978; McLean 1982b; Stohr and Khan 1985). The microstructure and thus the creep properties of in situ composites can be quite different from those of synthetic MMCs: the reinforcement often deforms by creep, the reinforcement shape can vary with the processing conditions (faceted or unfaceted fibers, ribbons or lamellae) or during creep (coarsening and spheroidization), and the fiber diameter and spacing can be small enough to induce additional microstructural strengthening. Dislocation tangles, networks, and cells were found to be stabilized at the fiber-matrix interface during deformation, resulting in a strengthening contribution sensitive to the fiber radius. Pile-ups of matrix dislocations were observed to break fibers by stress concentration, resulting in the onset of tertiary creep. Also, in some cases, matrix flow filled the cracks that formed between fiber fragments. Measured values of the stress exponent of in situ composites tended to be higher than those for the bulk matrix. Conversely, the activation energy (Equation 11.4) was either close to that of the fibers, indicating that the fibers determined the overall creep of the composite, or intermediate between the matrix and fiber activation energy, when both phases were rate-controlling (Proulx and Durant 1974; Lawley 1976; Dirnfeld and Zuta 1988). In most cases, in situ nickel- and cobalt-base composites tested in the longitudinal direction have higher rupture life than do advanced superalloys (Figure 11.2).

11.1.2 Discontinuous Reinforcement

11.1.2.1 Experimental Studies

Nardone and Strife (1987) studied the creep behavior in tension of Al 2124/20% SiC_w at temperatures between 150°C and 300°C. They found that the composite minimum strain rate could be fitted to Equation (11.4) with values of n_c and Q_c equal to 8.4 and 277 kJ/mol respectively, in the lower temperature range, and with

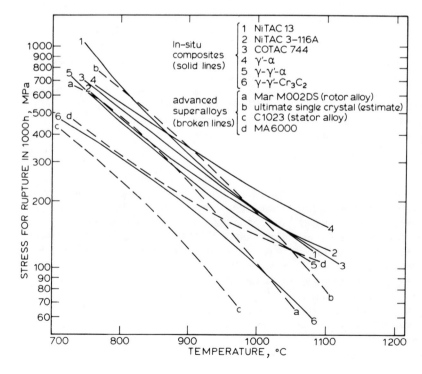

Figure 11.2. Stress to cause rupture in 1000 hours as a function of temperature for in situ composites and advanced superalloys (*From:* McLean, M., In A Numerical Study of High Temperature Creep Deformation in Metal-Matrix Composites and Metal. *Fifth International Conference on Composite Materials.* 37–51: TMS, 1985).

values equal to 21 and 431 kJ/mol respectively, in the higher temperature range. These values are much higher than those for the bulk matrix in the same temperature range ($n_o = 4$ for dislocation motion and $Q_{SD} = 146$ kJ/mol for self-diffusion). Nardone and Strife (1987) proposed to use an equation developed for particle-strengthened alloys that also exhibit high values of Q and n: the introduction of a threshold stress σ_R in Equation (11.4) leads to the equation

$$\dot{\varepsilon}_c = A'\left(\frac{\sigma-\sigma_R}{E}\right)^{n_0} \exp\left(-\frac{Q_{SD}}{RT}\right) \quad (11.5)$$

which is in good agreement with experimental data of the composite. Different possible microstructural mechanisms responsible for the threshold stress were discussed.

Nieh (1984) measured the creep properties in tension of Al 6061/20% SiC$_w$ at 232°C, 278°C, and 332°C. At all temperatures, the data could be fitted to a single power law equation (Equation 11.4) with $n_c = 20.5$ and $Q_c = 390$ kJ/mol (Figure 11.3). The same alloy reinforced with 30% SiC in the form of particles (SiC$_p$) exhibited the same high stress exponent, indicating that the rate-controlling creep process was similar. Its creep resistance was, however, much lower than that for the composite containing 20% oriented whiskers, a difference attributed to the greater load-bearing capability of the whiskers. Compression data on the same whisker composite between 300°C and 500°C yielded values of $n_c = 11.8$ and $Q_c = 297$ kJ/mol (Xiong et al. 1990).

Much higher strains could be reached by Nieh et al. (1988) and Xia et al. (1990) in subsequent studies in which Al 2124/20% SiC$_p$ was tested in shear between 200°C and 400°C. They found values of $Q_c = 360 \pm 50$ kJ/mol for the activation energy and $n_c = 9.5$ for the stress exponent. As the deformation temperature decreased, the composites did not exhibit a clearly defined second-stage creep, but tended to shift from primary to tertiary creep in a continuous manner. The authors did not observe a threshold stress, unlike Nardone and Strife (1987). Observations by transmission electron microscopy of deformed Al 2124/SiC$_p$ revealed interactions between particles and dislocations that were held up in arrays; however, neither subgrain boundaries nor voids were observed at the relatively large shear strain of 42%.

Pickard and Derby (1989) tested Al 1100/20% SiC$_p$ at 150°C, 200°C, and 350°C and also found a high value of n_c (about 20) for all temperatures, with no detectable threshold stress. Fracture was observed to occur by necking at all temperatures, initiating in the matrix far from the interface at 350°C; at 200°C, some voids were also found originating from the particle interface.

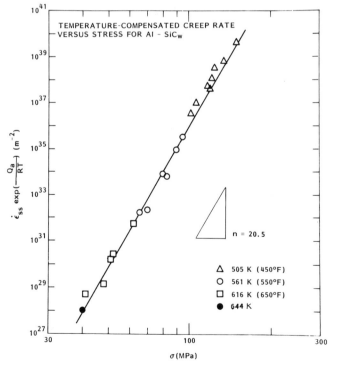

Figure 11.3. Temperature-compensated steady-state creep rate as a function of the applied stress for Al6061/SiC (Equation 11.4) (From Nieh 1984).

Mishra and Pandey (1990) reanalyzed the data of Nieh (1984) and Morimoto et al. (1988), and found a good fit with a phenomenological equation proposed by Sherby et al. (1977):

$$\dot{\varepsilon}_c = A'' \frac{D_L \lambda^3}{b^5} \left(\frac{\sigma - \sigma_R}{E} \right)^8 \quad (11.6)$$

where A'' is a constant, D_L the lattice diffusivity, λ the subgrain size, b the Burgers vector magnitude, and E Young's modulus. Equation 11.6 is valid for metals with a constant substructure deforming by lattice diffusion-controlled creep. Mishra and Pandey (1990) pointed out that the subgrain size in the composite could be controlled by the average spacing between the reinforcement, and that the second phase could stabilize the subgrains size by pinning. They determined the threshold stress from Equation 11.6, and found that it increases with decreasing temperature, a result they attributed to the temperature dependence of the elastic modulus.

Mishra (1992) introduced "dislocation creep mechanism maps," in which the normalized interparticle spacing λ/b is plotted against the inverse of the normalized effective stress, $E/(\sigma - \sigma_R)$. At high stresses, the material is in the field of power law breakdown. At lower stresses, three regions can be defined for subgrain forming matrices such as aluminum: (1) at high λ/b, a stress-dependent substructure field where $n = 5$, (2) at intermediate λ/b, a constant substructure field where $n = 8$ (Equation 11.6), and (3) at low λ/b, a dislocation-particle interaction field where $n = 5$. Existing data on MMC cluster in fields (1) and (2), while field (3) is more typical of oxide dispersion strengthened metals. This type of maps allows identification of the likely dominant mechanism for a given stress and reinforcement size and volume fraction.

Pandey et al. (1990) compared creep experiments in tension and compression on pure aluminum reinforced with 10% SiC particles. The minimum strain rate in tension at high stresses did not always correspond to the steady-state secondary strain rate because of the early onset of tertiary creep in tension, as also reported by Nieh et al. (1988), Xia et al. (1990), and Barth et al. (1990). Tests in compression or shear can thus be expected to yield more accurate values of the steady-state creep rate because the onset of tertiary creep is delayed as compared to tests in tension. Pandey et al. (1990) found a good correlation between the compression data and Equation 11.6, yielding a threshold stress of 15 MPa. Finally, they observed a critical stress below which fracture in tension is caused by cavitation and above which macroscopic necking is predominant.

Park et al. (1990) studied the creep behavior in shear of Al 6061/30% SiC$_p$. Their measurements of steady-state strain rates extend over seven order of magnitudes and exhibit two regions: at low stresses and strain rates, the composite stress exponent n_c is high and increases with decreasing stress (from about 9 to about 23), while at high stresses and strain rates, n_c tends towards a constant value of about 7.4. The apparent activation energy Q_c is also lower at high stress (270 kJ/mol) than at low stresses (494 kJ/mol). These observations are consistent with those of Nardone and Strife (1987). The applicability to MMCs of three models with threshold stress developed for dispersion strengthened alloys was assessed. In these models, the threshold stress originates from (1) Orowan bowing between particles, (2) a dislocation-climb mechanism resulting in a backstress, or (3) an attractive force between dislocations and particles, resulting from the partial relaxation of the strain field of the dislocations at the incoherent particle/matrix interface (Arzt and Wilkinson 1986). The data suggest that the threshold shear stress is equal to 8.1 MPa, resulting in a value for n_o of 5 in Equation 11.5, consistent with published results for unreinforced aluminum at high stress. The observed threshold stress is, however, much larger than predicted by any of these mechanisms when the SiC particles are considered obstacles. Because the composite was produced by powder metallurgy, the authors assumed that it contained a dispersion of fine oxide particles, as observed in other Al/SiC MMC processed by powder metallurgy (Pickard and Derby 1989; Liu et al. 1992) and in most solid-state processed aluminum alloys (Kim et al. 1985). Considering the oxide particles as obstacles yields a predicted value of the threshold stress in agreement with mechanism (3). Therefore, the observation reported by many investigators that MMCs and dispersion strengthened alloys have similar creep behavior could simply be explained by the presence in the matrix of fine oxide particles resulting from the powder metallurgical processing methods used.

This hypothesis does not, however, explain the high stress exponent values observed in three studies in which the composites were produced by squeeze casting, a process which leads to low oxide content in the matrix. Aluminum alloys reinforced with nonaligned alumina short fibers exhibited a stress exponent between $n_c = 12.2$ and $n_c = 15.5$, two to four times the value for the matrix alone (Komenda and Henderson 1991; Dragone and Nix 1992). Measurements at 150°C on zinc alloys reinforced with 20 vol % randomly oriented steel fibers yielded a stress exponent of $n_c = 12.5$, also much higher than the value of $n_m = 3.7$ found for the unreinforced matrix with the same processing history (Dellis et al. 1991). Lack of alignment of the fibers in these studies, however, introduces a complicating factor that is successfully modeled in the finite-element models of Dragone and Nix (1992), as reviewed later.

In agreement with this hypothesis, aligned composites with matrices having a lesser oxidation tendency than aluminum (zinc, lead, and silver) do not exhibit very large values of the stress exponent n_c, as will be reviewed next. In an extensive study, Dragone et al. (1991) investigated lead reinforced with short, aligned nickel fibers processed by diffusion bonding. The composites exhibited a stress exponent $n_c = 9.9$ and an activation energy $Q_c = 152$ kJ/mol, similar to the values found for the matrix processed in the same manner ($n_c = 8, Q_c = 150$ kJ/mol). The matrix values are higher than those reported for cast lead, a difference attributed to impurities and/or oxide particles in the lead resulting from the processing route used. In a study on cast lead reinforced with bronze fibers, Kelly and Street (1972a) also found that both matrix and composite, when processed in similar ways, had the same stress exponent of 14. The cast composite Ag/W investigated by Kelly and Tyson (1966) exhibited a low stress exponent at low stresses ($n_c = 3$, as compared to $n_m = 6$ for the matrix) which increased to $n_c = 14$ as the temperature and/or stress rose. At low temperature, the activation energy was also about equal to that of the matrix, but was found to be about five times larger at high temperature. A similar behavior was observed by Gulden and Shyne (1963) in indium containing high volume fractions of glass spheres 5 μm to 30 μm in diameter. This transition in composite stress exponent and activation energy was ascribed to a change of creep mechanism in the matrix, consistent with the concept of high local matrix stresses in the composite.

11.1.2.2 Superplasticity

Superplasticity is characterized by high fracture strains in tension and values of the strain-rate exponent m above about 0.5, with m defined by:

$$\sigma = K\dot{\varepsilon}^m \qquad (11.7)$$

where K is a constant.

Superplasticity has been observed for MMCs under both isothermal and nonisothermal conditions. The latter type—also called internal stress superplasticity—is treated later in this chapter (section 11.2.2.2). Isothermal superplasticity was mostly investigated in aluminum composites containing whiskers or particles. A summary of the studies published to date is given in Table 11.2. Superplastic composites can be separated into two groups according to the strain rate $\dot{\varepsilon}^*$, at which the maximum elongation ε_{max} occurs. High strain rate superplasticity ($\dot{\varepsilon}^* = 0.1$–10), which is observed at strain rates much higher than those encountered in conventional creep deformation, has been reviewed by Nieh and Wadsworth (1991). They concluded that this type of superplasticity cannot be explained by a single mechanism. The matrix grain size and composition, the matrix/reinforcement interface, and the possibility of localized melting during deformation are all important considerations. As in creep studies, threshold stresses and high activation energies were observed. Also, the unreinforced matrix was not always superplastic under the same experimental conditions of temperature and strain rate, unlike the four studies of composites showing superplasticity at low strain rates ($\dot{\varepsilon}^* = 10^{-4}$–$10^{-3}$) reviewed next.

Mahoney and Ghosh (1987b) deposited 10 to 15 vol % SiC whiskers on thin foils of 7475 Al that were diffusion bonded and thermomechanically processed to a fine-grained structure. Tests at 520°C revealed that the composite was superplastic, with a strain rate exponent $m = 0.5$, which is, however, lower than that of the equivalent unreinforced alloy. Mahoney and Ghosh (1987a) also processed by powder metallurgy both bulk and reinforced 7064 Al containing different volume fractions of SiC particles. After a fine grain producing thermal treatment, both materials were superplastic at 500°C. The composite exhibited higher stress values and lower m values than did the bulk alloy; this tendency increased as the volume fraction of SiC particles increased. Also, the strain rate sensitivity of the composite decreased more rapidly with strain than it did for the bulk alloy, increasing the composite tendency for necking. By superimposing a hydrostatic pressure of 4.14 MPa during deformation, the stress and strain to failure of the composite could be significantly increased, as a result of the inhibition of void formation and growth at the matrix/whisker interface. A model was then developed that compares the strain rate in the matrix constrained by the particles to that of the composite. This leads to a stress increase in the composite, in qualitative agreement with data. This increase is not, however, sufficient to explain the high strength of the composite at low strain rates. This may be indicative of an additional effect such as a threshold stress, below which no deformation takes place because of the reinforcement preventing grain-boundary sliding and diffusional creep.

Pilling (1989) investigated aluminum alloys 2014 and 7475 containing 15% SiC particles, heat treated to produce a fine-grained structure. The strain to failure and strain rate sensitivity were lower than those measured in the equivalent monolithic matrix Supral 220 and 7475. Two reasons were found for this difference. First, as also observed by Mahoney and Ghosh (1987a),

Table 11.2. Summary of Studies on the Superplasticity of MMC.

Matrix	Reinforcement	T [°C]	m_{max}[1]	$\dot{\varepsilon}_{m_{max}}$[2]	ε_{max}[3]	$\dot{\varepsilon}_{\varepsilon_{max}}$[4]	Reference
Slow Strain Rate Superplasticity							
2014	15vol.% SiC_p	480	0.4	$4\ 10^{-4}$	97	$4\ 10^{-4}$	(Pilling 1989)
		480			350*	$4\ 10^{-4}$	(Mahoney and Ghosh 1987a)
7064	10vol.% SiC_p	500	0.5	$2\ 10^{-3}-10^{-2}$	>450	$2\ 10^{-4}-10^{-3}$	(Pilling 1989)
7475	15vol.% SiC_p	515	0.42	$4\ 10^{-4}$	160	$4\ 10^{-4}$	
		515			310^5	$2\ 10^{-4}$	
7475	12vol.% SiC_w*	520	0.5	$5\ 10^{-4}-10^{-3}$	>450	—	(Mahoney and Ghosh 1987b)
High Strain Rate Superplasticity							
2124	20vol.% SiC_w	525	0.33	0.33	300	0.33	(Nieh et al. 1984)
2124	20vol.% SiCw	475	0.33	0.15	—	—	(Chokshi et al. 1988)
		550			250	0.33	
2124	αSi_3N_{4w}	525	0.38	0.18–1	130	0.6	(Imai et al. 1990a)
	βSi_3N_{4w}	525	0.5	0.035–0.9	225	0.1–0.2	(Imai et al. 1990b)
	20vol.% Si_3N_{4p}	515	0.5	0.1–1	830	0.04	(Mabuchi et al. 1992)
6061	20vol.% SiC_w	550	0.32	0.015–0.8	300	0.17	(Xiaoxu et al. 1991)
6061	20vol.% αSi_3N_{4w}	525	0.4	0.16–0.8	160	0.8	(Mabuchi et al. 1991b)
	20vol.% βSi_3N_{4w}	545	0.4	0.08–0.8	230	0.2	
6061	20vol.% βSi_3N_{4w}	525	0.5	0.16–0.35	225	0.16	(Mabuchi and Imai 1990)
		555			220	0.16	
6061	20vol.% Si_3N_{4w}	545	0.5	0.2–1.5	600	0.2	(Mabuchi et al. 1991b)
6061	20vol.% Si_3N_{4p}	545	0.45	0.4–10	450	0.1 and 1.5	(Mabuchi et al. 1991a)
		565	0.3–0.5	2–30	600	2	(Mabuchi et al. 1991a)
7064	αSi_3N_{4w}	525			250	0.27	(Imai et al. 1990a)
	βSi_3N_{4w}	525			250	0.15–0.3	
9021	15vol.% SiC_p	550	0.5	10–100	610	5	(Higashi et al. 1992)

All composites were produced by powder metallurgy and extruded (except where noted *), resulting in reinforcement alignment. (1) maximum strain rate exponent; (2) strain rate at which m_{max} is reached; (3) maximum elongation; (4) strain rate at which ε_{max} is reached; (5) under an hydrostatic pressure of 5.25 MPa; (*) not extruded.

the grain size grew during the deformation, thus decreasing the strain rate sensitivity and increasing the flow stress. Second, the undeformed composites exhibited pores at the Al/SiC interface resulting from processing. These pores tended to grow and coalesce into large voids, particularly where whiskers were clustered, resulting in early failure. As was the case for the study described above, the superimposition of a hydrostatic pressure of 5.25 MPa resulted in much higher strains to failure by inhibiting this cavitation effect. It was nevertheless possible to superplastically form a dome with the 7475 Al MMC without back pressure.

11.1.2.3 Analytic Studies

Most micromechanical models developed to date consider the idealized case of aligned short fibers in a creeping matrix. This permits simplification of the model to the case of a unit cell containing a single fiber loaded along its axis, which can be treated by the simple shear lag analysis. In one of the earliest analyses published, de Silva (1968) used this approach, assuming that a combination of three processes, shear-stress relaxation at the fiber ends, fiber creep, and load transfer from the matrix to the fibers due to matrix relaxation by creep, were responsible for the creep of the composite. Each process was examined individually and discussed in light of a parametric study.

Also using a shear-lag approach, Mileiko (1970) considered two geometrically simple unit cells (rigid plates or hexagonal fibers in a matrix deforming by power law creep), repeated in a regular or irregular manner, to describe a lamellar or fibrous composite. Upper and lower bounds for the steady-state creep rate of the composite were derived and found to be in reasonable agreement with experiments on Ag/W by Kelly and Tyson (1966). Matrix continuity was, however, not maintained in the model, resulting in the formation of pores at the end of the reinforcement during deformation.

Basing his analysis on the energy of deformation during creep and the shear lag analysis, McLean (1972) derived simple equations, similar to those developed by Mileiko (1970), for the creep stress of the composite. Void formation at the end of the fiber was avoided by the introduction of prismatic vacancy loops that shrink by vacancy diffusion as they glide away from the end of the fiber. The stress at the end of the fiber was found to be proportional to the fiber radius; thus, decohesion at the end of the fiber is more likely for thick fibers. The local strain rate was found to be highly sensitive to the local volume fraction. Because of this, fiber clustering was predicted to induce locally enhanced strain rates and early specimen fracture, although the overall strain rate was affected only slightly.

Lilholt (1978, 1982, 1984, 1985) refined the above models by taking into account the matrix region subjected to tensile stresses found between the ends of the fibers, explicitly avoiding the formation of pores during deformation. He also considered the case of creeping fibers and took into account the possibility of a matrix frictional stress, taken as the sum of the Orowan stress and a source-shortening stress (resulting from the decreasing spacing between fibers as deformation proceeds). The contribution of the frictional stress is expected to be important only when the spacing between the reinforcement is small, i.e. for dispersion strengthened alloys or for in situ composites with fine microstructures. Lilholt then developed a general formulation for composite creep that can accommodate different creep laws for the matrix. He pointed out that, while at a given overall strain rate the composite may obey a power law, the matrix in the composite is better described by an exponential law because it is subjected to high local creep stress for which the power law breaks down. This behavior results in a high composite stress exponent, as observed experimentally by many investigators (for example, Nieh 1984; Nardone and Strife 1987). Good quantitative agreement between models and experimental data on composites reinforced with continuous fibers (treated as short fibers with infinite aspect ratio) was found for the systems Cu/W, Ni/W, Ni-Ni$_3$Al/Cr$_3$C$_2$ (Figure 11.4), and some of the data on Ag/W. The model, however, overestimates the strength of Al 6061 and Al 2124 reinforced with SiC whiskers (SiC$_w$) (Nieh 1984; Lilholt and Taya 1987). Possible reasons for this overestimation are poor whisker alignment and distribution, uncertain materials properties, and microstructural synergistic effects between matrix and fibers.

Kelly and Street (1972b) also used the shear-lag analysis to develop a micromechanical model. In their analysis, the matrix is constrained at the fiber/matrix interface and deforms at a rate equal to the sum of the fiber creep rate and interface sliding velocity. Away from the interface, the matrix is unconstrained and deforms more rapidly. The difference in matrix velocity results in shear strains and stresses that are calculated assuming incompressibility, a Tresca-type yield criterion, and power-law deformation of the matrix. The load transferred by shear from the matrix to the fiber induces a longitudinal stress in the fiber, from which the composite behavior can be calculated by the rule of mixtures.

For the case of rigid fibers of length, l, and diameter, d, the composite stress σ_c is

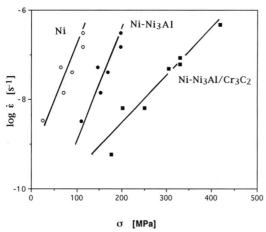

Figure 11.4. Creep data at 825°C for Ni, Ni-Ni$_3$Al, and Ni-Ni$_3$Al/Cr$_3$C$_2$ (McLean 1980b; McLean 1980a) and predictions of model by Lilholt (1982). (*From:* Taya, M., Dunn, M., Lilholt, H., Long Term Properties of Metal Matrix. Roskilde, Denmark: Risø National Laboratory, 1991.)

$$\sigma_c = \sigma_{m0}\left(\frac{\dot{\varepsilon}_m}{\dot{\varepsilon}_{m0}}\right)^{1/n_m} \cdot$$
$$\left[1 + v_f\left(\phi(1-\eta)^{1/n_m}\left(\frac{l}{d}\right)^{(n_m+1)/n_m} - 1\right)\right] \quad (11.8)$$

where the load transfer parameter, ϕ, is a function of the fiber volume fraction and the matrix stress exponent. The sliding parameter, η, is equal to zero for a perfect bond, and to unity for a completely debonded interface. Assuming that the creep rate of the composite is the same as that of the matrix (Dragone et al. 1991), the composite creep rate can be written as

$$\dot{\varepsilon}_c = \dot{\varepsilon}_{m0}\left(\frac{\sigma_c}{\sigma_{m0}}\right)^{n_m} \cdot$$
$$\left[1 + v_f\left(\phi(1-\eta)^{1/n_m}\left(\frac{l}{d}\right)^{(n_m+1)/n_m} - 1\right)\right]^{-n_m} \quad (11.9)$$

For a small value of the matrix stress exponent n_m, sliding has a large effect on the stress distribution at the interface and in the fiber; the effect is much less important for high values of n_m. The critical aspect ratio for which the fiber is loaded to rupture is also derived.

The case of creeping fibers is treated similarly. The fiber is assumed to deform by power law with a stress exponent n_f along its central portion, and to remain rigid at its ends. When the whole fiber is creeping without interface sliding, the composite creep rate can be reduced to Equation 11.2 for long fibers. The composite stress exponent is equal to that of the matrix n_m in the case of rigid fibers; for creeping fibers however, it is larger than that of the matrix when $n_f > n_m$, and is smaller than that of the matrix when $n_f < n_m$.

Creep data from lead reinforced with bronze fibers (Kelly and Street 1972a) can be satisfactorily explained with the model for creeping fibers; a lesser agreement is found with Equation 11.8 for rigid fibers (without sliding). In a recent study on lead containing nickel fibers (Dragone et al. 1991), Equation 11.9 was found to fit the data for a very high value of $\eta = 0.999$ (i.e., almost completely debonded interface), which corresponds, however, to a regime where the model is inaccurate. The interface was indeed found to be weak, as shown by full debonding of the fibers at the fracture surface.

McLean (1988, 1989) extended the model of Kelly and Street (1972b) to the case where fiber fracture takes place. He considers three cases: (1) rapid crack growth, in which the cracks formed by the broken fibers are unstable and propagate rapidly in the matrix, (2) ductile crack tearing, in which cracks grow in the matrix in a stable manner leading to a reduced internal load-bearing section, and (3) crack healing, in which cracks are filled by plastic flow of the matrix, resulting in a well-bonded composite with shorter fibers. Computed composite creep curves based on mechanisms (2) and (3) exhibit tertiary creep before fracture, while mechanism (1) leads to rapid fracture of the composite in the secondary creep region.

Taya and Lilholt (1986) also used a similar approach to Kelly and Street (1972b) to investigate the case of rigid fibers in a matrix assumed to creep according to an exponential law. An additional assumption is that load can be transferred by normal stresses at the ends of the fibers. For the case of a perfectly bonded rigid fiber, the composite strain rate becomes

$$\dot{\varepsilon}_c = (1-v_f)\dot{\varepsilon}_{m0}\exp\left(\frac{\sigma_c/\sigma_{m0}-c}{1+v_f(l/2d)}\right) \quad (11.10)$$

where c is a constant containing geometrical parameters.

The case of the debonded interface is treated by inserting into the above model a debonding parameter ξ, which takes a value of unity when the whole interface is debonded (a case assumed to occur at a threshold stress σ_2), and a value zero for no debonding (corresponding to a threshold stress σ_1). This modeling of debonding is different from that of Kelly and Street (1972b), who considered the interface as sliding over the full length of the fiber, with the degree of sliding being the adjustable parameter. In their model, Taya and Lilholt (1986) assumed that no load transfer can take place along the fully debonded interface at the ends of the fiber, while full load transfer occurs in the intact interface at the center of the fiber; in all cases, the fiber ends are debonded. The composite creep rate then becomes

$$\dot{\varepsilon}_c = (1 - v_f)\dot{\varepsilon}_{m0} \cdot$$
$$\exp\left(\frac{\sigma_c/\sigma_{m0} - c'}{1 - v_f + v_f(l/2d)(1-\xi)}\right) \quad (11.11)$$

where c' is a constant containing geometric parameters and the debonding parameter ξ. There is close agreement between Equation 11.11 and the data on Al 6061/SiC$_w$ by Morimoto et al. (1988) for values of the adjustable parameters σ_1 and σ_2 equal to 80 MPa and 130 MPa, respectively (Figure 11.5). The latter value compares favorably with the fracture stress of Al 6061 at the temperature of interest. Finally, the case of full debonding is proposed as a model for the tertiary creep of composites, during which void coalescence at the fiber/matrix interface has been experimentally observed by some investigators (Nieh 1984; Lilholt 1988).

Pachalis et al. (1990) also followed Kelly and Street (1972b) in a model considering creeping fibers with a sliding interface in a creeping matrix. The matrix region away from the fiber ends is also assumed to carry both shear and axial stresses. This more detailed geometric description of the composite thus allowed the modeling of different degrees of overlap between the ends of neighboring fibers. The effect of interface sliding, fiber volume fraction, and aspect ratios were also considered.

Lee et al. (1990) considered a case in which the fibers were elastic rather than rigid, while the matrix was deforming by power law creep. Using the shear-lag model, they derived equations for full bonding at the interface, except at both ends of the fibers; their results reduce to those of Kelly and Street (1972b) for the case of rigid fibers. The predicted composite stress exponent is equal to that of the matrix, in disagreement with the data of Nieh (1984). A good fit with these results was only reached after introduction of an empirical correction factor.

Following the model of Kelly and Street (1972b), Goto and McLean (1989, 1991b) developed equations for a composite with a third interfacial matrix zone having a distinct modulus and stress exponent. A strong, work-hardened interface extends the creep life of the composite by temporarily supporting the load being transferred from the creeping matrix to the elastic fiber; conversely, a weak interface has a considerable weakening effect on the composite. The model also allows a description of transient primary creep, which can represent an important fraction of the composite total strain. Comparing continuously and discontinuously reinforced composites, Goto and McLean (1989) found that a weak interface has a much larger effect on the latter. The strengthening effect of a work-hardened interface is however similar in both cases.

Finally, Rösler and Evans (1992) investigated the contribution of diffusion at the matrix-reinforcement interface to the relaxation of the mismatch between the two phases, responsible for the reduced creep of the composite. They compare the strain relaxation rate by diffusion to the creep rate so as to derive a critical reinforcement aspect ratio below which composite strengthening is eliminated.

Unlike the aforementioned one-dimensional models, the Eshelby method (Eshelby 1957; Mura 1982) allows a three-dimensional description of stresses and strains in composites. It was applied to creep by Zhu and Weng (1990) who examined the effect of the reinforcement aspect ratio (from thin disk to sphere to fiber) on the creep rate of the composite. Choi et al. (1991) also used the Eshelby method under the assumption that the internal stress and the strain rate depend linearly on each other, as opposed to the power or exponential relationships usually considered (Equations 11.1 and 11.3). Assuming that matrix and fibers deform by dislocation slip and diffusional flow respectively, the authors derived a simple expression for the composite creep rate. The composite was predicted to have a stress exponent of unity and a threshold stress below which

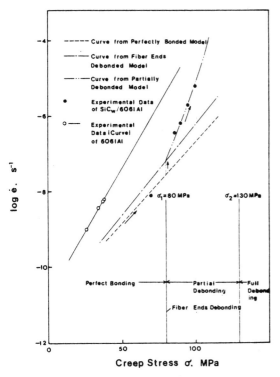

Figure 11.5. Creep data at 300°C for Al6061/SiC$_w$ (Morimoto et al. 1988) and predictions of Equations 11.7 and 11.8 (*From* Morimoto, T., et al. Second Stage Creep of SiC Whisker/6061 Aluminum Composite at 573 K. © Copyright 1988: 70–76. Transactions of the ASME—*Journal of Engineering Materials and Technology*.)

no creep occurs, two characteristics found experimentally for the Al-Al$_3$Ni system over the small range of strain rates investigated ($5 \cdot 10^{-9} - 5 \cdot 10^{-8}$ s^{-1}). The same authors (Wakashima et al. 1990) used a similar approach for the case of diffusional stress relaxation of the matrix. They equated the rate of work hardening by backstress accumulation to the rate of relaxation by diffusion in the matrix, and found a steady-state creep equation. The exact solution depends on the actual path of the matrix atoms diffusing around the fibers.

Taya et al. (1991) also used the Eshelby method to describe the redistribution of stresses from the creeping matrix (assumed to obey an exponential law) to the elastic fibers, some of which were assumed to be debonded. The resulting creep curve for the composite exhibits a transient primary-like region (despite a matrix considered to deform at steady state), and also exhibits a secondary-like region in which the creep rate decreases only slowly with time; no true steady-state creep rate is predicted. A three-parameter Weibull distribution for the matrix-fiber interfacial strength is used to describe fiber debonding that is responsible for the tertiary-like creep region. As with the model of McLean (1988, 1989), which was based on fiber fracture and damage accumulation, the above model can simulate the entire composite creep curve (Figure 11.6), with the exception of the value of the fracture strain. The predicted creep rates for the debonded case are in general agreement with the experimental data on aluminum reinforced with SiC whiskers, assuming a volume fraction of debonded whiskers of about 15%.

A third modeling route, in addition to the continuum mechanics models based on the shear-lag and Eshelby approaches, is by finite element analysis, allowing a precise description of the spatial distribution of stresses and strains in the composite for given experimental conditions. As also reported by other authors modeling low-temperature deformation of composites using the finite element method (Christman et al. 1989; Levy and Papazian 1990; see also Chapters 8, 9, 10, and 13 of this text), Dragone and Nix (1990a, 1990b) predicted sharp stress and strain concentrations at the ends of oriented fiber and plate reinforcements in MMCs deformed by creep. Figure 11.7 shows that, for the system Al/20% SiC, values of strains in the matrix region close to the fiber end can exceed ten times the average composite strain. Also shown in Figure 11.7 are the high compressive axial and hydrostatic stresses induced in the matrix by the constrained plastic flow, leading to a reduction of the composite creep rate. Large fiber aspect ratios, high volume fractions, and radial fiber clustering were found to result in smaller creep rates as opposed to asymmetry in the fiber distribution (longitudinal fiber offsetting), which increased the composite creep rate. Although the model could accurately predict the results on Ag/W by Kelly and Tyson (1966), a quantitative agreement with the data on Al/SiC$_w$ by Nieh (1984) was not found. This discrepancy was attributed to the formation of voids at the interface resulting from the predicted local stress concentration. Other configurations of fiber clustering were investigated by Sørensen (1991), who predicted increased creep strains for both radial and longitudinal clustering.

Dragone and Nix (1992) further used finite-element modeling to describe the creep behavior of an aluminum alloy containing randomly oriented alumina fibers. By considering load transfer to an interconnected network of fibers, and by taking into account damage to the network during deformation, their model successfully predicts the observed high values of stress exponent and

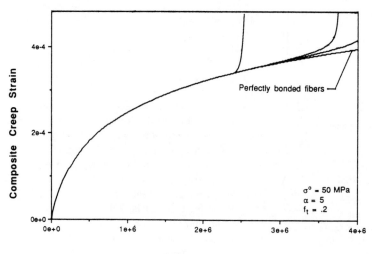

Figure 11.6. Predicted composite creep curve for Al6061/SiC for different values of the Weibull shape parameter. Temperature: 573 K, stress: 50 MPa, fiber aspect ratio: 5, fiber volume fraction: 0.2. (*From:* Lilholt, H. Relations between Matrix and Composite Creep Behavior. Roskilde, Denmark: Risø National Laboratory, 1991.)

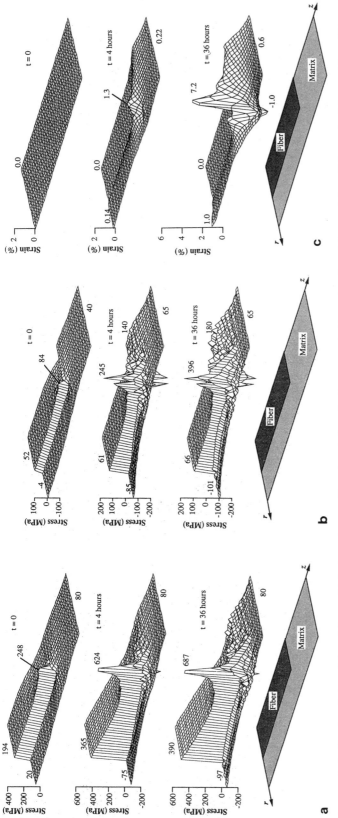

Figure 11.7. Predicted evolution of the axial stress (a), hydrostatic stress (b), and equivalent plastic creep strain (c) as a function of time, and the radial, r, and axial, z, position within the unit cell. Al6061/SiC, temperature: 561 K, stress: 80 MPa, fiber aspect ratio: 5, fiber volume fraction: 0.2. (*From:* Dragone, T.L. and Nix, W.D. Ceramic Matrix Composites: Processing, Modelling, Mechanical Behavior, pp. 367–380. *TMS*, 1990.)

activation energy found in their materials and many other MMCs, as described earlier. The activation energy for strain recovery upon unloading is however found to be similar to that for creep of the unreinforced alloy, indicating that no other strengthening mechanism than load transfer operates during creep of these composites.

11.2 Thermal Cycling

There are considerable differences between the coefficients of thermal expansion (CTE) of a large number of matrix and reinforcement combinations (see Table 11.1), leading to significant differential strains and internal stresses upon temperature change. During cooling to room temperature after fabrication at elevated temperature, this CTE mismatch between reinforcement and matrix may generate differential strains. The resulting internal stress distributions will subsequently influence local mechanical properties in a number of ways:

1. Significant differences between tensile and compressive yield strength have been measured on a range of in situ, long fiber, and discontinuous (elongated) reinforced metal-matrix composites, such as Cu/W and Fe/Fe$_2$B (de Silva and Chadwick 1969), Al/Al$_2$Cu (Pattnaik and Lawley 1971), and aluminum reinforced with SiC whiskers (Arsenault and Taya 1987).
2. Large dislocation densities have been observed in in situ carbide reinforced superalloys (Bibring 1973), Al/SiC MMC (Vogelsang et al. 1986), and Cu/W MMC (Chawla and Metzger 1972).
3. Yield strengths of particulate reinforced metal-matrix composites are significantly increased by quenching (Derby and Walker 1988; Humphreys 1988).
4. High levels of residual elastic strain have been measured by neutron diffraction in particle reinforced metal-matrix composites (Allen et al. 1987; Withers et al. 1987).

Thus, there is strong evidence for considerable internal stresses and plastic flow in MMCs during temperature changes after fabrication. It can be expected that these effects also occur during heating, leading to a cyclic internal stress regime during repeated thermal cycling and microstructural damage, as also observed in monolithic alloys.

Other materials classes also experience large internal stresses during changes in temperature:

1. Polycrystalline metals with anisotropic single crystal coefficients of thermal expansion (e.g., Zn and α-U) show thermal ratcheting (Burke and Turkalo 1952) and accelerated creep under load during thermal cycling (Roberts 1960, Lobb et al. 1972; Wu et al. 1987).
2. Similar behavior to that detailed in (1) has also been seen in materials undergoing a solid-state phase change (Stobo 1960; Guy and Pavlick 1961). This behavior has also been related to large internal stress levels caused by constrained volume change during transformation (Buckley et al. 1958; Greenwood and Johnson 1965).

The behavior under thermal cycling of continuous or long fiber (very high aspect ratio) MMC is quite different from that of discontinuous particle or very short aspect ratio fiber reinforced metals, because of the nature of the constraint imposed by the reinforcement on the deformation of the soft matrix. For long reinforcements, this constraint can only be relieved by the presence of a free surface, a crack in the reinforcement, or by sliding at the reinforcement/matrix interface. Short fibers, however, can relax by dislocation motion or by diffusional flow in the matrix during cycling, leading to different responses to thermal cycling than are observed with continuous MMC. Another major difference in behavior can be distinguished between the cases of composites cycled with or without externally applied load. This section will be subdivided into continuous fiber MMCs and discontinuous reinforced composites, and within each division, cycling under stress and without load will be considered separately.

11.2.1 Continuously Reinforced Composites

In the majority of synthetic and in situ composites, the reinforcing phase is much stronger and stiffer than the matrix. The variation during thermal cycling of internal stresses and strains of the composite has been modeled using simple one-dimensional mechanical models (de Silva and Chadwick 1969; Hoffmann 1973; Garmong 1974a, 1975, 1976). These assume that the reinforcement is sufficiently creep-resistant to deform elastically, and that all plastic and time-dependent relaxation strains are confined to the matrix. Extensive plastic flow is thus predicted in the matrix during thermal cycling. The total plastic strain is the sum of post-yield and relaxation strains, and is a function of the time available for relaxation: long-period cycling allows greater relaxation and larger internal strains than short-period cycling. Daehn (1989) recently extended these models by considering time-dependent processes in both matrix and reinforcement. His model can be used to predict thermal ratcheting similar to the behavior observed during the thermal cycling of thick monolithic sections (Bree 1967), in which thermal gradients induce the necessary internal strain differences. However, in most composite systems of current interest, the relaxation times are much greater for the reinforcement than they are for the matrix, and the simple models neglecting reinforcement creep are thus probably sufficient.

11.2.1.1 Microstructural Damage During Thermal Cycling

Potential damage mechanisms have been reviewed for in situ composites by Garmong (1974b). Macroscopic shape change and reinforcement fracture were identified by Bibring (1973), Breinan et al. (1973), and Garmong and Rhodes (1973). However, no grain boundary cavitation was reported, despite evidence for extensive matrix plastic flow and grain boundary sliding. Continuous fibre reinforced MMCs seem more damage-resistant than do in situ composites. There are few reports of significant fiber damage during cycling (Chawla 1976; Mackay 1990), and Mackay suggested that the majority of this damage occurred after cooling from fabrication rather than during thermal cycling. An accelerated degradation of boron fiber strength occurred after thermal cycling of Al/B alloy MMC (Anthony and Chang 1968; Wright 1975; Grimes et al. 1977). The boron fibers exhibited reduced strength when extracted from the matrix after thermal cycling, and this reduction was greater than that measured after subjecting the composite to an isothermal exposure of equivalent duration. Also, extensive matrix plastic flow resulted in microstructural damage accumulation. Surface roughening and matrix cracking were observed during the thermal cycling of boron reinforced Al 6061 and Al 2024 (Wright 1975) and Al 6061/Borsic (Chawla 1976). Surface roughening appears to be confined to Al alloys reinforced with large diameter fibers and is not reported in 10μm diameter C and SiC fiber reinforced Al alloys (Pepper et al. 1971; Kyono et al. 1986; Colclough et al. 1991). However, other matrix damage was reported in some small diameter fiber MMCs: fiber matrix debonding was found in Al/C (Kyono et al. 1986), and axial splitting was reported in Mg/C composites (Wolff et al. 1985). The lack of large-scale surface shape change can be attributed to the finer scale of the reinforcement distribution that suppresses slip band formation, leading to a more diffuse deformation. Chawla (1973a, 1973b) observed fine surface slip markings in Cu/W composites, as well as extensive grain boundary sliding and cavitation after cycles up to 600°C. Matrix cracking was reported in Ni-Cr/W MMC (Hoffmann 1973) Ti6Al4V/SiC (Thomin and Dunand 1993) and in a number of intermetallics reinforced with SiC and Al$_2$O$_3$ (Kim and Kleek 1990; Noebe et al. 1990). Kim and Kleek also reported damage to the carbon core of the SCS-6 SiC fibers used in their study.

11.2.1.2 Macroscopic Shape Change

Damage by extrusion can also lead to gross shape change of the composite. Hoffmann (1973) and Warren et al. (1982) reported the extrusion of W wires during thermal cycling from a series of reinforced nickel alloys and stainless steel. In other studies, the matrix is reported to extrude, including Al in Al/CuAl$_2$ (Garmong and Rhodes 1973) and Cu in Cu/W MMC (Yoda et al. 1978). In an earlier paper, Yoda et al. (1977) reported a shape change parallel to the growth direction of Al/Al$_3$Ni eutectics without extrusion; Garmong and Rhodes (1973) also studied Al/Al$_3$Ni and saw no extrusion. These contradictory results prevent the identification of a single mechanism for matrix or fiber deformation in the general case. A much more extensive shape change was reported by Yoda et al. (1978) for Cu/W MMCs that exhibited mean extensions parallel to the fibers (accompanied by matrix extrusion) in excess of 10% after 500 cycles between 400°C and 800°C. Such large extensions appear to be confined to the Cu/W system, with Garmong (1974b) and Wolff et al. (1985) reporting a stabilized hysteresis loop of thermal strain after repeated cycling of other MMCs. In Al/C composites, Kyono et al. (1986) found some growth perpendicular to the fiber direction after cycling, probably the result of extensive fiber matrix debonding and void growth that was reported, leading to specimen volume expansion. Colclough et al. (1991) found no extensions parallel or perpendicular to the fiber direction of Al/SiC composites, even when a load parallel to the fibers was imposed during the thermal cycles. However, loads normal to the fiber direction considerably enhanced creep during cycling. This behavior appears similar to that seen with discontinuously reinforced MMCs that, however, have much higher strains to failure (Wu and Sherby 1984; Pickard and Derby 1990) than the strain values of 1% to 2% found for long fiber MMCs.

To model the thermal cycle behavior of Cu/W MMCs, Yoda et al. (1978) considered thermally activated sliding at the fiber/matrix interface, which reduces the constraining effect of the reinforcement and allows matrix deformation. They also examined the influence of the composite microstructure on the thermal cycling deformation. Figure 11.8 shows that the linear expansion is proportional to the number of cycles after an initial incubation period in which the strain per cycle is reduced. For a given thermal cycle amplitude ΔT, the final extension per thermal cycle $\Delta \varepsilon^T$ was seen to depend solely on reinforcement volume fraction V_r: composites with $V_r = 10\%$ showed a value of $\Delta \varepsilon^T$ approximately double that with $V_r = 5\%$. The number of cycles in the incubation period decreases with (1) decreasing reinforcing fiber length, (2) increasing fiber

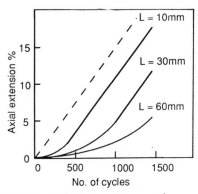

Figure 11.8. Relationship between mean reinforcement fiber length and extension during thermal cycling of a 10% V_r Cu/W MMC (Yoda et al. 1978). Note the constant slope after an initial incubation period; the dashed line is the zero incubation asymptote.

diameter, indicating that the fiber aspect ratio is the controlling parameter, and (3) increasing time spent at the higher temperature part of the cycle. With a constant fiber size and aspect ratio under identical thermal cycling conditions, the incubation periods for $V_r = 10\%$ and $V_r = 5\%$ MMCs were identical. The authors then developed a simple one-dimensional model in which the residual stresses in the soft matrix are relaxed by a time-dependent mechanism described as the viscous flow of a thin interphase region between the matrix and the fibers. The total strain is shown to depend only on reinforcement volume fraction and elastic properties, while the rate dependence of the relaxation depends on parameters such as reinforcement length and temperature. Colclough et al. (1991) later adapted this model to predict the behavior of their thermally cycled Al/SiC composites when subjected to external transverse loads.

11.2.1.3 Variation in Cycle Characteristics

Thermal cycling has been shown to cause considerable degradation in mechanical properties in a number of in situ and long fiber MMCs. In almost all of these cases, the decrease in mechanical properties could be correlated with a parallel degradation in the microstructure. In the case of boron and Borsic reinforcements, this degradation is a decrease in the fiber strength, as discussed earlier. In the majority of other cases, the degradation of the fiber-matrix interface was identified as the chief cause of strength reduction (Al/Al$_2$O$_3$: Kim et al. 1979; Al/SiC: Bhatt and Grimes 1983; Ti/SiC: Park and Marcus 1983; Al/C: Kyono et al. 1986). However, in a number of instances, no strength degradation was found after cycling similar MMC systems (Pepper et al. 1971; Gray and Sanders 1976; Wolff et al. 1985). Careful examination of reports of cycling damage reveals a number of common features. In particular, the higher the upper temperature of the cycle and the longer the cycle period (or time spent at the upper temperature), the greater the damage, whether microstructural or mechanical (Woodford 1976). These observations support the prediction of Garmong (1976) that increased matrix strain will lead to greater composite damage. This points to the importance of relaxation time and the importance of internal CTE mismatch in determining the response of continuous MMC microstructures to thermal cycling.

11.2.2 Discontinuously Reinforced Metal-Matrix Composites

The difference in geometry between discontinuously and continuously reinforced MMCs leads to a different response during thermal cycling. For discontinuous MMCs, the lack of reinforcement continuity relaxes the condition of equal strain in matrix and reinforcement along the major reinforcement axis. Matrix flow around discontinuous reinforcements is, in principle, much easier than sliding at fiber/matrix interfaces. Therefore, greater macroscopic shape changes and reduced damage might be expected during thermal cycling than is expected for continuous reinforced systems.

11.2.2.1 Thermal Cycling without Applied Stress

Patterson and Taya (1985) thermally cycled Al 2124 with partially aligned SiC whisker ($V_r = 15\%$). Two thermal cycle profiles with a period of about 3 minutes were used: from room temperature to 400°C, and from room temperature to 500°C. After 1000 cycles to 400°C, the specimen extended about 5% parallel to the alignment direction and shrank in its other dimensions to conserve volume. The cycles to 500°C resulted in a reduction in this cyclic strain. In a more detailed experiment, Hall and Patterson (1991) studied the influence of thermal cycle upper temperature and number of cycles. Highly anisotropic shape changes were observed, depending on the orientation with respect to the rolling direction. These strains increased with the upper cycle temperature and then reversed at upper temperatures higher than about 400°C. Similarly, the composite strength decreased with thermal cycles up to 300°C, while no effect was found for cycles above 400°C. Nakanishi et al. (1990) carried out a similar study with squeeze-cast aluminum containing randomly oriented SiC whiskers. Using a slightly longer thermal cycle of 8 minutes from 100°C to 400°C, very little specimen shape change was observed (< 0.1%) after 1000 cycles. Warwick and Clyne (1989) examined the thermal cycling behavior of a SiC whisker reinforced Mg-Li alloy produced by squeeze casting followed by extrusion. Thermal cycles of 100°C to 350°C over a period of about 800 minutes over about 100 cycles produced a positive strain per cycle in the extrusion direction of $\Delta \varepsilon^T = 4 \times 10^{-4}$. A series of slightly different thermal cycles were performed and showed only a slightly greater strain per thermal cycle when the upper temperature was raised to 435°C. If an extended plateau (at either the hot or cold limits of the cycle) was added to the original saw-tooth cycle, which had equal and opposite heating and cooling rates, a considerable increase in strain per thermal cycle was observed, despite the unaltered total cycle period.

Two effects must be considered in these experiments: the shape change concurrent with thermal cycling, and the changes in strength and ductility after cycling. Hall and Patterson (1991) carried out an extensive microstructural study of thermally cycled Al 2124 with and without SiC whiskers; they concluded that the dominant damage mechanisms were the overaging and dissolution of the precipitation-hardened microstructure, as would be expected from isothermal exposure (see Chapter 7). High upper-cycle temperatures did not reduce matrix or MMC strengths because complete dissolution occurred during cycling, followed by natural aging after cycling to restore its strength. At intermediate cycle ranges however, composites showed a greater strength degradation than did the unreinforced matrix in the lower temperature ranges. This effect was explained by the greater dislocation density expected in the MMC after cycling, the accelerating diffusion pathways, and the microstructural coarsening (Dutta and Bourell 1989).

The continuous straining of the composites along the direction of whisker alignment is probably caused by the combined plastic and diffusional relaxations of the CTE mismatch stresses, as modeled by many authors. Patterson and Taya (1985) proposed that the permanent strain of the composite after each complete cycle was caused by a hysteresis in the temperature-driven stress-strain behavior. This is, however, not consistent with the principles of plastic relaxation of the misfit strains that would predict a net zero dimension change on cycling. Even if a different temperature change for the initiation of plastic flow on heating and cooling is assumed (Pickard and Derby 1990), a fixed offset strain is predicted to be approached asymptotically, similar to that found in long fiber Al/C MMC by Wolff et al. (1985), rather than by a permanent incremental strain per cycle. Taya and Mori (1987) proposed a model using the Eshelby (1957) analysis to calculate a mean matrix residual stress that was then relaxed by power law creep. When compared to measurements by Patterson and Taya (1985), the model predicted strains of the appropriate sign and order, which were too large by a factor of about three. The model has two limitations, namely its high sensitivity to the exponent of the power law relaxation mechanism and its treatment of the temperature dependence of the yield stress of the matrix. The model assumes that the stress relaxation occurs with a creep exponent of 5, consistent with dislocation recovery creep of the unreinforced matrix. However, the results of Withers et al. (1988) suggest that diffusion rather than dislocation climb is the dominant relaxation mechanism. This would require a creep exponent of 1 in the Taya and Mori (1987) model, which would further reduce the agreement with experiment.

The strains per thermal cycle measured by Patterson and Taya (1985) and Hall and Patterson (1991) are in the range of $1-2 \times 10^{-5}$, while those measured by Warwick and Clyne (1989) are much larger (about 4×10^{-4}). These strains are considerably smaller than is the total thermal misfit strain for the cycles studied with $\Delta \alpha \Delta T = 7.2 \times 10^{-3}$ for Al/SiC (Patterson and Taya 1985) and $\Delta \alpha \Delta T = 7.5 \times 10^{-3}$ for Mg-Li/SiC (Warwick and Clyne 1989). The much higher strains per cycle seen in the latter case are almost certainly caused by the much longer cycle duration, which allows more complete relaxation of the residual elastic stresses. Warwick and Clyne (1989) found a dependence of strain per cycle on thermal cycle shape, which can be interpreted by assuming that the majority of the measured axial extension occurs by relaxation of residual stresses. The lower degree of relaxation in the Al/SiC composites is consistent with the measurements by neutron diffraction of relaxation made by Withers et al. (1988), who found relaxation time constants in the range of 40 to 400 minutes, much longer than the cycle times of 8 minutes used by Patterson and Taya (1985) and Hall and Patterson (1991). However, none of the current mechanism models can explain the sign reversal of cycling strain seen by Hall and Patterson at the largest thermal cycle amplitudes.

11.2.2.2 Thermal Cycling with Applied Stress

The original work on Al 2024/SiC$_w$ MMC by Wu and Sherby (1984) showed a considerably greater creep rate during thermal cycling between 100°C to 450°C than during identical isothermal tests at 450°C. A stress exponent of approximate unity and tensile elongations in excess of 300% prior to fracture were measured. Subsequent investigations by Le Flour and Locicéro

(1987) on a thermally cycled Al 7090/SiC$_p$ MMC demonstrated an increased creep rate that was found to depend on the stress and thermal cycle amplitude. However, these authors used much smaller temperature cycle amplitudes than did Wu and Sherby (up to 130°C), and they found a creep stress exponent of 4. Pickard and Derby (1988, 1990) examined the effect of the cycle duration, cycle amplitude and MMC microstructure in a study of the deformation under thermal cycling of Al 1100/SiC$_p$ MMC. Further data on SiC whisker reinforced Al alloys has been published by Hong et al. (1988), Daehn and González-Doncel (1989) and Chen et al. (1990).

The characteristics of the thermal cycle used have been seen to influence the enhanced creep rate of MMCs. Both Hong et al. (1988), and Pickard and Derby (1988, 1990) found a linear relation between the number of thermal cycles and the total creep strain under constant stress; therefore, the creep strain per thermal cycle $\Delta\varepsilon^T$ is constant. Neither a primary period of increased creep rate at the beginning of the creep test or during an increase in load, nor a period of zero creep or reduced creep rate on load reduction were seen. Such transients are observed in the isothermal creep of monolithic materials and MMCs. Strains of up to 300% were recorded by Wu and Sherby (1984) and Pickard and Derby (1990) under thermal cycling conditions (Figure 11.9). These large tensile elongations are consistent with the stress exponent of unity found in all the studies on the thermal cycling of MMCs, with the exception of the report by Le Flour and Locicéro (1987). The thermal cycle amplitude or temperature interval ΔT strongly affects the creep strain per cycle under constant stress. Fig. 11.10 shows that an almost linear relation exists between $\Delta\varepsilon^T$ and ΔT above some critical value of ΔT. In both studies reported in Figure 11.10, the cycle duration did not affect significantly the results.

The influence of MMC microstructure on the thermal cycle creep of Al 2024/SiC$_w$ was studied by Hong et al. (1988). They found that the alignment of the whiskers after extrusion had little effect on the thermal cycle creep rate, leading to very similar deformations parallel and perpendicular to the extrusion direction. They also reported that the creep rate increased with an increasing whisker volume fraction, a result confirmed by Pickard and Derby (1990) (Fig. 11.11a). On reducing the reinforcement size, the creep rate under identical conditions decreases (Fig. 11.11b), but on further increasing the volume fraction above 20% the creep rate remains effectively constant at $V_r = 30\%$ and decreases when $V_r = 40\%$.

As discussed earlier, there are numerous similarities between the observed thermal cycle creep enhancement in MMCs and both the behavior of anisotropic metals and transformation plasticity. Three types of model have been developed to explain the phenomena that can be grouped as internal stress driven enhanced plasticity. The first, by Roberts and Cottrell (1956) but put on a sounder analytic footing by Anderson and Bishop (1962), assumes that the internal plastic flow, which occurs after the internal stress exceeds the material yield point, is biased by the external applied stress. This bias generates an excess plastic strain in the direction of the applied stress after each complete thermal cycle. Using the Levy-Von Mises equations, the excess plastic strain $\Delta\varepsilon^T$ is given by Greenwood and Johnson (1965) as:

$$\Delta\varepsilon^T = \frac{5}{6}\frac{\Delta V}{V}\frac{\sigma}{\sigma_Y} \qquad (11.12)$$

where $\Delta V/V$ is the fractional constrained volume change during a thermal cycle, σ is the applied stress and σ_Y is the yield stress of the material. Using this approach, very similar relations were derived by Derby (1985) for whisker-reinforced MMCs, and by Daehn and Oyama

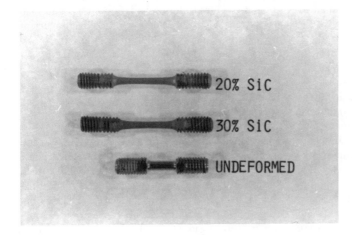

Figure 11.9. 150% tensile elongation in 20% and 30% V_r particle reinforced MMCs after thermal cycling 150°C–450°C under load. (Reprinted with permission from Derby, B. 1991. *In:* Metal Matrix Composites—Processing, Microstructure, and Properties. (N. Hansen, D.J. Jensen, T. Leffers, et al., eds.), 31–49, Roskilde, Denmark: Risø National Laboratory.)

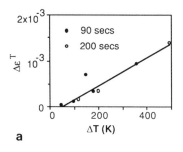
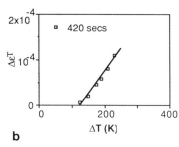

Figure 11.10. The influence of thermal cycle amplitude on strain per thermal cycle. (a) SiC$_w$/Al 2024 stressed to 10 MPa (Daehn and González-Doncel 1989). (b) SiC$_p$/Al 1100 stressed to 2.85 MPa (Pickard and Derby 1990). In both cases, there is a linear relation between $\Delta\varepsilon^T$ and ΔT above some critical ΔT.

(1988) who used a simplified two-dimensional model. Pickard and Derby (1990) developed the Levy-Von Mises model for the case of MMC microstructures and extended it further to include a temperature dependence of the yield stress and the presence of a critical temperature change before plastic flow occurs. Finally, a similar model based on biased plasticity was proposed by Zhang et al. (1990) who used a finite element technique and arrived at similar predictions.

The second type of model, which was also first developed by Anderson and Bishop (1962), considers the time-dependent creep deformation of the material by the combined action of the internal and external stresses. This model was developed further by Wu et al. (1987) who derived the following relation for the mean creep strain rate per cycle

$$\dot{\varepsilon} = nK\frac{D^*}{b^2}\left(\frac{\sigma_i}{E}\right)^{n-1}\frac{\sigma}{E} \qquad (11.13)$$

where n is the creep stress exponent, K is a creep pre-exponent, D^* is the mean diffusion coefficient averaged over one thermal cycle, b is the Burger's vector, E is the elastic modulus, and σ_i is the mean internal stress. Equation (11.13) can be used to predict a strain per cycle if the cycle duration is known, that is, $\Delta\varepsilon^T = \dot{\varepsilon}\Delta T$. This model, originally developed for anisotropic metals, was further extended by Daehn and González-Doncel (1989) for MMCs.

A third model proposed by Poirier (1982) uses a micromechanical approach to predict an increased dislocation density from the geometrically necessary dislocations needed to accommodate the internal volume change, leading to an accelerated creep rate. The model is formulated in a way more suited to accommodate geophysical time scales, but produces a relation identical to that of Greenwood and Johnson (1965).

Equations (11.12) and (11.13) both predict a linear relation between strain per cycle (or strain rate) and the applied stress for a given thermal cycle. However, there are significant differences in the model predictions. In Equation 11.12, the constrained volume expansion $\Delta V/V$ is related to the thermal misfit strain which is approximately equal to $3\Delta\alpha\Delta T$, and thus the observations of the influence of thermal cycle amplitude in Figure 11.10 are predicted. In contrast, in Equation 11.13, the thermal cycle controls the value of the internal stress σ_i; this should be constant once internal plastic flow begins and thus no relation between $\Delta\varepsilon^T$ and ΔT is expected. A further criticism of the enhanced creep models is that the internal stress σ_i should relax as creep proceeds. Neither Wu et al. (1987) nor Daehn and González-Doncel (1989) take into account this relaxation, even when considering deformation in the absence of an applied stress. Anderson and Bishop (1962) derived a relation for the maximum strain per thermal cycle during the relaxation of any internal stress during temperature cycling. This strain, which is *independent of relaxation mechanism*, is given by

$$\Delta\varepsilon_r^T = \frac{\sigma}{E}ln\left[\frac{\sigma_Y}{2\sigma}\right] \qquad (11.14)$$

where the internal stress σ is taken as having a maximum value equal to the the yield stress σ_Y. If an internal stress driven mechanism exists, then the constant strain per cycle independent of cycle duration can only be explained by a relaxation of the internal stress. This

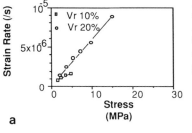
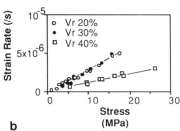

Figure 11.11. Apparent strain rate as a function of stress during 150°C–450°C thermal cycling of Al 1100 MMC with a different volume fraction of SiC$_p$ *(Data from Pickard and Derby 1990)*. (a) 10 μm SiC particles with V_r of 10% and 20%. (b) 2.3 μm particles with V_r of 20%, 30%, and 40%.

would limit the strain per cycle to a few multiples of the constrained thermal *elastic* strain, which is much lower than the observed strain per cycle. Thus, relaxation appears to be of little importance in determining the behavior of discontinuous MMCs when stressed during thermal cycling. Recently, Povirk et al. (1992) have applied continuum deformation modeling to the results of Hong et al. (1988) to show that these results, originally explained by an enhanced creep mechanism, could also be modeled using a plasticity model. Their model, similar to that of Pickard and Derby (1990), also predicts that increases in reinforcement volume fraction do not necessarily improve thermal cycle creep resistance.

Relaxation and recovery processes must operate during thermal cycling in order for the observed very high tensile elongations to occur. Superplastic extensions are clearly possible and these would not be expected if dislocation accumulation was important. TEM observations by Pickard and Derby (1988, 1990) show a substantially unchanged matrix dislocation density after extensive thermal cycle strain (Figure 11.12). The mechanisms that are responsible for dislocation generation during thermal cycling are easily formulated, but the mechanism that allows such rapid dislocation removal is, at the time of this writing, unknown.

Summary

The addition to a metal of a creep-resistant reinforcement results in a composite with lower creep rate. The case of continuously reinforced composites loaded parallel to the fibers or in an off-axis direction is well understood and can be modeled with continuum mechanics models. Composites containing discontinuous reinforcement (short fibers, whiskers, or particulates) exhibit a more complicated behavior and have been modeled using the shear-lag theory, Eshelby analysis, or by finite elements techniques. These composites typically show a stress exponent and activation energy higher than those of the matrix, and three possible explanations have been proposed to describe this behavior: (1) power law break-down of the matrix (Kelly and Tyson 1966; Lilholt 1985), (2) interface decohesion (Taya and Lilholt 1986), and (3) a threshold stress because of dislocation-obstacle interactions (Nardone and Strife 1987; Park et al. 1990).

Superplasticity has been observed at both low and very high strain rates. The latter case has not yet been completely explained and may be caused by localized melting as well as by interfacial phenomena. Superplasticity at low strain rate seems to be controlled by the superplastic behavior of the matrix.

In all MMC creep models, it is assumed that the properties of the matrix are time-independent. However, it has been shown (see Chapters 5, 6, and 7) that the reinforcement can alter the microstructure of the matrix during high-temperature exposure (by precipitate coarsening or recrystallization), with kinetics different from those prevalent in the unreinforced matrix. To date, no model exists that takes into account this effect, although this effect may significantly alter the creep properties of the MMC.

In all cases of thermal cycle damage, either at the microstructural level or by a macroscopic shape change, the onset of damage occurs above some critical thermal cycle amplitude. This applies both for stress-free cycling

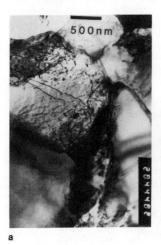
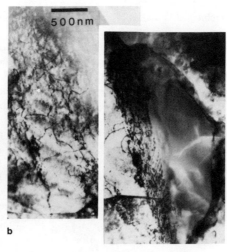

Figure 11.12. TEM micrographs of 20% V_r SiC$_p$/Al 1100 MMC microstructure in the as-received state (a) and after 90% thermal cycle strain (b). In both micrographs, an essentially similar dislocation substructure is visible. (Reprinted with permission from Derby, B. 1991. *In:* Metal Matrix Composites—Processing, Microstructure and Properties. (N. Hansen, D.J. Jensen, T. Leffers, et al., eds.), 31–49, Roskilde, Denmark: Risø National Laboratory.)

and for cycling under load. In general, damage is accentuated by holding at high temperatures and minimized with rapid thermal cycling.

With long fiber reinforced composites, dimensional changes and extensive matrix deformation remain confined to the material edges particularly where fiber ends intersect a free surface. There seems to be no clear guide regarding the nature of this damage, which is extremely system-dependent. The other important damage mechanism is degradation of the fiber/matrix interface. This degradation appears to be enhanced by the presence of dislocations in the matrix generated by local plasticity during cycling.

As for isothermal creep, the thermal cycling behavior of discontinuously reinforced MMCs is more complex. There is little evidence for microstructural damage other than the coarsening of age-hardened structures. In the absence of an applied stress, a shape change is observed only in materials with aligned elongated reinforcements. The total strain per thermal cycle is much smaller than is the thermal mismatch strain; it is also slightly smaller than the residual elastic strain expected after internal plastic flow, and so can be explained as a simple relaxation during the high temperature dwell of a cycle. In the presence of an applied stress, thermal cycling accelerates creep and very large tensile elongations are possible without failure. The stress exponent and strain rate sensitivity of the composites are close to unity, and the MMCs can then be considered to be superplastic under thermal cycling conditions. The mechanism of this accelerated creep appears to be controlled by the internal plastic strain during cycling and not by a relaxation of internal stresses.

The common thread linking the behavior of MMCs during thermal cycling is the CTE mismatch between reinforcement and matrix. Thus, if this mismatch could be eliminated, no concern would arise from MMC thermal cycling. However, the majority of reinforcement/matrix combinations of interest have significant CTE mismatches (see Table 11.1). Because in most cases the degree of thermal strain appears to be dominant in determining the damage mechanisms, matrices with high yield stresses should be most resistant.

Acknowledgments

DCD gratefully acknowledges support from AMAX in the form of a chair at MIT. Part of the research by BD was supported under the SERC Rolling Grant on metal-matrix composites, Grant No. GR/F68760.

References

Allen, A.J., Burke, M., Hutchings, M.T., et al. 1987. *In:* Residual Stresses in Science and Technology. (E. Macherauch and V. Hauck, eds.), 151–157, Oberursel, FRG: DGM Int. Verlag.

Anderson, R.G., and J.F.W. Bishop. 1962. *In:* Symp. Uranium and Graphite, Paper 3. 17–23, London: Institute of Metals.

Anthony, K.C., and W.C. Chang. 1968. *Trans. ASM.* 61:550–558.

Arsenault, R.J., and M. Taya. 1987. *Acta Metall.* 35:651–659.

Arzt, E., and D.S. Wilkinson. 1986. *Acta Metall.* 34:1893–1898.

Balis, C.D., Curran, D.R., and S.S. Wang. 1988. *In:* Fourth Japan-U.S. Conference on Composite Materials. 148–178, Lancaster, PA: Technomic Publishing.

Barth, E.P., Morton, J.T., and J.K. Tien. 1990. *In:* Fundamental Relationships Between Microstructure & Mechanical Properties of Metal-Matrix Composites (P.K. Liaw and M.N. Gungor, eds.), 839–846, Warrendale, PA: The Metallurgical Society.

Bhatt, R.T., and H.H. Grimes. 1983. *In:* Mechanical Behaviour of Metal Matrix Composites. (J.E. Hack and M.F. Amateau, eds.), 51–64, Warrendale, PA: TMS-AIME.

Bibring, H. 1973. *In:* Proc. Conf. on in situ composites. (F.D. Lemkey and E.R. Thompson, eds.), 1–69, Vol. 2, NMAB 308, Washington, DC: National Academy of Sciences-National Academy of Engineering.

Binienda, W.K., and D.N. Robinson. 1991. *J. Eng. Mech.* 117:624–639.

Blank, E. 1988. *In:* 9th Risø International Symposium on Metallurgy and Materials Science (S.I. Andersen, H. Lilholt, and O.B. Pedersen, eds.), 303–308, Roskilde, Denmark: Risø National Laboratory.

Bree, J. 1967. *J. Strain Anal.* 2:226–238.

Breinan, E.M., Thompson, E.R., and F.D. Lemkey. 1973. *In:* Proc. Conf. on in situ Composites. (F.D. Lemkey and E.R. Thompson, eds.), 201–222, Vol. 2, NMAB 308, Washington, DC: National Academy of Sciences-National Academy of Engineering.

Brown, L.M. 1982. *In:* Fatigue and Creep of Composite Materials. (H. Lilholt and R. Talreja, eds.), 1–18, Roskilde, Denmark: Risø National Laboratory.

Buckley, S.N., Harding, A.G., and M.B. Waldron. 1958. *J. Inst. Metals.* 87:150–154.

Bullock, E., McLean, M., and D.E. Miles. 1977. *Acta Metall.* 25:333–344.

Burke, J.E., and A.M. Turkalo. 1952. *J. Metals.* 4:651–657.

Chawla, K.K. 1973a. *Metallography.* 6:155–169.

Chawla, K.K. 1973b. *Phil. Mag.* 28:401–413.

Chawla, K.K. 1976. *J. Mater. Sci. Lett.* 11:1567–1569.

Chawla, K.K., and M. Metzger. 1972. *J. Mater. Sci.* 7:34–39.

Chen, Y-C., Daehn, G.S., and R.H. Wagoner. 1990. *Scripta Metall. Mater.* 24:2157–2162.

Choi, B.H., Wakashima, K., and T. Mori. 1991. *In:* Metal Matrix Composites—Processing, Microstructure and Properties. (N. Hansen, D. Jensen, T. Leffers, et al., eds.), 259–264, Roskilde, Denmark: Risø National Laboratory.

Chokshi, A.H., Bieler, T.R., Nieh, T.G., et al. 1988. *In:* Superplasticity in Aerospace. (H.C. Heikkenen and T. McNelley, eds.), 229–245, Warrendale, PA: The Metallurgical Society.

Christman, T., Needleman, A., Nutt, S., and S. Suresh. 1989. *Mater. Sci. Eng.* A107:49–61.

Colclough, A., Dempster, B., Farry, Y., and D. Valentin. 1991. *Mater. Sci. Eng.* A235:203–207.

Crossman, F.W., and R.F. Karlak. 1976. In: Failure Modes in Composites III. (T.T. Chiao, and D.M. Schuster, eds.), 260–287, Warrendale, PA: TMS-AIME.

Crossman, F.W., Karlak, R.F., and D.M. Barnett. 1974. In: Failure Modes in Composites II. (J.N. Fleck and R.L. Mehan, eds.), 8–21, Warrendale, PA: TMS-AIME.

Daehn, G.S. 1989. *Scripta Metall. Mater.* 23:247–252.

Daehn, G.S., and G. González-Doncel. 1989. *Metall. Trans.* 20A:2355–2368.

Daehn, G.S., and T. Oyama. 1988. *Scripta Metall.* 22:1097–1102.

de Silva, A.R.T. 1968. *J. Mech. Phys. Solids.* 16:169–186.

de Silva, A.R.T., and G.A. Chadwick. 1969. *J. Mech. Phys. Solids.* 17:387–403.

Dellis, M.A., Schobbens, H., Neste, M.V.D., et al. 1991. In: Metal Matrix Composites—Processing, Microstructure and Properties. (N. Hansen, D. Jensen, T. Leffers, et al. eds.), 299–304, Roskilde, Denmark: Risø National Laboratory.

Derby, B. 1985. *Scripta Metall.* 19:703–707.

Derby, B. 1991. In: Metal Matrix Composites—Processing, Microstructure and Properties. (N. Hansen, D.J. Jensen, T. Leffers, et al., eds.), 31–49, Roskilde, Denmark: Risø National Laboratory.

Derby, B., and J.R. Walker. 1988. *Scripta Metall.* 22:529–532.

Dirnfeld, S.F., and Y. Zuta. 1988. *Mater. Sci. & Eng.* A104:67–74.

Dragone, T.L., and W.D. Nix. 1990a. In: Metal & Ceramic Matrix Composites: Processing, Modeling & Mechanical Behavior (R.B. Bhagat, A.H. Clauer, P. Kumar, and A.M. Ritter, eds.), 367–380, Warrendale, PA: The Minerals, Metals & Materials Society.

Dragone, T.L., and W.D. Nix. 1990b. *Acta Metall. Mater.* 38:1941–1953.

Dragone, T.L., and W.D. Nix. 1992. *Acta Metall. Mater.* 40:2781–2791.

Dragone, T.L., Schlautmann, J.J., and W.D. Nix. 1991. *Metall. Trans.* 22A:1029–1036.

Dutta, I., and D.L. Bourell. 1989. *Mater. Sci. Eng.* A112:67–77.

Endo, T., Chang, M., Matsuda, N., and K. Matsuura. 1991. In: Metal Matrix Composites—Processing, Microstructure and Properties. (N. Hansen, D. J. Jensen, T. Leffers et al., eds.), 323–328, Roskilde, Denmark: Risø National Laboratory.

Eshelby, J.D. 1957. *Proc. Roy. Soc. Lond.* A241:376–396.

Evans, J.T., Ningyun, W., and H.W. Chandler. 1990. *Acta Metall. Et Mater.* 38:1565–1572.

Garmong, G. 1974a. *Metall. Trans.* 5:2183–2190.

Garmong, G. 1974b. *Metall. Trans.* 5:2199–2205.

Garmong, G. 1975. *Metall. Trans.* 6A:1179–1182.

Garmong, G. 1976. In: Conference on *in situ* Composites II. (M.R. Jackson, J.L. Walter, F.D. Lemkey, and R.W. Hertzberg, eds.), 137–153, Schenectady, NY: General Electric Corp. R&D.

Garmong, G., and C.G. Rhodes. 1973. In: Conference on *in situ* Composites. (F.D. Lemkey and E.R. Thompson, eds.), 251–264, Vol. 1, NMAB 308, Washington, DC: National Academy of Sciences-National Academy of Engineering.

Goto, S., and M. McLean. 1989. *Scripta Metall. Mater.* 23:2073–2078.

Goto, S., and M. McLean. 1991a. *Acta Metall. Mater.* 39:153–164.

Goto, S., and M. McLean. 1991b. *Acta Metall. Mater.* 39:165–177.

Gray, H.R., and W.A. Sanders. 1976. In: Conference on *in situ* Composites II. (M.R. Jackson, J.L. Walter, F.D. Lemkey, and R.W. Hertzberg, eds.), 201–210, Schenectady, NY: General Electric Corp. R&D.

Greenwood, G.W., and R.H. Johnson. 1965. *Proc. Roy. Soc. London.* 283A:403–422.

Grimes, H.H., Lad, R.A., and J.E. Maisel. 1977. *Metall Trans.* 8A:1999–2005.

Gulden, T.D., and J.C. Shyne. 1963. *Trans. AIME.* 227:1088–1092.

Guy, A.G., and J.E. Pavlick. 1961. *Trans AIME.* 221:802–807.

Hall, I.W., and W.G. Patterson. 1991. *Scripta Metall. Mater.* 25:805–810.

Higashi, K., Okada, T., Mukai, T. et al. 1992. *Scripta Metall. Mater.* 26:185–190.

Hoffman, C.A. 1973. *J. Eng. Mater. Tech.* 95:55–62.

Hong, S.L., Sherby, O.D., Divecha, A.P., et al. 1988. *J. Comp. Mater.* 22:102–123.

Humphreys, F.J. 1988. In: Mechanical and Physical Properties of Metallic and Ceramic Composites. (S.I. Andersen, H. Lilholt, and O.B. Pedersen, eds.), 51–65, Roskilde, Denmark: Risø National Laboratory.

Imai, T., Mabuchi, M., Tozawa, Y. et al. 1990a. In: Metal & Ceramic Matrix Composites: Processing, Modeling & Mechanical Behavior (R.B. Bhagat, A.H. Clauer, P. Kumar, and A.M. Ritter, eds.), 235–242, Warrendale, PA: The Minerals, Metals & Materials Society.

Imai, T., Mabuchi, M., Tozawa, Y., and M. Yamada. 1990b. *J. Mater. Sci. Lett.* 9:255–257.

Johnson, A.F. 1977. *J. Mech. Phys. Solids.* 25:117–126.

Kelly, A., and K.N. Street. 1972a. *Proc. Roy. Soc. Lond.* A328:267–282.

Kelly, A., and K.N. Street. 1972b. *Proc. Roy. Soc. Lond.* A328:283–293.

Kelly, A., and W.R. Tyson. 1966. *J. Mech. Phys. Solids.* 14:177–186.

Khan, T., Stohr, J.F., and H. Bibring. 1980. In: Superalloys 1980 (J.K. Tien, S.T. Wlodek, H. Morrow, et al., eds.), 531–540, Warrendale PA: ASM.

Kim, W.H., Koczak, M.J., and A. Lawley. 1979. In: New Developments and Applications in Composites. (D. Kuhlman-Wilsdorf and W.C. Harrigan Jr., eds.), 40–53, Warrendale PA: TMS-AIME.

Kim, Y.-W., and J.J. Kleek. 1990. In: Intermetallic matrix composites. (D.L. Anton, P.L. Martin, D.B. Miracle, and R. McMeeking, eds.), 315–321, Pittsburgh PA: MRS.

Kim, Y.W., Griffith, W.M., and F.H. Froes. 1985. *J. of Metals.* 37:27–33.

Komenda, J., and P.J. Henderson. 1991. In: Metal Matrix Composites—Processing, Microstructure and Properties. (N. Hansen, D. Juul Jensen, T. Leffers, et al., eds.), 449–454, Roskilde, Denmark: Risø National Laboratory.

Kyono, T., Hall, I.W., Taya, M., and A. Kitamura. 1986. *In:* Proceedings of the 3rd US Japan Conference on Composite Materials. (K. Kawata, S. Umekawa, and A. Umekawa, eds.), Tokyo: Society for Composite Materials.

Lawley, A. 1976. *In:* Conference on *In Situ* Composites-II. (M.R. Jackson, J.R. Walter, F.D. Lemkey, et al., eds.), 451–473, Schenectady, NY: General Electric R&D.

Le Flour, J.C., and R. Locicéro. 1987. *Scripta Metall.* 21:1071–1076.

Lee, Y.S., Batt, T.J., and P.K. Liaw. 1990. *Int. J. Mech. Sci.* 32:801–815.

Lee, Y.S., and H. Gong. 1987. *Int. J. Mech. Sci.* 29:669–694.

Lee, Y.S., Gungor, M.N., and P.K. Liaw. 1991. *J. Comp. Mater.* 25:536–555.

Levy, A., and J.M. Papazian. 1990. *Metall. Trans.* 21A:411–420.

Lilholt, H. 1978. *In:* Advances in Composite Materials. (G. Piatti, ed.), 209–233, London: Applied Science Publishers.

Lilholt, H. 1982. *In:* Fatigue and Creep of Composite Materials (H. Lilholt and R. Talreja, eds.), 63–76, Roskilde, Denmark: Risø National Laboratory.

Lilholt, H. 1984. *In:* Fundamentals of Deformation and Fracture. (B.A. Bilby, K.J. Miller, and J.R. Willis, eds.), 263–276, Cambridge, UK: Cambridge University Press.

Lilholt, H. 1985. *Composites Sci. and Tech.* 22:277–294.

Lilholt, H. 1988. *In:* 9th Risø International Symposium on Metallurgy and Materials Science (S.I. Andersen, H. Lilholt, and O.B. Pedersen, eds.), 89–107, Roskilde, Denmark: Risø National Laboratory.

Lilholt, H. 1991. *Mater. Sci. & Eng.* A135:161–171.

Lilholt, H., and M. Taya. 1987. *In:* Sixth International Conference on Composite Materials, ICCM 6. (F.L. Matthews, N.C.R. Buskell, J.M. Hodginson, and J. Morton, eds.), 2.234–2.244, London: Elsevier Applied Science.

Liu, Y.L., Juul Jensen D., and N. Hansen. 1992. *Met. Trans.* 23A:807–819.

Lobb, R.C., Sykes, E.C., and R.H. Johnson. 1972. *Metal Sci. J.* 6:33–39.

Lynch, C.T., and J.P. Kershaw. 1972. *In:* Metal Matrix Composites. 111–118, Cleveland: CRC Press.

Mabuchi, M., and T. Imai. 1990. *J. Mater. Sci. Lett.* 9:761–762.

Mabuchi, M., Higashi, K., Okada, Y., et al. 1991a. *Scripta Metall. Mater.* 25:2003–2006.

Mabuchi, M., Higashi, K., Tanimura, S., et al. 1991b. *Scripta Metall. Mater.* 25:1675–1680.

Mabuchi, M., Imai, T., Kubo, K., et al. 1991a. *In:* Advanced Composite Materials: New Developments and Applications. 259–266, Metals Park, OH: ASM.

Mabuchi, M., Imai, T., Kubo, K., et al. 1991b. *Mater. Let.* 11:339–342.

Mabuchi, M., Higashi, K., Wada, S., and S. Tanimura. 1992. *Scripta Metall. Mater.* 26:1269–1274.

Mackay, R.A. 1990. *Scripta Metall. Mater.* 24:167–172.

Mahoney, M.W., and A.K. Ghosh. 1987a. *Metall. Trans.* 18A:653–661.

Mahoney, M.W., and A.K. Ghosh. 1987b. *In:* Sixth International Conference on Composite Materials, ICCM 6. (F.L. Matthews, N.C.R. Buskell, J.M. Hodginson, and J. Morton, eds.), 2.372–2.381, London: Elsevier Applied Science.

McDanels, D.L., Signorelli, R.A., and J.W. Weeton. 1967. *NASA TN D-4173*.

McLean, D. 1972. *J. Mater. Sci.* 7:98–104.

McLean, M. 1982a. *In:* Fatigue and Creep of Composite Materials. (H. Lilholt, and R. Talreja, eds.), 77–88, Roskilde, Denmark: Risø National Laboratory.

McLean, M. 1982b. *In:* In Situ Composites IV. (F.D. Lemkey, H.E. Cline, M. McLean, eds.), 2–19, Amsterdam: Elsevier.

McLean, M. 1983. Directionally Solidified Materials for High Temperature Service. London: The Metals Society.

McLean, M. 1985. *In:* Fifth International Conference on Composite Materials ICCM V. (W.C. Harrigan, J. Strife, and A.K. Dhingra, eds.), 37–51, Warrendale, PA: TMS.

McLean, M. 1988. *In:* High Temperature/High Performance Composites. (F.D. Lemkey, S.G. Fishman, A.G. Evans, and J.R. Strife, eds.), 67–79, Pittsburgh: Materials Research Society.

McLean, M. 1989. *In:* Materials and Engineering Design: The Next Decade. (B.F. Dyson, and D.R. Hayhurst, eds.), 287–294, London: The Institute of Metal.

Mileiko, S.T. 1970. *J. Mater. Sci.* 5:254–261.

Miles, D.E., and M. McLean. 1977. *Metal Sci.* 11:563–570.

Mishra, R.S. 1992. *Scripta Metall. Mater.* 26:309–313.

Mishra, R.S., and A.B. Pandey. 1990. *Metall. Trans.* 21A:2089–2090.

Morimoto, T., Yamaoka, T., Lilholt, H., and M. Taya. 1988. *Trans. ASME-J. of Eng. Mater. and Tech.* 110:70–76.

Mura, T. 1982. Micromechanics of Solids, The Hague: Martinus Nijhoff.

Nakanishi, M., Nishida, Y., Matsubara, H., et al. 1990. *J. Mater. Sci. Lett.* 9:470–472.

Nardone, V.C., and J.R. Strife. 1987. *Metall. Trans.* 18A:109–114.

Nieh, T.G. 1984. *Metall. Trans.* 15A:139–146.

Nieh, T.G., and J. Wadsworth. 1991. *Mater. Sci. & Eng.* A147:129–142.

Nieh, T.G., Henshall, C.A., and J. Wadsworth. 1984. *Scripta Metall.* 18:1405–1408.

Nieh, T.G., Xia, K., and T.G. Langdon. 1988. *J. Eng. Mater. Techn.* 110:77–82.

Noebe, R.D., Bowman, R.R., and J.J. Eldridge. 1990. *In:* Intermetallic Matrix Composites. (D.L. Anton, P.L. Martin, D.B. Miracle, and R. McMeeking, eds.), 323–331, Pittsburgh: MRS.

Pachalis, J.R., Kim, J., and T.-W. Chou. 1990. *Composites Sci. Tech.* 37:329–346.

Pandey, A.B., Mishra, R.S., and Y.R. Mahajan. 1990. *Scripta Metall. Mater.* 24:1565–1570.

Park, K.T., Lavernia, E.J., and F.A. Mohamed. 1990. *Acta Metall. Mater.* 38:2149–2159.

Park, Y.H., and H.L. Marcus. 1983. *In:* Mechanical Behaviour of Metal Matrix Composites. (J.E. Hack and M.F. Amateau, eds.), 65–75, Warrendale PA: TMS-AIME.

Patterson, W.G., and M. Taya. 1985. *In:* Fifth International Conference on Composite Materials. (W.C. Harrigan Jr., J. Strife, and A. Dhingra, eds.), 53–66, Warrendale PA: TMS-AIME.

Pattnaik, A., and A. Lawley. 1971. *Metall. Trans.* 2:1529–1536.

Pepper, R.T., Upp, J.W., Rossi, R.C., and E.G. Kendall. 1971. *Metall. Trans.* 2:117–121.

Pickard, S.M., and B. Derby. 1989. *In:* Developments in the Science and Technology of Composite Materials (ECCM3).

(A.R. Bunsell, P. Lamicq, and A. Massiah, eds.), 199–204, Amsterdam: Elsevier.

Pickard, S.M., and B. Derby. 1988. *In:* Mechanical and Physical Properties of Metallic and Ceramic Composites. (S.I. Andersen, H. Lilholt, and O.B. Pedersen, eds.), 447–452, Roskilde, Denmark: Risø National Laboratory.

Pickard, S.M., and B. Derby. 1990. *Acta Metall. Mater.* 38:2537–2552.

Pilling, J. 1989. *Scripta Metall.* 23:1375–1380.

Poirier, J.P. 1982. *J. Geophys. Res.* 87:6791–6797.

Povirk, G.L., Nutt, S.R., and A. Needleman. 1992. *Scripta. Metall. Mater.* 26:461–466.

Proulx, D., and F. Durant. 1974. *In:* Failure Modes in Composites II. (J.N. Fleck, and R.L. Mehan, eds.), 188–196, Warrendale PA: TMS-AIME.

Roberts, A.C. 1960. *Acta Metall.* 8:817–819.

Roberts, A.C., and A.H. Cottrell. 1956. *Phil. Mag.* 1:711–717.

Rösler, J., and A.G. Evans. 1992. *Mater. Sci. Eng.* A153:438–443.

Sherby, O.D., Klundt, R.H., and A.K. Miller. 1977. *Metall. Trans.* 8A:843–850.

Shimmin, K.D., and I.J. Toth. 1972. *In:* Failure Modes in Composites. (I. Toth, ed.), 357–393, Warrendale PA: TMS-AIME.

Sørensen, N. 1991. *In:* Metal Matrix Composites—Processing, Microstructure and Properties. (N. Hansen, D.J. Jensen, T. Leffers, et al., eds.), 667–673, Roskilde, Denmark: Risø National Laboratory.

Stobo, J.J. 1960. *J. Nucl. Mater.* 2:97–109.

Stohr, J.F., and T. Khan. 1985. *In:* Introduction aux Matériaux Composites. (R. Daviaud and C. Filiatre, eds.), 265–313, Paris: CNRS.

Stoloff, N.S. 1978. *In:* Conference on In Situ Composites-III. (J.L. Walter, M.F. Gigliotti, B.F. Oliver, and H. Bibring, eds.), 357–375, Lexington, MA: Ginn Custom Publishing.

Taya, M. 1991. *In:* Metal Matrix Composites: Mechanisms and Properties. (R.K. Everett, and R.J. Arsenault, eds.), 189–216, Boston: Academic Press.

Taya, M., and R.J. Arsenault. 1989. Metal Matrix Composites: Thermomechanical Behavior. Oxford: Pergamon Press.

Taya, M., and H. Lilholt. 1986. *In:* Advances in Composite Materials and Structures. (S.S. Wang, and Y.D.S. Rajapakse, eds.), 21–27, New York: ASME.

Taya, M., and T. Mori. 1987. *In:* Thermomechanical Couplings in Solids. (H.D. Bui, and Q.S. Nquyen, eds.), 147–162, Amsterdam: Elsevier.

Taya, M., Dunn, M., and H. Lilholt. 1991. *In:* Metal Matrix Composites—Processing, Microstructure and Properties. (N. Hansen, D.J. Jensen, T. Leffers, et al., eds.), 149–171, Roskilde, Denmark: Risø National Laboratory.

Thomin, S.H., and D.C. Dunand. 1993. *In:* Mechanisms and Mechanics of Composite Fracture. (R.B. Bhagat, ed.), Warrendale PA: TMS-AIME.

Thompson, E.R., and F.D. Lemkey. 1974. *In:* Metallic Matrix Composites. (K.G. Kreider, ed.), 135–139, New York: Academic Press.

Vogelsang, M., Arsenault, R.J., and R.M. Fisher. 1986. *Metall. Trans.* 17A:379–389.

Wakashima, K., Choi, B.H., and T. Mori. 1990. *Mater. Sci. & Eng.* A127:57–64.

Warren, R., Larsson, L.O.K., Ekström, P., and T. Jansson. 1982. *In:* Progress in Science and Engineering of Composites. Proceedings of the Fourth International Conference on Composite Materials. (T. Hayashi, K. Kawata, and S. Umekawa, eds.), 1419–1426, Tokyo: Society for Composite Materials.

Warwick, C.M., and T.W. Clyne. 1989. *In:* Fundamental Relationships Between Microstructures and Mechanical Properties of Metal Matrix Composites. (M.N. Gungor, and P.K. Liaw, eds.), 209–223, Warrendale PA: TMS.

Withers, P.J., Juul Jensen, D., Lilholt, H., and W.M. Stobbs. 1987. *In:* Sixth International Conference on Composite Materials and Second European Conference on Composite Materials. (F.L. Matthews, J.M. Hodgkinson, and J. Morton, eds.), 2255–2263, London: Elsevier.

Withers, P.J., Lilholt, H., Juul Jensen, D., and W.M. Stobbs. 1988. *In:* Mechanical and Physical Properties of Metallic and Ceramic Composites. (S.I. Andersen, H. Lilholt, and O.B. Pedersen, eds.), 503–510, Roskilde, Denmark: Risø National Laboratory.

Wolff, E.G. 1989. *In:* Reference Book for Composites Technology. (S.M. Lee, ed.), 111–142, Lancaster, PA: Technomic.

Wolff, E.G., Min, B.K., and M.H. Kural. 1985. *J. Mater. Sci.* 20:1141–1149.

Woodford, D.A. 1976. *In:* Proc. Conf. on *in situ* Composites II. (M.R. Jackson, J.L. Walter, F.D. Lemkey, and R.W. Hertzberg, eds.), 211–221, Schenectady, NY: General Electric Corp. R&D.

Wright, M.A. 1975. *Metall. Trans.* 6A:129–134.

Wu, M.Y., and O.D. Sherby. 1984. *Scripta Metall.* 18:773–776.

Wu, M.Y., Wadsworth, J., and O.D. Sherby. 1987. *Metall. Trans.* 18A:451–462.

Xia, K., Nieh, T.G., Wadsworth, J., and T.G. Langdon. 1990. *In:* Fundamental Relationship Between Microstructure and Mechanical Properties of Metal Matrix Composites. (P.K. Liaw, and M.N. Gungor, eds.), 543–556, Warrendale PA: The Metallurgical Society.

Xiaoxu, H., Qing, L., Yao, C.K., and Y. Mei. 1991. *J. Mater. Sci. Lett.* 10:964–966.

Xiong, Z., Geng, L., and C.K. Yao. 1990. *Composites Sci. and Tech.* 39:117–125.

Yancey, R.N., and M.J. Pindera. 1990. *J. Eng. Mater. Techn.* 112:157–163.

Yoda, S., Kurihara, N., Wakashima, K., and S. Umekawa. 1977. *Metall. Trans.* 8A:2028–2030.

Yoda, S., Kurihara, N., Wakashima, K., and S. Umekawa. 1978. *Metall. Trans.* 9A:1229–1236.

Zhang, H., Daehn, G.S., and R.H. Wagoner. 1990. *Scripta Metall. Mater.* 24:2151–2155.

Zhu, Z.G., and G.J. Weng. 1990. *Mech. Mater.* 9:93–105.

PART IV
DAMAGE MICROMECHANISMS AND MECHANICS OF FAILURE

Chapter 12
Models for Metal/Ceramic Interface Fracture

ZHIGANG SUO
C. FONG SHIH

Metal-matrix composites are excellent candidate materials for advanced engineering systems. However, they have one major shortcoming that has limited their widespread use—their tendency to fracture easily. In many systems, the low ductility or brittleness of these composites is caused by microfailure processes that invariably begin at the interfaces. Thus, the mechanical behavior and the overall performance of metal-matrix composites are not limited by bulk properties or bulk phases, but by interface properties and toughness. Theories on interface fracture are reviewed in this chapter. With few exceptions, attention is limited to continuum mechanics considerations. Readers are referred to Rice et al. (1989, 1990, 1992) and references therein for atomistic and thermodynamic aspects of this subject. This article is concerned with recent advances within the confines of small-scale inelasticity and loading conditions, such that a major portion of the crack faces remain open. We review works regarding relatively brittle interfaces where the inelastic zone is small compared to the overall component. Large-scale bridging is reviewed by Bao and Suo (1992). Large scale contact has been treated by Hutchinson and Jensen (1990) within the context of fiber pullout against frictional sliding. The focus of this article is on theories. For a broader coverage of topics, the reader is referred to published proceedings of symposia on interfacial fracture. These include those edited by Suresh and Needleman (1989), Rühle et al. (1990), and Ashby et al. (1992). Experimental aspects of interface fracture and fatigue are reviewed by Evans et al. (1990), Kim (1991), and Cannon et al. (1992). Several aspects of fatigue are also discussed in a recent article (Woeltjen et al. 1992).

The mechanics of interface fracture has its root in the earlier works of Griffith (1921) and Irwin (1960) on the general theory of fracture, of Williams (1959) on the elastic stress distribution around an interface crack, of England (1965), Erdogan (1965), and Rice and Sih (1965) on explicit solutions for interface cracks, and of many practicing engineers on ingenious methods to measure adhesive strength of bonds. However, the subject did not take off until the 1980s. Advanced composites for high temperature engines, and layered materials for microelectronic and optical devices, have been the main technical driving force for new theoretic developments. Rapid advancement in high-resolution microscopes, high-speed computers, and the general theory of fracture have all provided tools for solving these challenging technical problems.

The classical fracture mechanics, as advanced by Irwin (1960), Rice (1968), and Hutchinson (1979), and as summarized in the textbook by Kanninen and Popelar (1985), is largely phenomenological. It enables us to predict, without a detailed description of the crack tip processes, crack growth in a structure by utilizing the observed crack growth behavior in a fracture specimen. This approach relies on the existence of stress intensity factor and on fracture resistance measured by mechanical testing. The advantage and the deficiency of this black-box approach both originate from the same fact: this approach requires little, nor does it generate much, knowledge of the physical process of fracture.

Mechanism-based fracture mechanics attempts to link the fracture resistance to the microstructural variables, providing guidelines for processing better and newer materials. This approach is as old, if not as fully developed, as the phenomenological approach. Indeed, in his original paper, Griffith prescribed a phenomenological fracture criterion as well as a physical mechanism of fracture resistance: rearranging atoms in a bulk into surfaces requires energy. It was eminently clear to Irwin that any form of heat dissipation that accompanies fracture, such as plastic flow in metals, contributes to fracture resistance. But just how atomic separation and irreversible atomic movements are interconnected remains an open question even today. Partial theories, each valid for a particular mechanism at a particular length scale, have been devised, largely in the earlier works by Cottrell (1963) and Rice and Johnson (1970) on hole growth, by Ritchie et al. (1973) on cleavage of alloys, and by Aveston et al. (1971) on fiber reinforced components. An example of the success of the mechanism-based approach is the thorough understanding of advanced ceramics toughened by ductile particles, or transforming particles, or strong fibers that has been achieved (Evans 1990). As a by-product of the mechanism-based approach, short cracks, small components, and nagging questions in the framework of classical fracture mechanics can be addressed with a unified, conceptually simple viewpoint (Stump and Budiansky 1989; Bao and Suo 1992). The dual approach, phenomenological and mechanism-based, is kept in mind throughout this review article.

The chapter is organized as follows. Section 12.1 reviews the energy consideration for a traction-free interface crack under small-scale inelasticity conditions, leading to the concepts of debond driving force, $\mathcal{G}$, and debond resistance, Γ. Supplemented by elasticity solutions of $\mathcal{G}$ for given components and loading conditions and experimentally measured Γ for given materials, this energy consideration is sufficient for most engineering applications. Section 12.2 reviews Williams' solution for an interface crack between two elastic half spaces. The near-tip stress field lends itself to a precise definition of mode mixity; the near-tip opening sets the condition for small-scale contact. Calibration of fracture specimen is also discussed. Crack-tip plasticity is reviewed in Section 12.3. The relevant mode mixity is the ratio of shear over tension on the interface immediately outside of the plastic zone. The plastic zone size is shown to depend on the mode mixity, as is the opening of the blunted crack tip. An analysis of a metal foil sandwiched between two ceramic plates is included, demonstrating the effect of constrained plastic flow on the fracture resistance. Section 12.4 reviews results on growing cracks and the concepts leading to a fundamental understanding of fracture resistance. Included are crack growth in elastic-plastic materials, crack bridging with or without background plasticity, and cleavage in the presence of plastic flow.

12.1 Energy Balance

The following energy arguments are essentially the same as those of Griffith (1921) and Irwin (1960). Cracks often run when a test-piece is still predominantly elastic, inelastic deformation being localized in thin layers beneath the crack surfaces. Taking advantage of this fact, one can partition the total energy supplied by the applied work into (1) elastic energy stored in the test-piece and (2) the heat dissipated by plastic flow and residual energy stored in the thin layers. From (1) comes a definition of debond driving force, $\mathcal{G}$, and from (2) comes debond resistance, Γ. Obviously, this partition becomes meaningless when the inelastic deformation spreads over a large part of the test-piece, either because the test-piece is small, or because the material is very ductile. These have been reviewed elsewhere (Stump and Budiansky 1989; Bao and Suo 1992).

12.1.1 Debond Driving Force

Consider an interface crack extending over an area A (Figure 12.1). Material near the crack front undergoes inelastic deformation; the interface is usually ill defined, containing misfit dislocations, an interdiffusion zone, or reaction compounds. However, these zones are typically small compared to the overall dimension of the test-piece, so that the crack front can be treated as a mathematical line, and the interface a mathematical plane. In computing the elastic energy stored in the test-piece, the two solids are taken to be (possibly nonlinearly) elastic. That is, each solid can be described by a strain energy density function $w(\epsilon_{11}, \epsilon_{12}, ...)$, such that stresses σ_{ij} are derived from

$$dw = \sigma_{ij} d\epsilon_{ij} \qquad (12.1)$$

Coupled with equilibrium and compatibility equations, these considerations define an elasticity problem. In particular, stress and strain are computed everywhere in the test-piece, down to the crack front and the interface; likewise the energy density w is computed everywhere. The elastic energy stored in the test-piece is an integral extended over the entire test-piece, such that

$$U = \int_V w \, dV \qquad (12.2)$$

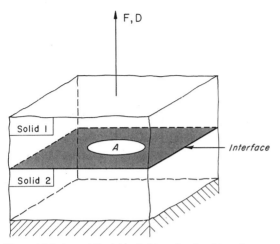

Figure 12.1. A partially debonded interface is subjected to applied load.

The test-piece in Figure 12.1 is loaded by displacement D, with work-conjugating force F. The elastic energy U depends on applied displacement and the crack size, thus,

$$U = U(D,A) \quad (12.3)$$

Note that U also depends on the geometry of the test-piece and the elastic moduli, but they remain constant during testing.

Upon loading, U varies as

$$dU = FdD - \mathcal{G}dA \quad (12.4)$$

With crack size held fixed ($dA = 0$), the above equation simply states that the energy increment equals the work applied. Because all the other quantities are defined, Equation 12.4 defines the quantity $\mathcal{G}$ when $dA \neq 0$. Just as F is the driving force for D, $\mathcal{G}$ is the driving force for crack size A. Explicitly, $\mathcal{G}$ is the decrease of elastic energy associated with a unit increment of crack area:

$$\mathcal{G} = -\frac{\partial U(D,A)}{\partial A} \quad (12.5)$$

Note that $\mathcal{G}$ has dimension energy/area.

The above concepts can be explained graphically. Figure 12.2(a) shows a load-displacement curve of the test-piece measured with fixed crack size ($dA = 0$). The curve should be straight for linear elastic materials. From Equation 12.4, U is the area under the load-displacement curve. Figure 12.2(b) shows two such load-displacement curves, measured in two independent tests with slightly different crack sizes, A and $A + dA$. The test-piece with the larger crack is more compliant; the shaded area is the energy decrease, dU,

associated with dA. In early days, this graphical interpretation was employed to experimentally determine $\mathcal{G}$ (Rivlin and Thomas 1953).

Standard thermodynamics manipulations apply to the present discussions. For load-controlled tests, for example, it is more convenient to work with the potential energy

$$\Pi = U - FD \quad (12.6)$$

which is indicated in Figure 12.2(a). The independent variables are now F and A. Upon loading, Equation 12.4 becomes

$$d\Pi = -DdF - \mathcal{G}dA \quad (12.7)$$

Therefore, an alternative definition of $\mathcal{G}$ is

$$\mathcal{G} = -\frac{\partial \Pi(F,A)}{\partial A} \quad (12.8)$$

The definitions (12.5) and (12.8) are of course equivalent.

For an interface along the x_1-axis, and with displacement and traction continuous across the interface, Rice's J-integral (1968)

$$J = \int (wn_1 - n_i\sigma_{ij}u_{j,1})\,ds \quad (12.9)$$

vanishes over contours not enclosing any singularity. For a traction-free crack on the interface, $\mathcal{G}$ equals the J-integral over any path that begins at a point on the

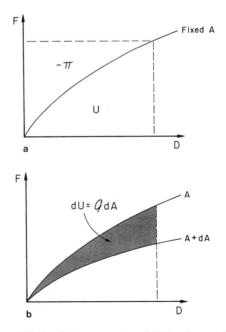

Figure 12.2. Graphic interpretation of (a) U and π and (b) $\mathcal{G}$.

lower crack face, and ends at another point on the upper crack face. This provides a tool for calculating $\mathcal{G}$ in finite element analysis (Moran and Shih 1987).

In general, $\mathcal{G}$ can be computed with an elasticity analysis of a given test-piece. Several illustrations requiring only elementary mechanics are given. Solutions to a wide range of geometries can be found in Hutchinson and Suo (1992) and the references therein.

Consider a fiber being pulled out of a matrix (Figure 12.3). The energy stored in the fiber can be estimated by regarding the fiber as a tensile bar, clamped at the debond front. The pullout displacement is $D = L\sigma/E$, so that $U = (\sigma^2/2E)(\pi R^2 L)$. The potential energy is

$$\Pi(\sigma,L) = -\frac{\sigma^2}{2E}\pi R^2 L. \quad (12.10)$$

Because the debond area is $A = 2\pi RL$, carrying out the differentiation in Equation 12.8 gives

$$\mathcal{G} = R\sigma^2/4E \quad (12.11)$$

The estimate, which ignores the compliance of the fiber-matrix junction, is accurate when the debond length is large compared to the fiber diameter. Observe that $\mathcal{G}$ does not depend on the debond length L. Once debond starts, it will run to the other end of the fiber without any increase in load.

Thin-film cracking of many patterns has inspired a new problem area (Evans et al. 1988; Hutchinson and Suo 1992). Figure 12.4 illustrates a circular interface crack emanating from the edge of a hole in a thin film, driven by a residual tensile stress in the film. The stress in the debonded film is partially relieved, leading to a reduction in the elastic strain energy. The debonded film may be treated as a ring in plane stress, clamped at the debond front. The energy release rate is found to be

$$\mathcal{G} = \frac{2hE\varepsilon_0^2}{1-\nu^2}\left[1 + \frac{1-\nu}{1+\nu}\left(\frac{a}{a_0}\right)^2\right]^{-2} \quad (12.12)$$

where ε_0 is the mismatch strain between the film and the substrate caused by thermal or epitaxial mismatch,

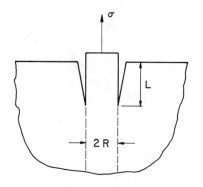

Figure 12.3. A fiber is being pulled out from a matrix.

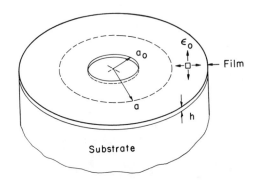

Figure 12.4. Thin-film decohesion emanating from a circular hole.

a_0 is the hole radius, a is the debond radius, and E and ν are the elastic constants of the film (Farris and Bauer 1988). Observe that $\mathcal{G}$ decreases rapidly as a increases, so that the debond is stable. Also note that $\mathcal{G}$ scales linearly with film thickness h: the thinner the film, the smaller the decohesion area. Debond can be practically suppressed if the film is sufficiently thin.

12.1.2 Debond Resistance

The essential idea of Griffith and Irwin is illustrated in Figure 12.5. Inelastic processes, such as atomic separation, twinning, phase transformation, and dislocation motion, require sufficiently high stress to activate, so they are confined to a region close to the crack tip where the stress is intensified. As the crack front extends, thin layers beneath the crack surface are left in the wake in which the atoms have undergone irreversible movements. The processes near the tip are complex and the quantification requires detailed knowledge of deformation mechanisms. Nonetheless, an effectively uniform deformation state along the x_1-axis is attained in the wake. Consider two cylinders of unit cross-sectional area normal to the interface, one far ahead of the crack front (A), and the other far behind (B). Let Γ be the energy spent to transform cylinder A to cylinder B. Obviously Γ depends on the deformation history that cylinder B underwent, including surface energy, heat dissipation, and elastic energy trapped in the wake.

The total energy variation, elastic as well as inelastic, is given by

$$FdD - \mathcal{G}dA + \Gamma dA \quad (12.13)$$

When D is held fixed, no work is externally applied to the test-piece and the total energy remains unchanged, so that

$$\mathcal{G} = \Gamma \quad (12.14)$$

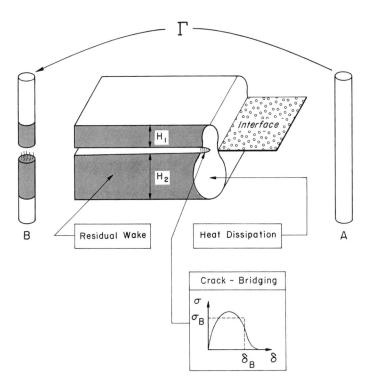

Figure 12.5. Inelastic processes accompanying debond.

The driving force $\mathcal{G}$ depends on the test-piece and can be evaluated by an elastic stress analysis. Fracture resistance Γ depends on the inelastic mechanisms. Equation 12.14 provides a connection between the macroscopic loading condition of a test-piece and the microscopic inelastic process associated with debonding.

Debond resistance can be measured phenomenologically. For example, this can be carried out using the fiber pullout experiment shown in Figure 12.3. The stress required to drive debond is measured, which can be translated to Γ using Equation 12.11 and noting Equation 12.14. This approach is purely phenomenological—no detailed knowledge of physical processes is required, nor is such knowledge generated. Nevertheless, the key quantity, debond resistance Γ, is measured and this can be used in device design. Debond resistance has been measured for a range of bimaterials for applications to thin films and fiber/matrix composites (Evans et al. 1990; Cannon et al. 1992).

In principle, test pieces of any geometry can be used to measure debond resistance. Several convenient geometries are sketched in Figure 12.6. It has been observed experimentally that debond resistance depends on the geometry of the test-piece. Specifically, debond resistance depends on the ratio of the sliding to normal loading parameterized by ψ:

$$\Gamma = \Gamma(\psi) \quad (12.15)$$

Mode mixity ψ will be elaborated upon later. The trend of the curve is shown in Figure 12.6. The double cantilever beam is predominantly opening mode ($\psi \approx 0°$) and the measured debond resistance is low. The fiber pullout is shear dominant ($\psi \approx 70°$), giving a high debond resistance. The other two, four-point bend and microindentation, produce nearly equal amounts of opening and shear ($\psi \approx 45°$), representative of the conditions in thin-film delamination and fiber/matrix debonding. Microindentation is particularly convenient for small samples (Davis et al. 1991). Other geometries have also been used to measure debond resistance (Argon et al. 1989; Kim 1991; Liechti and Chai 1992; O'Dowd et al. 1992a; Thouless 1990; Wang and Suo 1990).

The dependence of Γ on loading phase can be understood on the basis of inelastic mechanisms. For example, the fiber pullout experiment is dominated by shearing while friction adds to the debond resistance. This mechanism has been examined quantitatively by Hutchinson and Evans (1989). In metal/ceramic interfaces, shear-dominated loading produce larger plastic zones which increase debond resistance (O'Dowd et al. 1992b).

12.2 Williams' Singularity

This section collects mathematical details that refine the concept of mode mixity. The two solids are linearly

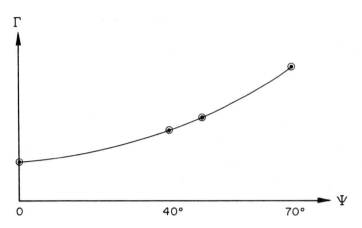

Figure 12.6. Convenient geometries to measure debond resistance.

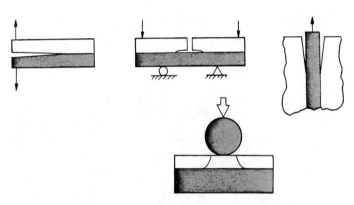

elastic and isotropic; corresponding results for anisotropic elasticity have been reviewed elsewhere (Suo 1990). The inelastic region is taken to be small compared to all other relevant dimensions of the crack geometry, so that the crack front is a mathematical line, the interface is a mathematical plane, and the crack is semi-infinite. The crack faces are traction-free. This eigenvalue problem was solved by Williams (1959). The essential features of the solution, stress oscillation and crack face contact, are described below. The following interpretation is largely due to Rice (1988).

12.2.1 Slow Oscillation in Stress Field

The tractions at a distance r ahead of the crack tip, on the interface, are found to be

$$\sigma_{yy} + i\sigma_{yx} = \frac{Kr^{i\varepsilon}}{\sqrt{2\pi r}} \quad (12.16)$$

In the above, x and y are Cartesian coordinates centered at the tip and $i = \sqrt{-1}$. The bimaterial constant ε is defined by

$$\varepsilon = \frac{1}{2\pi} \ln\left[\frac{(3-4\nu_1)/\mu_1 + 1/\mu_2}{(3-4\nu_2)/\mu_2 + 1/\mu_1} \right] \quad (12.17)$$

Here ν is Poisson's ratio, μ the shear modulus, and subscripts 1 and 2 refer to material 1 and 2, respectively; the constant ε is bounded, $|\varepsilon| < (1/2\pi)\ln 3 \approx 0.175$.

The complex-valued stress intensity factor, K, cannot be determined by the eigenvalue problem, but can be determined by solving the full boundary-value problem for a given test-piece. The magnitude of K scales with the applied stress, and the phase angle of K represents the relative amount of shear to tension. It can be seen from (12.16) that K has the dimensions

$$K = [\text{stress}] \, [\text{length}]^{1/2 - i\varepsilon} \quad (12.18)$$

Let $\hat{L}$ be an *arbitrary* length, and define $\hat{\psi}$ by

$$K = |K|\hat{L}^{-i\varepsilon}\exp(i\hat{\psi}) \quad (12.19)$$

The magnitude, $|K|$, has the dimension stress × length$^{1/2}$, which is independent of the choice of $\hat{L}$, because $|\hat{L}^{i\varepsilon}| = 1$. Indeed, $|K|$ is related to $\mathcal{G}$ (Malyshev and Salganik 1965) by

$$\mathcal{G} = \frac{1}{4}\left(\frac{1-\nu_1}{\mu_1} + \frac{1-\nu_2}{\mu_2}\right)\frac{|K|^2}{\cosh^2 \pi\varepsilon} \quad (12.20)$$

Consequently, $|K|$ and $\mathcal{G}$ are equivalent quantities characterizing the magnitude of the applied load.

Next examine the significance of $\hat{\psi}$. Combining Equations 12.16 and 12.19 yields

$$\sigma_{yy} + i\sigma_{yx} = \frac{|K|}{\sqrt{2\pi r}} \exp i[\hat{\psi} + \varepsilon \ln(r/\hat{L})] \quad (12.21)$$

The identity, $(r/\hat{L})^{i\varepsilon} \equiv \exp[i\varepsilon \ln(r/\hat{L})]$, is used in the above. From Equation 12.21, the ratio of shear stress to tensile stress at a distance r ahead of the crack tip is given by

$$\sigma_{xy}/\sigma_{yy} = \tan[\hat{\psi} + \varepsilon \ln(r/\hat{L})] \quad (12.22)$$

Observe that the traction ratio varies with position r, and that $\tan\hat{\psi}$ equals the ratio of the shear stress to the tensile stress at $r = \hat{L}$. This feature of interface cracks, caused by elastic mismatch, does not exist in mixed mode fracture in homogeneous materials. The result in Equation 12.22 is commonly referred to as the oscillatory singularity, and ε is referred to as the oscillation index. Contrary to a popular misconception, this oscillation is *not* rapid, because ε is small and because a logarithm is a slowly varying function. Thus, in specifying mode mixity, $\hat{L}$ need not be precisely defined, so long as it is broadly representative of the length scale of interest.

Up to this point, $\hat{L}$ has not been given any physical identity. Because Williams' elastic solution describes the stress state outside of the inelastic zone, it is sensible to specify $\hat{L}$ to be on the order of the inelastic zone size. For example, in discussing dislocation emission from an atomistically sharp crack tip, a natural choice of $\hat{L}$ is atomic spacing, so that $\hat{\psi}$ describes the stress state over several atomic spacing (Rice et al. 1990). For a metal/ceramic interface, where dislocation motion prevails over distances many times of the lattice constants, $\hat{L}$ should be chosen as the plastic zone size. Given two choices L and $\hat{L}$, the corresponding loading phases, ψ and $\hat{\psi}$, shift by

$$\psi - \hat{\psi} = \varepsilon \ln(L/\hat{L}) \quad (12.23)$$

Debond resistance Γ should depend on stress state surrounding the inelastic zone, which in turn is characterized by the local phase angle $\hat{\psi}$. Consequently, Equation 12.15 can be rewritten in a more rigorous form.

$$\Gamma = \Gamma(\hat{\psi}) \quad (12.24)$$

Because the size of the inelastic zone depends on fracture mechanisms ranging from nanometers to centimeters, it is meaningless to employ a single $\hat{L}$ for all bimaterials. For interfaces with debond resistance sensitive to mode mixity, the value of $\hat{\psi}$, together with $\hat{L}$, must be reported together with the value of Γ. Moreover, a common $\hat{L}$ must be used in the definition of $\hat{\psi}$ when comparing toughness values at different mode mixities.

12.2.2 Small-Scale Contact

In a homogeneous material, crack faces come into contact under compression. By contrast, interface crack faces may come into contact regardless of loading condition. The size of the contact zone depends on the mode mixity. In composites, the fiber and the matrix may remain in contact because of the residual compression or asperities, sliding against friction during pullout. This provides an example of large-scale contact (Hutchinson and Jensen 1990). In many other technical problems, such as thin-film decohesion, contact zone is small compared to the overall dimension. This section provides a criterion for small-scale contact.

Williams elastic solution shows that the displacement jump at a distance r behind the crack tip is

$$\delta_y + i\delta_x = \left(\frac{1-\nu_1}{\mu_1} + \frac{1-\nu_2}{\mu_2}\right) \cdot \frac{Kr^{i\varepsilon}}{(1+2i\varepsilon)\cosh\pi\varepsilon}\sqrt{\frac{2r}{\pi}} \quad (12.25)$$

From the above, the crack opening is

$$\delta_y = \delta\cos[\hat{\psi} + \varepsilon\ln(\hat{L}/r) - \tan^{-1}(2\varepsilon)] \quad (12.26)$$

where $\delta = (\delta_x^2 + \delta_y^2)^{1/2}$ is the magnitude of the displacement jump. If $\hat{L}$ is interpreted as the process zone size, and if the crack is required to remain open, i.e., $\delta_y > 0$, within $\hat{L} < r < 100\hat{L}$, the mode mixity must be confined within

$$-\pi/2 + 2\varepsilon < \hat{\psi} < \pi/2 + 2.6\varepsilon \quad \text{for } \varepsilon > 0$$
$$-\pi/2 - 2.6\varepsilon < \hat{\psi} < \pi/2 + 2\varepsilon \quad \text{for } \varepsilon < 0 \quad (12.27)$$

The number 100 is arbitrary, but the condition in Equation 12.27 is not sensitive to this number. When $\varepsilon = 0$, the above condition simply states that contact will not occur under tension, which is known for homogeneous materials.

12.2.3 Specimen Calibrations

For a given test piece, the complex stress intensity factor K can be solved by an elastic stress analysis. It has the generic form

$$K = YT\sqrt{L}L^{-i\varepsilon} \exp(i\psi) \quad (12.28)$$

where T is a representative stress magnitude, and L a characteristic crack dimension. Y and ψ are dimensionless real numbers that depend on elastic constants, geometric parameters, and loading. Note that ψ is by definition the phase of $KL^{i\varepsilon}$. Solutions have been compiled by Hutchinson and Suo (1992).

As an example, consider a Griffith crack of length L on the interface between two materials (Figure 12.7). The complex stress intensity factor is

$$K = (1 + 2i\varepsilon)\sqrt{\pi L/2}L^{-i\varepsilon}T \exp(i\omega) \quad (12.29)$$

where ω is the remote loading angle. Driving force $\mathcal{G}$ is obtained by substituting Equation 12.29 in Equation 12.20; the loading angle consistent with Equation 12.28 is $\psi = \omega + \tan^{-1}(2\varepsilon)$. The mode mixity at $r = \hat{L}$ is given by

$$\hat{\psi} = \omega + \tan^{-1}(2\varepsilon) + \varepsilon \ln(\hat{L}/L). \quad (12.30)$$

For a numerical illustration, consider an alumina/glass interface ($\varepsilon = -0.05$). The crack size $L = 1$ mm, and the process zone size is of the order $\hat{L} = 1$ nm. Under

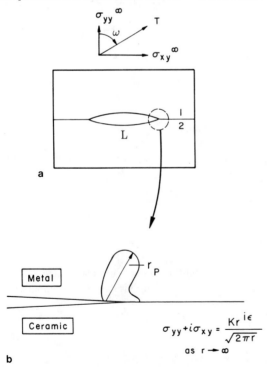

Figure 12.7. (a) A finite interface crack subjected to a remote tension T at angle ω; the plastic zone size is assumed to be small compared to the crack size. (b) A small-scaled yielding problem is posed.

remote tension, $\omega = 0°$, one finds $\hat{\psi} = 34°$, that is, a large shear component exists near the crack tip. Now change the crack size to $L = 1$ cm, and everything else being the same, the new phase angle becomes $\hat{\psi} = 40°$.

12.3 Crack Tip Plasticity

Plastic flow around the tip of an interface crack has been analyzed by Shih and Asaro (1988), Shih et al. (1991), and Zywicz and Parks (1989, 1992). Plasticity aspects of interface cracks are reviewed by Shih (1991). Consider a stationary interface crack between two materials, at least one of which is plastically deformable (Figure 12.7). Small-scale yielding conditions prevail—that is, the plastic zone size r_p is much smaller than the characteristic specimen dimension (e.g., crack size L for a finite crack in an infinite body). Stress distribution over distances $r \gg r_p$ is approximately determined by elasticity, as if near-tip plasticity were nonexistent. In particular, the stress field in the annulus, $r_p \ll r \ll L$, is given by Williams' singular solution discussed in Section 12.2. The boundary value problem thus consists of two semi-infinite materials bonded over $x_1 > 0$, but unbonded over $x_1 < 0$; Williams' stress distribution is applied as boundary conditions as $r \to \infty$, with a complex stress intensity factor as follows:

$$K = |K|L^{-i\varepsilon}\exp(i\psi) \quad (12.31)$$

Here L is the crack size, and ψ the load angle in the elasticity problem of finite crack. The elastic-plastic response is characterized by J_2 flow theory.

12.3.1 Plastic Zone Size

The problem contains two length scales, L and $(|K|/\sigma_Y)^2$, σ_Y being the lower yield stress of the two materials. Elementary considerations suggest that r_p scales with $(|K|/\sigma_Y)^2$, providing a natural length to define mode mixity. Define a dimensionless number by

$$\xi = \psi + \varepsilon \ln\left(\frac{|K|^2}{\sigma_Y^2 L}\right) \quad (12.32)$$

According to the interpretation in Section 12.2.1, $\tan \xi$ broadly represents the traction ratio σ_{xy}/σ_{yy} near $r = (|K|/\sigma_Y)^2$, or just outside of the plastic zone.

The plastic zone size is given by

$$r_p = \mathcal{R}(|K|/\sigma_Y)^2 \quad (12.33)$$

The dimensionless factor $\mathcal{R}$ depends weakly on material constants, but is sensitive to mode mixity, ranging from

0.15 to about 0.65 as $|\xi|$ increases from 0 to $\pi/2$. Furthermore, the shape of the plastic zone depends on the sign of ξ, which may lead to different debond resistance for loading with opposite shear directions. The above results are rigorously correct for deformation plasticity; numerical calculations have shown that they are quite accurate for flow theory.

12.3.2 Stress Distribution Around a Blunted Crack Tip

Next consider the stress distribution *within* the plastic zone. The region of interest is bounded by the plastic zone size r_p, and the crack tip opening displacement δ_t. The latter is given by

$$\delta_t = \mathcal{D} J/\sigma_Y \qquad (12.34)$$

The prefactor, $\mathcal{D}$, ranges from 0.5 to 0.7 for $|\xi| \leq \pi/6$ when the metal has low strain hardening ($N \leq 0.1$). The two lengths, r_p and δ_t, differ by a factor comparable to the yield strain.

The traction ratio, σ_{xy}/σ_{yy}, ahead of the blunted crack tip, within $\delta_t < r < r_p$, is shown in Figure 12.8. Note that $\tan \xi \approx \sigma_{xy}/\sigma_{yy}$ near $r = (|K|/\sigma_Y)^2$. For the opening mode $\xi \approx 0$, the traction ratio remains small over the range of distances shown. Under mixed mode loading $\xi \neq 0$, moderate variation in the traction ratio is observed. The trends displayed in Figure 12.8 are representative of metal/ceramic interfaces. Figure 12.9 shows the distribution of the hoop stress ahead of the crack tip. Focus on the curve for $\xi = 0$. The blunted crack tip relieves the constraint, leading to a low stress

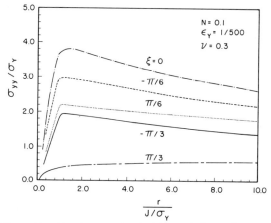

Figure 12.9. Tensile stress ahead of the crack tip, on the interface. Note that $\delta_t \sim J/\sigma_Y$. Metal/ceramic bimaterial; metal properties are $N = 0.1$, $\varepsilon_Y = 0.003$, and $\nu = 0.3$.

within $r < J/\sigma_Y$. The hoop stress reaches a maximum at distance $r \approx J/\sigma_Y$. The stiffer substrate provide additional constraint to plastic flow so that the stress for $\xi = 0$ is about 10% higher than the level for the corresponding homogeneous material (Shih et al. 1991). The constraint is partially relieved when the loading contains a large shear component.

In recent experiments with niobium diffusion bonded to alumina, O'Dowd et al. (1992b) found that debond resistance varies significantly with mode mixity; for example, $\Gamma(40°)/\Gamma(0°) \approx 10$. An attempt was made to correlate mixed mode debond resistance on the basis of a cleavage stress at a characteristic distance (for example, distance between triple point junctions in Al_2O_3), as an extension of the early work of Ritchie et al. (1973) on mode I fracture in mild steels.

The evolution of cyclic near-tip fields ahead of a stationary interface crack has been investigated by Woeltjen et al. (1992). Under monotonic loading to peak tensile load an essentially mode I near-tip field is observed over the major portion of the plastic zone, similar to the result in Figure 12.8 for $\xi = 0$. However, a mixed-mode field is generated near the tip upon removal of the tensile load. The development of strong shear tractions ahead of the interface crack tip has important implications for fatigue fracture mechanisms and fatigue life.

12.3.3 Constrained Plasticity

Reimanis et al. (1991) have carried out fracture experiments with gold foils that were diffusion bonded between sapphire plates. The foil thickness, h, is much smaller than the overall dimension of the specimen. The

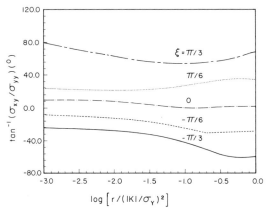

Figure 12.8. The ratio of shear over tension ahead of the crack tip, on the interface, in the range $\delta_t \leq r \leq r_p$. Metal/ceramic bimaterial; metal properties are $N = 0.1$, $\varepsilon_Y = 0.003$, and $\nu = 0.3$.

plastic zone is comparable to (or even larger than) h, but the total inelastic zone size is small compared to specimen dimensions. Therefore, the remote load can be prescribed by a stress intensity factor. Upon loading, partial debond develops at a distance several times the foil thickness ahead of the crack tip. These micro-debonds do not connect with the crack tip. With furthur loading, new debonds nucleate at a even larger distance ahead of the crack tip, as shown in Figure 12.11(d). The intact metal ligaments bridge the crack, leading to a rapidly rising resistance curve (R—curve). Here we focus on the initiation of the micro-debond, the precursor to bridging.

The above phenomenon is an extreme form of large-scale yielding (relative to the foil thickness h), in which the metal foil is highly constrained by the sapphire plates. A finite element analysis by Varias et al. (1991, 1992) reveals that the hydrostatic stress in the metal foil increases steadily as the applied load increases; this is in contrast to the stress distribution ahead of an interface crack between two substrates that cannot elevate above three to four times the yield stress. The behavior of the mean stress in the metal foil is shown in Figure 12.10. Near the tip ($r/h \ll 1$), the stress distribution is not affected by the constraint of the foil thickness, so the mean stress is about three times the yield stress, similar to the distribution in Figure 12.9. At a distance several times the foil thickness, the mean stress reaches the maximum, which increases with applied load; the location of the maxima shifts ahead as the load increases. These elevated stress maxima are responsible for micro-debonds.

12.4 Growing Cracks and Debond Resistance

Debonding rearranges the atoms that form the interface into two free surfaces, consuming the Griffith energy

$$\Gamma_G = \gamma_1 + \gamma_2 - \gamma_{int} \qquad (12.35)$$

Here γ_1 and γ_2 are the surface energies of material 1 and 2, respectively, and γ_{int} is the interface energy. The Griffith energy is small, since only a few layers of atoms participate in irreversible movements. To increase debond resistance, more atoms must be brought into the inelastic process, through mechanisms activated by stress lower than that required for atomic separation. For example, $\Gamma_G < 10$ J/m^2 for Al$_2$O$_3$, but even single crystal Al$_2$O$_3$ has fracture resistance exceeding 30 J/m^2. Some heat-dissipating, atomic-scale snapping processes might exist, involving atoms off the crack plane (K.-S. Kim, private communication). The fracture resistance of polycrystalline Al$_2$O$_3$ is further increased by grain-scale dissipating mechanisms such as pullout against friction (Vekinis et al. 1990).

Studies on crack growth resistance in metals were initiated by McClintock and Irwin (1965). They used small-scale yielding solutions for growing cracks in mode III, together with a growth criterion based on the attainment of a critical strain at a characteristic distance ahead of the tip. Later developments along this line were given by Drugan et al. (1982), and have been extended by Drugan (1991) and Ponte Castañeda and Mataga (1992) to cracks growing along bimaterial

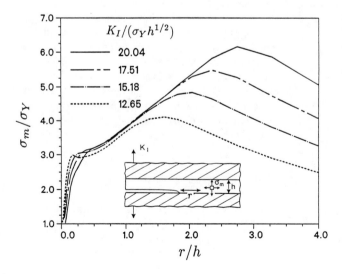

Figure 12.10. Inset: a metal foil bonded between two ceramic substrates, subjected to a remote Mode I stress intensity factor. The mean stress distribution ahead of the crack tip is plotted for several loading levels.

interfaces. Mechanism-based models to relate debond resistance to micromechanisms have been developed recently, and provide a focus for the subsequent presentation. We will limit our attention to predominantly opening mode of fracture.

12.4.1 Crack-Bridging

A solid will fall apart unless something holds it together. A far reaching, unifying idea, sufficiently rigorous for our purpose, is to represent "binding" by a relation between attractive stress, σ, and separation, δ. Such a relation is sketched in Figure 12.5, and is written as

$$\sigma/\sigma_B = \chi(\delta/\delta_B) \quad (12.36)$$

The dimensionless function χ describes the shape of the relation and the scale is set by σ_B and δ_B. The energy required to separate unit area of surfaces so bridged is

$$\Gamma_B = \sigma_B \delta_B \int_0^\infty \chi(\kappa) d\kappa \quad (12.37)$$

The dimensionless integral is of order unity. In practice, the shape function χ is difficult to determine precisely, but the quantities σ_B and δ_B are readily related to microstructural variables (Evans 1990). One can therefore estimate fracture resistance by

$$\Gamma_B \approx \sigma_B \delta_B \quad (12.38)$$

Sketched in Figure 12.11 are several bridging mechanisms, and Table 12.1 lists the representative values of σ_B, δ_B and Γ_B for these mechanisms. Atomic bond has high strength but small debond separation, resulting in a small fracture resistance. Ductile, crack-bridging ligaments give rise to a substantially higher fracture resistance; these ligaments are believed to operate in polycrystalline steels at lower shelf (Hoagland et al. 1972), and in a ceramic matrix containing metallic particles. In the latter, δ_B scales with the diameter of the particles. It remains unclear for polycrystalline steels whether δ_B is set by grain size or some other microstructural lengths. Holes can nucleate in ductile alloys around hard inclusions, or on metal/ceramic interface around pores or

Table 12.1. Illustrative Properties for Bridging Mechanisms

	$\sigma_B(N/m^2)$	$\delta_B(m)$	$\Gamma_B \approx \sigma_B \delta_B (J/m^2)$
Atomic bond	10^{10}	10^{-10}	1
Ductile ligament	10^8	10^{-5}	10^3
Hole growth	10^9	10^{-4}	10^5
Metal foil	10^7	10^{-5}	10^2

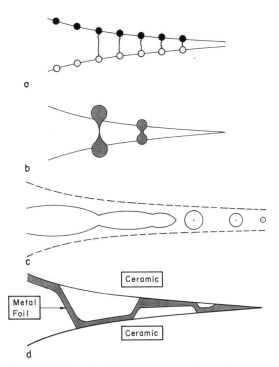

Figure 12.11. Crack-bridging mechanisms: (a) atomic adhesion, (b) ductile ligaments, (c) hole growth, and (d) alternating debonding.

triple-point junctions. Hole growth usually dissipates large amounts of energy. Thin metallic foils sandwiched between two ceramic substrates may debond along alternating interfaces, either because of periodic weak spots fabricated during bonding (Oh et al. 1988), or because of constrained plastic flow (Reimanis et al. 1991; Varias et al. 1991, 1992). More detailed review on crack-bridging concepts can be found elsewhere (Bao and Suo 1992; Suo et al. 1993).

12.4.2 Crack-Bridging and Background Plasticity

It is assumed in the previous section that inelastic deformation can be fully represented by a bridging law while the background material is elastic. In practice, several inelastic mechanisms can operate simultaneously. An interesting example involves a ceramic matrix containing both ductile and transforming particles. The ductile particles form bridges, while the transforming particles contribute to background dissipation. Bridging increases the height of the wake, transforming more particles and thereby dissipating more energy; transformation shields the bridging zone. Thus, the synergism (Amazigo and Budiansky 1988).

Consider the deformation history that a material at distance y off the interface experiences as the crack tip passes by. The energy density variation for the entire process is

$$W(y) = \int_0^\varepsilon \sigma_{ij} d\varepsilon_{ij} \qquad (12.39)$$

The integral is carried over the entire history, including the heat dissipation when the particle is in the active plastic zone, and the residual stress energy when the particle is in the wake. Let H_1 and H_2 be the depths of the inelastic layers in the two materials. The total energy expended in the background for the steady-state crack to move unit distance is

$$\Gamma_P = \int_{-H_2}^{H_1} W(y) dy \qquad (12.40)$$

The total fracture resistance, which includes Griffith energy, bridging energy, and stress work in the background, is given by

$$\Gamma = \Gamma_G + \Gamma_B + \Gamma_P \qquad (12.41)$$

Because Γ_P and Γ_B are typically much larger than Γ_G, it is sometimes assumed that Γ_G is an irrelevant parameter for fracture involving substantial plasticity. However, several authors have pointed out that if cleavage is the basic fracture mechanism, Γ_P or Γ_B must, in some way, depend on Γ_G — that is, the small quantity Γ_G serves as a "valve" for large dissipation Γ_B and Γ_P (Jokl et al. 1980). For example, in transformation-toughened ceramics, the matrix toughness sets the extent of the transformation zone and thereby Γ_P (McMeeking and Evans 1982; Budiansky et al. 1983).

A more familiar example is ductile fracture of alloys, where the near-tip mechanism of hole growth and coalescence serves as the valve for larger-scale plastic dissipation. This process has been analyzed by Needleman (1987, 1990), Varias et al. (1990), and Tvergaard and Hutchinson (1992). Consider a precut remotely loaded by a monotonically increasing $\mathcal{G}$. When $\mathcal{G} < \Gamma_B$, the bridging develops ahead of the crack tip, as does the plastic zone, while the crack remains stationary. The crack begins to grow or, rather, the bridges start to break when $\mathcal{G} = \Gamma_B$. In this sense, background plasticity does *not* provide any shielding prior to crack growth. This can be readily understood by the J–integral, and by the fact that plastic flow is proportional prior to crack growth. A reference length is defined by

$$R_B = \frac{1}{3\pi} \frac{\Gamma_B}{\sigma_Y \varepsilon_Y} \qquad (12.42)$$

This reference length scales with the extent of the plastic zone size when $\mathcal{G} = \Gamma_B$.

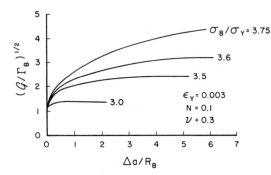

Figure 12.12. Fracture resistance curve resulting from background plasticity shielding. *(From Tvergaard and Hutchinson 1992.)*

As the crack grows, the bridging zone translates in the material: old bridges are broken in the wake, and new bridges are formed in the front. The background material also experiences elastic unloading and possibly reverse plastic loading. The complicated deformation shields the crack. The shielding ratio, $\mathcal{G}/\Gamma_B$, increases with the crack increment Δa, as shown in Fig 12.12. It is evident that the steady-state is established when the crack growth is greater than several times R_B. The steady-state fracture resistance, Γ_{SS}, depends on σ_B/σ_Y. The trend can be better seen in Figure 12.13. For a nonhardening material ($N = 0$), no contribution is derived from the background plasticity if $\sigma_B/\sigma_Y < 2$; conversely, the crack is "lock up," or has infinite fracture resistance when $\sigma_B/\sigma_Y \geq 3$. Similar trends are observed for strain-hardening materials.

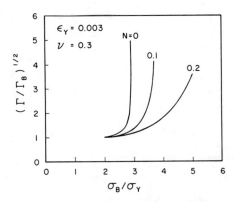

Figure 12.13. Steady-state shielding ratio as a function of bridging strength relative to yield strength. *(From Tvergaard and Hutchinson 1992.)*

12.4.3 Brittle Debonding in the Presence of Plastic Flow

It is known that a sharp, cleaving crack can propagate, slowly or dynamically, surrounded by substantial dislocation motion. For example, a sharp crack can grow slowly by cleavage along a gold/sapphire interface even though the gold deforms plastically; the measured fracture energy is much larger than Γ_G (Reimanis et al. 1991). Similar behavior is observed in copper/glass (Oh et al. 1987), copper/sapphire (Beltz and Wang 1992), niobium/alumina (O'Dowd et al. 1992b), and copper bicrystals contaminated by bismuth (Wang and Anderson 1991). This phenomenon cannot be explained by the models discussed in the previous sections. Atomic cohesive strength, σ_B, is known to be orders of magnitude higher than macroscopic yield strength, σ_Y. When σ_B/σ_Y exceeds about 4, crack-bridging models within the framework of continuum plasticity predict that the crack blunts, limiting the near-tip stress to several times σ_Y (Figure 12.9). Consequently, cleavage cannot proceed from the crack tip. Instead, one has to appeal to other fracture mechanisms, such as hole growth (Rice and Johnson 1970) and cleavage from a remote defect (Ritchie et al. 1973), both leading to rough fracture surfaces not observed in experiments cited in the previous paragraph.

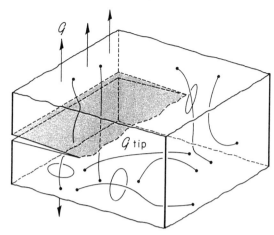

Figure 12.14. A decohesion front in a network of pre-existing dislocations. The diameter of the decohesion core is about 1 nm; the average dislocation spacing is more than 100 nm.

Figure 12.14 conveys the essentials of a theory proposed by Suo et al. (1993). The fundamental process for plastic flow is discrete, consisting of at least two length scales: the Burgers vector $b \sim 10^{-10}$ m, and dislocation spacing $D \sim 10^{-6}$ m. On one scale, atoms exhibit individuality ultimately governed by quantum mechanics. On the other scale, dislocations interact through continuum elasticity. Continuum plasticity applies when stress variation over a multiple of D is small compared to the macroscopic yield strength. The discreteness becomes important for events occurring between lengths b and D.

The theory is based on a single premise: the crack front does not emit dislocations. This happens, for example, for cleavable materials such as steel and silicon below the ductile-brittle transition temperature, or contaminated grain boundaries, or interfaces subjected to environmental degradation, or interfaces with a few atomic layers of brittle reaction compounds. As illustrated in Figure 12.14, so long as dislocation spacing D is much larger than the lattice constant, the probability for a pre-existing dislocation to blunt a major portion of the crack front should be extremely small. Consequently, a crack that does not emit dislocation will remain nanoscopically sharp, advancing by atomic decohesion. Within the cell, essentially free of dislocations that surrounds the crack front, the crystal is linearly elastic down to a nanometer. Near the crack tip, nonlinearity arises from partial atomic separation and nanoscopic shear bands. The size of the elastic cell, represented by D, is several orders of magnitude larger than the nonlinear zone size. Consequently, information regarding the nanoscopic nonlinearity is transmitted—to an observer outside the elastic cell—through a single quantity: the Griffith energy Γ_G. The elastic cell provides a medium through which the stress decays rapidly, matching the high atomic debond stress on one side, and the low macroscopic yield stress on the other. For example, with $b = 10^{-10}$ m and $D = 10^{-6}$ m, the stress decays approximately by a factor $\sqrt{D/b} = 100$ over a distance of 1 μm. The dislocation motion at the characteristic distance D away from the crack tip dissipates plastic energy, Γ_P, which is typically much larger than Γ_G. In summary, atoms around a crack front can be divided into three regions: nanoscopic decohesion zone, microscopic elastic cell, and macroscopic dislocation dissipative background.

The elastic cell is a nanomechanics concept with imprecise, if any, continuum description. The concept can be approximately understood in terms of spatially varying yield strengths. Sketched in Figure 12.15 is yield strength varying with the distance from a representative atom at the crack tip. The theoretical shear strength is approached near the crack tip; the strength decays to the macroscopic yield strength in the background. The shape of the decay function has not been investigated; dislocation cell models may provide some

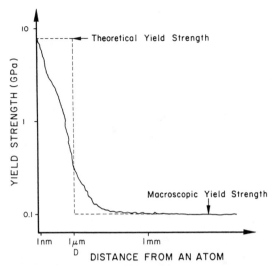

Figure 12.15. Yield strength as a function of the distance from an atom at the center of an elastic cell.

insight (Lubarda et al. 1993; Kubin et al. 1992). Nevertheless, the decay function *must* have a characteristic length comparable to the dislocation spacing D.

Consider a cleavable, rate-independent material with Griffith energy Γ_G, yield strength σ_Y and yield strain $\epsilon_Y = \sigma_Y/E$, E being Young's Modulus. The crack tip energy release rate, $\mathcal{G}_{tip}$, is shielded by background dislocation motion from the remotely applied energy release rate, $\mathcal{G}$. Dimensional analysis dictates that

$$\mathcal{G}/\mathcal{G}_{tip} = g(D\epsilon_Y\sigma_Y/\mathcal{G}_{tip}) \qquad (12.43)$$

The shielding ratio g also depends on crack increment and material constants such as ϵ_Y, Poisson's ratio ν and in particular, the shape of the decay function in Figure 12.15. For properties representative of metals (e.g. $D \sim 1$ μm, $\epsilon_Y \sim 10^{-3}$, $\sigma_Y \sim 10^8$ N/m^2, $\Gamma_G \sim 1$ J/m^2), the parameter $D\epsilon_Y\sigma_Y/\Gamma_G$ ranges from 10^{-2} to 10. The parameter can be understood in several ways; e.g., all else being fixed, an increase in elastic cell size D reduces the total energy dissipation. Under steady-state growth, $\mathcal{G}_{tip} = \Gamma_G$ and $\mathcal{G}$ equals the measured fracture energy Γ. The plastic dissipation Γ_P is given by $\Gamma = \Gamma_P + \Gamma_G$.

In the present theory, it is assumed that no low strength, long range bridges, such as tearing caused by cleavage plane reorientation between neighboring grains, operate in the crack wake. These bridges are responsible for the large "cleavage energy" reported for polycrystalline steels. When operating, the bridges may serve as a bigger valve than atomic decohesion. If this is the case, a bridging law may be used in the present model. Indeed, when $\sigma_B/\sigma_Y < 4$, the present model should reduce to a regular bridging model without an elastic cell.

Further simplifications are needed to make quantitative predictions (Figure 12.16). The decohesion zone is small compared to D so that the square root singular elasticity solution prevails in $b \ll r \ll D$. Detailed atomistic description of decohesion is unnecessary except for a prescription of a cleavage energy Γ_G. The shape of the elastic cell is unimportant because the plastic zone height is typically much larger than D; we use a strip to represent the elastic cell. A disc translating with the crack tip can be another convenient choice, but the difference is expected to be minor in so far as $\mathcal{G}/\mathcal{G}_{tip}$ is concerned. The background dislocation motion is represented by continuum plasticity. A refinement, if needed, may include individual dislocations or a dislocation network in the transition region between the elastic cell and the continuum plastic flow.

The crack starts to grow when $\mathcal{G} \geq \Gamma_G$; more load is required to maintain the growth, leading to a resistance curve. The plastic zone also increases as the crack grows, attaining a steady-state height H. The energy release rate reaches a steady-state value Γ_{SS}. The model geometry is analyzed in the steady-state using finite elements. Figure 12.17 shows that the shielding ratio increases rapidly as D or σ_Y decrease. The influence of strain hardening exponent, N, can also be seen. For nonhardening metals, the plastic dissipation completely shields the crack tip at a finite $D \epsilon_Y\sigma_Y/\mathcal{G}_{tip}$. In practice, D may be used as a fitting parameter to correlate experimental data. For example, a metal with $\sigma_Y = 10^8$ N/m^2, $\epsilon_Y = 3.3 \times 10^{-3}$ and $\Gamma_G = 2$ J/m^2 gives $\Gamma_G/\sigma_Y\epsilon_Y = 6$ μm. If the measured fracture energy $\Gamma_{SS} = 20$ J/m^2, one finds from Figure 12.17 that $D \approx 0.1$ μm.

In an experiment with a single crystal of copper diffusion bonded to a sapphire disc (Beltz and Wang

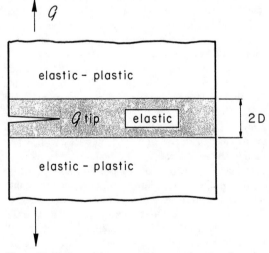

Figure 12.16. A model system with a step-function decay in yield strength.

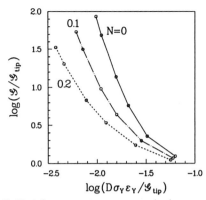

Figure 12.17. A fracture resistance curve: the fracture energy increases as the crack grows. Computed shielding ratio as a function of various parameters: N—hardening exponent, D—elastic cell size.

1992), interface debonding was driven in two crystallographic directions at slightly different energy release rates. The phenomenon was interpreted according to the Rice-Thomson model (1974): dislocations emit from the crack tip in one direction but not from the other. An alternative interpretation appears to be possible on the basis of the present theory: both crack tips do not emit dislocations and the different debond energies result from the different extent of background dislocation motion. Indeed, the micrographs show much denser slip lines in one case than they do in the other. Experiments at higher magnifications are needed to ascertain which of the two interpretations is appropriate for the copper/sapphire system. Calculations within the framework of the present theory, taking into account single crystal plasticity, are in progress to facilitate a direct comparison with such experiments.

Obviously, the competition between atomic decohesion and dislocation emission (Rice et al., 1992) cannot be addressed by the present theory. Instead, the consequences of the premise that dislocations do not emit from the crack front can be explored. Included in Suo et al. (1993) are slow cleavage cracking, stress-assisted corrosion, fast-running crack, fatigue cracking, constraint effects, and mixed mode fracture along metal/ceramic interfaces.

Acknowledgements

The work of Z.S. was supported by an NSF Young Investigator Award, by DARPA/URI Contract N00014-86-K-0753, and by a Visiting Associate Professor appointment at Brown University funded by NRC/ONR Grant N00014-90-J1380. The work of C.F.S. was supported by NRC/ONR Grant N00014-90-J13800, and by the Materials Research Group funded by NSF through Grant DMR-9002994.

References

Amazigo, J.C., and B. Budiansky. 1988. *J. Mech. Phys. Solids.* 36:581–595.
Aveston, J., Cooper, G.A., and A. Kelly. 1971. In: The Properties of Fiber Composites. Conference Proceedings, National Physical Laboratory (Guildford UK). 15–24, Teddington, UK: IPC Science and Technology Press, Ltd.
Argon, A.S., Gupta, V., Landis, H.S., and J.A. Cornie. 1989. *J. Mater. Sci.* 24:1406–1412.
Ashby, M.F., Rühle, M. (eds.). 1992. Proceedings of International Symposium on Metal/Ceramic Interfaces. To be published in *Acta Metall. Mater.*
Bao, G., and Z. Suo. 1992. *Appl. Mech. Rev.* In Press.
Beltz, G.E., and J.-S. Wang. 1992. *Acta Metall. Mater.* 40:1675–1683.
Budiansky, B., Hutchinson, J.W., and J.C. Lambropoulos. 1983. *Int. J. Solids Struct.* 19:337–355.
Cannon, R.M., Dalgleish, B.J., Dauskardt, R.H. et al. 1992. In: Fatigue of Advanced Materials (R.O. Ritchie, R.H. Dauskardt, and B.N. Cox, eds.) Edgbaston, UK: MCEP Publishing Ltd.,
Cao, H.C., and Evans, A.G. 1989. *Mech. Mater.* 7:295–305.
Cottrell, A.H. 1963. In: Tewksbury Symposium on Fracture. 1–27, Melbourne: University of Melbourne.
Davis, J.B., Cao, H.C., Bao, G., and A.G. Evans. 1991. *Acta Metall. Mater.* 39:1019–1024.
Drugan, W.J. 1991. *J. Appl. Mech.* 58:111–119.
Drugan, W.J., Rice, J.R., and T.-L. Sham. 1982. *J. Mech. Phys. Solids.* 30:447–473.
England, A.H., 1965. *J. Appl. Mech.* 32:400–402.
Erdogan, F. 1965. *J. Appl. Mech.* 32:403–410.
Evans, A.G. 1990. *J. Am. Ceram. Soc.* 73:187–206.
Evans, A.G., Drory, M.D., and M.S. Hu. 1988. *J. Mater. Res.* 3:1043–1049.
Evans, A. G., and J.W. Hutchinson. 1989. *Acta Metall. Mater.* 37:909–916.
Evans, A.G., Rühle, M., Dalgleish, B.J., and P.G. Charalambides. 1990. *Mater. Sci. Eng.* A126:53–64.
Farris, R.J., and C.L. Bauer. 1988. *J. Adhesion.* 26:293–300.
Griffith, A.A. 1921. *Phil. Trans. Roy. Soc. Lond.* A221:163–197.
Hoagland, R.G., Rosenfield, A.R., and G.T. Hahn. 1972. *Metall. Trans.* 3:123–136.
Hutchinson, J.W. 1979. Nonlinear Fracture Mechanics. Department of Solid Mechanics, Technical University of Denmark.
Hutchinson, J.W., and H.M. Jensen. 1990. *Mech. Mater.* 9:139–163.
Hutchinson, J.W., and Z. Suo. 1992. *Adv. Appl. Mech.* 29:63–191.
Irwin, G.R. 1960. In: Structural Mechanics. (J.N. Goodier and N.J. Hoff, eds.), 557–591, Oxford: Pergamon Press.
Jokl, M.L., Vitek, V., and C.J. McMahon, Jr. 1980. *Acta Metall.* 28:1479–1488.

Kanninen, M.F., and C.H. Popelar. 1985. Advanced Fracture Mechanics. Oxford: Oxford University Press.

Kim, K.-S. 1991. *Mat Res Soc Symp Proc.* 203:3–14.

Kubin, L.P., Canova, G., Condat, M. et al. 1992. In: Nonlinear Phenomena in Materials Science II (G. Martin and L. P. Kubin, eds.). In press.

Liechti, K.M., and Y.-S. Chai. 1992. *J. Appl. Mech.* 59:295–304.

Lubarda, V.A., Blume J.A., and A. Needleman. 1992. *Acta Metall. Mater.* 41:625–642.

Malyshev, B.M., and R.L. Salganik. 1965. *Int. J. Fract. Mech.* 5:114–128.

McClintock, F.A., and G.R. Irwin. 1965. ASTM-STP 381, 84–113, Philadelphia: ASTM.

McMeeking, R.M., and A.G. Evans. 1982. *J. Am. Ceram. Soc.* 65:242–246.

Moran, B., and C.F. Shih. 1987. *Eng. Fract. Mech.* 27:615–642.

Needleman, A. 1987. *J. Appl. Mech.* 54:525–531.

Needleman, A. 1990. *J. Mech. Phys. Solids.* 38:289–324.

O'Dowd, N.P., Shih, C.F., and M.G. Stout. 1992a. *Int. J. Solids Struct.* 29:571–589.

O'Dowd, N.P., Stout M.G., and C.F. Shih. 1992b. *Phil. Mag.* 66A:1037–1064.

Oh, T.S., Cannon, R.M., and R.O. Ritchie. 1987. *J. Am. Ceram. Soc.* 70:C352–C355.

Oh, T.S., Cannon, R.M., and R.O. Ritchie. 1988. *Acta Metall.* 36:2083–2093.

Ponte Castañeda, P., and P.A. Mataga. 1992. Submitted for publication.

Reimanis, I.E., Dalgleish, B.J., and A.G. Evans. 1991. *Acta Metall. Mater.* 39:3133–3141.

Rice, J.R. 1968. *J. Appl. Mech.* 35:379–386.

Rice, J.R. 1988. *J. Appl. Mech.* 55:98–103.

Rice, J.R., Beltz, G.E., and Y. Sun. 1992. In: Topics in Fracture and Fatigue. (A.S. Argon, ed.), 1–58, Berlin: Springer-Verlag.

Rice J.R., and M.A. Johnson. 1970. In: Inelastic Behavior of Solids. (M.F. Kanninen, W.F. Adler, A.R. Rosenfield, and R.I. Jaffee, eds.), 641–672, New York: McGraw-Hill.

Rice, J.R., and G.C. Sih. 1965. *J. Appl. Mech.* 32:418–423.

Rice, J.R., Suo, Z., and J.S. Wang. 1990. In: Metal-Ceramic Interfaces, Acta-Scripta Metallurgica Proceedings Series (M. Rühle, A.G. Evans, M.F. Ashby, and J.P. Hirth, eds.), Vol. 4, 269–294, Oxford: Pergamon Press.

Rice, J.R., and R.M. Thomson. 1974. *Phil. Mag.* 29:73–97.

Rice, J.R., and J.S. Wang. 1989. *Mater. Sci. Eng.* A107:23–40.

Ritchie, R.O., Knott, J.F., and J.R. Rice. 1973. *J. Mech. Phys. Solids.* 21:395–410.

Rivlin, R.S., and A.G. Thomas. 1953. *J. Polym. Sci.* 10:291–318.

Rühle, M., Evans, A.G., Ashby, M.F., and J.P. Hirth (eds.) 1990. Metal-Ceramic Interfaces, Acta-Scripta Metallurgica Proceedings Series, Vol. 4. Oxford: Pergamon Press.

Shih, C.F. 1991. *Mater. Sci. Eng.* A143:77–90.

Shih, C.F., and R.J. Asaro. 1988. *J. Appl. Mech.* 55:299–316.

Shih, C.F., Asaro, R.J., and N.P. O'Dowd. 1991. *J. Appl. Mech.* 58:450–463.

Stump, D.M., and B. Budiansky. 1989. *Acta Metall. Mater.* 37:3297–3304.

Suo, Z. 1990. *Proc. R. Soc. Lond.* A427:331–358.

Suo, Z., Shih, C.F., and A.G. Varias. 1993. A Theory for Cleavage Cracking in the Presence of Plastic Flow. *Acta Metall. Mater.* 41:1551–1557.

Suresh, S., and A. Needleman, eds. 1989. Interfacial Phenomena in Composites: Processing, Characterization and Mechanical Properties. London: Elsevier Applied Science.

Thouless, M.D. 1990. *Acta Metall.* 38:1135–1140.

Tvergaard, V., and J.W. Hutchinson. 1992. *J. Mech. Phys. Solids.* 40:1377–1397.

Varias, A.G., O'Dowd, N.P., Asaro, R.J., and C.F. Shih. 1990. *Mater. Sci. Eng.* A126:65–93.

Varias, A.G., Suo, Z., and C.F. Shih. 1991. *J. Mech. Phys. Solids.* 39:963–986.

Varias, A.G., Suo, Z., and C.F. Shih. 1992. *J. Mech. Phys. Solids.* 40:485–509.

Vekinis, G., Ashby, M.F., and P.W.R. Beaumont. 1990. *Acta Metall. Mater.* 38:1151–1162.

Wang, J.S., and P.M. Anderson. 1991. *Acta Metall. Mater.* 39:779–792.

Wang, J.S., and Suo, Z. 1990. *Acta Metall. Mater.* 38:1279–1290.

Williams, M.L. 1959. *Bull. Seismol. Soc. Am.* 49:199–204.

Woeltjen, C., Shih, C.F., and S. Suresh. 1993. "Near-Tip Fields for Fatigue Cracks along Metal–Metal and Metal–Ceramic Interfaces." *Acta Metall. Mater.* 41:2317–2335.

Zywicz, E., and D.M. Parks. 1989. *J. Appl. Mech.* 56:577–584.

Zywicz, E., and D.M. Parks. 1992. *J. Mech. Phys. Solids.* 40:511–536.

Chapter 13
Matrix, Reinforcement, and Interfacial Failure

ALAN NEEDLEMAN
STEVE R. NUTT
SUBRA SURESH
VIGGO TVERGAARD

The potential for widespread use of metal-matrix composites in high-performance structural applications is somewhat restricted by their poor ductility and low fracture toughness. Minimizing these limitations through microstructural design calls for a thorough understanding of the micromechanisms of failure processes. This chapter addresses the mechanics and micromechanisms of various failure mechanisms influencing the mechanical response of metal-matrix composites, with attention focused primarily on monotonic tensile deformation and on discontinuously reinforced metal-ceramic systems. However, whenever appropriate, reference is made to cyclic loading and crack growth resistance and to continuously reinforced metal-matrix composites. Further details on the micromechanisms of failure during quasistatic fracture and fatigue can be found in Chapters 14 and 15, respectively. The focus here is on room-temperature failure mechanisms. High-temperature behavior is discussed in Chapter 11.

13.1 Background

The basic failure mechanisms influencing the constitutive response and fracture resistance of metal-matrix composites can be broadly classified into three groups:

1. Ductile failure by the nucleation, growth, and coalescence of voids in the matrix.
2. Brittle failure of the reinforcement.
3. Debonding and fracture along the interface between the matrix and the reinforcement.

The degree to which these failure mechanisms individually and collectively influence the overall deformation and fracture resistance is strongly dictated by factors as diverse as

1. The size, shape, concentration, and spatial distribution of the reinforcement.
2. The concentration of impurities present in the constituent phases of the composites.
3. The processing and heat treatment procedures, including any aging treatments, to which the composite is subjected prior to mechanical loading.
4. The thermal and chemical environment in which mechanical properties are evaluated.
5. The coatings, if any, applied to the reinforcement with the specific objective of modifying the interfacial characteristics.

Descriptions are given of each of the failure processes commonly encountered in discontinuously reinforced metal-matrix composites. Section 13.2 focuses on failure of the matrix material. Here, experimental results for the effects of ductile matrix failure by void growth on the monotonic and cyclic deformation characteristics of the composite are examined. The experimental results are interpreted using finite element analyses, which employ a continuum model for a progressively cavitating matrix. Section 13.3 addresses the effects of reinforcement fracture on the overall elastic and plastic response of the composite. Experimental observations of particle fracture are described for several different composite systems and are compared with the predictions of analytic and numerical models. The dependence of particle fracture on the size, shape, and concentration of particles is also addressed. The issue of debonding and fracture along the interface is discussed in Section 13.4, which begins with a review of constitutive models for interfacial delamination and review of microstructural size scale issues. Experimental observations of imperfections and voids at interfaces along with the numerical modeling of this failure process are considered. A brief discussion of the role of interfacial frictional contact on deformation is also provided. The chapter concludes with a summary of the status of the field and recommendations for future investigations.

13.2 Matrix Failure

Ductile failure by the nucleation, growth, and coalescence of voids within the matrix is known to be a dominant failure process in many discontinuously reinforced aluminum matrix composites, such as 2024 and 2124 aluminum alloy matrices reinforced with Al_2O_3 and SiC particles (Davidson 1987; You et al. 1987; Christman and Suresh 1988a; Christman et al. 1989; Llorca et al. 1991). Aluminum alloys generally contain three types of particles: (1) constituent intermetallic particles (of diameter typically of the order of a μm or greater) that form as a result of the presence of impurities such as Fe, (2) dispersoids (of diameter in the range 0.1–3 μm), and (3) strengthening precipitates (of diameter in the range 100 nm to a fraction of 1 μm). The constituent particles with the largest size provide the primary nucleation sites for the formation of cavities, although the growth of such large cavities can be facilitated by voids formed around the smaller particles. For example, studies of 2xxx series aluminum alloys have shown that voids nucleate around intermetallic inclusions (3–10 μm in diameter), and that the coalescence of such voids is aided by the linkage of cavities nucleated around dispersoid particles (0.5–2.0 μm in diameter) (Tanaka et al. 1970; Van Stone et al. 1974). It has been found that approximately 10% of the larger inclusions fracture at a strain of 0.03, while as many as 45% of the void-nucleating particles crack at a strain of about 0.05. Furthermore, transmission electron microscopy analyses of 2124 aluminum alloys reinforced with SiC whiskers have identified the existence of $Cu_2Mn_3Al_{20}$ and $FeCu_2Al_{17}$ constituents, which are possible void nucleation sites (Christman and Suresh 1988b). Experimental results, discussed below, show clearly how void growth around constituent particles in the ductile matrix affects the monotonic flow stress and ductility. In addition, it is known that the fracture toughness and fatigue response of the composite can be strongly influenced by cavitation within the matrix (Liu et al. 1989; Bonnen et al. 1991; Suresh 1991; Llorca et al. 1992).

13.2.1 Model for a Progressively Cavitating Material

For a ductile material containing a certain volume fraction f of voids, Gurson (1975) has developed an approximate yield condition of the form $\Phi(\sigma, \sigma_M, f) = 0$ based on averaging techniques similar to those used by Bishop and Hill (1951) for polycrystalline aggregates. Here, σ is the average macroscopic Cauchy stress tensor, and σ_M is an equivalent tensile flow stress representing the actual microscopic stress state in the matrix material. Based on upper-bound rigid-plastic analyses for a spherical volume element with a concentric spherical void, Gurson (1975) derived an approximate yield condition, which is of the form

$$\Phi = \frac{\sigma_e^2}{\sigma_M^2} + 2q_1 f^* \cosh\left(\frac{3q_2\sigma_m}{2\sigma_M}\right) - 1 - q_1^2 f^{*2} = 0 \quad (13.1)$$

for $q_1 = q_2 = 1$ and $f^* = f$. Here, $\sigma_e^2 = \sqrt{\frac{3}{2}\sigma' : \sigma'}$ is the macroscopic Mises stress, $\sigma' = \sigma - \sigma_m I$ is the stress deviator, and $\sigma_m = \frac{1}{3}\sigma : I$ is the macroscopic mean stress. The parameters q_1 and q_2 were introduced by Tvergaard (1981, 1982) to bring predictions of the model at low volume fractions in closer agreement with full numerical analyses for periodic arrays of voids. The function $f^*(f)$ was proposed by Tvergaard and Needleman (1984) to model the loss of stress-carrying capacity associated with void coalescence. Modifying Φ by $f^*(f)$ is only relevant for void volume fractions larger than certain critical value f_c, and this function is taken as

$$f^* = \begin{cases} f & f \leq f_c \\ f_c + \frac{f_u^* - f_c}{f_f - f_c}(f - f_c) & f \geq f_c \end{cases} \quad (13.2)$$

where $f_u = 1/q_1$ and $f^*(f_f) = f_u^*$. It is noted that the material loses all stress carrying capacity as $f \to f_f$ (i.e., $f^* \to f_u^*$). Based on experimental and numerical studies, the two failure parameters in (Equation 13.2) have been chosen as $f_c = 0.15$ (or somewhat smaller) and $f_f = 0.25$.

It is noted that alternative yield conditions for porous ductile solids have been developed, mainly for the purpose of studying powder compacted metals, (Fleck et al. 1992). Also, kinematic hardening models for studying the mixed influence of porosity and the formation of a rounded vertex on the yield surface have been introduced. A discussion of such alternative models is given by Tvergaard (1990a).

A rate-sensitive version of the Gurson model (Pan et al. 1983) is obtained by assuming an expression of the form

$$d^p = \Lambda \frac{\partial \Phi}{\partial \sigma} \quad (13.3)$$

for the plastic part of the strain rate, and by using the function of Equation 13.1 as a plastic potential. The matrix plastic strain rate is taken to be given by the power law relation

$$\dot{\epsilon}^p_M = \dot{\epsilon}_0 \left(\frac{\sigma_M}{g(\epsilon^p_M)} \right)^{1/m} \quad (13.4)$$

where m is the strain-rate hardening exponent, and $\dot{\epsilon}_0$ is a reference strain rate. If strain hardening follows a power law with exponent N and initial yield strength $\sigma_0 = E/\epsilon_0$, the function $g(\epsilon^p_M)$ may be expressed as

$$g(\epsilon^p_M) = \sigma_0 (1 + \epsilon^p_M/\epsilon_0)^N \quad (13.5)$$

By setting the macroscopic plastic work rate equal to the matrix dissipation,

$$\sigma : d^p = (1 - f) \sigma_M \dot{\epsilon}^p_M \quad (13.6)$$

the plastic flow proportionality factor, Λ in Equation 13.3, is found to be

$$\Lambda = (1 - f) \sigma_M \dot{\epsilon}^p_M \left[\sigma : \frac{\partial \Phi}{\partial \sigma} \right]^{-1} \quad (13.7)$$

The total strain-rate is taken to be the sum of an elastic part and the plastic part, $d = d^e + d^p$. Then, assuming small elastic strains, the elastic incremental stress-strain relationship is of the form $\hat{\sigma} = \mathcal{R} : d^e$, and the constitutive relations for the viscoplastic material can be written as

$$\hat{\sigma} = \mathcal{R} : (d - d^p) \quad (13.8)$$

Here, $\hat{\sigma}$ is the Jaumann (corotational) rate of the Cauchy stress tensor.

The rate of increase of the void volume fraction is taken to be given by

$$\dot{f} = (1 - f) d^p : I + \mathcal{B}[\dot{\sigma}_M + \dot{\sigma}_m + \mathcal{D} \dot{\epsilon}^p_M] \quad (13.9)$$

where the first term represents the growth of existing voids and the last two terms represent stress-controlled nucleation and strain-controlled nucleation, respectively (Needleman and Rice 1978). Void nucleation is taken to follow a normal distribution (Chu and Needleman 1980). Thus, for strain-controlled nucleation, the coefficients in Equation 13.9 are given by

$$\mathcal{D} = \frac{f_N}{s_N \sqrt{2\pi}} \exp\left[-\frac{1}{2} \left[\frac{\epsilon^p_M - \epsilon_N}{s_N} \right]^2 \right] \quad \mathcal{B} = 0 \quad (13.10)$$

where ϵ_N, s_N and f_N are the mean nucleation strain, the standard deviation, and the volume fraction of void nucleating particles, respectively.

13.2.2 Numerical Results

The basic material behavior represented by the porous ductile material model is seen in a uniformly strained solid, where voids will nucleate and grow until failure occurs by coalescence with neighboring voids, thus showing overall softening behavior in the later stages, even though the matrix material hardens. However, the effects of porosity and ongoing nucleation also significantly promote the onset of plastic flow localization in shear bands, as has been shown in many investigations (Saje et al. 1982, for example). Thus, the model also represents the commonly observed void sheet failure, where coalescence occurs in a thin slice of highly deformed material. These different features of the material model are illustrated by an analysis of the cup cone fracture in the neck of a tensile test specimen (Tvergaard and Needleman 1984), where coalescence in the central part of the neck develops in a rather uniform strain field, while a void sheet forms in the outer parts.

For discontinuously reinforced aluminum matrix composites, an analysis of ductile matrix failure has been carried out by LLorca et al. (1991). Strain-controlled nucleation has been assumed, and based on quantitative studies for the unreinforced aluminum alloy, the nucleation parameters in Equation 13.10 are taken to be $\epsilon_N = 0.05$, $s_N = 0.01$ and $f_N = 0.05$, with no voids initially.

The effect of reinforcement shape on matrix failure has been studied by axisymmetric numerical cell model analyses representing materials with periodic arrays of transversely aligned whiskers or particles. Figure 13.1 shows a comparison of the uniaxial stress-strain curves predicted for particles modeled as cylinders with aspect ratio one for spherical particles, and for whiskers with aspect ratio five. The SiC volume fraction was 12.5% in all three cases. The stress-strain curves predicted by the Gurson model, accounting for matrix failure, are compared with curves for no matrix failure and with the stress-strain curve for the unreinforced aluminum. The strain hardening and the flow strength are maximum for the whiskers and minimum for the spheres, which is a consequence of the geometry-dependence of the hydrostatic stress build-up associated with constrained plastic flow. However, the whiskers also give rise to higher strain concentrations at the sharp edge, leading to earlier void nucleation and rapid void growth in the high hydrostatic tension field.

The distributions of void volume fraction, effective plastic strain, and hydrostatic tension are illustrated in Figure 13.2 at different levels of average axial strain for the three reinforcement geometries, showing clearly the strong effect of sharp edges on whiskers or particles. For the whiskers the load decays abruptly at $\epsilon_{max} \approx 0.013$ (see Figure 13.1), but Figure 13.2a shows that matrix failure is much like a fiber debonding, leading to gradual fiber pull-out with a cone of undamaged matrix material attached to the fiber end. Therefore, based on previous studies of whisker debonding and pull-out (Tvergaard 1990b), it is expected that a lower load plateau on the stress-strain curve of the whisker-reinforced MMC would be reached, if the computation was continued beyond the stage shown in Figure 13.1.

The effect of reinforcement distribution on matrix failure has been studied by LLorca et al. (1991) based on plane strain models of materials with periodic distributions of whiskers. Plane strain models allow for studies of distribution effects such as fiber clustering or nonalignment; but the results are qualitative, because only idealized planar fiber geometries are considered. Figure 13.3 shows contours of void volume fractions, effective plastic strains, and hydrostatic tension, for uniformly distributed plane strain whiskers and for axially clustered whiskers, where the two whiskers are moved in the axial direction inside the same unit cell. The corresponding stress-strain curves are shown in Figure 13.4. It is seen in Figure 13.3(a) that the distribution of failure near the whisker ends predicted by the plane strain model is rather similar to that predicted by the analogous axisymmetric model of Figure 13.2(a). Note also that the corresponding stress-strain curves in Figures 13.1 and 13.4 are rather similar. The axial clustering gives less hardening (Figure 13.4), because of less constraint on

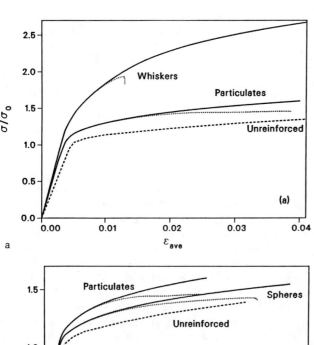

Figure 13.1. Finite element predictions of the effect of reinforcement shape on the stress-strain curve and ductility of 2124 Al-12.5 vol% SiC composites in the peak-aged condition. (a) Whiskers and particles. (b) Particles and spheres. *(From LLorca et al. 1992.)*

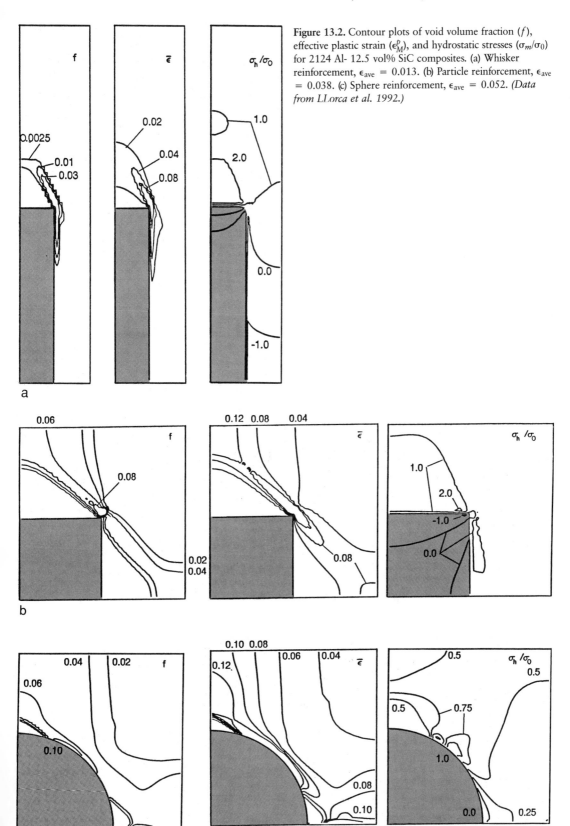

Figure 13.2. Contour plots of void volume fraction (f), effective plastic strain (ϵ_M^p), and hydrostatic stresses (σ_m/σ_0) for 2124 Al- 12.5 vol% SiC composites. (a) Whisker reinforcement, $\epsilon_{ave} = 0.013$. (b) Particle reinforcement, $\epsilon_{ave} = 0.038$. (c) Sphere reinforcement, $\epsilon_{ave} = 0.052$. *(Data from LLorca et al. 1992.)*

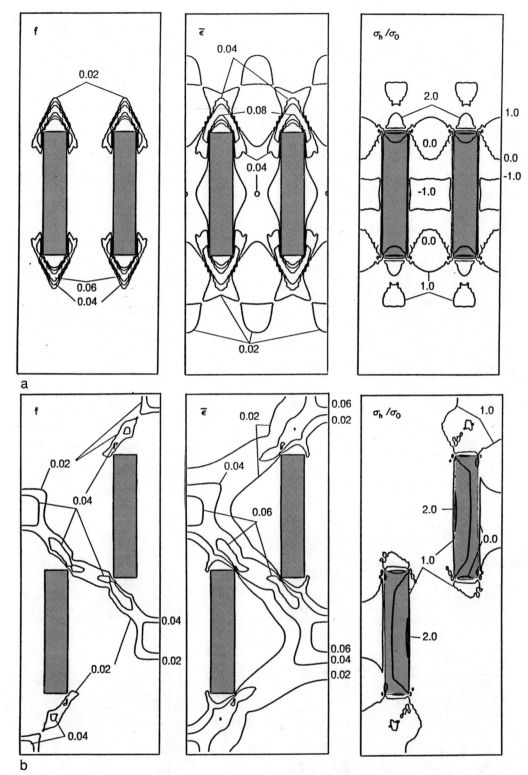

Figure 13.3. Contour plots of void volume fraction (f), effective plastic strain (ϵ_M^p), and hydrostatic stress (σ_m/σ_0) for 2124 Al-12.5 vol% SiC whisker-reinforced composite, $\epsilon_{ave} = 0.02$. (a) Uniform distribution. (b) 50% vertical clustering. *(Data from LLorca et al. 1992.)*

plastic flow, and it is seen in Figure 13.3(b) that here failure develops in a shear band so that matrix fracture by a void-sheet mechanism appears to develop.

In the highly constrained plastic flow that tends to develop in metal/ceramic systems, the rate of void growth may be significantly higher than predicted by the Gurson model, which relies on rigid-plastic studies. In fact, for sufficiently high stress triaxiality in an elastic-plastic solid cavitation instabilities will occur (Huang et al. 1991), which means that a void will grow very large without further overall straining. Such behavior is also relevant to voids growing in the metal at a metal-ceramic interface. For such voids initiating at initially unbonded spots of the interface, the growth under tensile loading normal to the interface has been studied by Tvergaard (1991). Figure 13.5 shows the development of the volume V, the average tensile stress T_a and the average mean stress σ_m, for various area fractions $(R_i/R_o)^2$ of initially hemispherical interface voids in an elastic-perfectly plastic material with $\sigma_y/E = 0.003$. It is seen in Figure 13.5 that for small area fractions of interface cavities, the stress triaxiality grows very large. Also, the growth rate $dV/d\epsilon_a$ increases significantly for decreasing area fraction, and for $R_i/R_o \leq 0.03$, the growth rate reaches infinity at a certain critical value of ϵ_a. It is noted that studies of fiber debonding curves like those shown in Figure 13.5 give a clear indication of the maximum stress levels that can be reached at the interface if failure occurs by a ductile mechanism.

The nucleation and growth of voids to coalescence is an important failure mechanism in structural alloys in general, and the Gurson model has provided qualitative and quantitative agreement with experiments in a num-

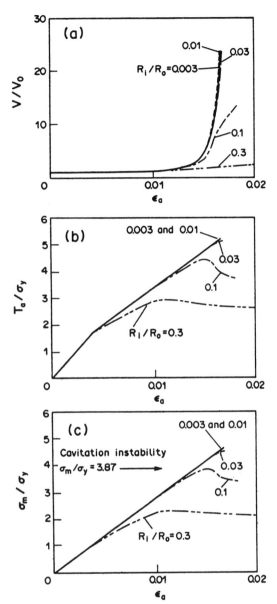

Figure 13.5. Axisymmetric cell model results for void growth along a metal-ceramic interface. R_i is the initial void radius and R_0 is the cell radius. (a) Void volume versus applied strain. (b) Average traction versus applied strain. (c) Mean stress in the material far away from the interface versus applied strain. (Reprinted with permission from Tvergaard 1991 *Acta Metall. Mater.* 39:419–426.)

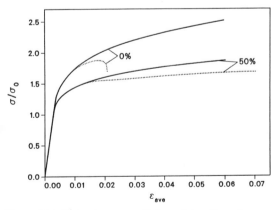

Figure 13.4. Finite element predictions of the effect of vertical clustering of the reinforcement on the stress-strain curve and ductility of 2124 Al reinforced with 12.5 vol% SiC whiskers. The solid lines show the response when the composite matrix is a fully dense Mises material and the dotted lines show composites with a Gurson matrix. *(Data from LLorca et al. 1992.)*

ber of cases (Becker et al. 1988, 1989). However, a limitation of the applicability of porous ductile material models to the study of ductile fracture in metal-matrix composites stems from the size of the matrix void nucleating particles relative to the size of the fibers or particles. In a phenomenological porous plastic constitutive relation the porosity is treated as a continuous variable, which is only appropriate when the matrix voids are significantly smaller than the reinforcement. The matrix void nucleation sites in the whisker reinforced 2124 aluminum alloy matrix in LLorca et al. (1991) are of the order of 0.4 μm, which is comparable to the whisker diameter. Hence, a limitation on the applicability of the porous plastic constitutive relation for this material is due to discrete void effects. In the particle reinforced Al-Cu alloy in LLorca et al. (1991) (3-4 μm mean particle size), matrix void nucleation sites are of a similar size, which is sufficiently small for the assumption of a continuous porosity distribution to be a better approximation.

Numerically predicted stress-strain curves for SiC particle-reinforced Al-Cu alloy composites are plotted in Figure 13.6 along with corresponding experimental measurements. It is seen that the predicted values of elastic modulus, yield strength, and strain hardening exponent closely match the experimental results. Both the numerical and the experimental results are terminated at strain values representative of complete failure. The numerical results also predict a reduction in ductility with increasing volume fraction of the reinforcement, a trend consistent with experimental results. However, the predicted values of strain-to-failure are lower than those found experimentally. One possible reason for the observed differences stems from uncertainty regarding the characterization of void nucleation within the Al matrix. Furthermore, in addition to the occurrence of void formation in the ductile matrix, there was some evidence of reinforcement particle fracture, especially for the higher volume fractions of the reinforcement.

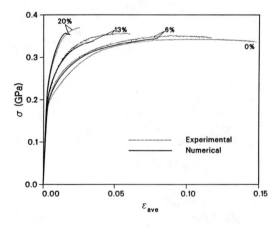

Figure 13.6. Numerical predictions and experimental results for Al-Cu matrix composites reinforced with various volume fractions of SiC particles. *(After LLorca et al. 1992.)*

13.3 Particle Fracture

The higher elastic moduli of metal-matrix composites, as compared to those of unreinforced metals, make them attractive for a variety of structural and electronic applications. In many discontinuously reinforced metals, the enhancement in modulus caused by brittle particle reinforcement can be fully or partially annulled (or more than counterbalanced) by the cracking of the particles themselves during mechanical and/or thermal loading. In addition, particle fracture reduces the flow stress, strain hardening exponent, and ductility of reinforced metals during monotonic tension loading. Under cyclic loading conditions, it is known (see Chapter 15 of this volume for further details) that failure of the reinforcing phase decreases the total life in low-cycle fatigue tests (Bonnen et al. 1991) and increases the apparent crack propagation rates in high-cycle fatigue tests (Kumai et al. 1992; Sugimura and Suresh 1992).

Although the fracture of the reinforcing particle, by itself, has a detrimental effect on the overall load-bearing capacity of the composite, the ensuing damage can also influence other concomitant failure processes. For example, sharp microcracks that develop as a consequence of particle fracture can enhance localized plastic flow within the ductile matrix and aid in such failure phenomena as ductile separation by void growth or shear banding. In some reinforced metals, the propensity for particle fracture can also increase the sensitivity of the overall deformation of the composite to such geometric factors as the size and spatial distribution of the reinforcing phase because, as discussed later, these factors have a strong bearing on the extent of particle fracture.

13.3.1 Experimental Observations

In recent years, quantitative experimental information has emerged on the factors influencing reinforcement fracture and its effects on the constitutive response of metal-matrix composites. Hunt et al. (1991) used an Al-Si-Mg model composite system to investigate the effects of particle size and volume fraction on development of damage and deformation. In this model composite system, Si particles are formed as a consequence of a eutectic reaction, and the volume fraction and size of the Si particles as well as the matrix strength can be varied drastically by a proper manipulation of the processing parameters. Hunt et al. (1991) studied the effects of particle fracture occurring during deformation on the change in elastic moduli of the composite for six different combinations of Si particle size and volume fraction. Figure 13.7 shows Young's modulus

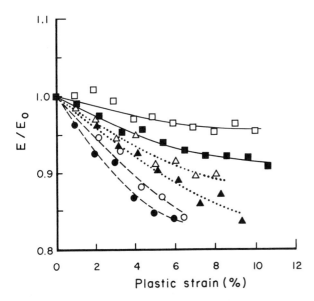

Figure 13.7. Young's modulus of the composite (E), normalized by Young's modulus of the undamaged composite (E_0), as a function of plastic prestrain in various Al-Si-Mg alloys. *(After Hunt et al. 1991.)*

of the composite E (normalized by Young's modulus of the undamaged composite, E_0) as a function of the plastic strain applied previously to induce particle fracture for three different volume fractions of Si, each of which contained two different average sizes of Si particles. Here, the rate of decrease of Young's modulus with plastic strain (and hence the rate of damage accumulation by particle fracture, as corroborated by microscopic observations) increased with the volume fraction of Si particles. Furthermore, it was observed that an increase in particle size resulted in a faster rate of decrease of Young's modulus with plastic strain.

Similar experimental results have also been observed in other composite systems. Mochida et al. (1991) conducted in situ tensile tests in a scanning electron microscope (SEM) on a 6061 aluminum alloy reinforced with 15 vol % Al_2O_3 particles. Their experiments revealed a greater propensity for particle fracture for the larger size particles. Identical trends, based on in situ SEM tests, have also been reported by Brechet et al. (1991) for SiC-reinforced aluminum alloys. Systematic studies of the evolution of particle fracture in response to the application of uniaxial tensile plastic strains have also been conducted by Shen et al. (1993) for an Al-3.5 wt% Cu alloy reinforced with 20 vol% SiC particles. They found a rapid increase in the fraction of fractured SiC particles beyond an average plastic strain of 0.25%. At an overall plastic strain of 2.5%, as much as 50% of the reinforcing SiC particles were seen broken (see Figure 13.8). The results clearly show that larger particles fracture more readily than smaller ones, at fixed values of plastic strain.

13.3.2 Modeling of Particle Fracture

13.3.2.1 Elastic Deformation

Numerical modeling of the effective elastic response of metal-matrix composites with a known fraction of broken reinforcing particles has been carried out by recourse to finite element simulations within the context of axisymmetric and plane strain unit cell formulations (Hunt et al. 1991; Shen et al. 1993), Eshelby's equivalent inclusion method (Mochida et al. 1991), and a novel three-phase damage model (Bao 1992).

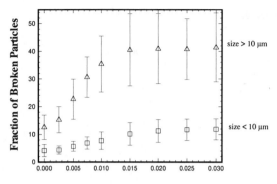

Plastic Strain

Figure 13.8. Fraction of broken SiC particles (of two size groups) as a function of plastic strain in an Al-3.5% Cu alloy reinforced with 20 vol% SiC particles. The composite was solutionized at 510°C for 2 hours, and then artificially aged at 190°C for 18 hours to provide peak strength in the matrix. *(After Shen et al. 1993.)*

Figure 13.9 shows the numerically predicted effects of particle fracture when all of the particles are fractured on the overall elastic moduli of the metal-matrix composites for different volume fractions of the reinforcing phase and several different combinations of the elastic moduli for the matrix and reinforcement. Also shown for comparison are the analytic predictions of changes in moduli in an unreinforced matrix alloy containing various volume fractions of penny-shaped microcracks. This figure reveals that when Young's modulus of the reinforcing particle E_p is three times that of the matrix E_m, the increase in effective modulus arising from reinforcing the matrix with particles is essentially offset by the fracture of the entire population of particles. When $E_p > 3E_m$, the overall Young's modulus is raised as a result of reinforcement even if all the particles fracture.

The aforementioned methods for modeling particle fracture invoke continuum or averaged descriptions of the constituent phases with one or several particles within the "model cell" used to simulate the overall response. Consequently, the effects of geometric factors such as particle size or spatial distribution seen in real composites can only be simulated for certain highly idealized cases. However, it is known from experiments that the propensity for reinforcement fracture is higher for (1) larger particles, (2) regions of the composite where the particles are clustered, and (3) larger volume fractions of the particles. The greater extent of particle fracture for larger particle sizes may arise from (1) the greater probability of finding a pre-existing crack or other defects within the particle, and (2) the probability of the occurrence of a larger sized flaw (as compared to the case of a smaller particle) which can trigger particle fracture at a lower far-field tensile stress. Furthermore, local clustering and high volume fractions of particles significantly elevate the local hydrostatic stress. Particle fracture is one of the mechanisms by which high triaxial stresses that evolve within the composite can be relieved.

Systematic studies of the effects of reinforcement distribution on the elastic response have been conducted using plane strain unit cell formulations for particle-reinforced metals (e.g., Shen et al. 1993) and using three-dimensional or generalized plane strain models for fiber-reinforced metals (Brockenbrough and Suresh 1990; Brockenbrough et al. 1991; Nakamura and Suresh 1992; Chapter 10 of this volume). These studies have shown that reinforcement distribution (at a fixed reinforcement concentration) can alter the elastic moduli (by up to 25% at a volume fraction of 46%) in metal-matrix composites. Further work is necessary to develop a quantitative understanding of the effective elastic moduli of metal-matrix composites with realistic criteria for reinforcement fracture, wherein the effects of such geometric parameters as reinforcement size, shape, concentration, and spatial distribution on the propensity for particle fracture are considered.

13.3.2.2 Plastic Deformation

As noted previously, fracture of reinforcing particles during mechanical loading can lead to pronounced changes in the constitutive response of the metal-matrix composites. The effects of particle failure on the plastic

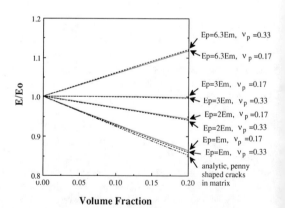

Figure 13.9. Numerically predicted change in elastic modulus (normalized by the modulus of the undamaged composite) as a function of the volume fraction of the reinforcement for different combinations of matrix and reinforcement elastic properties. Also shown are analytic predictions of matrix elastic modulus for different volume fractions of penny-shaped cracks. Further details are given in the text. *(After Shen et al. 1993.)*

response of reinforced metals has been investigated by Bao (1992) and Shen et al. (1993). Bao developed a micromechanical model for particle fracture in a metal-matrix composite with uniformly distributed spherical or cylindrical particles by considering a three-phase damage cell. The unit cell consisted of a cracked particle in a cylindrical cell of an elastic–perfectly plastic matrix material, which was embedded within an undamaged composite cell. The ratio of the volume of the broken particle and matrix cell to that of the entire unit cell was adjusted to be the volume fraction of the broken particle. For the case of spherical particles, the decrease in composite limit flow stress caused by particle fracture varied linearly with the fraction of fractured particles. It was also found that the sharp corners in the particles and the aspect ratio of the particles had a strong influence on the overall flow stress of the composite.

The full extent of the effect of particle fracture on the plastic deformation characteristics of a reinforced metal can be recognized by recourse finite element analyses within the context of axisymmetric unit cell models. Consider, for example, the case of an Al–3.5 wt% Cu alloy reinforced with 20 vol% of SiC particles. This composite can be modeled as a circular cylindrical cell of the matrix material embedded at the center of which is a cylindrical whisker of aspect ratio unity. The matrix is modeled as an elastic–isotropically hardening plastic solid, the reinforcement is elastic and the interface between the matrix and the reinforcement is taken to be perfectly bonded. The topmost curve in Figure 13.10 is the numerically predicted stress–strain response of the metal-matrix composite where the reinforcement is intact. If the cylindrical particle is now broken into two equal halves, with the plane of fracture being normal to its axis (with no contact between the broken halves), the numerically predicted stress–strain response is illustrated by the lowest curve in Figure 13.10. (Note that this model pertains to the situation where all the particles in the composite are fractured.) It is seen that particle fracture results in a noticeable reduction in Young's modulus and flow stress; however, the strain hardening exponent is essentially unaltered by particle failure. The dashed line shows the experimentally measured stress–strain response for the composite. (The experimental curve is terminated at the point of catastrophic fracture in the tensile specimen.) The experimental results fall between the predictions for the perfect and broken particle cases for average strains of up to 1%. Beyond this strain, the experimental data fall below the numerical prediction for particle fracture, presumably because of the instigation of other failure mechanisms, such as ductile failure by void growth within the matrix.

13.4 Interfacial Failure

Control of interfacial failure is one of the most potent approaches for altering composite properties. The properties of the interface are, in most cases, amenable to modification through the application of coatings and surface treatments for the reinforcements, and through post-processing heat treatment of the composite. Unfortunately, these properties are also susceptible to degradation through chemical reaction, interdiffusion, and environmental attack. The following section presents some of the microstructural factors governing interface failure and a methodology for describing these processes within model calculations.

13.4.1 Microstructural Observations

Composite materials are inherently heterogeneous by definition, and this characteristic often leads to additional deleterious effects at interfaces because of solute segregation, enhanced local dislocation density, precipitation reactions, and reinforcement clustering. These processes can lead to imperfections at interfaces that facilitate nucleation of cavities and cracks. Furthermore, reinforcements typically exhibit microscopic or macroscopic surface roughness, and their shapes are seldom perfectly round or cylindrical. The sharp corners constitute geometric imperfections that can promote damage initiation and growth. Plastic flow is concentrated near such regions and a local region of large

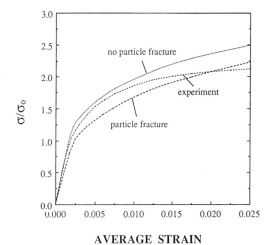

Figure 13.10. Numerically predicted stress-strain response of an Al-3.5% Cu alloy reinforced with 20 vol% SiC particles, in which the constitutive response is simulated with and without particle fracture. Also shown are experimental results for the composite. (*After Shen et al. 1993.*)

244 DAMAGE MICROMECHANISMS AND MECHANICS OF FAILURE

tensile hydrostatic stresses develops. Here, some interface imperfections encountered in aluminum-matrix composites and their effects on damage mechanisms are described.

One of the mechanisms that appears to be involved in the premature tensile failure of Al-SiC whisker composites is void nucleation via interface decohesion at fiber ends. The characteristic fiber geometry includes a sharp corner profile (caused by cleavage on {111} planes) that concentrates stress during loading. Consequently, composites deformed in tension often exhibit voids at fiber ends that typically nucleate at the fiber corners (Nutt and Duva 1986: Nutt and Needleman 1987). Some examples are shown in Figure 13.11, electron micrographs acquired from regions located 10 μm to 50 μm below the fracture surface of a tensile fracture specimen. In Figure 13.11(a), the dark contrast in the vicinity of the fiber end is caused by a high dislocation density resulting from intense plastic strain. A void 20 nm to 30 nm in diameter has nucleated at the sharp corner formed by the whisker end. Similar voids of different sizes were frequently observed at distances up to 50 μm below the fracture surface. In most instances, voids appeared to have nucleated at the fiber corner and to have grown toward the center of (and away from) the fiber end, as shown in Figure 13.11(b). Here the twin voids have nucleated at the fiber corners, and there is a local depression between the two voids, evidenced as region of light contrast. The appearance suggests the existence of a third void that nucleated at a corner outside the specimen plane or perhaps at the fiber center. A third, less frequently observed pattern of void growth, is depicted in Figure 13.11(c). Debonding appeared to nucleate at the site indicated by the arrow, and decohesion propagated rapidly across a segment of the whisker end, implying a weak or brittle mode of debonding.

Growth and coalescence of multiple voids at fiber ends sometimes resulted in single equiaxed voids that were approximately the same diameter as the fiber.

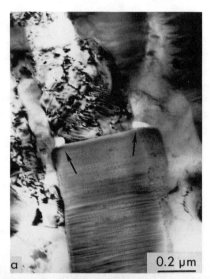

Figure 13.11. Patterns of void initiation in tensile fracture specimen of 6061 Al-SiC composite. (a) Void nucleation at corners of the fiber end (arrows), with intense strain in the matrix. (b) Void coalescence at the fiber end. Voids initiate at corners and grow toward the center of the fiber end. (c) Interface decohesion at the fiber end. Arrow indicates the probable site of void nucleation.

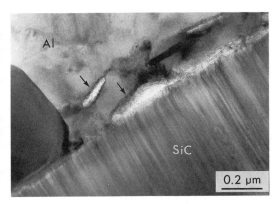

Figure 13.12. Debonding on a whisker flank during tensile deformation. Arrows indicate sites where debonding has occurred at oxide clusters in P/M processed Al-SiC composite.

These voids were manifest as microdimples on the composite fracture surface. Voids and cracked whiskers were generally confined to regions near the fracture surface. However, when the test temperature was increased to 200°C to 300°C in hot tension tests and tensile creep experiments, void nucleation at fiber ends generally occurred throughout the gauge section, following patterns similar to those shown in Figure 13.11. Damage was extensive and mechanisms included void nucleation at fiber ends and transverse cracking in the whiskers. The composites were much more tolerant of microstructural damage at high test temperatures than at room temperature.

In addition to the significant effects of fiber geometry on interface failure modes, precipitate phases are often present and a description of some of these common imperfections and the effects on interfacial failure follows. One of the processes commonly used to fabricate metal-matrix composites with whisker reinforcements is powder metallurgy (PM), in which the products often contain fine oxide particles. These oxides are normally present on the powder particles prior to compaction and densification, and the particles often end up at whisker interfaces in the densified composite. The particles are not necessarily detrimental to the mechanical properties provided they are well-dispersed, which is usually accomplished by thermomechanical working. However, hot-working of composites does not always succeed in dispersing the oxide particles, and oxide clusters sometimes remain at interfaces. These clusters constitute a serious imperfection that can facilitate interface decohesion during composite loading (Nutt 1986). The micrograph shown in Figure 13.12 illustrates how oxide clusters at the fiber-matrix interface can lead to decohesion and damage accumulation. The composite was subjected to uniaxial tension, and the whisker shown is misoriented to the tensile axis by about 20 degrees. Slit-like openings appear at and near the interface where decohesion has taken place within the oxide clusters (arrows). This damage mechanism was rarely observed in uniaxial tensile tests, but the mechanism is expected to be more prevalent in transverse tensile tests and in situations in which the tensile component of stress normal to the interface is stronger.

Precipitation of matrix solute often occurs at composite interfaces and contributes to failure modes in several ways. Precipitate-free zones (PFZs) are commonly observed in fiber-reinforced age-hardenable aluminum alloys, a consequence of localized depletion of solute atoms and/or vacancies through migration to the incoherent fiber-matrix interface. An excess vacancy concentration is normally frozen into the matrix when the material is quenched from the solutionizing temperature. The interface is an effective vacancy sink, and excess vacancies from the adjoining matrix are often depleted during quenching. A reduction in excess vacancy concentration near the interface inhibits the homogeneous nucleation of GP zones and contributes to the formation of PFZs, as shown in Figure 13.13 (Nutt 1989). The image shows an interface region of a carbon fiber–aluminum composite (CFAL) in which the matrix is A201 aluminum (Al, 4.6-Cu, 0.7-Ag). The PFZ contains several dislocations extending away from the fiber, and these are attributed to thermal stresses that develop during cooling as a result of thermal expansion mismatch. These PFZs along interfaces are expected to be softer than the age-hardened matrix and thus furnish compliant zones around the fibers.

Interface PFZs in composites are often accompanied by precipitation reactions at the fiber interface. Figure 13.14 shows a $CuAl_2$ precipitate nucleated on a fiber

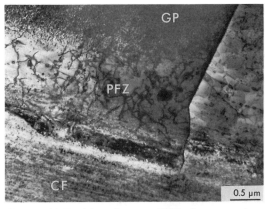

Figure 13.13. Precipitate-free zone (PFZ) adjacent to carbon fiber (CF) in naturally aged A201 Al matrix (Al, 4.6-Cu, 0.7-Ag). GP zones are indicated.

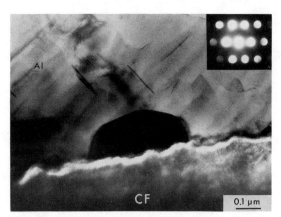

Figure 13.14. Heterogeneous nucleated $CuAl_2$ precipitate at carbon fiber-aluminum interface in A201 matrix. Inset microdiffraction pattern is $[0\bar{2}1]$ $CuAl_2$.

interface in an artificially aged carbon fiber aluminum (CFAL) composite. The specimen is a transverse section and provides an edge-on view of the interface, revealing surface roughness on a scale of 30 nm to 50 nm. The fiber is coated with a thin layer of TiB_2 to inhibit interfacial reaction and interdiffusion. During ion thinning of the specimen, the coating usually debonds from the fiber surface, leaving a narrow gap (Nutt 1989). Composite fabrication is achieved by continuous casting of composite wires, followed by hot-pressing and conventional aging treatment. Solutes tend to segregate to the fiber interface during heat treatment, often leading to the heterogeneous nucleation of precipitates. In this case, copper is the diffusing species, and the precipitate phase that nucleates is $CuAl_2$. In some composite materials, interface precipitation can lead to the formation of a continuous interfacial layer. Composites of 6061 aluminum reinforced with SiC whiskers often showed a thin continuous layer of polycrystalline oxide, as seen in Figure 13.15. The layer was ≈ 3 nm thick, and microanalysis led to an identification of MgO. Heterogeneous nucleation of precipitate phases at interfaces is an almost unavoidable consequence of the inherent heterogeneous nature of composites, which provides large areas of incoherent interface that are ideal low-energy nucleation sites for precipitation reactions.

Geometric imperfections on fiber flanks were also present, contributing to stress concentration and inhibiting frictional sliding. The example shown in Figure 13.15 illustrates the microscopic roughness often observed in SiC whisker reinforcements. A line of bright contrast caused by Fresnel diffraction separates the oxide layer from the SiC whisker and highlights the interface profile. The profile is microscopically rough, a consequence of surface energy anisotropy that causes microfaceting on close-packed {111} planes. The facets sometimes result in an irregular sawtooth profile consisting of notches and teeth that coincide with planar defects within the whisker. Although the planar defects are not detrimental to the whisker strength, they terminate in surface notches that tend to be circumferential and in severe cases may constitute strength-limiting defects. In moderate cases, such as that shown in Figure 13.15, the roughness can be expected to increase friction and to inhibit both sliding and fiber pullout, both of which are desirable toughening mechanisms.

In some composite systems, the fiber and the matrix react during high-temperature processing, leading to the formation of reaction products that are deleterious to composite properties. These interfacial reactions are sometimes unavoidable, even when protective coatings are applied to the reinforcements. For example, SiC monofilaments are grown with a carbon-rich surface layer (denoted SCS) designed to improve fiber strength and inhibit interface reactions at elevated temperatures (Ning and Pirouz 1991; Nutt and Wawner 1985). When aluminum-matrix composites are fabricated by hot-pressing, the matrix reacts with the carbon-rich coating, forming small crystallites of Al_4C_3. Carbon diffuses from the coating, and carbide platelets grow into the matrix. Although small amounts of reaction have negligible effects on the mechanical properties of the composite, aluminum carbide is water-soluble. Thus, the interface is susceptible to environmental attack, and exposure to humid atmospheres is likely to cause interface failure and degradation of mechanical properties.

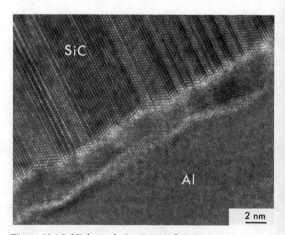

Figure 13.15. High-resolution image showing a polycrystalline oxide layer (MgO) along an SiC whisker flank in a 6061 Al matrix. Planar defects in the SiC whisker are a consequence of the growth process. *(From Nutt 1988.)*

13.4.2 Modeling of Interfacial Failure

A predictive model for interfacial decohesion needs to incorporate a separation criterion that characterizes the interfacial fracture process. A continuum framework has been developed where constitutive relations are written for the matrix, the reinforcement and for the interface (Needleman 1987, 1992). Consider a multiphase solid under quasistatic loading and suppose that one phase is distributed in regions having one dimension much smaller than any other characteristic dimension of the solid. Furthermore, this phase is presumed to be distributed along the boundaries of other phases. In the limit of vanishing phase boundary thickness, the principle of virtual work can be written as,

$$\int_V s : \delta F dV - \int_{S_{int}} T \cdot \delta \Delta dS = \int_{S_{ext}} T \cdot \delta u dS \quad (13.11)$$

where $\Delta = u^+ - u^-$ is the displacement jump across the interface and the sign convention is such that T is the restoring traction.

Although the interface thickness does not explicitly appear in Equation 13.11, an interface characteristic length enters implicitly through the interface constitutive relation. The material constitutive relation, which relates stress and strain, has a parameter having units of stress. The interface constitutive relation, which specifies the dependence of the interface tractions on the displacement jump across the interface and, possibly, on other "internal variables," contains a parameter having units of stress × length. By solving appropriate boundary value problems based on Equation 13.11, the effect of the evolution of the interface traction-bearing capacity on aggregate behavior can be assessed.

There are a few general requirements that it seems reasonable to impose on an interface constitutive relation which include: (1) The interface response is independent of any superposed rigid body motion, and (2) The interface response is dissipative, which requires the work expended in a closed process (one that begins and ends with the same value of Δ) to be non-negative. This is expressed by

$$-\oint T \cdot d\Delta \geq 0 \quad (13.12)$$

Satisfaction of Equation 13.12 means that work cannot be extracted from the interface.

An elastic constitutive relation states that the interface traction is a function of the displacement jump so that

$$T = T(\Delta) = -\frac{\partial \phi}{\partial \Delta} \quad (13.13)$$

where the existence of the potential ϕ for the interface traction vector is a consequence of Equation 13.12.

Various potentials have been used for analyses of interfacial decohesion (Nutt and Needleman 1987; Tvergaard 1990b; Povirk et al. 1991; Xu and Needleman 1992). As an example, a potential with a well-defined separation point that has been used in several analyses of void nucleation in metal-matrix composites, is given in two dimensions by

$$\phi(\Delta_n, \Delta_t) = \frac{27}{4}\sigma_{max}\delta\left\{\frac{1}{2}\left(\frac{\Delta_n}{\delta}\right)^2\left[1 - \frac{4}{3}\left(\frac{\Delta_n}{\delta}\right) + \frac{1}{2}\left(\frac{\Delta_n}{\delta}\right)^2\right]\right.$$
$$\left. + \frac{1}{2}\alpha\left(\frac{\Delta_t}{\delta}\right)^2\left[1 - 2\left(\frac{\Delta_n}{\delta}\right) + \left(\frac{\Delta_n}{\delta}\right)^2\right]\right\} \quad (13.14)$$

for $\Delta_n \leq \delta$ and $\phi \equiv \phi_{sep}$ for $\Delta_n > \delta$, where ϕ_{sep} is the work of separation. The subscripts n and t denote quantities normal and tangential to the interface, respectively, and the characteristic length, δ, is given by $\delta = 16\phi_{sep}/9\sigma_{max}$. The interfacial tractions (force per unit initial area) are

$$T_n = \frac{-27}{4}\sigma_{max}\left\{\left(\frac{\Delta_n}{\delta}\right)\left[1 - 2\left(\frac{\Delta_n}{\delta}\right) + \left(\frac{\Delta_n}{\delta}\right)^2\right]\right.$$
$$\left. + \alpha\left(\frac{\Delta_t}{\delta}\right)^2\left[\left(\frac{\Delta_n}{\delta}\right) - 1\right]\right\} \quad (13.15)$$

$$T_t = \frac{-27}{4}\sigma_{max}\left\{\alpha\left(\frac{\Delta_t}{\delta}\right) \cdot \left[1 - 2\left(\frac{\Delta_n}{\delta}\right) + \left(\frac{\Delta_n}{\delta}\right)^2\right]\right\} \quad (13.16)$$

for $\Delta_n \leq \delta$ and $T_n = T_t = 0$ for for $\Delta_n > \delta$.

Because the tangential response of the interface is taken to be linear, use of Equations 13.15 and 13.16 is restricted to small values of Δ_t/δ. The key parameters for tensile-dominated decohesion are the interface strength, σ_{max}, and the work of separation per unit area, ϕ_{sep}. As long as the reinforcement dimensions are more than a few hundred times δ, the decohesion predictions are not sensitive to the actual form of the potential. For example, with characteristic reinforcement dimensions of the order of one micrometer, this requires a value of δ of the order of a few nanometers or smaller. Inelastic interface constitutive relations, accounting for frictional sliding, have been presented in Tvergaard (1990b) and in Povirk and Needleman (1993). At present, interface constitutive relations have not been developed that reflect inelastic mechanisms occurring in a thin layer near the interface that differ from the dominant bulk inelastic mechanism. For example, observations of interfaces have shown that a thin interfacial layer can form that differs structurally, and presumably, mechanically from the phases on either side (Nutt et al. 1990; Nutt and Carpenter 1985).

The effect of interfacial decohesion on composite response has been analyzed within the context of axisymmetric unit cell models. Figure 13.16 shows finite element meshes in the deformed configuration depicting

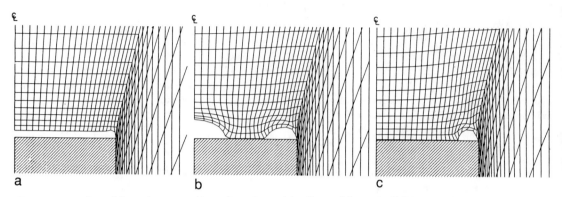

Figure 13.16. Deformed finite element meshes in the vicinity of the fiber end for 6061 Al-SiC composites. (a) $\sigma_{max}/\sigma_0 = 3.0$ at $\epsilon_a = 0.0151$, (b) $\sigma_{max}/\sigma_0 = 5.0$ at $\epsilon_a = 0.0231$, and (c) $\sigma_{max}/\sigma_0 = 6.0$ at $\epsilon_a = 0.0291$. The SiC inclusion is shaded and, for reference, the axis of symmetry is marked. The fiber axis is aligned with the tensile axis. *(Reprinted with permission from Nutt S.R., and A. Needleman 1987. Scripta Metall. 21:705-710.)*

the mode of decohesion obtained from two calculations modeling a 20 vol% SiC whisker reinforced 6061 aluminum matrix (Nutt and Needleman 1987). All geometric and material parameters are the same except for the interfacial strength; in Figure 13.16(a) the interfacial strength is three times the matrix flow strength, whereas in Figure 13.16(b) it is six times the matrix flow strength. In both cases, a void nucleates near the fiber corner because of the high stress and strain concentrations there. For the lower strength interface, there is crack-like propagation across the fiber end following a little void growth. Conversely, for the higher strength interface, decohesion remains confined near the fiber corner. For example, the analyses in Nutt and Needleman (1987) indicate that the mode of failure observed in Figure 13.11(c) implies a relatively weaker interface than the more frequently observed type of failure shown in Figure 13.11(a). Variation of interface strength, with fixed matrix material properties, leads to predicted patterns of void evolution that are virtually identical to those observed experimentally. Within the composite, interfacial strengths are expected to vary locally and will be affected by the presence of interfacial phases such as oxides and intermetallics (Nutt and Carpenter 1985). Thus, the calculations suggest the possibility that, with a sufficiently high interfacial strength, complete decohesion of the fiber end is substantially delayed, if not precluded altogether, even though void formation occurs at the fiber corner at relatively small strains.

Summary

The basic failure mechanisms in metal-matrix composites involve: (1) void nucleation, growth, and coalescence in the matrix, (2) brittle fracture of the reinforcement, and (3) debonding along the interface between the matrix and the reinforcement. Continuum models suitable for use in strength calculations are available for the first and third of these mechanisms, as discussed in Sections 13.2 and 13.4. Progress is being made in developing a predictive model for reinforcement fracture (Section 13.3). Once such a model is in place, predictions of the effects of reinforcement distribution, concentration, and shape on composite ductility can be made by incorporating the possible effects of all three failure mechanisms. However, as of this writing, failure analyses have been based on two-dimensional axisymmetric or plane strain models. Full three-dimensional analyses are needed, particularly to quantitatively account for distribution effects, as discussed for deformation response in Chapter 10. The growing capability and availability of massively parallel supercomputers could make three-dimensional failure analyses feasible in the not-too-distant future.

The failure models require input parameters such as a void nucleation strain (or stress) for the matrix, a particle fracture criterion, an interfacial strength, and work of separation. In order for the continuum models to serve as predictive tools, methods for the direct measurement of such quantities in the composite need to be developed, or lower-level models for the direct calculation of these quantities, such as atomistic analyses of the mechanics and chemistry of interfaces, are needed.

The models discussed in this chapter are based on classical isotropic elastic and plastic (or viscoplastic) constitutive characterizations. Continuum analyses that account for anisotropic matrix hardening caused by slip on discrete crystallographic slip systems predict behavior that generally is in qualitative and quantitative agreement with corresponding isotropic hardening predictions for the deformation response (see Chapter 8). There are differences in local fields that affect failure predictions and for the continuum slip matrix characterization, the discreteness of slip systems promotes shear localization in the matrix. On a smaller scale, the effects of discrete dislocations will come into play so that the continuum descriptions (even those accounting for the discreteness of slip systems) would no longer be appropriate. For example, a discrete dislocation model of plastic flow would give rise to a size dependence on the local stress and deformation fields in the composite as well as to a dependence on reinforcement volume fraction, geometry, and distribution. Additionally, it is possible that discrete dislocation contributions to the stress concentration at the reinforcement-matrix interface could affect the predicted onset of decohesion. As also mentioned in Chapters 8, 9, and 10 and in Section 13.2.2 of this chapter, various size scale limitations to continuum models remain to be quantified.

Nevertheless, phenomenological continuum theories appear to provide remarkably accurate qualitative and quantitative descriptions of a broad range of deformation and failure processes in metal-matrix composites. The development of continuum models for each of the principal composite failure mechanisms provides a basis for addressing the interactions between the various failure mechanisms. This, in turn, will permit consideration of the design of metal-matrix composites for optimum ductility.

Acknowledgments

Alan Needleman, Steve R. Nutt and Subra Suresh are pleased to acknowledge the support of the Materials Research Group at Brown University for *Micromechanics of Failure-Resistant Materials*, which is funded by a grant from the U.S. National Science Foundation (Grant No. DMR-9002994).

References

Bao, G. 1992. *Acta Metall. Mater.* 40:2547-2555.
Becker, R., Needleman, A., Richmond, O., and V. Tvergaard. 1988. *J. Mech. Phys. Solids.* 36:317-351.
Becker, R., Needleman, A., Suresh, S., et al. 1989. *Acta Metall.* 37:99-120.
Bishop, J.F.W., and R. Hill. 1951. *Phil. Mag.* 42:414-427.
Bonnen, J., Allison, J., and J.W. Jones. 1991. *Metall. Trans.* 22A: 1007-1019.
Brechet, Y., Embury, J.D., Tao, S., and L. Luo. 1991. *Acta Metall. Mater.* 39:1781-1786.
Brockenbrough, J.R., and S. Suresh. 1990. *Scripta Metall. Mater.* 24:325-330.
Brockenbrough, J.R., Suresh, S., and H.A. Weinecke. 1991. *Acta Metall. Mater.* 39:735-752.
Christman, T., and S. Suresh. 1988a. *Mater. Sci. Eng.* A102:211-216.
Christman, T., and S. Suresh. 1988b. *Acta Metall.* 36:1691-1704.
Christman, T., Needleman, A., and S. Suresh. 1989. *Acta Metall.* 37:3029-3050.
Chu, C.C., and A. Needleman. 1980. *J. Eng. Mater. Technol.* 102:249-256.
Davidson, D.L. 1987. Southwest Research Institute Report No. N00014-85-C-0206 06-8602/3.
Fleck, N.A., Kuhn, L.T., and R.M. McMeeking. 1992. *J. Mech. Phys. Solids.* 40:1139-1162.
Gurson, A.L. 1975. Ph.D. Thesis. Division of Engineering, Brown University, Providence, RI.
Huang, Y., Hutchinson, J.W., and V. Tvergaard. 1991. *J. Mech. Phys. Solids.* 39:223-241.
Hunt, W.H., Brockenbrough, J.R., and P.E. Magnusen. 1991. *Scripta Metall. Mater.* 25:15-20.
Kumai, S., King, J.E., and J.F. Knott. 1992. *Fatigue Fract. Eng. Mater. Struct.* 15:1-11.
Liu, L.S., Manoharan, M., and J.J. Lewandowski. 1989. *Metall. Trans.* 20A:2409-2420.
LLorca, J., Needleman, A., and S. Suresh. 1991. *Acta Metall. Mater.* 3:2317-2335.
LLorca, J., Suresh, S., and A. Needleman. 1992. *Metall. Trans.* 23A:919-934.
Mochida, T., Taya, M., and D.J. Lloyd. 1991. *Mater. Trans. Japan Inst. Metals.* 32:931-942.
Nakamura, T., and S. Suresh. 1992. Submitted for publication.
Needleman, A. 1987. *J. Appl. Mech.* 54:525-531.
Needleman, A. 1992. *Ultramicroscopy.* 40:203-214.
Needleman, A., and J.R. Rice. 1978. In: Mechanics of Sheet Metal Forming. (D.P. Koistinen and N.-M. Wang, eds.), 237-265, New York: Plenum Press.
Ning, X.J., and P. Pirouz. 1991. *J. Mater. Res.* 6:2234-2248.
Nutt, S.R. 1986. In: Interfaces in Composites, 157-167, Metals Park, OH: TMS of AIME.
Nutt, S.R. 1989. In: Treatise on Materials Science and Technology, Vol. 31. (A.K. Vasudevan and R.D. Doherty, eds.), 389-408, New York: Academic Press.
Nutt, S.R., and R.W. Carpenter. 1985. *Mater. Sci. Engr.* 75:169-177.
Nutt, S.R., and J.M. Duva. 1986. *Scripta Metall.* 20:1055-1058.
Nutt, S.R., Lipetzky, P., and P.F. Becker. 1990. *Mater. Sci. Eng.* A126:165-172.
Nutt, S.R., and A. Needleman. 1987. *Scripta Metall.* 21:705-710.
Nutt, S.R., and F.E. Wawner. 1985. *J. Mater. Sci.* 20:1953-1960.
Pan, J., Saje, M., and A. Needleman. 1983. *Int. J. Fract.* 21:261-278.
Povirk, G.L., and A. Needleman. 1993. *J. Eng. Mater. Technol.* 115:286-291.
Povirk, G.L., Needleman, A., and S.R. Nutt, S.R. 1991. *Mater. Sci. Eng.* A132: 31-38.
Saje, M., Pan, J., and A. Needleman. 1982. *Int. J. Fract.* 19:163-182.
Shen, Y.-L., Finot, M., Needleman, A., and S. Suresh. 1993. Manuscript in preparation.
Sugimura, Y., and S. Suresh. 1992. *Metall. Trans.* 23A:2231-2242.

Suresh, S. 1991. Fatigue of Materials. Cambridge: Cambridge University Press.

Tanaka, J.P., Pampillo, C.A., and J.R. Low. 1970. *In:* Review of Developments in Plane Strain Fracture Testing, ASTM STP 463, Philadelphia: ASTM, 191–214.

Tvergaard, V. 1981. *Int. J. Fract.* 17:389–407.

Tvergaard, V. 1982. *Int. J. Fract.* 18:237–252.

Tvergaard, V. 1990a. *Adv. Appl. Mech.* 27:83–151.

Tvergaard, V. 1990b. *Mater. Sci. Eng.* A125:203–213.

Tvergaard, V. 1991. *Acta Metall. Mater.* 39:419–426.

Tvergaard, V., and A. Needleman. 1984. *Acta Metall.* 32:157–169.

Van Stone, R.H., Merchant, R.H., and J.R. Low. 1974. *In:* Fatigue and Fracture Toughness—Cryogenic Behavior, ASTM STP 556, Philadelphia: ASTM, 93–120.

You, C.P., Thompson, A.W., and I.M. Bernstein. 1987. *Scripta Metall.* 21: 181–186.

Chapter 14
Fracture Behavior

BRIAN DERBY
PAUL M. MUMMERY

The use of metal-matrix composites in structural applications is attractive because of their exceptionally good stiffness-to-weight and strength-to-weight ratios. These properties suggest many possible uses in weight-sensitive components for aerospace or land transportation. However, although it is possible to manufacture metal-matrix composites of exceptionally high tensile failure strengths, these materials often have disappointingly low toughness and tensile ductility. It is often these fracture-related properties that limit the further use of metal-matrix composites.

Metal-matrix composites are not new materials. Thin metal wires, < 100 μm in diameter, have been available for some time and work on W wire reinforced Cu and steel wire reinforced Al alloys dates back about 25 years. Preliminary studies were very much based on an experimental mechanics approach. The later development of B fibers for the space program produced composites with a brittle reinforcement phase and subsequently, the fracture behavior of long fiber reinforced metals began to be explored. This early work produced many of the concepts of fracture that are still used for modern understanding of composite behavior. This work was extensively reviewed by Morley (1976). Current interest in metal-matrix composites has arisen because of a parallel development of new reinforcements and new, better understood processing methods, which have resulted in cheaper and more reliable material.

Modern metal-matrix composites contain reinforcements made from hard ceramic materials. These can either be in the form of long fibers made from graphite, SiC, B, or Al_2O_3, or in the form of discontinuous reinforcements in platelet, particle, and whisker form, chiefly of SiC and Al_2O_3. Both classes of composites have relatively poor fracture properties. Tables 14.1 and 14.2 show the measured fracture properties for a number of composites. In all cases, there is a substantial decrease in these properties when the reinforcing phase in both continuous and discontinuous form is introduced. However, there is no dramatic change in the fracture mechanism; Figures 14.1 and 14.2 show the fracture surfaces from broken composite specimens. Failure occurs by ductile rupture between the reinforcement, which fails in a brittle manner. Despite these similarities in fracture mechanism, continuous and discontinuous reinforcements are treated differently mainly because long fiber composites are more amenable to simpler mechanical analyses and, therefore, their fracture behavior will be considered in separate sections of this chapter.

The goal in writing this chapter is to present a sound historical background regarding the development of the understanding of fracture processes in metal-matrix composites and to show how it acts as a foundation for current models. This presentation will be developed to show the importance of matrix, reinforcement, and any interfacial phases present within the microstructure. Metal-matrix composites will be shown to display an unusual combination of brittle and ductile phenomena. Understanding their fracture behavior presents a new challenge.

Table 14.1. Toughness and Ductility Measurements of Some Fiber Reinforced Metal-Matrix Composites and Unreinforced Matrix Alloys

Material	Fracture Toughness (K_c) (MPa$\sqrt{m}$)	Impact Fracture Energy (kJm^{-2})	Elongation to Failure (%)	Data Source
Zn-Al alloy (ZA8)	29.9	—	—	1
ZA8 + 20% steel fibers	22.5	—	—	1
ZA8 + 20% Saffil fibers	9.6	—	—	1
Al-3.5%Cu	—	—	19.5	2
Al-3.5%Cu + 20% Saffil fibers	—	—	2.3	2
Pure Al + 5% Saffil fibers	38.2	4.8	—	3
Pure Al + 10% Saffil	24.5	42	—	3
Pure Al + 20% Saffil fibers	23.1	27	—	3
Pure Al + 30% Saffil fibers	18.7	15	—	3
Al 7010	30	—	10.5	4
Al 7010 + 15% Saffil fibers	13	—	< 0.2	4
Al-5%Mg	46	—	13.8	4
Al-5%Mg + 15% Saffil fibers	14	—	2.0	4

[1]Dellis et al. (1991); [2]Harris et al. (1988); [3]Keyhoe and Chadwick (1991); [4]Musson and Yue (1991).

Table 14.2. Toughness Measurements of Some Particle and Whisker Reinforced Metal-Matrix Composites and Unreinforced Matrix Alloys

Material	Fracture Toughness (K_c) (MPa$\sqrt{m}$)	Elongation to Failure (%)	Data Source
Al 1100	—	35	1
Al 1100 + 20% SiC	—	10–15	1
Al 6061	—	12	1
Al 6061 + 20% SiC	—	5–6	1
Al 2024	—	10	1
Al 2024 + 20% SiC	—	3–4	1
Al 7075	—	12	1
Al 7075 + 20% SiC	—	3–4	1
Al 8090	—	6	1
Al 8090 + 20% SiC	—	2–3	1
Al 2024	—	21.0	2
Al 2024 + 2% Al$_2$O$_3$	16.5	12.0	2
Al 2024 + 5% Al$_2$O$_3$	15.8	9.0	2
Al 2024 + 20% Al$_2$O$_3$	12.2	4.3	2
Al 2024	21.0	—	3
Al 2024	36.0	—	3
Al 2024 + 15% SiC	14.7	—	3
Al 2024 + 15% SiC	17.7	—	3

[1]Humphreys 1988; [2]Kamat et al. 1989; [3]Davidson 1989a.

14.1 Long Fiber Reinforced Composites

14.1.1 Tensile Fracture

The ultimate failure stress, as distinct from the toughness, of long fiber reinforced metal-matrix composites (LFMMC) has been the subject of study for some time. In the absence of large defects, fracture is controlled by the fibers and can be modeled using a simple rule of mixtures approach (Cooper 1971), with the composite UTS given by a relation

$$\sigma_{UTS} = \sigma_F' V_F + \sigma_F'(1 - V_F)\frac{E_M}{E_F} \quad (14.1)$$

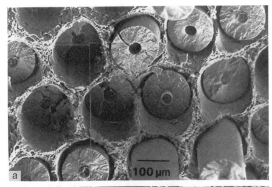

Figure 14.1. Fracture surfaces of long fiber reinforced metal-matrix composites. (a) Textron SCS-2 SiC fiber reinforced pure Al. (b) Nicalon SiC fiber reinforced pure Al. *(Reprinted with permission of B. Roebuck, National Physical Laboratory, Crown Copyright, National Physical Laboratory, U.K.)*

where σ_F' is the fiber failure stress, V_F is the volume fraction of fibers present, and E_M and E_F are the elastic moduli of the matrix and fiber, respectively. This simple model acts as an upper bound to the UTS of long fiber composites. This also assumes that the composite behaves in an ideal elastic manner to failure, and that no degradation of the fiber or matrix has occurred during

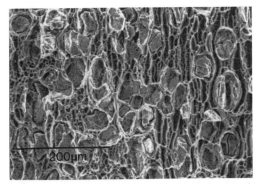

Figure 14.2. Fracture surface of an SiC particle reinforced Al-5050 matrix composite.

manufacture of the composite. Experimentally determined values tend to lie slightly below this value even when no degradation has occurred (Brindley et al. 1990; Masson et al. 1991).

For the case of ductile metal matrices, the ultimate strength of these materials is relatively insensitive to matrix behavior. When the influence of the matrix is seen, it is generally either by the effect of a degradation caused by fiber/matrix interactions or by cracking leading to local stress concentrations. However, the sequence of events prior to fracture depends on the matrix. The important variables are the matrix ductility and the fiber/matrix interface strength. In the case of highly ductile matrices, fiber fracture is still likely to begin before global plastic yield of the matrix because of the high stiffness of the fibers and the long lengths of fiber strained (reported fiber strengths are often over comparatively short gauge lengths). Even if the matrix exceeds its yield stress prior to fiber failure, the total matrix plastic strain before fiber failure will be much less than 10^{-2}. Once fibers begin to fail, local matrix yield at the broken fiber ends relaxes the stresses and prevents failure from progressing to adjacent fibers, as seen in Figure 14.3(a). In the case of more brittle matrix behavior the failure of a single fiber may propagate through the entire composite as seen in Figure 14.3(b). This catastrophic crack growth can be prevented if the fiber/matrix interface is weak. In this event, the crack can be deflected along the interface and the ends of the broken fibers will relax. (See Figure 14.3(c).) This weakness at the interface introduces considerable anisotropy into the behavior of fiber reinforced metal-matrix composites as can be seen from the data in Table 14.3.

The influence of single fiber fracture has been considered in the literature. The load carried by the fiber prior to fracture must be redistributed and is chiefly carried by adjacent intact fibers. These fibers are, therefore, more likely to fail if the composite is strained further. The degree of sequential local fiber failure will depend on the degree of coupling between the fibers

Table 14.3. Fracture Strengths of Fiber Reinforced Metal-Matrix Composites Measured Parallel (0°) and Perpendicular (90°) to Their Fiber Direction

Material	0° Strength (MPa)	90° Strength (MPa)	Data Source
Al + B	1400	140	1
Al + SiC	270	76	2
Al + C	800	18	3
Al-Li + Al$_2$O$_3$	500	175	4

[1]Kreiger and Prewo 1974; [2]Tsangarakis et al. 1987; [3]Nayeb-Hashemi and Seyyedi 1987; [4]Schulte and Minoshima 1991.

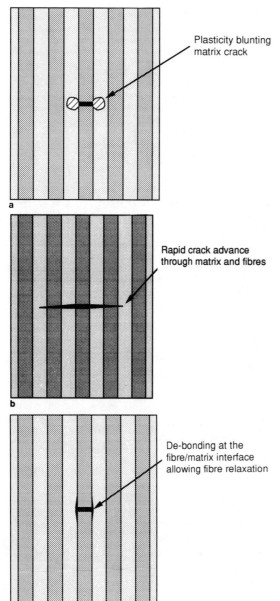

Figure 14.3. Possible crack tip processes in LFMMC fracture. (a) Crack blunting in a ductile matrix.
(b) Catastrophic crack growth through fibers and matrix.
(c) Debonding and crack deflection by fiber/matrix sliding.

and the matrix. At the limit of no coupling, the composite will behave like a bundle of fibers. The strengths of fiber bundles as a function of the distribution of fiber strengths have been known for some time (Coleman 1958). If the fibers and the matrix are strongly bonded, fracture propagates across both components as discussed above. The implications of groups of failed fibers have been considered extensively in the polymer-matrix composite literature and fracture is proposed to occur by the cumulative failure of fiber bundles. The effect of distributions of broken fibers has been calculated (Smith et al. 1983), and the critical number of adjacent broken fibers required for crack nucleation has been estimated (Zweben and Rosen 1970; Barry 1978; Harlow and Phoenix 1981).

We now consider the growth of large cracks during tensile straining parallel to the fiber direction. For the crack to propagate, failure must occur in both the fibers and the matrix. The nature of crack advance is strongly influenced by the nature of the fiber/matrix interface. In the case of material showing a very strong fiber/matrix interface, the fibers are likely to break immediately ahead of a matrix crack; matrix crack propagation occurs behind the fiber crack front. This should lead to a relatively planar crack front even if matrix fracture occurred by a ductile mechanism. This is shown in Figure 14.4(a). If the matrix is weakly bonded, the stress field at the matrix crack tip will introduce a debonding driving force that leads to crack blunting (Erdogan 1965; Rice and Shi 1965), as shown in Figure 14.4(b). In this case, fiber fracture will not necessarily be in the same plane as the matrix crack and fiber pull-out will be necessary to achieve full crack surface separation. An important result is that for both mechanisms, theoretic analysis shows that the volume of matrix that deforms during fracture, and hence the energy absorbed or toughness, will strongly depend on the fiber spacing and so, for a given volume fraction, on fiber size. Therefore, larger diameter reinforcing fibers will lead to a tougher composite. This result was first achieved by Cooper and Kelly (1967) who predicted a linear relation between the plastic work contribution of the matrix, Γ_M, to fracture surface energy and the fiber diameter, d, for the case when the fiber is strongly bonded to the matrix, with

$$\Gamma_M = \frac{(1-V_F)^2}{V_F}\sigma_M' \varepsilon_M' d_F \qquad (14.2)$$

where σ_M and ε_M' are the matrix ultimate failure stress and strain respectively, and d_F is the mean fiber diameter. When the fibers are weakly bonded to the matrix, toughening is controlled by frictional relaxations and work caused by fiber/matrix displacement during fracture. The contribution that fiber pull-out makes to toughening, $\Gamma_{\text{pull-out}}$, also depends on fiber size (Kelly 1970).

$$\Gamma_{\text{pull-out}} = \frac{V_F d_F \sigma_F'^2}{24\tau} \qquad (14.3)$$

Here, τ is the fiber/matrix interface shear strength. Finally, Morley (1983) has shown that if the fibers deflect a crack, and if the crack grows and leaves the

Fracture Behavior

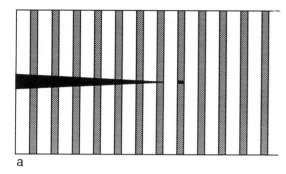

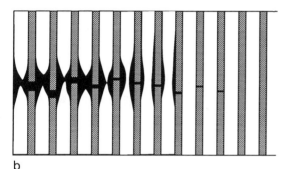

Figure 14.4. Crack propagating through LFMMC with strong fiber/matrix interface. (a) A planar crack occurs with brittle fracture in the matrix and highly constrained ductile matrix failure. (b) Weak fiber/matrix interface. Ductile flow in the matrix leads to extensive decohesion and nonplanar fracture.

fibers unbroken and thus bridging a crack, the crack tip driving force will be reduced and result in ultimate crack arrest. Further straining results in the accumulation of a family of stable bridged cracks in the matrix. Final failure occurs when there is sufficient load to initiate fiber failure or when the crack density reaches a level sufficient to induce global failure of the matrix.

The contribution of matrix plasticity to composite toughening also depends critically on the fiber/matrix adhesion. The toughening increment from matrix plasticity given by Equation 14.2 scales with the volume of material that must flow plastically during fracture. The plastic flow of the matrix will be constrained by the presence of the unyielding fibers, thus the matrix stress/strain response will be very different from that monitored by conventional tensile testing of the matrix alone. The influence of constraint of plastic flow has been studied using a model glass-lead system (Ashby, et al. 1989; Bannister and Ashby 1991). In these experiments, two model systems were investigated: a lead-filled glass tube, and a thin lead sheet constrained by parallel glass blocks. In both cases, cracks were introduced into the brittle glass and arrested at the glass/lead interfaces, the flow behavior of the matrix during straining was then studied. In these experiments, the energy absorbed during extension was found to be strongly dependent on the degree of glass/lead debonding. Thus, it would be expected that the extent of fiber/matrix adhesion would strongly influence the toughness in long fiber metal-matrix composites, with strongly bonded fibers resulting in smaller regions of plastic deformation in the matrix and consequently lower toughness.

Unfortunately, there have been too few systematic studies of the relationship between interface adhesion and fracture in a range of long fiber metal-matrix composites to fully test this debonding model. Such systematic studies, as have been published (Yang et al. 1991; Jeng et al. 1991a, 1991b) cannot separate changes in fiber/matrix decohesion behavior with other changes in the matrix or fiber/matrix interface.

14.1.2 Fracture During Cyclic Loading

Long fiber reinforced Al alloys show much greater fatigue resistance than do their unreinforced matrix alloys (Schulte and Minoshima 1991). This is not surprising considering that (1) the higher modulus ceramic fibers will bear the majority of the cyclic load and (2) ceramics are known to be highly resistant to fatigue failure. Schulte and Minoshima (1991) found fatigue cracks nucleating only after $> 10^6$ cycles, with a stress range of 600 MPa in a 40%V_F SiC fibre reinforced Al 1070 alloy. The crack nucleation was clearly a matrix-dominated event. Nucleation sites were at defects and in matrix-rich regions where shear band formation was possible. In these examples, fatigue life is nucleation controlled. However, there has recently been considerable interest in studying the growth of fatigue cracks through long fiber metal-matrix composites.

Cox and Marshall (1991) reviewed the considerable similarities that occur between fatigue cracks in metal-matrix composites and tensile cracks in highly brittle matrix composites. Crack propagation occurs mainly within the matrix in both cases, and three distinct modes of failure are identified for the case of a notch introduced perpendicular to the fiber axis:

1. Longitudinal crack growth. If the fiber/matrix interface is very weak, cracks propagate at 90 degrees to the notch along the fiber/matrix interface. See Figure 14.5(a).
2. Mode I (noncatastrophic). Here the interface is stronger and only partial debonding of the fibers occurs. A matrix crack can then propagate and leave

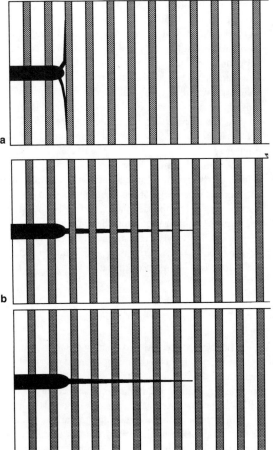

are complex but are controlled by microstructural parameters such as fiber size, volume fraction, and fiber/matrix interface strength. The transition from longitudinal to noncatastrophic Mode I failure is also controlled by the loading mode, with bending, such as is found in a single-edge notch specimen, tending to promote longitudinal failure. Cox and Marshall have developed a model for noncatastrophic Mode I failure in which there is a bridging zone of fibers behind the crack tip. Their shielding action is modeled as a distribution of crack closure forces on the crack surface (Marshall et al. 1985; Marshall and Cox 1987). Similar models have been developed by McCartney (1989). Cardona et al. (1993) developed a model using discrete closure loads to represent the fiber fractions. These models show that noncatastrophic failure is favored by: high fiber strength, large fiber radius, large matrix modulus, small fiber/matrix interfacial friction stress, and small fiber modulus.

Ibbotson et al. (1991) have presented a thorough experimental study confirming the nature of crack bridging in SiC fiber reinforced Ti-6Al-4V (Figure 14.6). A series of tests was carried out on an unidirectional composite. The transition from noncatastrophic to catastrophic Mode I propagation was observed and the number of intact fibers bridging the crack was measured. This allowed a rudimentary fracture map to be constructed that plots the fracture toughness at transition as a function of the number of fibers bridging a crack. Further experimental work of this nature will be needed to validate the crack growth models.

Figure 14.5. Fatigue crack propagating from a notch in an LFMMC. (a) Very weak fiber/matrix interface results in a longitudinal splitting mode. (b) Intermediate strength interface leads to stable Mode I cracking with a zone of bridging fibers behind the crack tip. (c) High strength fiber/matrix interface results in rapid Mode I crack growth.

intact fibers bridging the specimen. (See Figure 14.5(b).) More than one crack may grow and lead to a family of parallel cracks in the composite. This is essentially the fracture behavior predicted by Aveston et al. (1971) for the behavior of brittle matrix composites in tension.
3. Mode I (catastrophic). Here the fiber/matrix interface is so well bonded that a catastrophic crack propagation occurs, leading to immediate tensile fracture. See Figure 14.5(c).

The conditions that determine which of these failure modes is dominant at a given stress level in a composite

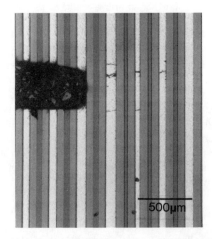

Figure 14.6. Fatigue crack propagating in an SCS-6 SiC reinforced Ti-6Al-4V LFMMC, zone of crack bridging is evident. *(Reprinted with permission from P. Bowen, University of Birmingham, U.K.)*

14.1.3 The Fiber/Matrix Interface

This is clearly one of the most important regions of composite microstructure. First, as discussed in detail in the previous two sections, it directly controls matrix behavior by imposing local plastic constraint or by affecting crack paths. Second, the interface region can itself be a source of defects for fracture nucleation.

Most metal-matrix composites are not in thermodynamic equilibrium and there is considerable driving force for chemical change at the metal/ceramic interface. Kinetic conditions suitable for reaction product formation can occur either during fabrication or during high-temperature service. Some possible interface interactions are:

1. $4Al + 3 SiC \rightarrow Al_4C_3 + 3Si$
2. $8Ti + 3SiC \rightarrow 3TiC + Ti_5Si_3$
3. $5Ti + Al_2O_3 \rightarrow 3TiO + 2Ti Al$

All of these interactions result in brittle interfacial products. The brittle interphases will be stressed whether by volume changes during reaction or by thermal residual strains upon cooling. The thickness of any interfacial reaction zone is important for controlling fracture behavior. If we assume that the strength of a fiber is limited by the presence of intrinsic defects, then the maximum reaction layer thickness should be no greater than the size of these, or premature fiber fracture will occur. The intrinsic defect size, c_{crit}, can be defined using a Griffiths formulation, with

$$c_{crit} = \frac{1}{Y}\left(\frac{K_{Fc}}{\sigma_F'}\right)^2 \quad (14.4)$$

where Y is a dimensionless stress intensity factor and K_{Fc} is the fracture toughness of the fiber. For an SiC monofilament, $K_{Fc} \approx 4 MPa\sqrt{m}$ and $\sigma_F' \approx 4 GPa$, which gives a critical reaction layer thickness of around $1\mu m$. Metcalfe (1967) used a similar argument to produce a slightly different result with

$$c_{crit} = \left(\frac{E_F}{10B\sigma_F'}\right)^2 r \quad (14.5)$$

where r is the crack tip radius in the reaction layer and $1 < B < 2$ is a dimensionless crack shape constant. Using $E_F \approx 400$ GPa for SiC, taking $B = 1$ and $r \approx 10^{-9}$m gives $c_{crit} \approx 0.1\mu m$. Thus, even very small reaction layer thicknesses can have serious implications on composite behavior (Metcalfe and Klein 1974; Ochiai et al. 1980; Fishkis 1991).

Equation 14.2 shows the importance of the fiber/matrix interface in controlling the contribution from fiber pull-out to fracture energy. Low values of the interfacial shear stress, τ, lead to high values of $\Gamma_{pull-out}$.

Although the contribution from matrix plasticity (Equation 14.1) is expected to dominate in most ductile matrix composites, the pull-out term will be important in composites with brittle intermetallic matrices such as Ti_3Al. It is documented that such intermetallic matrix composites show better mechanical properties if they possess weak, low shear strength, fiber/matrix interfaces (Cox et al. 1989). Even when Equation 14.1 dominates, the extent of plasticity will depend on local constraints that are also affected by the fiber/matrix adhesion. In order to characterize the adhesion between fiber and matrix, a number of mechanical test techniques have been developed. These can be grouped into three broad classes distinguished by their mechanical loading:

Type (1) Single fibers loaded in compression, either pushed through thin slices of composites or pushed into the bulk.

Type (2) Single fibers loaded in tension and pulled out of the composite.

Type (3) Fragmentation of fibers within the composite during tensile loading of the material.

These are illustrated schematically in Figure 14.7.

Tests of Type (1) were pioneered by Marshall and co-workers (Marshall 1984; Marshall and Oliver 1987). The technique has been applied to measure interfacial sliding resistance in Ti_3Al matrix composites (Eldridge and Brindley 1989; Yang et al. 1990b) and for Ti-6Al-4V composites (Watson and Clyne 1992). In these tests, thin slices of the composite, about 300 to 500 μm thick, have their fibers loaded in compression using a hardness indenter. The load at which displacement occurs is assumed to be balanced by the interfacial shear stress τ, which can be calculated if the thickness of the specimen is known.

Pull-out tests of Type (2) to determine τ, are common in the field of fiber reinforced polymers. Their use in testing interfacial strength in metal-matrix composites has been limited by the ability to fabricate suitable specimens in which a single fiber can be pulled from a matrix. Marshall et al. (1992) used micromachining to make suitable specimens. They argued that most pushing tests, because they use a fixed load, measure a peak stress for the initiation of interface fracture and not the more important shear stress, which resists sliding. The nature of the pull-out tests allows a better record of the load/displacement behavior and a more accurate measure of τ. Recent developments in tests of Type (1) have allowed complete load/displacement records to be monitored and more accurate values for τ to be obtained (Warren 1992). However, it can be argued that the loading of tests of Type (2) is closer to the real fiber loading during fracture.

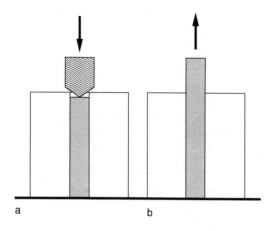

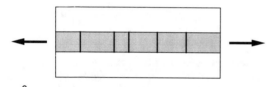

Figure 14.7. Schematic representation of the loading regimes used in determining fiber/matrix interfacial shear stress (τ). (a) Push-down or push-through configuration with an individual fiber loaded in indentation. (b) Pull-out configuration, essentially a reversal of (a). (c) Fiber fragmentation during tensile loading of the composite.

Table 14.4. Interfacial Shear Stress τ as Measured in a Range of Long Fiber Reinforced Metals.

System	τ (MPa)	Test Type Used	Data Source
ATi-6Al-4V/SiC(AVCO)	750	(3)	1
Ti-6Al-4V/SiC(BP)	345	(3)	1
Ti-6Al-4V/B(B4C)	240	(3)	1
Ti-6Al-4V/SiC(C)	180	(3)	1
Ti$_3$Al/SiC(C)	40–66	(2)	2
Ti-6Al-4V/SiC(C)	80	(1)	3
Ti-24Al-11Nb/SiC(C)	120	(1)	4
Ti-25Al-10Nb/SiC(C)	58	(1)	5
Ti-6Al-4V/SiC(C)	160	(1)	5
Ti-15V-3Al-3Cr/SiC(C)	120	(1)	5
Ti-6Al-4V/B(B$_4$C)	275	(1)	6

SiC(C) is the AVCO SCS6 C coated SiC fiber. [1]Le Petitcorps et al. 1989; [2]Marshall et al. 1992; [3]Watson and Clyne 1992; [4]Eldridge and Brindley 1989; [5]Yang et al. 1990b; [6]Yang et al. 1991.

14.2 Discontinuously Reinforced Composites

14.2.1 Tensile Fracture

The fracture surface morphology of discontinuously-reinforced metal-matrix composites (DRMMCs) has features characteristic of a matrix ductile rupture mechanism (Figure 14.2). This failure process can be conveniently split into three stages: void nucleation, growth, and coalescence. Each of these will be discussed. An alternative description, based on the initiation and accumulation of damage, will also be considered briefly.

14.2.1.1 Void Nucleation

The extension of simple models developed to describe the ductile fracture of monolithic alloys to DRMMCs may imply that the onset of void nucleation is the dominant process controlling the ductility in these materials. These models propose that void growth continues until a critical void size or volume fraction is attained. This is generally determined by some geometric consideration. If we consider the simplest formulation for the ductile rupture process, presented by Brown and Embury (1973), voids coalesce by a shear failure in the matrix between the voids when the plastic constraint for flow localization is removed. This occurs when the void height is equal to the intervoid spacing. This condition is satisfied upon void nucleation by increasing the volume fraction of void nucleating sites to 0.16. Thus, for the high volume fractions of reinforcements found in the more commercially attractive DRMMCs, one might expect the nucleation process to dominate if void

Fiber fragmentation tests of Type (3) have been developed to measure τ in polymer-matrix composites. (Fraser et al. 1975). In this test, a composite containing fibers of low failure strain are stressed parallel to the fibers until multiple cracking of the fibers occurs. The interfacial shear strength and fiber strength are related to the mean spacing of the cracks by an analysis identical to that used by Aveston et al. (1971) for matrix crack spacing in brittle matrix composites. This solution was extended by LePetitcorps et al. (1988) to account for statistical variations in fiber strength.

Table 14.4 shows a range of fiber/matrix interfacial shear strengths measured by the workers listed above. Note how the unprotected fibers tend to give large values of τ and note the general similarity of τ values derived by the three techniques when used on similar composites. The shear stresses measured are about one to two orders of magnitude greater than those normally found for ceramic- and polymer-matrix composites.

nucleation is at the reinforcing phase. It is, therefore, important to isolate and identify the void nucleation process and to consider the microstructural parameters that influence it. The prediction that the rate of nucleation can be identified with the ductility can then be tested.

Most authors (for example, Nair et al. 1985; Nieh et al. 1985; Flom and Arsenault 1989) assume that fracture in DRMMCs follows the same sequence as dispersion-strengthened alloys, namely, nucleation at the second phase particles followed by failure in the matrix through void coalescence. You et al. (1987), however, with some experimental support from Roebuck (1987), contended that the increased levels of stress and high levels of plastic constraint, imposed by the reinforcing particles on the matrix, lead to void nucleation in the matrix as the initiation step, with the final stage of fracture being the decohesion or cracking of the particles. In situ experiments, in which the deformation of a composite was studied while straining within a scanning electron microscope (SEM), have isolated and identified the void nucleation stage (Mummery and Derby 1989; Manoharan and Lewandowski 1989a). These experiments showed void nucleation occurring at the reinforcing particles and not in the matrix.

Two dominant void nucleation modes have been observed in DRMMCs (Figures 14.8 and 14.9): particle cracking (You et al. 1987; Roebuck 1987; Lloyd 1991; Davidson 1987), and decohesion at the particle/matrix interface (Crowe et al. 1985; Stephens et al. 1988; Manoharan and Lewandowski 1990a). The mode of void nucleation can be identified by simple fractography, although as Roebuck (1987) pointed out, a unique determination can only be made by matching areas from both halves of a fracture surface. The mode is sensitive to a number of microstructural parameters, such as size and volume fraction of reinforcement, and in studies where a systematic variation of these parameters has been made, transitions between these modes have sometimes been observed (Mummery and Derby 1991; Vasudevan et al. 1989; Yang et al. 1990a).

The two parameters that have the greatest influence on the mode of nucleation are the size of reinforcing phase and the interfacial bond strength. A change from interfacial decohesion to particle cracking has been observed on increasing the particle size. The reinforcing particles are brittle ceramics and, on adopting a Griffiths approach, this transition has been attributed to a reduction in mean failure stress of the particle with increasing size. The particle/matrix interfacial bond strength has been altered by a number of methods. In the first method, a change in matrix alloy composition was used (Man et al. 1991; Manoharan and Lewandowski 1989b; Strangwood et al. 1991). Here, the segregation of an alloying element to the interface, or its reaction

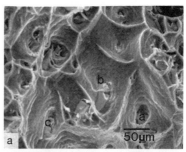
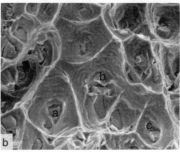

Figure 14.8. A matched fracture pair, (a) and (b), from an Al-1070 matrix composite containing a 5% volume fraction of 30 μm SiC particles. The presence of particles on both halves shows that the void nucleation mechanism was particle fracture. The meeting of the voids at a sharp lip and the serpentine glide on the inside of the voids are indicative of extensive void growth and coalescence through shear failure in the matrix. *(Reprinted with permission from Mummery, P.M., Derby, B., Buttle, D.J., and C.B. Scruby. Micromechanisms of Fracture in Particle-Reinforced Metal-Matrix Composites: Acoustic Emission and Modulus Reduction. Proc. Euromat 91. Institute of Materials, London.)*

with the reinforcements, have been used to vary the bond strength. The second method changed the surface properties of the reinforcement by baking it in a furnace (Ribes et al. 1990). This encouraged the formation of a brittle phase at the interface through which failure proceeded. A different reinforcement type within the same matrix has also been used by Stephens et al. 1988. Other parameters that have been shown to affect the nucleation mode are volume fraction (Mummery 1991), aspect ratio of the reinforcement (Whitehouse et al. 1991), matrix heat treatment (Manoharan and Lewandowski 1990a), and strain rate (Pickard et al. 1988). The influence of the mode of void nucleation on the ductility of the composite will be considered in Section 14.2.1.2.

We can now test the prediction that the void nucleation process dominates the ductility of DRMMCs. Their ductility has been found to be a function of a number of microstructural parameters. For a given matrix alloy, elongation to failure is reduced by increasing volume fraction (as shown in Table 14.2 (McDanels 1985; Lloyd 1991; Miller and Humphreys 1991) and

260 DAMAGE MICROMECHANISMS AND MECHANICS OF FAILURE

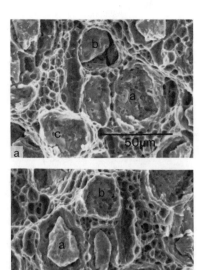

Figure 14.9. A matched fracture pair, (a) and (b), from an Al-5050 matrix composite containing a 5% volume fraction of 30 μm SiC particles. The particles are matched with voids, showing that the void nucleation mechanism was by decohesion at the particle/matrix interface. The presence of extensive local voiding in the matrix reduces the constraint on matrix flow and reduces macroscopic ductility. *(Reprinted with permission from P.M. Mummery and B. Derby. 1991. Metal-Matrix Composites: Processing, Microstructure and Properties. 12th Risø Symposium on Materials Science.)*

the size of the reinforcement (Kamat et al. 1989; Yang et al. 1990; Stephens et al. 1988; Liu et al. 1989). Composites of high-strength alloys have lower ductility than those of low-strength alloy matrices (Nair et al. 1985; Girot et al. 1987; England and Hall 1986) with decreasing ductility on aging to peak matrix strength (McDanels 1985; Lewandowski et al. 1989a, 1989b; Papazian and Adler 1990; Lloyd 1991). In non age-hardening alloys, the ductility has also been reduced on quenching (Miller and Humphreys 1989; Mummery and Derby 1989). The effect of the matrix parameters has been modeled as increasing the rate of void nucleation at the reinforcing phase by the suppression of strain relief mechanisms at the interface (Humphreys 1988). The models can now be tested because the rate of void nucleation should increase as the composite ductility is decreased.

Mummery et al. (1993) have monitored the acoustic emissions on tensile straining of a series of composite materials in which there has been a systematic variation in the microstructural parameters just outlined above (Figure 14.10). They showed that each emission corresponded to one nucleation event, and they also found that there was void nucleation from the onset of plastic deformation, which continued throughout the plastic region for some 10% to 20% additional strain. This indicates that there must be considerable void growth before failure. The rates of void nucleation (the number of emissions at a given far-field strain) increased with particle size and volume fraction, correlating with a decrease in ductility, but were independent of matrix yield strength and

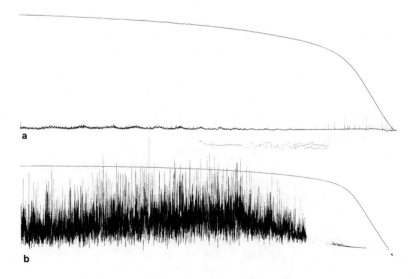

Figure 14.10. Combined load/displacement and r.m.s. acoustic emission traces from Al matrix composites with (a) 10 μm SiC particles, and (b) 30 μm SiC particles. The emissions were also recorded digitally to provide quatitative data. *(Reprinted with permission from Mummery, P.M., Derby, B., Buttle, D.J., and C.B. Scruby. Micromechanisms of Fracture in Particle-Reinforced Metal-Matrix Composites: Acoustic Emission and Modulus Reduction. Proc. Euromat 91. Institute of Materials, London.)*

heat treatment. The matrix must, therefore, affect void growth and coalescence but not nucleation.

Thus, the ductility of DRMMCs is not simply related to the rate of void nucleation. We can, however, increase the elongation to failure of the composites by suppressing void nucleation at the reinforcing phase. The composite's ductility is governed by matrix processes, that will be affected by the presence of the reinforcements. This is evidenced by the decrease in ductility while increasing reinforcement volume fraction.

14.2.1.2 Void Growth and Coalescence

Void growth and coalescence have been much neglected in the study of DRMMCs because they have been considered unimportant and difficult to monitor experimentally. However, as stated earlier, the ductility of DRMMCs cannot be uniquely correlated with the void nucleation rate at the reinforcing particles. For this reason, these additional processes must be considered.

Detailed microstructural information on void growth and coalescence can be obtained from sections through failed tensile specimens. If local deformation has taken place, distances away from the fracture surface correspond to lower strains and earlier stages in the failure process. However, if there is limited macroscopic ductility, most of the damage processes are confined to regions very close to the final fracture surface and no stable voids are seen (You et al. 1987; Lloyd et al. 1989; Lewandowski et al. 1989b). Studies of composites with high ductility matrices, such as pure aluminum, have provided some information on the nature of void growth and have indicated the influence of the nucleation mechanism on subsequent void coalescence (Mummery et al. 1992a; Mummery and Derby 1991).

Figure 14.11(a) shows a region near the fracture surface of a composite containing 20% volume fraction of 30 μm silicon carbide particles in an Al-1070 matrix. Most of the particles are cracked through their center with the plane of the crack normal to the applied stress in Mode I. Figure 14.11(b) shows a similar region, near the fracture surface of a composite, of 30 μm particles in the same Al-1070 matrix but of only 5% volume fraction. Extensive void growth can be seen in the tensile direction although no lateral growth is evident, indicating the great constraining effect that the still well-bonded interfaces have on matrix deformation. Linkage between adjacent voids is rarely seen. Figure 14.12 shows the 30 μm particles in an Al-1%Mg matrix. In this case, the voids have nucleated by decohesion at the particle ends, as has been observed elsewhere (Nutt and Duva 1986). Note the lateral growth and the very small amount of void growth in the tensile direction when compared with the Al-1070 matrix compos-

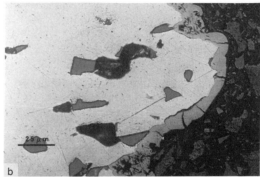

Figure 14.11. Section through failed tensile specimen of an Al-1070 matrix composite containing (a) 20% volume fraction of 30 μm SiC particles, and (b) 5% volume fraction of 30 μm particles. Note the extensive, constrained void growth. *(Reprinted with permission from Mummery, P.M., Derby, B., Buttle, D.J., and C.B. Scruby. Micromechanisms of Fracture in Particle-Reinforced Metal-Matrix Composites: Acoustic Emission and Modulus Reduction. Proc. Euromat 91. Institute of Materials, London.)*

ite. The elongation to failure of each composite in an Al-Mg matrix is approximately half that of the equivalent composite in an Al-1070 matrix (Mummery et al. 1992b), despite having the same void nucleation rate. Similarly, Manoharan and Lewandowski (1990a) found a significant reduction in crack initiation and growth toughness by inducing a change in void nucleation mechanism from particle fracture to interfacial decohesion on overaging. Thus, it appears that void nucleation by particle fracture hinders coalescence by constraining matrix flow, leading to increased elongation to failure and toughness.

Of greater influence on the strain at which coalescence occurs are local failure processes within the matrix. As voids are nucleated within the matrix, the constraint on plastic flow is reduced allowing void coalescence at lower strains. This is the principal cause of the lower ductility of composites of high yield strength matrices and not an increased void nucleation rate, as has been suggested (Humphreys 1988). This is consistent with both experimental and modeling work, which

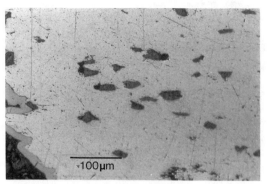

Figure 14.12. Section through failed tensile specimen of an Al-5050 matrix composite containing 5% volume fraction of 30 μm particles. Note the small growth and relatively easy coalescence. *(Reprinted with permission from Mummery, P.M., Derby, B., Buttle, D.J., and C.B. Scruby. Micromechanisms of Fracture in Particle-Reinforced Metal-Matrix Composites: Acoustic Emission and Modulus Reduction. Proc. Euromat 91. Institute of Materials, London.)*

considered the influence of a second population of small voids on the growth of large voids and subsequent onset of flow localization (Ohno and Hutchinson 1984; Melander 1980; Dubensky and Koss 1987). The effect of local matrix failure can be seen as a change in fracture surface morphology. Figures 14.8(a) and (b) are matched fracture halves from the same Al-1070 matrix composite discussed earlier. Note the simple appearance of the fracture surface with the matrix necking to a knife edge between the particles. Note also the serpentine glide on the dimples, which is indicative of a shear coalescence mechanism. Figures 14.9(a) and (b) are from a composite containing the same size and volume fraction of reinforcing phase but are now in an Al-1%Mg matrix. The more complicated fracture surface is the result of local matrix failure. The presence of the matrix voids has led to a reduction in the size of the voids associated with the particles and reduced macroscopic ductility.

The influence of matrix processes on void coalescence in DRMMCs has been shown in other studies. Lloyd (1991) took specimens of an Al-6061 matrix reinforced with silicon carbide, strained them to varying degrees, and then unloaded them. The material was then resolutionized to recover the matrix defect damage such as dislocations. This process did not recover particle fracture or cavitation. The specimens were then reloaded and the yield stress and strain to failure were recorded. It was found that the sum of the prestrain and subsequent strain to failure was equal to the strain to failure of the uninterrupted test, some 7% to 8%, for prestrain levels up to 5.5%. Although the ductility was recovered, the yield stress consistently fell below that of the uninterrupted test. Similarly, the suppression of void nucleation and the hindrance of void growth by performing tensile tests under an applied hydrostatic pressure has led to substantial increases in the elongation to failure (Liu et al. 1989; Zok et al. 1988; Lewandowski et al. 1991). It is, however, difficult to obtain a detailed knowledge of the void growth and coalescence processes as the tensile ductility of most DRMMCs is limited. Failure must be postponed artificially so that damage may develop away from the final fracture plane.

14.2.1.3 Damage Mechanics

An alternative approach that does not rely on a detailed microstructural knowledge of the failure process is to consider the change in some physical property of the material that mirrors its progression to failure. This damage mechanics approach is very attractive because failure can be monitored without the need for the development of real-time fracture experiments and because it can circumvent the problem of low ductility. In this case, failure is said to follow when some damage parameter reaches a critical value. This value will be a function of both the failure processes and the known microstructural variables as discussed earlier. Damage progresses in two stages: initiation and accumulation. Damage initiation must be deconvoluted from damage accumulation to allow the relative importance of each process to be assessed. Damage initiation can be measured directly by the acoustic emission technique so a method for monitoring damage accumulation is required.

In the case of DRMMCs that follow a ductile rupture mechanism and for which failure occurs when the void volume fraction reaches a critical level, a number of parameters have been suggested. Lloyd (1991), following work on a model composite system of a modified Al-Si eutectic (Hunt et al. 1991), has considered the reduction in elastic modulus as a function of strain. As the reinforcing particles are cracked, their load-bearing contribution is reduced. The total void volume fraction can be estimated by substitution into models for the modulus of porous media. A more traditional approach has been followed by Whitehouse et al. (1991) where the change in density on straining has been monitored. Embury and his co-workers (Embury et al. (1991); Brechet et al. (1991)) have considered the change in size of the particulates during compression or during tensile loading under applied hydrostatic pressure. During straining, the particles fracture and the variation in mean particle diameter is a measure of the damage initiation and accumulation. For these damage measures to be used in a model of DRMMC failure, much further work must be carried out on a range of systems with different microstructural parameters. There is much in common in this approach with the development of damage models used in other fracture problems (Lemaitre 1992).

14.2.1.4 Crack Propagation Under Monotonic Loading

There are two approaches to the study of monotonic crack propagation: real time and post failure. The real time techniques study the crack as it propagates by deforming a specimen inside a scanning electron microscope (SEM), inside a transmission electron microscope (TEM), or under an optical microscope.

The in situ SEM studies can themselves be subdivided into tensile straining (Wu and Arsenault 1989; Da Silva et al. 1988; Manoharan and Lewandowski 1989a, 1990b; Kryze et al 1991), and loading in other geometries (Mummery and Derby 1989; Mummery and Derby 1991). The observations of crack propagation were consistent in both techniques and across a range of high- and low-strength aluminum alloy matrices reinforced with silicon carbide and alumina particles. Voids were nucleated at the reinforcing phase before the onset of matrix failure. Upon further straining, a macroscopic crack was formed. Microcracked areas were found ahead of and near the crack tip that was associated with a region of intense deformation. The regions extended large distances ahead of the crack tip, typically some tens of interparticle spacings. Crack propagation occurred by a process whereby some of the microcracks join via matrix failure ahead of the original crack tip, concurrent with the initiation of additional microcracks ahead of the crack. This propagation sequence led to other experimental observations such as crack bridging and branching associated with discontinuities in the crack on the specimen surface. The results of the in situ TEM studies (Doong et al. 1989) showed broad agreement with the nucleation processes.

These in situ techniques necessarily study either a free surface or a thin foil; in both cases the fracture process is observed in plane stress. Neither of these may be representative of the failure processes within the bulk of a material which is in plane strain. Indeed, a number of the in situ observations were inconsistent with other fractographic and theoretic results. Wu and Arsenault (1989) and Mummery and Derby (1991) noted that during their in situ experiments, the crack avoided particles that had fractured earlier in the straining process, preferring to propagate through the matrix and away from the interface. Here fractography had previously shown the crack to pass through the particles. In addition, if an analogy between reinforcement failure in DRMMCs and carbide fracture in steels is made, the extent of damage ahead of the crack tip is far larger than that expected for the composites (Ritchie et al. 1973). Post-fracture studies on bulk specimens should be used to validate the in situ findings.

Mummery and Derby (1991) have loaded bulk specimens in a constant displacement, double cantilever arrangement. Sections were taken through stable cracks introduced into the material. Nucleation events can be seen ahead of the crack tip (Figure 14.13) and the crack profile shows regions of greater void growth linked by regions of lesser growth (Figure 14.14). This showed that the crack discontinuities on the surface observed during in situ straining do continue into the bulk. The variation in void growth near the crack tip was indicative of the simultaneous microcrack nucleation and ligament failure mechanism seen earlier. However, there was no evidence of crack branching and the damaged region ahead of the crack tip was greatly reduced. Because the plastic zone ahead of a crack tip is larger in plane stress than it is in plane strain, such a reduction in the damage region implies that the particles are loaded by shear forces at the interface caused by matrix deformation. Similar results have been obtained by Shang and Ritchie (1989a). As they increased ΔK in their fatigue tests, and so increased the size of the plastic zone associated with the crack tip, they found an increased damage region and crack bridging and branching. These post-fracture studies showed that the basic mechanism of crack propagation observed during in situ tests is valid, but care must be taken in their interpretation.

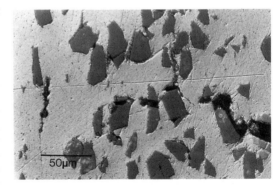

Figure 14.13. Detail of crack tip processes in an Al-1070 matrix composite containing 20% volume fraction of 30 μm SiC particles. The extent of damage ahead of the crack tip is confined to 1–2 interparticle spacings. *(Reprinted with permission from P.M. Mummery and B. Derby. 1991. Metal-Matrix Composites: Composites: Processing, Microstructure and Properties. 12th Risø Symp. on Material Science.)*

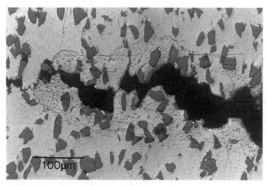

Figure 14.14. Section through stable crack in an Al-1070 matrix composite containing 20% volume fraction of 30 μm SiC particles. The nonuniform profile indicates a discontinuous growth mechanism. *(Reprinted with permission from P.M. Mummery and B. Derby. 1991. Metal-Matrix Composites: Processing, Microstructure and Properties. 12th Risø Symposium on Materials Sciences.)*

14.2.2 Fracture During Cyclic Loading

Despite having very inferior fracture toughnesses, DRMMCs often exhibit improved fatigue thresholds and crack growth rates over their constituent monolithic alloys. Many results have been reported on specific composite systems where there has been no variation in the microstructural parameters (Pao et al. 1989; Levin et al. 1989; Manoharan and Lewandowski 1989c; Biner 1989; You and Allison 1989). This approach does not aid the understanding of the role that the microstructural parameters play. In addition, the composites have often been compared with the unreinforced matrix alloy, which has had a different thermomechanical and processing history (Logsdon and Liaw 1986; McDanels 1985).

The behavior of DRMMCs in fatigue appears to be extremely sensitive to microstructural changes. In tests where a variation in the parameters has been made, unique conclusions cannot be obtained for their influence on fatigue fracture resistance without additional knowledge of the mode of crack growth (Crowe et al. 1985; Logsdon and Liaw 1986; Davidson 1987; Yau and Mayer 1986). For example, although some studies find an increase in fatigue growth resistance on increasing reinforcement volume fraction (Christman and Suresh 1988; Shang and Ritchie 1989; Davidson 1989a, 1989b), others find an opposing trend (Kumai et al. 1990; Suresh 1991). However, a number of processes that can be affected by the choice of size and volume fraction of reinforcement have been shown to be of importance.

The influence of particle size (Shang and Ritchie 1989a, 1989b; Shang et al. 1988) and volume fraction (Sugimura and Suresh 1992) of reinforcement on fatigue crack growth have been studied. These studies have shown that crack closure increases the threshold for crack initiation and produces lower crack growth rates in the composites when compared with the unreinforced matrix at low load ratios. The crack closure has been associated with crack deflection around the particles and asperity wedging caused by fracture surface roughness. Thus, a greater improvement in fatigue properties can be observed on increasing particle size. However, this mechanism for increasing growth resistance seems to be applicable only when the mode of crack propagation does not involve significant fracturing of the reinforcing phase. Sugimura and Suresh (1992) suggest that the crack front that passes through a particle is smoother microscopically than is the ductile path through the matrix, which has striations.

With increasing ΔK, there is increased fracture of the particles and the influence of crack closure on growth resistance is essentially removed. In this regime, the composites and their unreinforced matrix alloys have comparable crack growth rates. However, Shang et al. (1988) and Shang and Ritchie (1989a) have reported other crack tip shielding mechanisms responsible for some improvement in growth resistance, namely bridging induced by the presence of uncracked ligaments behind the crack tip. It appears to result from fracture events triggered ahead of the crack tip or from general nonuniform or discontinuous advance of the crack front. Two forms of bridging were observed: coplanar ligaments, where the intact ligament was in the same plane as the advancing crack, and overlapping ligaments, where there was significant development of the crack out of a single plane of advance. The shielding caused by coplanar uncracked ligaments was found to be negligible as compared with the substantial shielding of the overlapping ligaments. Another mechanism that retards crack growth is trapping of the crack tip at the reinforcing phase (Shang and Ritchie 1990).

As ΔK increases even further, approaching the critical stress intensity factor K_{IC}, crack growth rates in the composites become significantly higher than in the unreinforced material. This is shown in tests performed under strain control in which crack growth rates or reversals to failure are poorer in the composites than they are in the unreinforced matrices (Bonnen et al. 1991).

Summary

The fracture properties of both LFMMC and DRMMC are controlled by the behavior of the matrix when the matrix is ductile. However, the behavior of the matrix is modified by the presence of the more rigid reinforcing phase, with the nature of the metal/ceramic interface playing a key part. With brittle matrix material, such as intermetallics, the fiber/matrix interface plays a more important role controlling crack deflection, crack bridging, and pull-out toughening contributions.

In LFMMC, the ultimate tensile strength is determined by a rule of mixtures, with the high-strength ceramic fibers dominating. However, fracture energy is dominated by the plastic work of fracture of metal ligaments that occurs between fractured fibers. The toughness of these composites will be greater if large volumes of the matrix can deform during fracture. The extent of matrix plasticity will be strongly affected by the constraint imposed by the presence of fibers bonded to the matrix. Work with model constrained systems has clearly demonstrated that a lowering of interface adhesion greatly increases the volume of matrix deforming. However, there is insufficient systematic work on the influence of the fiber/matrix interface on the fracture of LFMMCs with ductile matrices to confirm this trend.

Fiber/matrix interfacial mechanical properties have been investigated by a range of different techniques. These include techniques that test a single fiber and those that can make multiple fiber measurements. These tests produce measurements of the interfacial friction stress in the range of 50 MPa to 100 MPa for a range of ceramic fibers within Ti alloy matrices (Table 14.4), with reasonable consistency between different testing techniques used on similar LFMMCs. These energies are about one order of magnitude greater than those found with reinforced glass and glass/ceramic-matrix composites. However, this does not indicate a lower toughness with LFMMCs than is found with these ceramic-matrix composites because the monofilament fibers used with Ti matrices are about an order of magnitude larger, and cancel out the influence of the stronger interfaces (Equation 14.2).

With DRMMCs is found a lower level of understanding of the mechanisms that control composite fracture. Fracture is, again, a ductile phenomenon; however, with low toughness and ductility levels, this is believed to be the result of reduced plastic zone sizes ahead of growing cracks. There have been a number of in situ studies of crack propagation in these materials. Fracture is seen to occur by a conventional void nucleation and growth mechanism. Because of the very high volume of brittle reinforcement phases in DRMMC, and because the belief that these were all potential void nucleation sites, it was initially believed that the ductility and toughness of these materials was controlled by the onset of void nucleation. Many studies of void nucleation showed that both reinforcement fracture and reinforcement/matrix decohesion occur, although there was little correlation between nucleation mechanism and fracture properties. However, recent results indicate that void nucleation occurs very early in deformation, probably beginning immediately after the onset of plastic flow. Thus, any predictive model of DRMMC fracture should include both nucleation and growth components.

Studies of systematic variations in microstructural parameters (reinforcement size, volume fraction, and matrix composition) indicate that changes in void nucleation mechanism occur with changes in microstructure. There is some evidence that there is a slightly greater ductility when void nucleation occurs by particle fracture, and that this is related to the extent of constraint imposed by the particle on the growth of the neighboring void. Hence, some influence of reinforcement/matrix interface properties on the bulk fracture behavior of DRMMC would be expected. However, unlike the case of LFMMC, it is difficult to make direct measurements of interface strength within the composite. It is unlikely that strengths calculated using model systems can be used because of the strong influence of interface composition on strength in most interphase systems.

Acknowledgments

We would like to thank the Science and Engineering Research Council for supporting this work under Grant No. GR/F/87660. Paul M. Mummery would also like to thank the SERC for a post-doctoral fellowship.

References

Ashby, M.F., Blunt, F.J., and M. Bannister. 1989. *Acta Metall.* 37:1847–1857.

Aveston, J., Cooper, G.A., and A. Kelly. 1971. *In:* The Properties of Fibre Composites, 15–26, London: National Physical Laboratory, IPC Science and Technology Press.

Bannister, M., and M.F. Ashby. 1991. *Acta Metall. Mater.* 39: 2575–2582.

Barry, P.W. 1978. *J. Mater. Sci.* 13:2177–2187.

Biner, S.B. 1989. *In:* Fundamental Relationships between Microstructures and Mechanical Properties of Metal Matrix Composites. (M.N. Gungor and P.K. Liaw, eds.), 825–838, Warrendale, PA: TMS.

Bonnen, J.J., Allison, J.E., and J.W. Jones. 1991. *Metall. Trans. A.* 22A:1007–1021.

Brechet, Y., Embury, J.D., Tao, S., and L. Luo. 1991. *Acta Metall. Mater.* 39:1781–1786.

Brindley, P.K., Draper, S.L., Nathal, M.V., and J.I. Eldridge. 1990. *In:* Fundamental Relationships between Microstructures and Mechanical Properties of Metal Matrix Composites, (M.N. Gungor and P.K. Liaw, eds.), 387–401, Warrendale, PA:TMS.

Brown, L.M., and J.D. Embury. 1973. In: Proceedings of 3rd International Conference on Strength of Metals and Alloys 164–170, London: Institute of Metals.

Cardona, D.C., Knott, J.F., and P. Bowen. 1993. Composites. 24:122–128.

Christman, T., and Suresh, S. 1988. Mater. Sci. and Eng. 102: 211–224.

Coleman, B.D. 1958. J. Mech. Phys. Solids 7:60–70.

Cooper, G.A. 1971. Rev. Phys. Tech. 2:49–63.

Cooper, G.A., and A. Kelly. 1967. J. Mech. Phys Sol. 15:279–289.

Cox, B.N., James, M.R., Marshall, D.B. et al. 1989. In: Proceedings of the 10th International SAMPE Conference, Birmingham, UK.

Cox, B.N., and D.B. Marshall, D.B. 1991. Fatigue. Fract. Eng. Mater. Struct. 14: 847–861.

Crowe, C.R., Gray, R.A., and D.F. Hasson. 1985. In: Proceedings of the 5th International Conference on Composite Materials (W.C. Harrigan Jr., J. Strife, and A.K. Dhingra, eds.), 843–849, Warrendale, PA:TMS.

DaSilva, R., Caldemaison, D., and T. Bretham. 1988. In: Mechanical and Physical Behavior of Metallic and Ceramic Composites. Proceedings of the 9th Risø International Symposium of Materials Science. (S.I. Anderson, H. Lilholt and O.B. Pedersen, eds.), 333–341, Roskilde, Denmark: Risø National Laboratory.

Davidson, D.L. 1987. Metall. Trans. A 18A:2115–2128.

Davidson, D.L. 1989a. J. Mater. Sci. 24:681–687.

Davidson, D.L. 1989b. Eng. Fract. Mech. 33:965–983.

Dellis, M.-A., Schobbens, H., Van Den Neste, M. et al. 1991. In: Metal Matrix Composites—Processing, Microstructure and Properties. Proceedings of the 12th Risø International Symposium of Materials Science. (N. Hansen, D. Juul Jensen, T. Leffers et al., eds.) 299–304, Roskilde, Denmark: Risø National Laboratory.

Doong, S.H, Lee, T.C., Robertson, I.M., and H.K. Birnbaum. 1989. Scripta Metall. 23:1413–1418.

Dubensky, E.M., and D. A. Koss. 1987. Metall. Trans. A 18A: 1887–1895.

Eldridge, J.I., and P.K. Brindley. 1989. J. Mater. Sci. Lett. 8:1451–1454.

Embury, J.D., Newell, J., and S. Tao. 1991. In: Metal Matrix Composites—Processing, Microstructure and Properties. Proceedings of the 12th Risø International Symposium on Materials Science. (N. Hansen, D. Juul Jensen, T. Leffers et al. eds.) 317–322, Roskilde, Denmark: Risø National Laboratory.

England, J. and I.N. Hall. 1986. Scripta Metall. 20:697–705.

Erdogan, F. 1965. Proc. Phys. Soc. Lon. B66:793–801.

Fishkis, M. 1991. J. Mater. Sci. 26:2651–2661.

Flom, Y. and R.J. Arsenault. 1989. Acta Metall. 37:2413–2423.

Frazer, W.A., Ancker, F.H., and A.T. Di Benedetto. 1975. In: 30th Anniversary Tech. Conf. Soc. of Plastics Ind. Inc. Proc. 22A:1–13.

Girot, F.A., Quenisset, J.M. and R. Naslain. 1987. Compos. Sci. Tech. 30:155–165.

Harlow, D.G., and S.L. Phoenix. 1981. Int. J. Fracture 17:601–630.

Harris, S.J., Dinsdale, K., Gao, Y., and B. Noble. 1988. In: Mechanical and Physical Behaviour of Metallic and Ceramic Composites. Proceedings of the 9th Risø International Symposium on Materials Science (S.I. Andersen, H. Lilholt, and O.B. Pedersen, eds.), 373–382, Roskilde, Denmark: Risø National Laboratory.

Humphreys, F.J. 1988. In: Mechanical and Physical Behaviour of Metallic and Ceramic Composites. Proceedings of the 9th Risø International Symposium on Materials Science. (S.I. Andersen, H. Lilholt, and O.B. Pedersen, eds.), 51–74, Roskilde, Denmark: Risø National Laboratory.

Hunt, W.H., Jr., Brockenbrough, J.R., and P.E. Magnusen. 1991. Scripta Metall. Mater. 25:15–20.

Ibbotson, A.R., Beevers, C.J., and P. Bowen. 1991. Scipta Metall. 25:1781–1786.

Jeng, S.M., Yang, J-M., and C.J. Yang. 1991a. Mater. Sci. Eng. A138:169–180.

Jeng, S.M., Yang, J-M., and C.J. Yang. 1991b. Mater. Sci. Eng. A138:181–190.

Kamat, S.V., Hirth, J.P., and R. Mehrabian. 1989. Scripta Metall. 23:523–528.

Kelly, A. 1970. Proc. Roy. Soc. Lon. A319:95–116.

Keyhoe, F.P., and G.A. Chadwick 1991. Mater. Sci. Eng. A135: 209–212.

Kreiger, K.G., and K.M. Prewo. 1974. In: Composite Materials. (L.H. Brontman, and R.H. Crock, eds.), 399–410, New York: Academic Press.

Kryze, J., Breban, P., Baptiste, D., and D. Francois. 1991. In: Metal Matrix Composites—Processing, Microstructure and Properties. Proceedings of the 12th Risø International Symposium on Materials Science (N. Hansen, D. Juul Jensen, T. Leffers et al. eds.), 455–460, Roskilde, Denmark: Risø National Laboratory.

Kumai, S., King, J.E., and J.F. Knott. 1990. Fatigue. Fract. Eng. Mater. Struct. 13:511–522.

Lemaitre, J. 1992. A Course on Damage Mechanics. Berlin: Springer-Verlag.

LePetitcorps, Y., Pailler, R., Lahaye, M., and R. Naslaim. 1988. Compos. Sci. Tech. 32:31–55.

Levin, M., Karlsson, B., and J. Wasen. In: Fundamental Relationships between Microstructures and Mechanical Properties of Metal Matrix Composites, (M.N. Gungor, and P.K. Liaw, eds.), 421–440, Warrendale, PA: TMS.

Lewandowski, J.J., Liu, C., and W.H. Hunt, Jr. 1989a. In: Powder Metallurgy Composites. (M. Kumar, K. Vedula, and A.M. Ritter, eds.), 117–139, Warrendale, PA: TMS.

Lewandowski, J.J., Liu, C., and W.H. Hunt, Jr. 1989b. Mater. Sci. Eng. A107:241–255.

Lewandowski, J.J., Liu, D.S., and C. Liu. 1991. Scripta Metall. Mater. 25:21–26.

Liu, C., Rickett, B.I., and J.J. Lewandowski. 1989. In: Fundamental Relationships between Microstructures and Mechanical Properties of Metal Matrix Composites, (M.N. Gungor, and P.K. Liaw, eds.), 145–160, Warrendale, PA, TMS.

Lloyd, D.J. 1991. Acta Metall. Mater. 39:59–71.

Lloyd, D.J., Lagace, H., McLeod, A., and P.L. Morris. 1989. Mater. Sci. Eng. A107:73–80.

Logsdon, W.A., and P.K. Liaw. 1986. Eng. Fract. Mech. 24:737–751.

McCartney, L.N. 1989. Proc. Roy. Soc. Lon. A425:215–244.

McDanels, D.L. 1985. *Metall. Trans. A.* 16A:1105–1115.

Man, C.F., Mummery, P.M., Derby, B., and M.L. Jenkins. 1991. *In:* Interfacial Phenomena in Composite Materials. (I. Verpoest and F. Jones, eds.) 175–178, Oxford: Butterworth-Heinemann.

Manoharan, M., and J.J. Lewandowski. 1989a. *Scripta Metall.* 23:1801–1805.

Manoharan, M., and J.J. Lewandowski. 1989b. *Scripta Metall.* 23:301–306.

Manoharan, M., and J.J. Lewandowski. 1989c. *In:* Fundamental Relationships between Microstructures and Mechanical Properties of Metal Matrix Composites. (M.N. Gungor and P.K. Liaw, eds.), 471–478, Warrendale, PA, TMS.

Manoharan, M., and J.J. Lewandowski. 1990a. *Acta Metall. Mater.* 38:489–496.

Manoharan, M., and J.J. Lewandowski. 1990b. *Scripta Metall. Mater.* 24:2357–2362.

Marshall, D.B. 1984. *Comm. Amer. Ceram. Soc.* 67:C259-C260.

Marshall, D.B., and B.N. Cox 1987. *Acta. Metall.* 25:2607–2619.

Marshall, D.B., Cox, B.N., and A.G. Evans. 1985. *Acta Metall.* 33:2013–2021.

Marshall, D.B., and W.C. Oliver. 1987. *J. Amer. Ceram. Soc.* 70: 542–548.

Marshall, D.B., Shaw, M.C., and W.L. Morris. 1992. *Acta Metall.* 40:443–454.

Masson, J.J., Weber, K., Miketta, M., and K. Schulte. 1991. *In:* Metal Matrix Composites—Processing, Microstructure and Properties. Proceedings of the 12th Risø International Symposium on Materials Science. (N. Hansen, D. Juul Jensen, T. Leffers, et al. eds.), 509–514, Roskilde, Denmark: Risø National Laboratory.

Melander, A. 1980. *Acta Metall.* 28:1799–1804.

Metcalfe, A.G. 1967. *J. Compos. Mater.* 1:356.

Metcalfe, A.G., and M.J. Klein. 1974. *In:* Interfaces in Metal Matrix Composites (K.G. Kreider, ed.) 310–330, New York: Academic Press.

Miller, W.S. and F.J. Humphreys. 1989. *In:* Fundamental Relationships between Microstructures and Mechanical Properties of Metal Matrix Composites. (M.N. Gungor and P.K. Liaw, eds.), 517–542, Warrendale, PA: TMS.

Miller, W.S., and F.J. Humphreys. 1991. *Scripta Metall. Mater.* 25:33–38.

Morley, J.G. 1976. *Int. Met. Reviews* 268:153–170.

Morley, J.G. 1983. *J. Mater. Sci.* 18:1564–1576.

Mummery, P.M. 1991. D. Phil. Thesis. Department of Materials. University of Oxford. Oxford, UK.

Mummery, P.M., and B. Derby. 1989. *In:* Fundamental Relationships between Microstructures and Mechanical Properties of Metal Matrix Composites, (M.N. Gungor and P.K. Liaw, eds.), 161–172, Warrendale PA: TMS.

Mummery, P.M. and B. Derby. 1991. *In:* Metal Matrix Composites—Processing, Microstructure and Properties. Proceedings of the 12th Risø International Symposium on Materials Science. (N. Hansen, D. Juul Jensen, T. Leffers, et al. eds.), 535–542, Roskilde, Denmark: Risø National Laboratory.

Mummery, P.M., Derby, B., Buttle, D.J., and C.B. Scruby. 1992a. *In:* Proceedings of the 2nd European Conference on Advanced Materials and Processes. T.W. Clyne and P.J. Withers eds. 441–447, London: Institute of Materials.

Mummery, P.M., Derby, B., Cook, J., and J.H. Tweed. 1992. *In:* Proceedings of the 2nd European Conference on Advanced Materials and Processes. T.W. Clyne and P.J. Withers, eds. 92-99. London: Institute of Materials.

Mummery, P.M., Derby, B., and C.B. Scruby. 1993. *Acta Metall. Mater.* 43:1431–1445.

Musson, N.J., and T.M. Yue. 1991. *Mater. Sci. Eng.* A135:237–242.

Nair, S.V., Tien, J.K., and R.C. Bates. 1985. *Int. Met. Rev.* 30: 275–290.

Nayeb-Hashemi, H., and J. Seyyidi. 1987. *Metall. Trans. A* 20A:727–739.

Nieh, T.G., Rainen, R.A., and D.J. Chellman. 1985. *In:* Proceedings of the 5th International Conference on Composite Materials. (W.C. Harrigan Jr., J. Strife, and A.K. Dhingra, eds.) 825–842, Warrendale, PA: TMS.

Nutt, S.R., and J.M. Duva. 1986. *Scripta Metall.* 20:1055–1058.

Ochiai, S., Urakawa, S., Ameyama, K., and Y. Murakami. 1980. *Metall. Trans.* 11A:525–530.

Ohno, N., and J.W. Hutchinson. 1984. *J. Mech. Phys. Sol.* 32: 63–85.

Pao, P.S., Gill, S.J., Pattnaik, A., et al. 1989. *In:* Fundamental Relationships between Microstructures and Mechanical Properties of Metal Matrix Composites. (M.N. Gungor, and P.K. Liaw, eds.), 405–420, Warrendale, PA: TMS.

Papazian, J.M., and P.N. Adler. 1990. *Metall. Trans.* 21A:401–411.

Pickard, S.M., Derby, B., Harding, J., and M. Taya. 1988. *Scripta Metall.* 22:601–606.

Ribes, H., DaSilva, R., Suery, M., and T. Brethau. 1990. *Mater. Sci. Tech.* 6:621–628.

Rice, J.R., and G.C. Shi. 1965. *ASME J. Appl. Mech.* 32:418–423.

Ritchie, R.O., Knott, J.F., and J.R. Rice. 1973. *J. Mech. Phys. Sol.* 21:395–410.

Roebuck, B. 1987. *J. Mater. Sci. Let.* 6:1138–1141.

Schulte, K., and K. Minoshima. 1991. *In:* Metal Matrix Composites—Processing, Microstructure and Properties. Proceedings of the 12th Risø International Symposium on Materials Science. (N. Hansen, D. Juul Jensen, T. Leffers, et al., eds.), 123–147, Roskilde, Denmark: Risø National Laboratory.

Shang, J.K. and R.O. Ritchie. 1989a. *Acta Metall.* 37:2267–2278.

Shang, J.K. and R.O. Ritchie. 1989b. *Metall. Trans. A.* 20A:897–908.

Shang, J.K., Yu, W., and R.O. Ritchie. 1988. *Mater. Sci. Eng.* 102A:181–192.

Smith, R.L., Phoenix, S.L., Greenfield, M.R., et al. 1983. *Proc. Roy. Soc. Lon.* A388:353–391.

Stephens, J.J., Lucas, J.P., and F.M. Hosking. 1988. *Scripta Metall.* 22:1307–1312.

Strangwood, M., Hippsley, C.A., and J.J. Lewandowski. 1991. *Scripta Metall. Mater.* 24:1483–1487.

Sugimura, Y., and S. Suresh. 1992. *Metall. Trans. A.* 23A:2231–2242.

Suresh, S. 1991. Fatigue of Materials, Cambridge, UK: Cambridge University Press.

Taya, M., and A. Daimerce. 1983. *J. Mater. Sci.* 18:3105–3116.

Tsangarakis, N., Andrews, B.O., and C. Cavallaro. 1987. *J. Comp. Mater.* 21:481–492.

Vasudevan, A.K., Richmond, O., Zok, F., and J.D. Embury. 1989. *Mater. Sci. Eng.* A107:63–74.

Warren, P. 1992. *Acta Metall. Mater.* In press.

Watson, M.C., and T.W. Clyne. 1992. *Acta Metall. Mater.* 40:141–148.

Whitehouse, A.F., Shahani, R.A., and T.W. Clyne. 1991. *In:* Metal Matrix Composites—Processing, Microstructure and Properties. Proceedings of the 12th Risø International Symposium on Materials Science. (N. Hansen, D. Juul Jensen, T. Leffers, et al. eds.), 741–748, Roskilde, Denmark: Risø National Laboratory.

Wu, S.B., and R.J. Arsenault. 1989. *In:* Fundamental Relationships between Microstructures and Mechanical Properties of Metal Matrix Composites. (M.N. Gungor, and P.K. Liaw, eds.), 241–254, Warrendale, PA: TMS.

Yang, J., Cady, C., Hu, M.S., et al. 1990. *Acta Metall. Mater.* 38:2613–2619.

Yang, C-J, Jeng, S.M., and J-M. Yang. 1990. *Scripta Metall. Mater.* 24:469–474.

Yang, J-M., Jeng, S.M., and C.J. Yang. 1991. *Mater. Sci. Eng.* A138:155–167.

Yau, S.S., and G. Mayer. 1986. *Mater. Sci. Eng.* 82:45–58.

You, C.P. and J.E. Allison. 1989. *In:* Proceedings of the 7th. International Conference on Fracture, (K. Salama, K. Ravi-Chandar, D.M.R. Taplin, and P. Rama Rao, eds.), 3005–3012. Oxford: Pergamon Press.

You, C.P., Thompson, A.W., and I.M. Bernstein. 1987. *Scripta Metall.* 21:181–185.

Zok, F., Embury, J.D., Ashby, M.F., and O. Richmond. 1988. *In:* Mechanical and Physical Behaviour of Metallic and Ceramic Composites. Proceedings of the 9th Risø International Symposium on Materials Science. (S.I. Andersen, H. Lilholt, and O.B. Pedersen, eds.), 517–526, Roskilde, Denmark: Risø National Laboratory.

Zweben, C., and B.W. Rosen. 1970. *J. Mech. Phys. Solids* 18:189–206.

Chapter 15
Fatigue Behavior of Discontinuously Reinforced Metal-Matrix Composites

JOHN E. ALLISON
J. WAYNE JONES

Applications for discontinuously reinforced metal-matrix composites (MMCs) are anticipated to grow significantly in the next decade (see Chapter 16). To ensure the structural integrity of components fabricated from these materials, it is necessary to understand the many potential failure modes to which they may be susceptible. The resistance to fatigue failure is an especially critical design requirement for many of these applications. Fatigue failures can be defined in several ways, including significant changes in geometry, initiation of small fatigue cracks, or final separation of a component into two or more pieces. This chapter reviews the effect of discontinuous reinforcement of metals on cyclic deformation, fatigue crack initiation and early growth, and fatigue crack propagation. The focus will be on discontinuously reinforced aluminum-based composites because of the technological significance of these materials to a broad spectrum of industries and because investigations of these materials provide the primary literature in this field. This chapter reviews fatigue at room temperature in ambient environments. Effects on cyclic deformation and fatigue behavior due to environment (Crowe and Hasson 1982; Hasson et al. 1984), elevated temperature (Seitz, et al. 1991; Healey and Beevers 1991), thermomechanical loading (Karayaka and Sehitoglu 1991) and notch root radii (Biner 1990) will not be covered.

Improvements in fatigue behavior are often suggested as an advantage of metal-matrix composites. However, fatigue is a complex process and such generalizations are overly simplistic. An objective of this review will be to compare the fatigue behavior of reinforced materials with their unreinforced counterparts to understand situations in which composite materials offer improvement and situations in which their resistance to cyclic loading is diminished. To provide a general understanding of fatigue behavior in these materials, underlying principles and mechanisms will be emphasized. An attempt is made to review the current state of knowledge about fatigue behavior of these materials (some of which is currently unpublished). It must be stressed that the scientific understanding of this complex process is not only far from complete, but is also complicated because of the lack of systematic investigation and the evolving nature of the metal-matrix composites industry.

15.1 Cyclic Deformation

When a material is subjected to a given cyclic strain (or stress), the resultant stresses (or strains) may change with continued cycling. The cyclic stress-strain *response* or cyclic hardening/softening behavior is a measure of this transient response. Under constant strain cycling conditions, the resultant stresses may be invariant during cycling, i.e., they are stable, or they may increase (harden) or decrease (soften) with continued cycling. For a material that cyclically hardens or softens, a stable stress, called the saturation stress, is generally reached during cycling. A cyclic stress-strain *curve* is used to characterize this steady-state behavior and is obtained by plotting these stabilized or saturated stresses versus the applied strain. Both the cyclic hardening/softening behavior and the cyclic stress-strain curve of a material can be affected by the presence of reinforcement particles (Williams and Fine 1985; Vyletel et al. 1991; Llorca et al. 1992).

The cyclic stress-strain response of composite materials has only recently become the subject of serious investigation (Vyletel et al. 1991, 1993; Seitz et al. 1991; Srivatsan et al. 1991; Srivatsan 1992; Liu 1991). Because the cyclic stress-strain response is highly sensitive to matrix microstructure, when comparing the cyclic stress-strain response of reinforced metals to their unreinforced counterparts, it is important to ensure that the microstructure of the reinforced matrix is similar to that of the unreinforced material or at least that the differences that may exist are fully understood. If care is not taken, similar aging treatments can lead to differences in the structure and volume fraction of precipitates formed (see Chapter 7). Similar issues exist relative to grain size, texture, and alloying. In particular, the slip character (planar versus wavy slip) and cyclic stress-strain response of a material are strongly affected by the structure of the strengthening precipitates. In general, when these issues are carefully controlled, the cyclic stability of reinforced aluminum is similar in character to that of the unreinforced matrix. An example of this is shown in Figure 15.1(a) and (b) for an Al-6Cu-0.3Mn* alloy (2219 aluminum) reinforced with 15% spherical TiC particles (2219Al/TiC/15$_p$) from the work of Vyletel et al. (1991). In both of these materials, θ' precipitates were formed by standard solution treatment and aging at elevated temperatures (artificial aging). These θ' precipitates serve as hard barriers to dislocations and lead to the development of a stable, homogeneous dislocation substructure. It is well known that precipitates of this type lead to a cyclic stress-strain response that is stable in unreinforced aluminum alloys (Calabrese and Laird 1974a), although some hardening may be experienced at higher stresses. As can be seen in Figure 15.1(a), when artificially aged 2219 aluminum is reinforced it remains cyclically stable, however, to

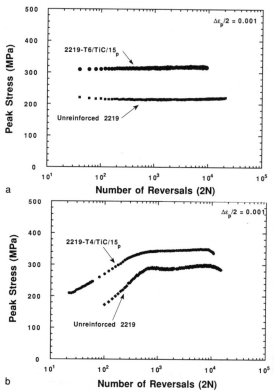

Figure 15.1. Cyclic stress-strain response in 2219 Al and 2219/TiC/15p. (a) Artificially aged microstructure. (b) Naturally aged microstructure. *(After Vyletel et al. 1991.)*

achieve a certain plastic strain, a higher applied stress is required in the composite.

In contrast with artificially aged aluminum, naturally aged aluminum, whether strengthened by GP zones or θ'' precipitates, is generally not cyclically stable and exhibits an initial hardening behavior. Depending on the alloy and the stress level, after this initial hardening stage, a final softening stage is often observed prior to fatigue fracture (Calabrese and Laird 1974b; Sanders et al. 1975; Fine and Santner 1975). The initial hardening is generally attributed to an initial increase in dislocation density and subsequent dislocation interactions. As straining continues, deformation bands or persistent slip bands form because of the shearing of precipitates. The softening phase is generally attributed to dissolution of GP zones from the repeated shearing that occurs within the persistent slip band (Calabrese and Laird 1974b). As seen in Figure 15.1(b), the presence of a reinforcement does not appear to change the general cyclic stress-strain response of 2219 aluminum strengthened by shearable precipitates, other than that the stresses required to enforce a given amount of plastic strain are higher for the composite material (Vyletel et al. 1991). In this

---

*Compositions are given in weight %.

particular alloy, an intermediate stable region (sometimes referred to as "secondary hardening") is observed and may be attributed to activation of secondary slip within the PSBs as well as a competition between the hardening and softening phenomena.

If the character of the cyclic stress-strain response is similar between an alloy and its composite, then it might be anticipated that no fundamental differences exist in the dislocation structures developed during cyclic deformation. Although not yet the subject of exhaustive investigation, this appears to be accurate. Figure 15.2(a) and (b) shows dislocation substructures developed in the 2219Al/TiC/15_p composite described above. These dislocation substructures are generally similar to those observed for the unreinforced material (Vyletel et al. 1993). In the artificially aged condition, both materials showed dislocation accumulation at the matrix/θ' precipitate interface and dislocations extending into the matrix between θ' precipitates. Because of the shearable nature of GP zones, the dislocation structures that developed in naturally aged conditions were quite different than those observed in the artificially aged condition. In the naturally aged materials, dislocation-free deformation bands were observed with extensive dislocation debris along the sides of these bands. A lower level of dislocation debris was also reported in matrix areas not associated with these bands. In the composite material, these bands traverse the entire width of certain grains but were not observed in all grains. Although there was no fundamental difference between deformation structures observed in the artificially aged composite and its unreinforced counterpart, the unreinforced materials did exhibit a much higher density of deformation bands, with numerous bands existing within each grain.

Figure 15.3 shows an example of the influence of SiC reinforcement on both the monotonic and cyclic stress-strain curves in a peak-aged Al-3.5Cu alloy (Llorca et al. 1992). As shown in Figure 15.3, the reinforcement produces a large increase in the work hardening of the material under both monotonic and cyclic conditions and further, this increase in work hardening is most pronounced under cyclic conditions. The increase in work hardening is important for understanding the effect reinforcement has on cyclic deformation and fatigue-life behavior under a variety of situations. To adequately understand the cyclic deformation of composite materials, it is necessary to consider the highly localized stresses that develop in the matrix around the reinforcement particles, and the influence these concentrated stresses have on the work hardening behavior of the composite. (For a more detailed examination of this issue refer to Chapter 9). Goodier (1933) and others have shown that in the matrix regions immediately adjacent to the reinforcements, the stresses at the "poles" (along the stress axis) are significantly increased relative to stresses remote from the reinforcement. This concentration of stresses results from the constrained deformation in the matrix that occurs because of the significant differences in the elastic modulus of the reinforcement and the matrix. This constraint also leads to a triaxial stress state and steep stress gradients in the matrix around reinforcement particles. In contrast, Eshelby (1957) has shown that the stresses within the reinforcement are uniform, at least for spherical and ellipsoidal reinforcements embedded in an infinite elastic matrix.

Elastic-plastic finite element analyses of monotonic (Hom and McMeeking 1991; Llorca et al. 1991; Davis and Allison 1993) and cyclic loading (Llorca et al. 1992) have shown that these highly localized stresses lead naturally to highly localized plastic deformation. As material is strained, a small region at the poles of the reinforcement particle begins to deform. Straightforward theoretical calculations of this process show that for low volume fraction composites in an elastic-perfectly plastic material, the proportional limit is well

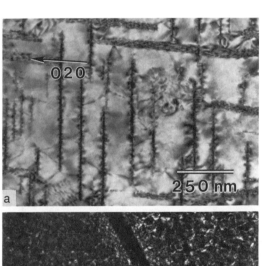

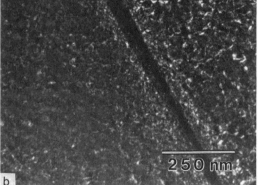

Figure 15.2. Dislocation structures produced in fatigue of 2219/TiC/15_p. (a) Artificially aged microstructure. (b) Naturally aged microstructure. *(After Vyletel et al. 1991.)*

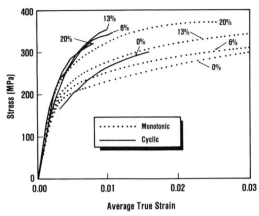

Figure 15.3. Comparison of cyclic stress-strain curves for peak-aged Al-3.5 Cu composites reinforced with SiC$_p$. *(After Llorca et al. 1992.)*

below the flow strength of the unreinforced matrix. Although the material begins to flow locally, the majority of the composite is strained elastically. As the stress is increased, the volume of material that is plastically deformed increases. Eventually, because the reinforcement is elastically loaded and shares in the straining, the material reaches an ultimate strength that is higher than the unreinforced matrix. The result is an increase in the measured work hardening rate although this is actually only an "apparent" work hardening. It merely reflects a change in the volume of material that has been plastically deformed. The term *apparent work hardening* is adopted to indicate that, while on a macroscopic basis the composite material demonstrates a higher resistance to plastic deformation than does an unreinforced material, the mechanisms of work hardening and, thus, the *local* hardening rate, are not necessarily different between a composite material and an unreinforced material. The thermal expansion coefficient of the ceramic reinforcement is typically much lower than that for the aluminum matrix. Thus, significant residual stresses can be produced in the composite during thermomechanical processing. Although the influence of these residual stresses is often assumed to be negligible, numerical studies (Zahl and McMeeking 1991; Davis and Allison 1993) have shown that the influence that these residual stresses have on initial yielding can be quite pronounced. The impact that these residual stresses have on cyclic deformation and fatigue behavior has not been previously considered in any detail. It is reasonable to speculate, however, that important effects do exist.

As previously described, the cyclic saturation stress is higher in composite materials than it is in unreinforced materials. This results directly from the high apparent work hardening that these materials exhibit. Recent theoretical efforts have suggested that the level of this saturation stress is determined by a competition between the high matrix work hardening behavior and softening that arises from the development of voids in the matrix regions experiencing high levels of plastic deformation (Llorca et al. 1992). However, the amount of hardening predicted by this model was considerably higher than that actually exhibited by the Al-3.5Cu/SiC$_p$ composites shown in Figure 15.3. Numerical models also predict an increase in the apparent work hardening as the volume fraction of reinforcement increases under both monotonic (Hom and McMeeking 1991; Llorca et al. 1991) and cyclic conditions (Llorca et al. 1992). The experimental observations shown in Figure 15.3 indicate that in an Al-3.5Cu composite reinforced with volume fractions of SiCp varying from 6% to 20% (Llorca et al. 1992), this is, in fact, the case for monotonic work hardening but not for cyclic work hardening. The cyclic work hardening of a 6% SiC$_p$ composite was significantly higher than that of the unreinforced material, but further increases in volume fraction up to 20% had only a minimal effect on the cyclic work hardening. This has been attributed to the reinforcement cracking that is commonly observed in cyclically loaded SiC$_p$ composites (Shang et al. 1988; Bonnen et al. 1991; Hall et al. 1993b). It is interesting to note that although the composites investigated by Llorca et al. 1992 all cyclically harden, the unreinforced material experiences a crossover in the monotonic and cyclic stress-strain curves. This indicates that in the peak aged condition the unreinforced Al-3Cu will cyclically harden at strains less than about 0.005, while above this strain level the material will cyclically soften (Figure 15.3). This is not a completely general result because the monotonic and cyclic stress strain curves diverge for many Al-Cu and Al-Cu-Mg alloys strengthened by shearable precipitates (Calabrese and Laird 1974b; Fine and Santner 1975; Landgraf 1970).

To date, observations on particle size or shape effects on cyclic deformation have been limited. Based on continuum theory, particle size should not influence cyclic hardening. However, because decreasing particle size leads to a reduced interparticle spacing (at a constant reinforcement volume fraction), it is reasonable to speculate that deformation would be more homogeneous. The effect this has on the cyclic deformation is unclear. Particle shape effects have been subjected to limited experimental investigation (Llorca et al. 1992) the results of which indicate that composites reinforced with SiC whiskers cyclically harden more than those reinforced by SiC particles. This observation has been attributed to the low percentage of fractured reinforcements observed in the SiC$_w$ composite. Detailed numerical analysis of monotonic and cyclic hardening in these composites correlated well with the above observations.

One distinction between whisker reinforced composites and particle/spherical reinforced composites is that the former has significantly reduced tensile and cyclic ductilities (Llorca et al. 1991; Llorca et al. 1992).

Many materials display a reduction in the compressive flow strength after being deformed in tension and vice versa. This phenomenon is called the Bauschinger effect and can lead to pronounced shifts in mean stresses. It is thus important for understanding and predicting the cyclic stress-strain response and effects such as cyclic creep. Although Bauschinger effects have been investigated in discontinuously reinforced aluminum composites by a number of investigators (Arsenault and Wu 1987; Taya et al. 1990; Llorca et al. 1990; Johannesson et al. 1991; Liu 1991; Mouritz and Bandyopadhyay 1991), there is much controversy on this topic (see Withers et al. 1989; Arsenault and Taya 1989). Further, the lack of standardized parameters for describing this phenomenon makes it difficult to compare the results of different investigations. Despite these difficulties, a consistent picture has recently begun to emerge. The Bauschinger effect appears to become more pronounced with increasing applied plastic strain, to be higher in composite materials than unreinforced materials, to increase with increasing volume fraction, and to increase in overaged microstructures. As an example, Figure 15.4 shows recent results for Al-3.5Cu-1.7Mg (MB85 or × 2080) alloy composites reinforced with SiC$_p$ (Liu 1991). Bauschinger stress in this figure is defined as the difference between the maximum stress in the forward direction and the yielding stress in the reverse direction (stress differential). As can be seen, increasing the volume fraction of SiC$_p$ leads to an increase in the Bauschinger stress. Moreover, overaged (OA) microstructures tend to have higher Bauschinger stress compared to underaged (UA) microstructures. Depending on the applied strain level and material of investigation, the reported stress differential varies between 0 MPa and 200 MPa, while the permanent softening is generally lower and between 0 MPa and 60 MPa. Taya et al. (1990) have observed that the differential stress is higher if the initial load is in tension and lower if the initial load is in compression, while others have reported the inverse (Arsenault and Wu 1987; Mouritz and Bandyopadhyay 1991; Johannesson et al. 1991). The Bauschinger effect has been attributed to a back stress developed in composite materials (Taya et al. 1990), and has been correlated with residual stresses caused by thermal expansion mismatch either determined theoretically or measured by X-ray diffraction (Mouritz and Bandyopadhyay 1991; Liu 1991).

15.2 Fatigue Crack Initiation and Early Growth

For many engineering applications, fatigue life data, often developed from smooth samples, is the primary information used for assuring resistance to failure caused by cyclic loading. Thus, understanding the influence that composite reinforcement has on this property is of obvious importance. Developing a mechanistic interpretation from the current information is complicated by a number of factors and, for composite materials, a lack of systematic study. One primary issue relates to the quality of composite materials, that is, the presence of imperfections in the form of clusters of reinforcement particles, isolated large reinforcement particles, or exogenous defects such as tramp particles in powder metallurgy processes or dross or refractory crucible-related defects in ingot metallurgy composites. Of all the mechanical properties, axial, smooth bar fatigue behavior is especially sensitive to the occurrence of such imperfections. Because the discontinuously reinforced composite industry has been an evolving industry, quality has also been evolving and the smooth bar fatigue data that are affected by this quality have shown significant improvements. This is illustrated graphically in Figure 15.5, which shows data generated in 1992 for a 6061/Al2O3/20$_p$ composite compared with data generated on the same composite manufactured by the same producer in 1990 (Klimowicz 1992). The improvements in fatigue lives reflect process improvements made during this time period and illustrate the difficulty of developing a fundamental, general understanding of the fatigue process on an evolving material. Similar improvements related to processing have also been observed in SiC whisker reinforced 2124 aluminum (Williams and Fine 1985). In many cast materials, casting-related defects such as porosity are an inherent feature of the microstructure and these defects

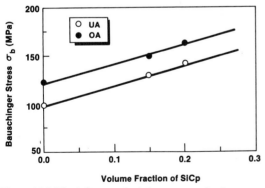

Figure 15.4. The influence of reinforcement and aging on the Bauschinger stress in 2xxxAl/SiC$_p$ composites. (*After Liu 1991.*)

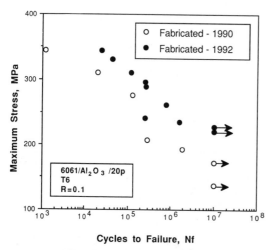

Figure 15.5. Effect of process improvements on the fatigue life behavior of 6061 composites. *(After Klimowicz 1992.)*

have an especially pronounced effect on their fatigue properties (Zhang et al. 1991; Masuda and Tanaka 1992). Much of the following generalities regarding fatigue behavior will deal with data generated on wrought alloys in which processing imperfections have presumably been minimized. However, understanding and improving the fatigue life behavior of cast materials represents a new challenge.

Testing practices also confound attempts to develop general principles. In the smooth bar fatigue sample, there is both a crack initiation period and a crack growth period. Separating these two distinct aspects, although technologically possible, is difficult in practice and thus not often done. Differences in total fatigue life for composite materials may then be attributed to differences in resistance to fatigue crack initiation as suggested by Williams and Fine (1985) or to differences in the short fatigue crack propagation behavior, or may be attributed to some combination of both. The use of both axial and rotating bending test apparati for developing fatigue life data is an additional complication. Axial fatigue data are known to be more conservative, therefore, to allow comparison between these two test types, data from bending tests are typically multiplied by a "load constant." Values for this load constant for steels can range from 0.75 to 1.0 (Juvinall 1967), although 0.8 is most commonly used. Such factors have not been determined for metal-matrix composites. Moreover, because the rotating bending sample emphasizes surface regions, there is much less susceptibility to the material imperfections that occur randomly throughout the material. The relative differences in the importance of the crack growth period between the axial and rotating bending test is unclear, although there have been suggestions that the initiation period pre-

dominates in the rotating bending test (Llorca et al. 1991). Given these various complicating factors and the lack of systematic investigations, the limitations of this review should thus be realized.

To rationalize the phenomenological effects described below, it is instructive to be reminded of the influence of the individual elastic and plastic strain components as given by Equation 15.1:

$$\Delta\varepsilon_T = \Delta\varepsilon_e + \Delta\varepsilon_p \qquad (15.1)$$

where $\Delta\varepsilon_T$ is the total applied strain range, $\Delta\varepsilon_e$ is the elastic strain range, and $\Delta\varepsilon_p$ is the plastic strain range. Although the plastic strain is the predominant factor influencing fatigue damage, elastic strains are also important and differences in elastic and thus total strains must be considered. In composite materials, the stresses and strains are not uniform and deformation can be highly concentrated. As discussed in the previous section, this has been shown to lead to high local plastic strains and high apparent work hardening rates. It is important to distinguish between *average* strains, which represent the response of the overall composite, and *local* strains, which are highly concentrated.

15.2.1 Phenomenology of Fatigue

15.2.1.1 Stress–Life Behavior

Many investigators have found that when stress-controlled fatigue experiments are conducted, the fatigue lives of discontinuously reinforced metals are generally longer than those of unreinforced metals (Crowe and Hasson 1982; Hasson et al. 1984; Harris and Wilks 1986; Harris 1988; Sharp et al. 1990; Bonnen et al. 1991; Hunt et al. 1991; Hall et al. 1993b; Llorca et al. 1991). Examples of the differences between unreinforced and reinforced materials are shown in Figure 15.6 (a and b) for an Al-3.5Cu-1.8Mg alloy (x2080) reinforced with 15% SiC particles (Bonnen et al. 1991). As this example shows, the improvements in fatigue resistance are most pronounced at intermediate and low stresses, depending on the stress ratio, R. At high stresses, these differences are minimized and, at a given stress level, the fatigue lives are generally similar. At positive or near-zero stress ratios, the most pronounced improvement is at the lowest stresses, that is, in the high-cycle fatigue regime. Under fully reversed ($R = -1$) axial loading, the main improvements in fatigue life occur at intermediate stresses. There are indications (Bonnen et al. 1991; Klimowicz 1992) that for fully-reversed axial tests, the fatigue lives converge under low stress, high-cycle fatigue conditions. An example of this is shown in Figure 15.6(a).

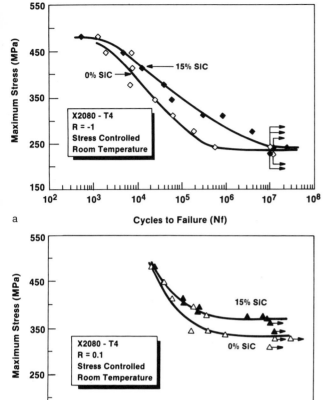

Figure 15.6.. Stress versus fatigue life behavior of X2080-T4 aluminum and X2080-T4/SiC/15$_p$. (a) For stress ratio, R, of -1. (b) For stress ratio, R, of 0.1. *(After Bonnen et al. 1991.)*

Under stress-controlled conditions at low and high stresses, a schematic of stable loops for a typical aluminum-copper alloy and its composite are shown in Figure 15.7(a) and (b). At a fixed level of cyclic stress, the composite material experiences a total strain that is much lower than that in the unreinforced material. This is true at both low stresses (see Figure 15.7(a)), for which elastic strains predominate and at higher stresses (see Figure 15.7(b)) for which plastic strains predominate. Two primary factors are responsible for this: (1) a reduced elastic strain that results from higher modulus of the composite and, at most stresses, (2) a reduced plastic strain that results from the increased apparent work hardening of the composite. The superior fatigue resistance of the composite under constant stress conditions then results from the lower total strain and most importantly, the lower plastic strain at which it is cycled. This is consistent with the observations of Williams and Fine (1985), who found that under constant stress cycling of SiC$_w$ reinforced 2124-T6, the crack initiation period was delayed and lives were a factor of 10 longer compared to unreinforced 2124-T6. This delayed initiation was presumably a manifestation of the lower plastic strain in the composite material. A similar delay in fatigue crack initiation has been observed in 6061 reinforced with 15% SiC$_p$ (Hall et al. 1993b).

The fatigue lives of composite materials at high stresses generally converge with unreinforced counterparts. This can be thought of in terms of a ductility "exhaustion" concept. At high cyclic plastic strains (at short lives) the cyclic ductility of a material limits the amount of plastic strain and thus the number of cycles that can be accumulated prior to fracture (Landgraf 1970). Because composite materials have lower tensile and cyclic ductilities (Llorca et al. 1992), the number of cycles that can be accumulated in the low cycle fatigue regime is limited. Therefore, at high stresses, although the plastic strains are lower in the composite material, so too is the cyclic ductility and thus resistance to these cyclic strains. This leads to the convergence of lives for the reinforced and unreinforced materials at high stresses.

As described in the previous section, continuum theory suggests that composite materials should exhibit a proportional limit that is lower than a corresponding

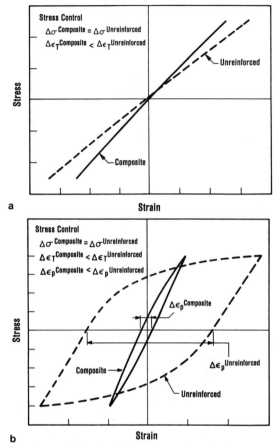

Figure 15.7. Stabilized cyclic stress-strain loops for discontinuously reinforced composite compared with an unreinforced counterpart under stress-control conditions (schematic). (a) At low stress, high cycle fatigue ($N_f = 1 \times 10^7$), (b) At intermediate stress, intermediate cycle fatigue (1×10^4).

in fatigue lives are observed as shown in Figure 15.6(b). The more interesting case is that of fully reversed loading (Figure 15.6(a)). Under these conditions, stable stress-strain loops form and the (potentially) higher plastic deformation in the composite is retained with subsequent cycling. Thus, under fully reversed loading at low stresses, the damage resulting from plastic deformation may be *greater* in the composite than it is in the unreinforced matrix. The lack of significant improvements in stress-life response at low stresses under fully reversed loading conditions may be related to the (theoretically) higher plastic strains present in the composite at these low stresses.

In a stress-controlled test, strain rachetting or the development of a tensile (or compressive) mean strain is commonly observed. This is because of the differences in the cyclic (and monotonic) strengths in tension and compression, and is related to the Bauschinger effect described in the previous section. These differences have been shown to increase with increasing plastic strain and with the introduction of reinforcement (Liu 1991). Again, counteracting forces exist in the form of a higher plastic strain in the unreinforced material and an increased Bauschinger effect in the composite. Both of these effects can lead to an increase in the amount of mean strain rachetting. Limited observations in X2080-T6 and X2080-T6/SiC/15_p composite indicate that, for a given stress condition, the increase in mean strain and Bauschinger effect is much more pronounced in the unreinforced material. The development of such a mean strain has an important effect on the dimensional stability of a structure and probably fatigue life. This has not, however, been the subject of extensive or systematic investigations.

It should be noted that improvements in the stress-controlled fatigue behavior with reinforcement are not universally observed. In a study of squeeze-cast 2618 aluminum, Harris (1988) has shown that reinforcement with short δ-Al_2O_3 fibers lead to a degradation of fatigue resistance, while Bloyce and Summers (1991) have reported that no significant differences could be observed between the fatigue behavior of a squeeze-cast A357 aluminum alloy with or without 20% SiC particles. Both of these studies were on cast materials exhibiting rather low tensile ductilities, which suggests that the fatigue data for these materials was not indicative of intrinsic fatigue response but rather of the influence of processing-related imperfections.

15.2.1.2 Total Strain–Life Behavior

Fatigue behavior can also be considered on the basis of total strain versus fatigue life curves, either where total strain is the controlled variable or where total strain is

unreinforced matrix. An interesting aspect of the stress-controlled test is the impact that this lower proportional limit may have on fatigue lives at low stresses under fully reversed loading ($R = -1$). At low stresses, the amount of plastic strain is generally quite low and fatigue life correlates better with the elastic strain. However, because plastic strains are of primary importance in producing fatigue damage, differences in these strains must be considered. Because the composite has a proportional limit which, at least theoretically, should be lower than that of the reinforced material, at low stress levels the *average* plastic strain in the composite may be *greater* than that experienced by the unreinforced material. Under positive mean stresses, plastic deformation will occur in the first cycle but subsequent cycling will be generally elastic. Because the elastic strains (and thus total strains) are lower in the composite, improvements that reached after stabilization of a stress-controlled test has occurred (e.g., specimen half-life). Without exception, when total strain-life behavior is investigated, the fatigue resistance of discontinuously reinforced materials is inferior to that of the unreinforced matrix (Hurd 1988; Shang and Ritchie 1989c; Bonnen et al. 1991; Klimowicz 1992; Liu 1991). Examples of this behavior are shown in Figure 15.8. Figure 15.8(a) shows stabilized strains (from stress-controlled tests) versus life for the same material (x2080) shown in Figure 15.6 (Bonnen et al. 1991). Figure 15.8(b) compares total strain-fatigue life behavior for 6061-T6 aluminum and 6061-T6/Al$_2$O$_3$/20$_p$ (Klimowicz 1992). As is evident, for a given total strain, the fatigue resistance of the composite material is generally inferior to that of the unreinforced material. The differences are most striking at high strains, that is, in the low-cycle fatigue regime.

A comparison of stable stress-strain loops for a typical aluminum-copper alloy and its composite under total strain control is shown in Figure 15.9. Under total strain-controlled conditions, the elastic strains are much lower in the composite material than they are in the unreinforced material. This is because of the higher modulus and lower proportional limit of the composite material. Thus, for a given total strain, the average plastic strain is significantly elevated in the composite material relative to the unreinforced metal as shown by Equation 15.1. This increased plastic strain dominates the development of fatigue damage and results in the inferior fatigue behavior of the composite material. This effect is most pronounced at high strains where the lower cyclic ductility of the composite material leads to a decreased ability to resist plastic strain accumulation. Similar to the mean strain effects produced under constant stress conditions, mean stresses may develop under constant strain conditions. These have not been examined to any significant extent, but undoubtedly play a role in influencing fatigue lives.

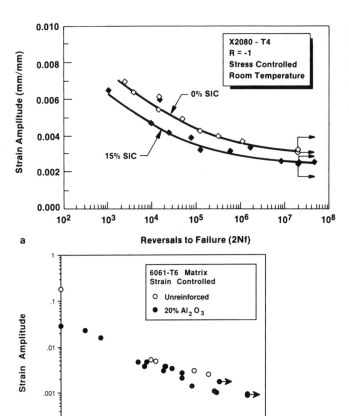

Figure 15.8. Total strain versus fatigue life (reversals) behavior of discontinuously reinforced composite compared with an unreinforced counterpart. (a) For X2080-T4 and X2080-T4/SiC/15$_p$ *(After Bonnen et al. 1991.)* (b) For 6061- and 6061/Al$_2$O$_3$/20$_p$ *(After Klimowicz 1992.)*

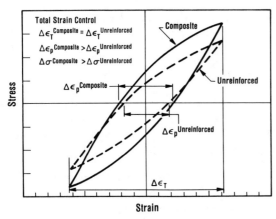

Figure 15.9. Stabilized cyclic stress-strain loops of discontinuously reinforced composite compared with an unreinforced counterpart under total strain-control conditions (schematic).

15.2.1.3 Plastic Strain–Life Behavior

Plastic strain is frequently used as the control variable for experiments conducted in the low cycle fatigue regime. It has not, however, been widely used for evaluating composite materials. In the only known study where average plastic strain was used as the control variable throughout the entire cyclic life, Vyletel et al. (1993) have shown that reinforcement of 2219Al with 15% spherical TiC particles leads to a factor of 2.5 to 10 decrease in fatigue life at a given level of plastic strain in an artificially-aged condition as shown in Figure 15.10. In a naturally-aged condition, the decrease in fatigue life was less, generally a factor of 1.5 to 6. In both cases, the difference in life increased as plastic strain increased. Under plastic strain cycling, stable loops similar to those shown in Figure 15.11 develop. Although the differences in total strain are not pronounced, composite materials generally achieve higher stresses than do unreinforced materials because of the high apparent work hardening of the composite materials. This does not, however, provide an explanation for the lower fatigue lives. A primary factor contributing to these lower lives is undoubtedly related to the concentration of plastic strains. Although the composite is deformed to an equivalent average plastic strain, local regions of the matrix at the poles of the reinforcement particle are subjected to considerably higher plastic strains. Thus, a higher degree of fatigue damage is induced in these locations than would be anticipated from the average plastic strain. In addition, there is limited experimental evidence (Vyletel et al. 1993) that, for a given plastic strain, the area inside the stabilized stress-strain loop is appreciably larger in the composite materials. This implies that to develop a given cyclic plastic strain (as measured at zero stress), the composite material requires more plastic "energy" or plastic work than does the unreinforced material. Finally, there is some ambiguity in the determination of the gauge length that should be assumed in calculating strains. Because the reinforcement particles are not plastically deforming, one can logically question whether the volume they fill should be considered in determining the gauge length used for calculating the plastic strain (Corbin and Wilkinson 1991). Thus, the calculated average plastic strain is decreased by the volume fraction, a small but systematic correction.

A limited number of studies have considered either *initial* plastic strain versus life behavior or the plastic strain at specimen half-life versus life behavior, however these were from experiments conducted under either total strain (Seitz et al. 1991; Srivatsan 1992; Liu 1991)

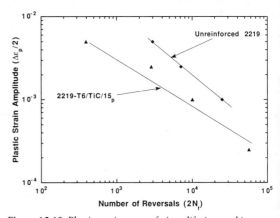

Figure 15.10. Plastic strain versus fatigue life (reversals) behavior of 2219-T6 compared with 2219-T6/TiC/15$_p$ *(Data from Vyletel et al. 1992.)*

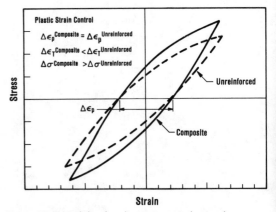

Figure 15.11. Stabilized cyclic stress-strain loops of a discontinuously reinforced composite compared with an unreinforced counterpart under plastic strain-control conditions (schematic).

or total stress control (Bonnen et al. 1991). At a given level of initial or half-life plastic strain, these investigations have reported increased fatigue resistance in composite materials (Bonnen et al. 1991; Seitz et al. 1991), no difference (Liu 1991), or a decrease in fatigue resistance compared to unreinforced material (Liu 1991). In view of the previously described differences in the cyclic softening-hardening response of composite materials compared with unreinforced matrices, these results should be viewed with some caution.

15.2.2 Mechanisms of Fatigue Crack Initiation

As described in the previous section, the dislocation structures that form during fatigue of composite materials appear to be quite similar to those that form in unreinforced materials subjected to similar plastic strains. In aluminum composites aged to produce shearable precipitates, persistent slip bands form, although they are appreciably less well defined than they are in corresponding unreinforced alloys. In composites with nonshearable precipitates, deformation is more homogeneous in nature and dislocations reside at precipitate-matrix interfaces and extend into the matrix between the precipitates. How these features develop into fatigue cracks has not been the subject of any investigation.

In particle reinforced composites, fatigue cracks tend to initiate well into the matrix (Hall et al. 1993b) or at the particle-matrix interface (Liu 1991) when initiation is not related to imperfections. An example of this for a 6061-T6 composite reinforced with SiC particles is shown in Figure 15.12. In both particle (Liu 1991) and whisker reinforced aluminum alloy (Williams and Fine 1985), this initiation is observed in the matrix at the poles of the reinforcement. This is consistent with analytic predictions (Hom and McMeeking 1991; Llorca et al. 1992) that indicate that high stress and high plastic strain levels occurs in this region.

One topic that has received extensive study is the manner in which the propagating crack in a smooth bar sample interacts with the reinforcement particle. Bonnen et al. (1991) and Hall et al. (1993a, 1993b) have noted a high number of fractured SiC particles on the fatigue surface after failure. The number of cracked particles on the fracture surface has been observed to not be a function of applied stress, however, the number of these fractured particles increases as the crack progresses away from the initiation site, which suggests a dependence on stress intensity. Similar correlations between particle fracture and increasing stress intensity have been reported on long crack fatigue crack growth samples (Shang et al. 1988; You and Allison 1989). Debonded particles are generally not observed in smooth bar samples, which distinguishes these fracture features from those in long crack samples where debonded particles are observed at low ΔK (Bonnen et al. 1990). Hall et al. (1993a, 1993b) have reported that the propensity for fractured SiC particles increases with increasing average particle size. The fracture of SiC is presumed to be related to the presence of faults that are known to be present in these particles, thus promoting easy fracture. Therefore, larger particles will have a higher propensity to have large faults and will fracture at a lower stress than will small SiC particles. Similar results have been reported for aluminum reinforced with silicon particles (Lukasak 1992). This subject will be discussed in more detail in the section on fatigue crack propagation.

Although a distinctive feature in composites reinforced with moderate to large SiC_p, particle fracture is by no means ubiquitous. In a 2219 composite reinforced by spherical TiC particles, Vyletel et al. (1993) have observed no particle cracking on the fatigue fracture surface. This may be attributed to the defect-free nature of the single crystal TiC particles, which leads to a rather high particle fracture strength. The small size (3 µm) of the particle is presumably also a factor as observed in SiC_p composites (Hall et al. 1993b). TiC particles with thin layers of aluminum have occasionally been observed on fatigue fracture surfaces, indicating matrix cracking near the reinforcement particle. This is taken to reflect the build-up of stresses in the matrix.

A common observation, especially in composites produced by casting, is that fatigue cracks initiate at imperfections (Zhang et al. 1991; Masuda and Tanaka 1992). Imperfections such as shrinkage porosity and reinforcement clustering are generally considered to be inherent in cast materials, however, recent work (Hoover 1991) has suggested that die casting of composite materials may serve to minimize these imperfections because of the higher solidification rates of die casting and the thixotropic nature of the composite materials (which minimizes gas entrapment during metal flow). For these types of materials, there is a need to quantify the levels of porosity and particle clustering

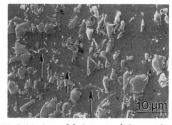

Figure 15.12. Initiation of fatigue crack in matrix of a 6061/SiC/15$_p$ composite. *(Courtesy of J. Hall.)*

that can occur, and to predict the effect of these imperfections. Approaches based on fracture mechanics have been advanced (Couper et al. 1990) to account for this situation. Cast composites, especially those using short fiber or whisker preforms can lead to reinforcements that are not well bonded to the matrix because of the lack of complete infiltration. In these materials, debonded reinforcements are frequently observed and associated with fatigue fracture initiation. These have been most often reported for short fiber composites fabricated using pressure infiltration techniques (Hurd 1988; Harris 1988), suggesting that improvements in these processes might lead to enhanced fatigue performance.

Imperfections are also observed in composites of wrought alloys produced by powder metallurgy. In an early vintage X2080/SiC/15$_p$ powder metallurgy composite, Bonnen et al. (1991) found that fatigue cracks initiated at a number of recognizable imperfections, which included large reinforcement particles, clusters of particles, and exogenous defects introduced by the powder metallurgy process. In a number of samples, initiation could not be attributed to pre-existing defects, indicating that these cracks initiated in the matrix. Interestingly, fatigue lives of samples in which fatigue cracks initiated from pre-existing defects were no different than those in which no initiation feature could be identified. Similar results have been obtained by Hall et al. (1993b), in which initiation was related to coarse intermetallic phases in addition to large particles and particle clusters. Again, imperfections had minimal effect on fatigue lives in this material suggesting that the effect of these defects on fatigue life diminishes as strength levels increase.

15.2.3 The Effects of Reinforcement Volume Fraction, Size and Shape, and Matrix Microstructure on Fatigue Life Behavior

Because of the many metallurgical variables that can be controlled in composites, these materials can be tailored or engineered to develop the balance of properties that is most suitable for a particular application. These variables include reinforcement volume fraction, size and shape, as well as matrix microstructure. This section will review the manner in which these factors affect fatigue lives.

Increasing the volume fraction of reinforcement has been observed to increase the fatigue life under stress-controlled conditions in wrought aluminum alloys as well as a magnesium alloy. As previously described, this can be attributed to the decreased elastic and plastic strains that result from the increasing modulus and apparent work hardening, both of which increase with increasing volume fraction. A confounding factor in this rationale is that different investigators have found qualitatively different behavior. That is, while some investigators have reported significant improvements in fatigue strength with increasing reinforcement volume fraction (Xia et al. 1991; Llorca et al. 1991; Couper and Xia 1991; Hall et al. 1993b) other data suggests that fatigue strength is only weakly increased by increasing volume fraction (Hunt 1993; Klimowicz 1992). It should be noted that a strong correlation exists between these different observations and the test technique used. The two studies that show a large effect of volume fraction were based on data generated using the rotating bending technique, while the data from the two studies that show a weak effect of volume fraction were developed using axial loading. This suggests that the rotating bending experiments are more affected by volume fraction than are the axial experiments. This is further supported by the observation that, in investigations using rotating bending, the influence of volume fraction increases with increasing stress, that is, the data diverge at high stresses. In contrast, for axial loading, the influence of volume fraction appears to be diminished at high stresses. The reason that the rotating bending tests are more influenced by volume fraction is unclear, although it may be related to differences in elastic modulus, work hardening, susceptibility to preexisting flaws, or small crack propagation behavior. These differences are apparently not due to differences in alloying, because an identical alloy (6061-T6) was characterized by both Xia et al. (1991) and Klimowicz (1992). When data from these two studies for the unreinforced 6061 alloy are compared, the ratio of fatigue strengths (at 10^5 cycles) for the axial experiments to that from rotating bending experiments is 0.78, which is consistent with "load factors" developed in steels (Juvinall 1967). Similar ratios for the composites appear to be lower and vary from 0.62 to 0.7.

Reinforcement particle size is also a potential variable that can be engineered. The contribution to composite work hardening and strengthening predicted by continuum mechanics is not scale dependent and thus, for an equivalent volume fraction, particle size should not have an effect on these factors or, in turn, on fatigue life. However, experimentation on wrought aluminum alloys clearly indicates that particle size does indeed have a significant influence. As shown in Figure 15.13, Holcomb (1992) has reported that in 2124-T6 reinforced by SiC$_p$, the fatigue strength (at 1×10^7 cycles) increased as the mean particle size decreased from 35 μm to 3 μm. Similar results have been reported by Hall et al. (1993b) for 6061-T6 reinforced by SiC$_p$ with mean particle sizes ranging from 5 μm to 19 μm. At the

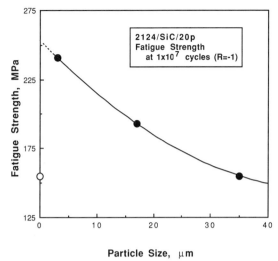

Figure 15.13. Influence of particle size on fatigue strength at 1×10^7 cycles in a 2124-T6/SiC/20$_p$ composite. *(After Holcomb 1992.)*

time of this writing, this effect of particle size is not well understood. It may be related to a refinement of slip length or to the increased propensity for particle cracking of the larger particles. It is interesting to note that composites manufactured using powder metallurgy techniques generally have smaller reinforcement particles and superior stress-controlled fatigue properties as compared with composites fabricated using ingot metallurgy techniques.

It is known that for a constant volume fraction, decreasing the particle size leads necessarily to a decrease in the interparticle spacing (Kamat et al. 1991; Martin 1980). Because the reinforcement particle is a barrier to slip, a decrease in the particle size can lead to a potential decrease in the slip length. Moreover, reinforcement particles act to pin grain boundaries (Vyletel et al. 1992), which leads to refinement of the grain size and reduction of the slip length through this additional path. Thus, in microstructures containing shearable precipitates, small reinforcement particles can lead to a reduction in slip length, either through grain size refinement or by the presence of impenetrable reinforcement particles that are closely spaced. Whether the strengthening precipitates present in the specific 6061-T6 or 2124-T6 composites investigated by Hall et al. (1993b) or Holcomb (1992) are shearable and thus support this rationale is not known. An important factor may be the propensity for increasing particle fracture with increasing particle size (Hall et al. 1993a), because particle fracture could lead to premature fatigue crack initiation in the matrix. In addition, it has been suggested that in SiC$_p$ reinforced composites, particle fracture leads to cyclic softening (LLorca et al. 1992) that, under constant stress experiments, would lead to an increase in the plastic strain.

Although reinforcement size is generally stated as an average value, reinforcement size distribution is also an important parameter. Decreasing the range of particle sizes present in a composite can have a beneficial influence on fatigue life by eliminating fracture-prone large particles. Couper and Xia (1991) have demonstrated this in 6061-T6 reinforced with 15% polycrystalline alumina spherical particles. Using a single batch of spherical particles, these investigators eliminated large particles using screening and classifying to arrive at three size ranges 0–100 μm, 0–45 μm, and 0–30 μm in ingot metallurgy fabricated composites. The average particle size in all three cases was reported to be 20 μm. Although the average fatigue strengths were identical for all three composites, the scatter in fatigue life increased significantly with increasing size ranges. This was attributed to the premature cracking of large particles that led to reduced fatigue lives. Thus, the *minimum* fatigue properties were significantly improved by narrowing the particle size distribution.

Particle shape has a pronounced influence on monotonic and cyclic flow strength and ductility of composite materials (Llorca et al. 1992) and recent analytical models have provided a basis for understanding this influence (Llorca et al. 1991; Llorca et al. 1992). These models have predicted that increasing aspect ratio should lead to increased flow strength and apparent work hardening as well as reduced tensile ductility. The strength increase is attributed to the geometric dependence of the build-up of hydrostatic stresses that govern flow. As previously described, these hydrostatic stresses build up because of the differences in the elastic modulus between the reinforcement and the matrix and the resultant constrained plastic flow. The increase in apparent work hardening competes, however, with void nucleation and growth, which ultimately leads to fracture. Therefore, although the high aspect ratio whiskers have high constraint and thus high flow strengths, these high hydrostatic stresses also promote void formation and lead to lower cyclic and tensile ductilities. This suggests that increasing the reinforcement aspect ratio should lead to increased fatigue lives at low stresses for which cyclic flow strength dominates life. In contrast, at high stresses for which cyclic ductility dominates life, high aspect ratio particles should lead to decreased fatigue lives. Limited experimental evidence supports this notion, at least in the low stress region. Crowe and Hasson (1982) compared whisker reinforcement to particle reinforcement in 6061-T6 reinforced with 20 volume fraction of SiC. They reported that whisker reinforcement led to a pronounced improvement in fatigue compared with particle reinforcement. It should be noted that different processing methods were used to manufacture these composites and that, given their relatively early vintage, processing differences cannot be discounted. In fact, recent fatigue studies on 6060-T6 reinforced with 15% SiC particles (Hall et al. 1993b) exhibit fatigue resistance that is superior to the 20% SiC whisker composite. Thus, while reinforcement aspect ratio effects are predicted to be significant, experimental work is required for verification of this hypothesis.

In addition to effects caused by the reinforcement, the matrix may have an influence on fatigue life behavior of composite materials. Matrix effects would include differences in alloying, type of precipitate structure, grain size, or dispersoids/intermetallic content. Grain size effects have not been examined explicitly, however the superior fatigue resistance of powder metallurgy composites of Al-Cu alloys (Bonnen et al. 1991; Hall et al. 1993b; Hunt et al. 1991) compared with ingot metallurgy composites with similar compositions (Klimowicz 1992) may be related, at least in part, to the fine subgrain and grain sizes of the powder metallurgy composites. Investigations that have explicitly explored matrix effects have been in Al-Cu and Al-Cu-Mg alloys in which the precipitate type can be varied widely. Vyletel et al. (1991) have shown that, despite the significant effect precipitate type has on cyclic hardening and flow strength, it has essentially no effect on either stress-controlled or plastic strain-controlled fatigue. As shown in Figure 15.14, they found that when a 2219/TiC/15$_p$ composite was naturally aged to produce GP zones, the fatigue life behavior was identical to that of an artificially aged condition aged to produce θ' precipitates. The only major differences in fatigue life behavior was in stress-controlled testing at high stresses. This was attributed to the cyclic instability (hardening) of the naturally aged condition, which produced large plastic strains and fatigue damage in the first few cycles and thus, short fatigue lives. This example shows that matrix strength is also not a major contributor to fatigue behavior as the naturally aged condition had a much lower yield and ultimate strength compared to the artificially aged condition. Similar results have been reported by Liu (1991) for strain-controlled experiments conducted in 2024 aluminum reinforced with Al_2O_3.

15.3 Fatigue Crack Propagation

15.3.1 Overview

Fatigue crack propagation represents, in many cases, the controlling phase of fatigue life and, in all cases, represents the final damage accumulation process that ultimately leads to service failure under cyclic loading conditions. Therefore, an understanding of fatigue crack propagation behavior is of obvious importance to fatigue life predictions and damage tolerance considerations. Fatigue crack propagation is best characterized by the rate of crack advance per cycle, da/dN, as a function of applied stress intensity range, ΔK, in which ΔK is the difference between the maximum, K_{max}, and the minimum, K_{min}, applied stress intensity factor. Crack propagation data, when plotted as log (da/dN) versus log (ΔK), generally exhibit the sigmoidal behavior shown schematically by the solid curve in Figure 15.15. In region I, crack growth rate rapidly decreases as the stress intensity range decreases and becomes difficult to measure below an apparent threshold, ΔK_{th}, in stress intensity range. In region II, at intermediate levels of ΔK and da/dN, crack growth rate is characterized by the Paris relation:

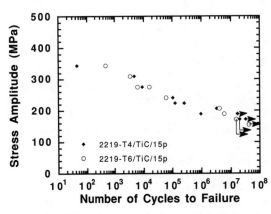

Figure 15.14. Effect of aging on the fatigue-life behavior of 2219/TiC/15$_p$ composite. (*After Vyletel et al. 1991.*)

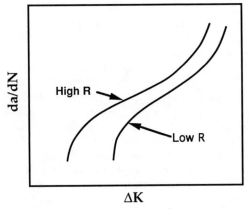

Figure 15.15. Schematic of da/dN versus ΔK including an illustration of the effect of R.

$$da/dN = C\Delta K^m \quad (15.2)$$

where C and m (approximately 2–4 for many alloys) are material sensitive parameters. In region III, a rapid acceleration in crack growth rate is observed as the applied maximum stress intensity, K_{max}, approaches the cyclic fracture toughness, ΔK_c, of the material.

In the near-threshold and near-ΔK_c regimes, crack growth rate can be very sensitive to microstructure and load ratio, $R = K_{min}/K_{max}$. The effect of increasing R is shown schematically by the dashed curve in Figure 15.15. Microstructure and load ratio exert a much smaller influence on crack growth rate in the Paris regime. Fatigue crack growth behavior is, therefore, defined by the Paris constants, the crack growth threshold, ΔK_{th}, and fracture toughness.

Numerous studies of the fatigue crack propagation behavior of discontinuously reinforced aluminum alloy composites, and especially those reinforced with particles, have been conducted since 1982, with most studies occurring since 1987. Questions regarding the role of matrix microstructure, load ratio, reinforcement size, and volume fraction on crack growth, especially in the near-threshold and Paris regimes, have been addressed in much of this work. Of particular interest is whether the incorporation of a reinforcement phase into an alloy improves or degrades the fatigue crack propagation behavior. The answer to this question is not a simple one and contradictory results are often observed from different investigators studying similar alloys. These comparisons are complicated by the many microstructural, compositional, processing, and reinforcement variables associated with MMCs. An additional complication arises because of the intrinsic difficulty in interpreting fatigue crack growth data generated by different test procedures. Despite these difficulties, considerable progress has been made in understanding the manner in which reinforcement causes fatigue crack propagation to differ from the matrix (if indeed it does) and in identifying the mechanisms responsible for fatigue crack propagation.

The fatigue crack propagation behavior of a wide variety of aluminum alloy-based MMCs have been examined and, in a few cases, reviewed (Nair et al. 1985; Kumai et al. 1991b; Shang and Ritchie 1989c; Davidson 1993). Although a few studies have been performed on whisker-reinforced composites (Logsdon and Liaw 1986; Yau and Mayer 1986; Musson and Yue 1991; Hirano 1991), the majority of studies in recent years have focused on particulate strengthened alloys. It is this latter group that forms the basis of this review of fatigue crack propagation. Although experimental parameters varied considerably in previous studies, it is fortunate for purposes of comparison that in almost all cases at least a portion of the crack growth data was generated under constant R at values less than 0.3. It has often been assumed that at low values of R, the introduction of a reinforcement phase results in higher thresholds, a steeper Paris region, and a transition to unstable fracture at lower ΔK levels. This latter behavior is characteristic of all discontinuously reinforced MMCs and is a direct consequence of the lower fracture toughness in composites compared to unreinforced alloys. However, the influence of reinforcement on the near-threshold and midrange growth behavior is more complicated than that implied by the generalization described above. In reality, threshold levels, as well as near-threshold and midrange growth rates in composites may be higher than, lower than, or roughly equivalent to, the values observed in unreinforced alloys. Although it is not possible to neatly categorize all of the observations, a majority of them can be reasonably described by one of the three following situations:

1. An increase in ΔK_{th} and an increased dependence of growth rate on ΔK in the midrange growth regime (larger value of m in the Paris equation).
2. Lower ΔK_{th} and higher near-threshold and midrange growth rates.
3. Lower or equivalent midrange growth rates and lower ΔK_{th}.

An example of the first type of crack growth behavior is shown in Figure 15.16 from the work of Lukasak and Bucci (1992) on X2080-T6 aluminum reinforced with 15 vol % SiC particles with an average diameter of 16 μm. Cyclic loading, at an R of 0.3, resulted in a modest increase in apparent threshold and a steeper slope to the midrange growth regime. The second type of behavior is illustrated by the work of Sugimura and Suresh (1992), who examined the influence of volume fraction on the near-threshold crack growth behavior in a peak aged, Al-3.5 Cu ingot metallurgy alloy reinforced with 2 μm to 4 μm average diameter SiC particles. As shown in Figure 15.17, when the reinforced alloys are compared to an unreinforced alloy with a 150 μm grain size, threshold values at $R = 0.1$ were lower and crack growth rates were higher in the reinforced alloys for the three reinforcement volume fractions examined. Finally, the third type of behavior is illustrated by the results of Shang and Ritchie (1989a), who have examined the crack propagation behavior of an overaged Alcoa MB78 composite containing coarse (16 μm) SiC particles. Their results for fatigue crack growth tests performed on a 15 vol % composite with $R = 0.1$ (Figure 15.18) show that, although ΔK_{th} is decreased only slightly with reinforcement, the midrange growth rates are decreased significantly in the reinforced alloy. Naturally not all

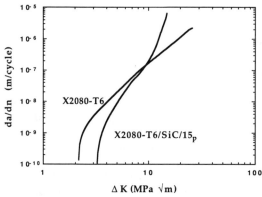

Figure 15.16. Crack growth behavior in X2080-T6 aluminum reinforced with 15 v/o SiC particles with an average diameter of 16 μm. *(After Lukasak and Bucci 1992.)*

results fit these simple categorizations. However, it is clear that the fatigue crack growth behavior of discontinuously reinforced MMCs cannot be described by an overly simple paradigm.

15.3.2 Crack Tip Phenomena

In order to explain what, at first view, might seem to be random variation in the influence of discontinuous reinforcement on fatigue crack propagation behavior in aluminum alloys, it is important to examine the observed behavior in the context of the crack tip phenomena normally associated with the existence of a threshold and the near-threshold crack growth behavior in these alloys. For discontinuously reinforced composites, the primary crack tip phenomena are crack closure,

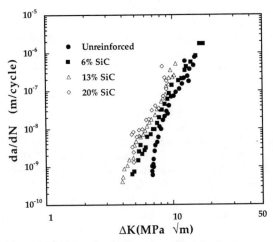

Figure 15.17. Near-threshold crack growth behavior in a peak aged, Al-3.5 Cu ingot metallurgy alloy reinforced with 2 μm to 4 μm average diameter SiC particles *(After Sugimura and Suresh 1992.)*

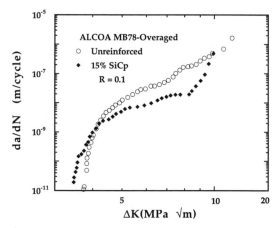

Figure 15.18. Fatigue crack propagation behavior in an overaged X2080 composite containing coarse (16 μm) SiC particles. *(After Shang, Yu, and Ritchie 1988.)*

crack bridging, crack deflection, and crack trapping. Through an understanding of the manner in which particle or whisker reinforcement affects these phenomena, a more consistent view will be attained of how matrix microstructure, composition, and reinforcement variables (volume fraction, size, shape) affect crack growth in these composites.

15.3.2.1 Fatigue Crack Closure

As described in the preceding section, the driving force for crack advance during fatigue is the stress intensity range, ΔK, experienced at the crack tip. Thus, phenomena that shield the crack tip from the full influence of the applied ΔK strongly influence crack growth behavior. Crack closure, the contact of the crack faces in the vicinity of the crack tip under positive applied loads, is the most important shielding mechanism for fatigue crack growth at low load ratios in aluminum alloys and their composites. Crack closure occurs over a range of applied stress intensity, and it is traditional, although somewhat simplistic, to attempt to define a single value, K_{cl}, the closure stress intensity, above which the crack is fully open and below which the crack faces are in contact. Crack closure can arise from a number of phenomena, including crack tip plasticity, oxide formation, and contact between asperities that are present on the crack surface (Suresh and Ritchie 1984). This latter behavior is called roughness-induced closure and plays an important role in fatigue crack growth in composites. Roughness-induced closure arises because the irregular surface topography of a propagating fatigue crack generates asperities that contact at stress intensity levels above K_{min}, thus leading to crack closure. An example of such an asperity contact is shown in Figure 15.19 (You and Allison 1989). K_{cl} is usually measured

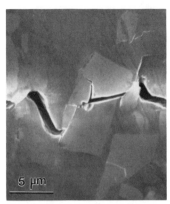

Figure 15.19. Asperity contact in 2124/SiC/20$_p$. *(Reprinted with permission from You, C.P. and J.E. Allison. 1989. In: Advances in Fracture Research. Vol. 4. Oxford, UK: Pergamon Press.)*

by noting changes in bulk specimen compliance as a function of K_{max}, and, in many cases, the phenomena responsible for crack closure can be inferred from the variation of K_{cl} with K_{max} (Allison 1988). You and Allison (1989) have shown that for a variety of aging conditions in 2124/SiC/20$_p$ K_{cl} does not vary as a function of K_{max}, as shown in Figure 15.20. This suggests that the operating closure mechanism is roughness-induced closure. Crack closure effectively reduces the stress intensity range experienced at the crack tip to an effective value:

$$\Delta K_{eff} = K_{max} - K_{cl} \quad \text{(where } K_{cl} > K_{min}\text{)} \quad (15.3)$$

The contribution of crack closure to crack tip shielding and, therefore, to fatigue crack growth behavior, decreases at high R values at which large crack tip openings occur. With increasing load ratio, a condition is eventually reached where K_{cl} is less than K_{min} and ΔK is equal to ΔK_{eff}. This explains the strong dependence of threshold and near-threshold growth rate on R, illustrated in Figure 15.15, and why, in many cases, fatigue crack growth data at different R values can be consolidated into a single curve by plotting against ΔK_{eff}.

Microstructure can strongly influence crack closure behavior in aluminum alloys and, therefore, aging condition is an important variable in the near-threshold fatigue crack propagation in these materials. In unreinforced, precipitation hardening aluminum alloys, such as those in the 2xxx, 6xxx, 7xxx, and 8xxx series, changes in microstructure as a result of aging can lead to significant changes in near-threshold behavior. Underaging in these alloys promotes planar slip within the crack tip plastic zone, while overaging homogenizes crack tip deformation. Planar slip leads to faceted crack growth and significant roughness-induced closure at low values of R. Underaged aluminum alloys, therefore, exhibit higher threshold levels and lower near-threshold crack growth rates than those observed in overaged alloys, where crack propagation produces relatively smooth fracture surfaces. The coupling of slip behavior and crack closure levels also explains the fact that in the underaged condition, increasing grain size, which corresponds to longer slip lengths, produces greater facet sizes and leads to improved near-threshold properties. (Suresh et al. 1984; Carter et al. 1984; Zaiken and Ritchie 1985). Because these microstructural effects on fatigue crack growth are manifested via mechanical crack closure, they are considered extrinsic in nature.

The results of numerous studies have indicated that the differences in near-threshold behavior between reinforced and unreinforced alloys are controlled by the

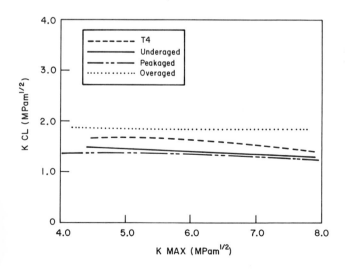

Figure 15.20. K_{cl} versus K_{max} as a function of aging in 2124/SiC/20$_p$. *(Reprinted with permission from You, C.P. and J.E. Allison. 1989. In: Advances in Fracture Research. Vol. 4. Oxford, UK: Pergamon Press.)*

significant effect of reinforcement on crack closure levels. In several cases, the influence of reinforcement on closure has been confirmed by direct measurement of crack closure levels. For example, Shang et al. (1988) examined crack closure and its effect on the near-threshold behavior in a peak-aged Alcoa MB78 alloy reinforced with 20 v/o fine (5 μm) and 20 v/o coarse (16 μm) SiC particles. As shown in Figure 15.21, near-threshold crack closure ratios, K_{cl}/K_{max}, showed a strong dependence on particle size with substantially higher closure levels, equivalent to the those of the unreinforced alloy, observed for the coarser particle composite. Shang and Ritchie (1989b) also observed that closure levels were essentially independent of volume fraction for the small range (15 v/o versus 20 v/o) examined. Sugimura and Suresh (1992), in their study of peak-aged Al-3.5Cu reinforced with 6, 13, and 20 v/o SiC of relatively small (2 μm to 4 μm) particles, found that crack closure levels were significantly reduced by the presence of reinforcement, as shown in Figure 15.22. Similar to the observations of Shang and Ritchie, they also found that the closure ratio increased significantly at low ΔK levels.

These results, which are representative of most of the closure measurements made in similar composites, indicate that closure levels increase with increasing particle size. Note that reductions in closure levels are observed for small particle sizes in both studies cited above. These observations can be best explained by understanding the influence of reinforcement on roughness-induced closure. Levin and Karlsson (1991), although not measuring closure levels, reported increases in crack surface roughness in 6061/SiC/15$_p$ and attributed decreased near-threshold growth rates to enhanced roughness-induced closure. At low R values in aluminum MMCs, crack tip opening displacements

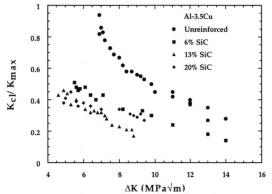

Figure 15.22. Crack closure levels in Al-3.5Cu as a function of SiC volume fraction. *(After Sugimura and Suresh 1992.)*

at near-threshold ΔK levels are approximately two orders of magnitude smaller than the reinforcement particle size (Shang and Ritchie 1989b). As described in more detail later, it is also generally observed that crack deflection around reinforcement particles, rather than particle fracture, occurs at low ΔK. Thus, crack closure levels are greater for larger particle sizes as a direct result of the increased levels of surface roughness.

It is essential to note that the relative changes in roughness-induced closure levels that result from the introduction of reinforcement depend significantly on microstructure and, therefore, aging condition. This is because closure levels in unreinforced alloys strongly depend on microstructure while closure levels in reinforced alloys do not. You and Allison (1989) have shown, for example, that changes in microstructure produced by various aging treatments have little effect on closure levels in a 2124 Al reinforced with 20 v/o of 5 μm SiC particles. This was because the size and spacing of reinforcement particles effectively control the fracture surface roughness and diminish or eliminate the effect of the matrix microstructure on roughness. This effect on fracture surface roughness occurs because of the reduction in slip length brought about by the presence of the hard particles and the tendency of the crack to be diverted by them. Thus, even though a particular microstructure may be susceptible to planar slip in the unreinforced condition, reinforcement effectively reduces the influence of slip character on fracture surface roughness. This has the significant consequence of producing very similar levels of surface roughness in composites in both the underaged and overaged conditions, and is responsible for the absence of a large effect of aging on crack closure in composites.

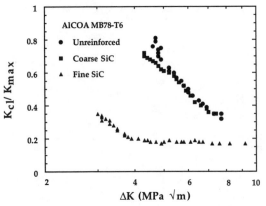

Figure 15.21. Crack closure levels as a function of particle size in peak aged X2080/SiC. *(After Shang et al. 1988.)*

15.3.2.2 Crack Trapping and Bridging

Crack tip processes other than crack closure may also be important in determining the influence of reinforcement on near-threshold crack growth behavior. At low ΔK, the reinforcement phase may act as a direct impediment to crack advance by "trapping" the crack front in the vicinity of the matrix/particle interface. Shang and Ritchie (1989b) developed a model which describes the situation in which reinforcement particles in the crack tip vicinity reduce the crack tip stress intensity and thus serve as crack traps. Other evidence supports the contention that reinforcement particles impede crack advance. Others have demonstrated or speculated that crack growth in the vicinity of particles can be retarded (Davidson 1987; Kobayashi et al. 1991; Levin and Karlsson 1991). You and Allison (1989) and Shang et al. (1988) have noted that the incidence of particle decohesion is greater at near-threshold growth rates than at higher growth rates. These observations provide support for crack trapping and/or crack deflection at near-threshold conditions. Shang and Ritchie (1989b, 1989c) have predicted that the influence of crack trapping on near-threshold behavior will be proportional to matrix yield stress and to the square root of particle size. Thus, larger particles should be more effective in reducing near threshold growth rates in composites, which is consistent with much of the published literature in this area.

At higher levels of ΔK, crack growth may be reduced in composites by the formation of uncracked ligaments ahead of the main crack front, as observed by Shang and Ritchie (1989a) in an Al-Zn-Mg-Cu alloy reinforced with either fine or coarse SiC particles. They attribute this behavior to a crack bridging mechanism that results from the fracture of large SiC particles ahead of the main crack. In the fine-particle composite, damage ahead of the fatigue crack results in a coplanar array of bridges, while particle fracture in the coarse-particle composite yields a region of overlapping cracks. Models of these process have shown that coplanar bridging is a much less effective crack tip shielding mechanism than overlapping cracks, which is consistent with experimental observations.

15.3.2.3 Particle Fracture

From the preceding discussion it is clear that the influence of reinforcement on crack tip phenomena will be significantly affected by the occurrence or absence of particle fracture during crack propagation. Crack trapping, for example, requires crack deflection around reinforcement particles, rather than particle fracture, while in crack bridging particle fracture is a requirement. Excessive particle fracture should also be expected, in some instances, to increase the rate of crack growth by providing an easy path for crack advance. It is therefore important to describe the nature of particle fracture in particle reinforced aluminum alloys during fatigue crack propagation. Particle fracture during fatigue has been shown to depend on particle size (Shang et al. 1988; Hall et al. 1993a) and particle volume fraction (Sugimura and Suresh 1992), and to increase with increasing crack tip stress intensity (Bonnen et al. 1991; Sugimura and Suresh 1992; Hall et al. 1993b). As noted by Shang et al. (1988), the probability of particle fracture during fatigue will depend on crack tip stresses and particle strength, with the latter varying inversely with particle size. This dependence has been quantified recently by Hall et al. (1993a, 1993b) who measured the frequency of particle fracture as a function of particle size, matrix yield strength and stress intensity. In their study, area fractions of particle fracture were determined for fatigue cracks that initiated in smooth axial specimens and propagated to failure. The probability of particle fracture, taken here as the area fraction of the fracture surface occupied by fractured particles normalized by the reinforcement volume fraction, strongly depends on particle size. K_{max} also depends on volume fraction. In Figure 15.23, the frequency of particle fracture for 6061 and 2124 aluminum alloys reinforced with various sizes and volume fractions of SiC is plotted as a function of $(K_{max})^2$. In the near-threshold regime, essentially no particle fracture occurs. The frequency of particle fracture increases with maximum stress intensity and with an increase in particle size. These observations and the reasonable correlation with $(K_{max})^2$ indicate that a fracture criterion which depends on crack tip stress and particle fracture strength is required to account for the probability of reinforcement particle fracture. It is interesting to note that essentially no particle fracture was observed in the fatigue study of Levin and Karlsson (1991) for 6061/SiC/15$_p$ for particle sizes in the same size range in which Hall et al. (1993a) observed particle fracture. The Levin and Karlsson results notwithstanding, the general propensity of particle fracture increases with increasing particle size and volume fraction. This effect and its influence on fracture surface roughness and subsequent crack closure must be understood before the influence of particle fracture on fatigue crack growth can be understood.

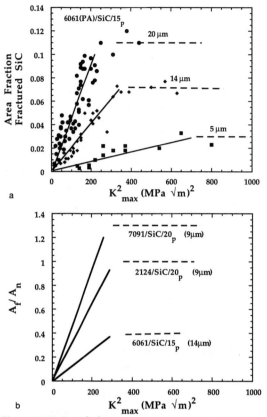

Figure 15.23. Particle fracture in various composites as a function of stress intensity. Dashed lines denote fast fracture region. (a) The influence of particle size on particle fracture in 6061/SiC/15p. (b) The influence of matrix on particle fracture (A_f is the area fraction of fractured SiC particles; A_n is numerically equivalent to reinforcement volume fraction.) (After Hall et al. 1993a.)

15.3.3 Near-Threshold Behavior

15.3.3.1 Effect of Particle Size and Volume Fraction

In the near-threshold region, fatigue crack growth at low R is dominated by crack closure effects. It is not surprising that, in this region, much of the fatigue crack growth behavior of aluminum alloy matrix composites, including the effects of reinforcement particle size and volume fraction, can be attributed to the influence of reinforcement on crack closure, as described in Section 15.3.2.1. For example, Shang et al. (1988) determined ΔK_{th} and near-threshold growth rates in a peak-aged Alcoa MB78 alloy reinforced with either 20 v/o fine (5 μm) or 20 v/o coarse (16 μm) SiC particles. As shown in Figure 15.24, in the alloy reinforced with the coarse particles ΔK_{th} was greater and near-threshold growth rates were lower than in the alloy with the fine reinforcement. This behavior paralleled the effect of particle size on crack closure. Sugimura and Suresh (1992) also observed a decrease in ΔK_{th} and an increase in near-threshold crack growth rates (Figure 15.18) in peak-aged, Al-3.5Cu reinforced with SiC particles with an average diameter of 2 μm to 4 μm. In their study, the magnitude of crack closure in the near-threshold region was much lower in the composites than it was in the unreinforced alloy, as seen in Figure 15.22. The near-threshold behavior observed by Sugimura and Suresh (1992), Shang et al. (1988), and Shang and Ritchie (1989b) are actually quite comparable for the composite materials when similar volume fractions and particle sizes are compared. Both studies show that reinforcement with small particles reduces closure levels and has a detrimental effect on ΔK_{th} and near-threshold growth rates. However, for larger particle sizes, crack closure levels are high and equivalent to those of the unreinforced alloy, and correspondingly, near-threshold fatigue crack growth resistance is equivalent to that of the unreinforced alloy. Downes et al. (1991) report a similar particle size effect for an 8090 alloy reinforced with 3 μm SiC and 21 μm SiC. Thus, in situations where roughness-induced closure dominates near-threshold behavior, larger particle sizes promote higher levels of closure and improved near-threshold fatigue behavior.

The effect of volume fraction on fatigue crack growth behavior has been examined by several investigators. Bonnen et al. (1990) have reported that in 2124 aluminum reinforced with fine SiC particles, increasing the volume fraction from 20% to 30% leads to an increase in fatigue crack growth resistance for both underaged and overaged microstructures (Figure 15.25). In a peak-aged Al-Zn-Mg-Cu composite reinforced with either 15 v/o or 20 v/o SiCp, Shang and Ritchie (1989a) observed no influence of volume fraction on threshold for composites reinforced with coarse particles and only a modest improvement in the case of

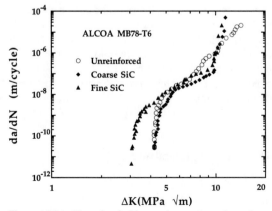

Figure 15.24. Near-threshold growth rates in peak-aged MB85 X2080 as a function of SiC particle size. (After Shang and Ritchie 1988.)

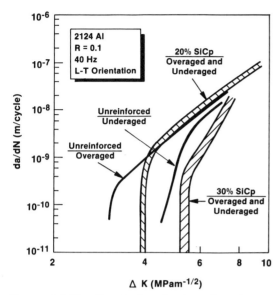

Figure 15.25. The effect of volume fraction and matrix microstructure on near-threshold crack growth in 2124/SiC/20$_p$. *(After Bonnen et al. 1990.)*

Bonnen et. al. 1990; Christman and Suresh 1988; Davidson 1989; Knowles and King 1991; Downes et al. 1990; Knowles and King 1990). Unlike crack propagation in unreinforced aluminum alloys, crack propagation behavior in aluminum composites is only weakly dependent on matrix microstructure (Figure 15.25). As described in Section 15.3.2.1, this is because aging has a weak influence on fracture surface roughness and, thus, a weak influence on roughness-induced crack closure. For microstructures that produce faceted crack growth in the near-threshold region in unreinforced alloys, reinforcement produces a significant reduction in faceted fatigue crack growth. The result is that near-threshold behavior is relatively independent of aging in the reinforced aluminum alloys.

The pronounced influence of microstructure on near threshold crack growth behavior in unreinforced aluminum alloys, and the relatively small effect in reinforced alloys, requires that careful attention be paid to microstructural condition when comparisons of threshold levels and near-threshold crack growth rates are made between unreinforced and reinforced alloys. For example, the introduction of reinforcement into an underaged microstructure has an effect that is similar to that produced by changing the matrix microstructure to an overaged condition, at least in terms of its influence on fracture surface roughness and the resultant roughness induced closure. This is illustrated from the work of Bonnen et al. (1990) (Figure 15.25), in which near threshold behavior is shown for a 2124 Al alloy in the overaged and underaged condition with and without 20 v/o of 5 μm diameter SiC particles. The large differences in the effect of microstructure on ΔK_{th} in the unreinforced condition leads to the conclusion that in one instance ΔK_{th} is raised by the addition of reinforcement and in another instance ΔK_{th} is decreased by the addition of the same reinforcement. Additionally, because of the smaller dependence of microstructure on near-threshold crack growth rates in reinforced alloys, threshold levels and near-threshold growth rates span a much larger range in unreinforced alloys than they do in their reinforced counterparts. This is illustrated rather convincingly in the work of Kumai et al. (1991b), who show that the variation in fatigue crack growth behavior for unreinforced 8090 aluminum is substantially greater than the variation seen for several reinforced alloys with differing volume fractions and particle sizes.

When comparing crack growth behavior in composites and in their unreinforced counterparts, the influence of grain size and, therefore, processing must not be overlooked. Many of the composites that have been examined have been produced by powder processing. For these materials, small grain sizes are normal in both the reinforced and unreinforced material and the influence of reinforcement on near-threshold behavior may

a fine particle size composite. Sugimura and Suresh (1992) examined a wider range of volume fractions in peak-aged Al-3.5Cu/SiC$_p$ composites and found only modest differences in near threshold behavior when volume fractions ranged from 6 v/o to 20 v/o, with 6 v/o and 20 v/o yielding almost identical threshold values and 13 v/o producing the lowest threshold value (Figure 15.17). Although only limited data on the effects of volume fraction are available, it appears that the influence of volume fraction on fatigue crack growth behavior is consistent with trends expected from roughness-induced closure, especially at low growth rates. At higher growth rates, static fracture modes may also be important and must be considered. Hall et al. (1993b) have shown that particle fracture during fatigue crack propagation increases with increasing volume fraction. Several investigators (Sugimura and Suresh 1992; Knowles and King 1991) have concluded that particle cracking and other static modes can lead to accelerated fatigue crack propagation rates. This suggests that increasing volume fraction of reinforcement may promote higher, rather than lower, growth rates at high stress intensity, where static modes are more important.

15.3.3.2 Microstructural Effects

Several studies of fatigue crack growth in discontinuously reinforced aluminum alloy composites have incorporated matrix precipitate structure as a variable. Materials have been examined in the underaged, peak aged and overaged conditions (You and Allison 1989;

be small. However, in ingot-metallurgy-produced materials, grain size in the unreinforced material can be significantly larger than that in the composite. Thus, in an unreinforced material with a microstructure susceptible to planar slip, the ingot metallurgy alloy with the larger grain size will have superior crack growth resistance and the effect of reinforcement will be pronounced (Sugimura and Suresh 1992). Further illustrating this point, Tanaka (1991) has observed threshold levels in an unreinforced cast 6061 heat treated to the T6 condition to be substantially higher that for a SiCw reinforced 6061 with the same heat treatment. Although no grain sizes are given, he notes that the matrix grain size is an order of magnitude smaller in the composite and that faceted slip observed in the unreinforced alloy is absent in the composite.

15.3.3.3 Load Ratio Effects and Intrinsic Threshold

Thus far, we have been concerned with near-threshold crack growth behavior under conditions of low load ratios. Under these conditions, the introduction of reinforcement alters near threshold behavior by influencing the level of roughness-induced crack closure in peak-aged and underaged microstructures. When closure effects are removed by conducting tests at high R levels, a significant lowering of ΔK_{th} is observed in these composites (Healy and Beevers 1991; Knowles and King 1991; Kumai et al. 1990a; Kumai et al. 1991a; Logsdon and Liaw 1986; Lukasak and Bucci 1992; Shang and Ritchie 1989b; Sugimura and Suresh 1992). $\Delta K_{th,eff}$ values for both reinforced and unreinforced alloys are naturally lower than ΔK_{th} measured at low R values, and in many cases, $\Delta K_{th,eff}$ values for the composites are greater than those for the unreinforced alloy. In these instances, nonclosure phenomena such as crack trapping (Shang and Ritchie 1989b) and crack deflection (Suresh 1991) may be responsible for the enhanced intrinsic threshold in the composites. Davidson (1988, 1991) has proposed a model similar to that of Shang and Ritchie, which is based on a mean free path for slip at the crack tip. The mean free path is presumed to be controlled by particle spacing and $\Delta K_{th,eff}$ can be expressed as a function of particle size and volume fraction. This model predicts an increase in intrinsic threshold with increasing particle size. It has also been argued that increased intrinsic thresholds arise because of the increased elastic modulus in the composites, which reduces crack tip opening displacements (Bonnen et al. 1990; Kumai et al. 1991a).

Caution should be exercised concerning the absolute values of $\Delta K_{th,eff}$ because of the difficulty in accurately determining absolute crack closure levels in aluminum alloys. You and Allison (1989) have shown that the values of closure load determined from global compliance measurements can differ significantly from the value determined by constant K_{max}, increasing R testing (CKIR). This test, which is conducted by increasing K_{min} while holding K_{max} constant, yields an independent measurement of the closure stress intensity as well as closure-free data (for $K_{max} > K_{cl}$). This independent determination of K_{cl} is important as closure loads determined from compliance techniques can vary significantly with the numerical procedure used to extract the closure load from the compliance data (You and Allison, 1989). This has been further illustrated by Sugimura and Suresh (1992), who also used a CKIR test procedure. Their results show that $\Delta K_{th,eff}$ based on compliance data was slightly greater in the composite than in the unreinforced material, but identical values were observed for both materials when the CKIR method was used. It should be mentioned that in many cases, the differences in intrinsic threshold between reinforced and unreinforced materials is small. Absolute values usually range from slightly less than 1 MPa m$^{1/2}$ to approximately 2 MPa m$^{1/2}$.

15.3.4 Midrange Growth Rates

Midrange fatigue crack growth rates in discontinuously reinforced aluminum MMCs have been shown in numerous investigations to have a stronger dependence on ΔK (a higher exponent, m, in the Paris equation) than do their unreinforced counterparts. For example, Davidson (1989) examined crack growth behavior in twelve different composites and found midrange crack growth rates to be very similar, with values of m ranging from 4.2 to 6.0, considerably above what would be expected for unreinforced alloys. This behavior has been explained, in several instances, as the result of an increased role of static modes of fracture, including particle fracture (Sugimura and Suresh 1992) and matrix void formation (Kumai et al. 1990a; Davidson 1989) during crack propagation. Davidson (1989) also argues from a practical viewpoint that the higher Paris exponents in composite materials are a natural consequence of the higher apparent thresholds and lower cyclic fracture toughnesses that are often observed.

Similar to the behavior in unreinforced aluminum alloys, most studies find little effect of aging condition on midrange growth rates. One notable exception is the work on the lithium-containing 8090 alloy reinforced with SiC (Knowles and King 1990, 1991), in which an apparent embrittlement at or near the matrix/particle interface and on matrix grain boundaries occurs during overaging.

Although particle fracture has been usually assumed to contribute to more rapid crack growth rates in the Paris regime, there is evidence that growth rates, in some cases, may actually be decreased by particle fracture. Shang and Ritchie (1989a), in a study of overaged Alcoa MB78/SiCp composites, noticed that in a composite with coarse particles, midrange growth rates are considerably lower than they are in unreinforced material. The effect was less pronounced when a composite with fine particles was examined. They argueed that crack bridging, as described in Section 15.3.2.2, is responsible for the decreased crack growth rates.

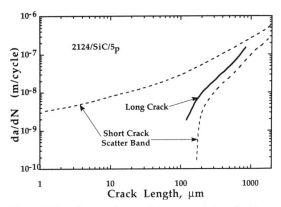

Figure 15.26. Short crack growth in 2124 reinforced with 5% SiC. *(After Downes et al. 1991.)*

15.3.5 Short Crack Behavior

In numerous materials, fatigue crack growth rates for physically short cracks are substantially higher than the growth rate of long cracks fatigued at the same ΔK and R (see Suresh 1991 for a review of short crack behavior). At low R values, short cracks have been found to grow at ΔK levels well below ΔK_{th} for long cracks in the same material. The potential for lower short crack thresholds and higher near-threshold growth rates requires that short crack growth behavior in MMCs be understood if accurate life predictions are to be made. Investigations of short and long crack growth behavior in a series of SiC reinforced aluminum alloys, have shown that a short crack growth effect does indeed exist in these materials, with crack growth possible well below the apparent long crack, low R threshold levels. Although exact behavior varies from material to material, it appears that, similar to unreinforced materials, the short crack effect can be related to the absence of crack closure for the short cracks (Downes et al. 1991). Short crack growth rates have been observed to merge with long crack growth rates for a crack length of approximately 200 μm in 2124 reinforced with 5% SiC and approximately 90 μm in a 6061 alloy reinforced with 15% SiC (Figure 15.26). Particle fracture has been observed during short crack growth in some instances (Kumai et al. 1990b; Biner 1990), and crack deflection near particle interfaces has been observed in others (Kumai et al. 1992).

Although these initial investigations on short crack growth behavior have shown that the effect does exist and that it appears to behave in a conventional fashion, the presence of reinforcement particles complicates the issue. Fractured particles may provide an easy crack propagation path and may accelerate growth rate; unfractured particles may act to impede crack growth by crack trapping or crack deflection.

Summary

The reinforcement of metals by ceramic particles or whiskers can have a significant impact on cyclic deformation as well as on the initiation and growth of fatigue cracks. This influence can be beneficial and can allow use of matrix alloys such as aluminum in applications for which they would not otherwise be considered. It can also be detrimental, therefore, certain limitations must be taken into account to properly design components fabricated from metal-matrix composites. Understanding the complex manner in which reinforcement influences fatigue behavior is limited by (1) a lack of systematic studies, and (2) because fatigue properties are highly sensitive to material quality, which has been continually changing. Despite these limitations, some central principles have emerged that may ultimately allow truly tailored or engineered materials to become a reality.

The incorporation of reinforcement generally leads to an increase in work hardening rate and a decrease in ductility under both monotonic and cyclic loading. Recent advances in modeling the concentrated deformation that occurs around reinforcements has lead to significant improvements in our understanding of this cyclic deformation process. In turn, an improved understanding of the role of reinforcement particles in influencing the resistance to fatigue crack initiation and early growth (fatigue-life) has also been possible. Fatigue life behavior can be improved or degraded by the incorporation of reinforcement, depending on the loading mode. This can be understood by careful consideration of the influence of reinforcement on modulus and work hardening. Fatigue life behavior has not, however, been intensively investigated, and issues such as volume fraction effects and particle shape and size effects on fatigue life are worthy of increased attention.

Relative to fatigue-life behavior, considerably more research has been focused on the influence of reinforcement on fatigue crack propagation. It has been concluded by many researchers that crack tip phenomena such as roughness-induced crack closure and crack trapping have a direct and significant impact on fatigue crack growth rates. This has provided an insight into otherwise contradictory results related to particle size and volume fraction effects. Similar to unreinforced metals, increases in fatigue crack growth rates have been observed in short cracks relative to long cracks. This has been attributed to a lack of crack closure in the crack wake.

For many metals, matrix precipitate structures can have a significant influence on fatigue behavior. In discontinuously reinforced aluminum composites, a common observation for both fatigue-life behavior and fatigue crack propagation is the reduced influence of matrix microstructure. Generally, qualitative changes in mechanisms have not been observed, which suggests that both properties are dominated by the macroscopic composite mechanics. Although this limits the material designer's options to modify fatigue properties, it has potential advantages in that it provides materials that may be resistant to heat treatment variations.

Engineering materials for fatigue resistance is a complex and often paradoxical problem. In composite materials, as in many unreinforced metals, factors that improve resistance to the initiation of a crack lead to a decrease in the resistance to the growth of a crack. An example of this is the influence of reinforcement on the fatigue behavior of aluminum alloys. In underaged aluminum alloys, reinforcement leads to an improvement in stress-controlled fatigue behavior but produces an increase in fatigue crack propagation rates. In contrast, in overaged aluminum alloys, reinforcement not only improves the stress-controlled fatigue behavior but also the resistance to fatigue crack propagation. Under stress-controlled high cycle fatigue conditions, a high volume fraction of fine reinforcement particles is preferred for fatigue crack initiation resistance. For fatigue crack growth resistance, a high volume fraction of coarser particles is preferred.

Acknowledgments

The authors acknowledge fruitful discussions with many of their students and colleagues. We are especially appreciative of permission to include information previously unpublished. We make special note of the contributions of G. Vyletel and J. Hall and collaborations with Professor D. C. Van Aken and Dr. C. P. You.

Finally, the continued support of Ford Motor Company (J.E.A. and J.W.J.) and General Motors Corporation (J.W.J.) is most gratefully acknowledged.

References

Allison, J. E. 1988. In: Fracture Mechanics: Eighteenth Symposium, ASTM STP 945. (D.T. Read and R.P. Reed, eds.), 913–933. Philadelphia, PA: ASTM.

Arsenault, R.J., and M. Taya. 1989. Mat. Sci. and Eng. 108:285–288.

Arsenault, R.J., and S.B. Wu. 1987. Mat. Sci. and Eng. 96:77–88.

Biner, S B. 1990. Fatigue Fract. Engng. Mater. Struc. 13:637–646.

Bloyce, A., and J.C. Summers. 1991. Mat. Sci. and Eng. A135:231–236.

Bonnen, J.J., Allison, J.E., and J.W. Jones 1991. Metall. Trans. A. 22A:1007–1019.

Bonnen, J.J., You, C.P., Allison, J.E., and J.W. Jones. 1990. In: Proceedings of the Fourth International Conference on Fatigue and Fatigue Thresholds, 887–892, Honolulu, HI.

Calabrese, C., and C. Laird. 1974a. Mat. Sci. and Eng. 13:159–174.

Calabrese, C., and C. Laird. 1974b. Mat. Sci. and Eng. 13:141–157.

Carter, R.D., Lee, E.W., Starke, E.A. and C.J. Beevers. 1984. Metall. Trans. A15:555.

Christman, T., and S. Suresh. 1988. Mater. Sci. Eng. 102:211–216.

Corbin, S.F,. and D.S. Wilkinson. 1991. In: Metal Matrix Composites—Processing, Microstructure and Properties. (N. Hansen, et al., eds.), 283–290, Roskilde, Denmark: Risø National Laboratory.

Couper, M.J., and K. Xia. 1991. In: Metal Matrix Composites—Processing, Microstructure and Properties. (N. Hansen, et al., eds.), 291–298, Roskilde, Denmark: Risø National Laboratory.

Couper, M.J., Neeson, A.E., and J. Griffiths. 1990. Fatigue Fract. Engng. Mater. Struc. 13:213–227.

Crowe, C.R., and D.F. Hassen. 1982. In: Strength of Metals and Alloys, ICSMA Vol. 2. 6. (R.C. Gifkins, ed.) 859–865, Oxford, UK: Pergamon Press.

Davidson, D.L. 1987. Metall. Trans. A 18A:2115–2128.

Davidson, D.L. 1988. Acta Met. 36:2275–2282.

Davidson, D.L. 1989. Engineering Fracture Mechanics. 33:965–977.

Davidson, D.L. 1991. Metall. Trans. A 22A:97–112.

Davidson, D.L. 1993. Composites, 24:248–255.

Davis, L.C., and J.E. Allison. 1993. Metal Trans., in press.

Downes, T.J., Knowles, D.M., and J.E. King. 1990. In: ECF 8 Fracture Behaviour and Design of Materials and Structures. (D. Firrao, ed.) Cradley Heath, UK: Engineering Materials Advisory.

Downes, T.J., Knowles, D.M., and J.E. King, 1991. In: Fatigue of Advanced Material. (R.O. Ritchie, R.H. Dauskardt, and B.N. Cox, eds.), 395–407, Birmingham, UK: Materials and Component Engineering Publications Ltd.

Eshelby, J.D. 1957. Proc. R. Soc. A, 241:376–396.

Fine, M.E., and J.S. Santner. 1975. *Scripta Met.* 9:1239–1241.
Goodier, J.N. 1933. *Journal of Appl. Mech.* 55-7:39–44.
Hall, J., Jones, J.W., and A. Sachdev. 1993a. *Fatigue 93.* (J.P. Bailon, and J.I. Dickson, eds.), 1129–1135, Cradley Heath, UK: Engineering Material Advisory Services.
Hall, J., Jones, J.W., and A. Sachdev. 1993b. *Materials Sci. Eng.* Submitted.
Harris, S.J. 1988. *Mat. Sci. and Tech.* 4:231–239.
Harris, S.J., and T.E. Wilks. 1986. *I. Mech. Eng.* C37(86):19–28.
Hasson, D.F., Crowe, C.R., Ahearn, J.S., and D.C. Cook. 1984. *In:* Failure Mechanisms in High Performance Materials, (J. Early, R. Shives, and J. Smith, eds.), 147–156, Cambridge, UK: Cambridge University Press.
Healy, J.C., and C.J. Beevers. 1991. *Mat. Sci. Engng.* A142:183–192.
Hirano, K. 1991. *In:* Mechanical Behavior of Materials—IV. Proceedings of the 6th International Conference (M. Jono, and T. Inoue, eds.), Vol. 3, 93–100, Oxford, U.K.: Pergamon Press.
Holcomb, S. 1992. M.S. Thesis, Department of Materials Engineering, University of Southern California, Los Angeles, CA.
Hom, C.L., and R.M. McMeeking. 1991. *Int. J. of Plasticity* 7:255–274.
Hoover, W.R. 1991. *In:* Metal Matrix Composites—Processing, Microstructure and Properties. (N. Hansen, et al., eds.), 387–392, Roskilde, Denmark: Risø National Laboratory.
Hunt, W.H. Jr., Cook, C.R., and R.R. Sawtell. 1991. Cost Effective High Performance PM Aluminum Matrix Composites for Automotive Applications, SAE Paper 910834, Warrendale, PA: Society of Automotive Engineers.
Hunt, W.H., Jr. 1993. Private communication.
Hurd, N.J. 1988. *Mat. Sci. and Tech.* 4:513–517.
Johannesson, B., Ogin, S.L., and P. Tsakiropoulos. 1991. *In:* Metal Matrix Composites—Processing, Microstructure and Properties. (N. Hansen, et al., eds.), 411–416, Roskilde, Denmark: Risø National Laboratory.
Juvinall, R.C. 1967. *Engineering Considerations of Stress, Strain and Strength,* New York: McGraw-Hill Book Co., p. 228.
Kamat, S.V., Rollet, A.D., and J.P. Hirth, J.P. 1991. *Scripta Met.* 25:27–32.
Karayaka, M., and H. Sehitoglu. 1991. *Metall. Trans. A.,* 22A:697–707. Klimowicz, T. (ed.) 1992. DURALCAN USA, Fatigue Data Handbook. San Diego, CA.
Knowles, D.M., and J.E. King. 1990. *In:* Proceedings of the Fourth International Conference on Fatigue and Fatigue Thresholds. 641–646, Honolulu, HI.
Knowles, D.M., and J.E. King. 1991. *Acta Metall. Mater.* 39:793–806.
Kobayashi, T., Niinomi, M., Iwanari, H., and H. Toda. 1991. *In:* Science and Engineering of Light Metals (RASEL '91), 543–548, Tokyo, Japan: Japan Institute of Light Metals.
Kumai, S., Higo, Y., and S. Nunomura. 1991a. *In:* Science and Engineering of Light Metals (RASEL '91), 489–494, Tokyo, Japan: Japan Institute of Light Metals.
Kumai, S., King, J.E., and J.F. Knott. 1990a. *Fatigue Fract. Engng. Mater. Struc.* 13:511–524.
Kumai, S., King, J.E., and J.F. Knott. 1990b. *In:* Proceedings of the Fourth International Conference on Fatigue and Fatigue Thresholds, 869–874, Honolulu, HI.

Kumai, S., King, J.E., and J.F. Knott 1991b. *Mater. Sci. Engng.,* A146:317–326.
Kumai, S., King, J.E., and J.F. Knott. 1992. *Fatigue Fract. Engng. Mater. Struc.* 13:1–11.
Landgraf, R.W. 1970. *In:* Achievement of High Fatigue Resistance in Metals and Alloys, ASTM STP 467. 3–36, Philadelphia: ASTM.
Levin, M., and B. Karlsson. 1991. *Mater. Sci. Tech.,* 7:596–607.
Liu, C. 1991. Ph.D. Thesis, Department of Materials Science and Engineering, Case Western Reserve University, Cleveland, OH.
Llorca, J., Needleman, A., and S. Suresh. 1990. *Scripta Met.* 24:1203–1208.
Llorca, J., Needleman, A., and S. Suresh. 1991. *Acta. Metall. Mater.* 39(10):2317–2335.
Llorca, J., Suresh, S., and A. Needleman. 1992. *Metall Trans A,* 23A:919–934.
Llorca, N., Bloyce, A., and T.M. Yue. 1991. *Mat. Sci. and Eng.* A135:247–252.
Logsdon, W.A., and P.K. Liaw. 1986. *Eng. Frac. Mech.* 24:737–751.
Lukasak, D.A., and R.J. Bucci. 1992. Alloy Technology Div. Rep. No. KF-34. Alcoa Technical Center, Alcoa Center, PA.
Lukasak, D.A. 1992. Alloy Technology Div. Rep. No. KF-37, Alcoa Technical Center, Alcoa Center, PA.
Martin, J.W. 1980. Micromechanisms in Particle-Hardened Alloys, Cambridge, UK: Cambridge University Press.
Masuda, C., and Y. Tanaka. 1992. *J. Mater. Sci.,* 27:413–422.
Mouritz, A.P., and S. Bandyopadhyay. 1991. Proceedings of a Forum on Metal Matrix Composites. University of New South Wales, Sydney, Australia.
Musson, N.J., and T.M. Yue. 1991. *Mat. Sci. Eng.* A135:237–242.
Nair, S.V., Tien, J.K., and R.C. Bates. 1985. *Int. Metals Reviews* 30: 275–290.
Sanders, T.H., Staley, J.T., and D.A. Mauney. 1975. *In:* Fundamental Aspects of Structural Alloys Design, 487, New York: Plenum Press.
Seitz, T., Baer, J., and H.-J. Gudladt. 1991. *In:* Metal Matrix Composites—Processing, Microstructure and Properties (N. Hansen, et al., eds.), 649–654, Roskilde, Denmark: Risø National Laboratory.
Shang, J.K., Yu, W., and R.O. Ritchie. 1988. *Mater. Sci. Eng.* A102: 181–192.
Shang, J.K., and R.O. Ritchie, 1989a. *Metall. Trans. A* 20A:897–908.
Shang, J.K., and R.O. Ritchie, 1989b. *Acta Metall.* 37:2267–2278.
Shang, J.K., and R.O. Ritchie. 1989c. *In:* Metal Matrix Composites (R.J. Arsenault, and R.K. Evertt, eds.), Boston, MA: Academic Press.
Sharp, P.K., Parker, B.A., and J.R. Griffiths, 1990. *In:* Proceedings of the Fourth International Conference on Fatigue and Fatigue Thresholds (H. Kitagawa and T. Tanaka, eds.), 875–880, Honolulu, HI. Birmingham, UK: MCE Publications.
Srivatsan, T.S., Auradkar, R., Prakash, A., and E.J. Lavernia. 1991. *In:* International Conference on Fracture of Engineering Materials and Structures, Singapore: FEFG/ICF.

Srivatsan, T.S. 1992. *Int. J. Fatigue.* 14:173–182.

Sugamuri, Y., and S. Suresh. 1992. *Metall. Trans. A,* 23A:2231–2242.

Suresh, S., and R.O. Ritchie. 1984. *In:* Fatigue Crack Growth Threshold Concepts (S. Suresh, and D.L. Davidson, eds.), 227, Warrendale, PA: TMS-AIME.

Suresh, S., Vasudevan, A.K., and P.E. Bretz. 1984. *Metall. Trans. A,* A15:369–379.

Suresh, S. 1991. Fatigue of Materials, 138, Cambridge, UK: Cambridge University Press.

Tanaka, K. 1991. *In:* Mechanical Behavior of Materials—IV, Proceedings of the 6th International Conference (M. Jono and T. Inoue, eds.), 414–420, Oxford, UK: Pergamon Press.

Taya, M., Lulay, K.E., Wakashima, K., and D.J. Lloyd. 1990. *Mat. Sci. and Eng.* 124:103–111.

Vyletel, G.M., Krajewski, P.K., Van Aken, D.C., Jones, J.W., and J.E. Allison. 1992. *Scripta Metall. Mater.,* 27(5):549–554.

Vyletel, G., Van Aken, D.C., and J.E. Allison. 1991, *Scripta Metall. Mater.* 25:2405–2410.

Vyletel, G., Van Aken, D.C., and J.E. Allison. 1993. *Metall. Trans A.* In press.

Williams, D.R., and M.E. Fine. 1985. *In:* International Conference on Composite Materials, IV, Proceedings. (W.C. Harrigan, J. Strife, and A.K. Dhingra, eds.), 639–669, Warrendale, PA: TMS.

Withers, P.J, Pedersen, O.B., Brown, L.M., and W.M. Stobbs. 1989. *Mat. Sci. and Eng.* 108:281–284.

Xia, K., Couper, M.J., and J. Griffiths. 1991. Presented at TMS Symposium on Fatigue and Creep of MMCs. New Orleans, LA.

Yau, S.S., and G. Mayer. 1986. *Mat. Sci. Eng.,* 82:45–57.

You, C.P., and J.E. Allison. 1989. *In:* Advances in Fracture Research, Vol. 4. Oxford, UK, Pergamon Press, 3005–3012.

Zahl, D.B., and R.M. McMeeking. 1991. *Acta Metall. Mater.* 39:1117–1122.

Zaiken, E., and R.O. Ritchie. 1985. *Mater. Sci. Eng.* 70: 151.

Zhang, R.J., Wang, Z., and C. Simpson. 1991. *Mat. Sci. Eng.* A148:53–66.

PART V
APPLICATIONS

Chapter 16
Metal-Matrix Composites for Ground Vehicle, Aerospace, and Industrial Applications

MICHAEL J. KOCZAK
SUBHASH C. KHATRI
JOHN E. ALLISON
MICHAEL G. BADER

Metal-matrix composites (MMCs) have emerged as a class of materials capable of advanced structural, aerospace, automotive, electronic, thermal management, and wear applications. These alternatives to conventional materials provide the specific mechanical properties necessary for elevated and ambient temperature applications. In many cases, the performance of metal-matrix composites is superior in terms of improved physical, mechanical, and thermal properties (specific strength and modulus, elevated temperature stability, thermal conductivity, and controlled coefficient of thermal expansion), although substantial technical and infrastructural challenges remain.

The performance advantage of metal matrix composites is their tailored mechanical, physical, and thermal properties that include low density, high specific strength, high specific modulus, high thermal conductivity, good fatigue response, control of thermal expansion, and high abrasion and wear resistance. In general, the reduced weight and improved strength and stiffness of the MMCs are achieved with various monolithic matrix materials. However, material liabilities for continuous fiber systems include low transverse and interlaminar shear strength, foreign object impact damage, mechanical/chemical property incompatibility, and high fiber and processing costs.

Continuous fiber metal-matrix composites are primarily utilized for high specific stiffness and high-temperature applications. The intimate contact between the fiber and the matrix, coupled with the elevated processing and service temperatures, promotes an interfacial reaction that can result in fiber degradation. As a result, the properties of MMCs such as specific fiber/matrix systems must be considered individually, and cannot be generalized as a class of reinforced composites.

The ability to transition a metal-matrix composite from an advanced composite material to a cost-effective application for the commercial market involves several factors, including a large material production capacity, reliable static and dynamic properties, cost-effective processing, and a change in design philosophy based on experience and extensive durability evaluation. A comparison of relative costs of reinforcement, competing materials, and the cost of a 2000 cc automobile is detailed in Table 16.1 (Funatani 1986; Koczak 1989). The structural potential, particularly for aluminum MMC materials, has been demonstrated, although the result has rarely proven to be cost effective.

In application-driven environments, MMCs are gaining rapid prominence in the future automotive, electronic, and aerospace sectors. Higher operating temperatures of materials and wear resistance are mandated by

Table 16.1. Relative Fiber and Materials Costs Compared to the Cost of a 2000 cc Automobile

	Fiber Material Cost				
	$3/lb ¥1/g	$30/lb ¥10/g	$300/lb ¥100/g	$3,000/lb ¥1,000/g	$30,000/lb ¥10,000/g
Fibers	Glass Fiber 3. 0.5	Potassium Titanate Fiber 2 4 Aramid Fiber 5 8	Carbon Fiber (PAN) 10 50	Si_3N_4 Fiber 200 300 Boron Fiber 180 300 Al_2O_3 Fiber 80 120	
	Al_2O_3-SiO_2 Short Fiber 1 3	Al_2O_3 Short Fiber 7 · 10	SiC Whisker 50 120		
Other Materials	Al 0.3 .05 Cu 0.3 0.4 Fe • 0.07	Ti 1.5 Si_3N_4 2.5 10 Engineering Plastics 0.7 1.3 Plastics 0.3 0.7	Shape Memory Alloy 15 30 ZrO_2 PSZ 15 20	Ge 250 270 Si Wafer 150	GaAs Single Crystal 6,000 7,000 Pt • 2,500
Products		2,000 cc Automobile 1.4 1.7			

(Data from Yamada 1986.)

the increasingly stringent performance requirements of the aerospace and automotive industry. Although titanium alloys are used at the intermediate temperature ranges and result in cost and weight increases, development of aluminum and titanium based, cost-effective composites that would be viable at elevated temperatures ($T > 250°C$) would be beneficial to design engineers for automotive and aerospace applications (Table 16.2). Apart from the development of high performance aerospace products, several industrial efforts have materialized and have provided a low cost, cast metal-matrix composite. These developments provided the necessary base capacity for the first high volume production applications of MMCs. Notable production advances include:

1. Development of reinforced pistons and engine blocks using preform infiltration technology on a mass production basis by Toyota and Honda, respectively (Toyota 1991; Honda 1991; Ebisawa 1991).
2. Spray deposition processes for producing MMC ingots and products by Alcan (Alcan 1990, 1991).
3. Production by Lanxide via direct metal oxidation and pressureless infiltration of wear-resistance composites and electronic packaging components (Lanxide 1991).
4. Development of large-scale production capacity for commercial particulate composites by both foundry and powder routes. (Alcan; Duralcan Co.; Hydro-Aluminum Co.; BP; Alcoa.)

Table 16.2. Goal of FRM in Jisedai Project

Fiber Composite System	V (%)	Tensile Strength (MPa) R.T.	450C	Weibull Probability
PCS SiC NICALON/1050 Aluminum alloy F	50	1320	1190	20
CVD SCS SiC/1050 Aluminum alloy F	50	1600	1175	NA
SiC (Whisker)/1050 Aluminum alloy F	30	686	NA	20
SiC Whiskers/Al 4032 Powder	28	460	NA	NA
SiC (CVD)/Ti alloy	35	1800	1530	NA
SiC (CVD)/Ti powder	30	1470	1323	NA

Data from Minoda (1986).

Table 16.3. Comparison of Monolithic Materials and Composite

Material	Density g/cm^3	Modulus GPa	Strength MPa	E/ρ 10^8 cm	σ/ρ 10^8 cm
4340 Steel	8.0	200	1760	2.5	2.2
7475 Aluminum	2.7	70	570	2.6	2.1
Ti-6Al-4V	4.4	110	1290	2.5	2.9
INCO 718	8.2	210	1430	2.5	1.8
Al-50v/o SiC	2.6	140	1500	5.4	5.7

Data from Textron (1991).

Table 16.4. Characteristics of Several Fiber-Reinforcement Materials

Material	Specific Gravity	Tensile Strength (psi × 10^6) (MPa × 10^6)	Modulus of Elasticity (psi × 10^6) (MPa × 10^3)
Whiskers			
Graphite	2.2	3 (20)	100 (690)
Silicon carbide	3.2	3 (20)	0 (480)
Silicon nitride	3.2	2 (14)	55 (380)
Aluminum oxide	3.9	2–4 (14–28)	60–80 (415–550)
Fibers			
Aramid (Kevlar 49)	1.4	0.5 (3.5)	19 (124)
E-Glass	2.5	0.5 (3.5)	10.5 (72)
Carbon	1.8	0.25–0.80 (1.5–5.5)	22–73 (150–500)
Aluminum oxide	3.2	0.3 (2.1)	25 (170)
Silicon carbide	3.0	0.50 (3.9)	62 (425)
Metallic Wires			
High-carbon steel	7.8	0.6	30
Molybdenum	10.2	0.2 (1.4)	52 (360)
Tungsten	19.3	0.62	58

Data from Koczak et al. (1988).

By selecting appropriate reinforcing constituents for matrix material (volume fraction, shape, and size), it is possible to design alloys that will enhance strength and stiffness. Metal-matrix composites incorporate a wide variety of metal systems (for example, Al, Mg, Ti, Cu, Fe, Ni alloys, see Table 16.3) using several types of reinforcements in the form of whiskers (SiC), monofilament (SiC, B, W), continuous fiber tows (SiC, alumina, graphite), and particulates (TiC, SiC, see Table 16.4). In terms of fibers cost and performance, the present fibers for elevated temperatures are considered to be inadequate. There is a critical need to develop lower cost fibers comparable to carbon. Chemically pure fibers such as alumina were considered desirable because of their reduced reactivity with metallic matrices. Of the currently available continuous fibers suitable for aerospace applications, the larger-diameter fibers are considered preferable. The multifilament, SiC-based fibers such as Nicalon (Nippon Carbon, Tokyo, Japan) and Tyranno (Tonen, Tokyo, Japan), were inadequate in several areas, especially with regard to strength and high-temperature stability. The alumina-based fibers, Saffil and Saffimax (ICI, Wilmington, DE), were considered to be very promising, although both have limited specific strength and stiffness properties. The performance of current continuous, large diameter Al_2O_3 fibers does not justify their cost. There appears to be a continuing interest in metallic fibers such as W-based alloys and intermetallics in both continuous and high-aspect, ratio-discontinuous forms. High-performance ductile fiber may have considerable potential. Likewise, there is continued interest in the exploitation of high-aspect ratio whiskers, despite potential health hazards.

During the past decade, specific property material requirements for advanced applications have escalated where conventional alloy systems are not suitable. A range of specific strengths versus temperatures for the application for reinforced metal- and ceramic-matrix composites are required, see Figures 16.1 and 16.2 (Driver 1989; Minoda 1986, 1989). These synthetic metal-matrix composites are a thermodynamically unstable, artificial mix of fiber and matrix, and reflect the high cost of fiber reinforcements and processing. For successful progress and application of metal-matrix composites, major fundamental scientific challenges lie in the materials processing, synthesis, kinetic understanding, thermodynamic control, reinforcement stability, and strength. By controlling alloy matrix design and interface reactions, a stable metal-matrix composite may be generated.

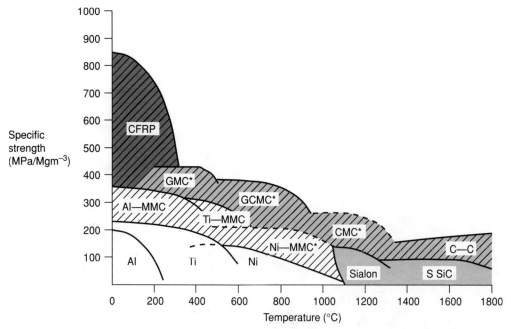

Figure 16.1. Specific strengths versus temperature for metal- and ceramic-matrix composites. CFRP, carbon fiber reinforced polymers; GMC, glass-matrix composites; GCMC, glass-ceramic matrix composites; CMC, ceramic-matrix composites; MMC, metal-matrix composites; and C-C, carbon-carbon composites. GMC, GCMC, CMC, and MMC are research materials *(Courtesy of Rolls Royce)*.

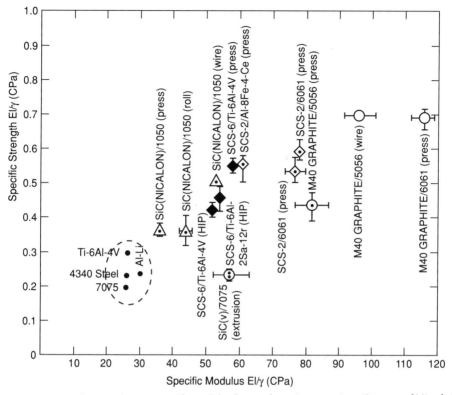

Figure 16.2. Specific strength versus specific modulus for metal-matrix composites. *(Courtesy of Minoda.)*

Commercial motivation is directed toward the development of a cost-effective, aluminum metal-matrix composite. The enhancement of the power to weight ratio demonstrates the viability of aluminum base engines and MMC cylinder engine blocks. Replacing the gray cast iron liners with an aluminum MMC enables a weight savings of 4.5 kg, with an increase in engine displacement and increased cooling efficiency. See Figure 16.3 (Ebisawa et al. 1991; Honda 1991). Aluminum-matrix composite alloys have been used extensively in aerospace structural applications at ambient temperature because of their high specific strength (high strength-to-weight ratio). There has been a concerted research effort to develop aluminum alloys for applications at above-ambient temperatures. The development of these alloys would make it economically viable to replace existing alloys in certain aerospace structural and engine components. In addition to synthetic fiber/particulate systems, in situ metal-matrix composites processed via liquid metal and gas reactions as well as other in situ directional solidification processes, have developed reinforced alloys with high strength and modulus as well as good elevated temperature stability. Metal-matrix composites may be utilized to develop high volume fraction, thermally stable, wear-resistant

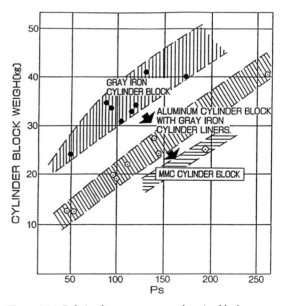

Figure 16.3. Relation between power and engine block weight. *(Courtesy of Honda.)*

alloys, as well as alloy composites with tailored or graded physical properties such as acoustic, wear, coefficient linear expansion, and thermal conductivities.

This chapter reviews ground vehicle, aerospace, and industrial applications for MMCs. The ground vehicle applications currently include engine applications such as the Honda Prelude cylinder liners and the Toyota diesel pistons, as well as near-term applications that include chassis and drive trains (driveshafts, brake rotors, and calipers) (Honda 1991; Toyota 1991). Engine applications are considered for cylinder liners, pistons, connecting rods, valve lifters, and piston pins. Aerospace applications simultaneously require higher performance, weight savings, high operating temperatures, and competitive life cycle costs. Candidate components for aerospace engine applications include exhaust nozzles, links, blades, cases, shafts, and vanes (Driver 1989; Kelly 1988; Harris 1988, 1990). Structural aerospace applications include aft fuselage structures, landing and arresting gears, stiffeners, drag brakes, and compression and torque tubes. Industrial applications include wear-resistant, hard materials for cutting tools, drill bits, valves, and gates. The control of thermal conductivity and thermal expansion permits the use of MMCs for electrical connectors, substrates, packing materials, battery plates, and superconducting wires. For materials in high thermal and stress gradients, graded composite structures are considered. Recreational applications include golf club heads, structural tubes for sail masts, and bicycle frames.

16.1 Applications in the Automotive Industry

The automotive industry is currently facing substantial technical challenges as it seeks to improve fuel economy, reduce vehicle emissions, increase styling options, and enhance performance. Improved technology will be required to meet these challenges. Thus, significant opportunities exist for the use of advanced materials. The increased use of lightweight metals is being given serious consideration for use in high production volume applications. Examples of this include aluminum in auto body structures, chassis components (Komatsu et al. 1991), and engine blocks (Ebisawa et al. 1991), the application of titanium alloys to engine valves (Allison et al. 1988) and connecting rods (Kimura et al. 1991), the development of titanium-aluminides valves (Dowling et al. 1992), and the development of rapidly solidified aluminum compressor vanes (Arnhold and Mueller-Schwelling 1991). Of the new materials being considered, metal-matrix composites are particularly promising because of their superior properties and our ability to tailor these properties for particular applications. Although these opportunities exist, barriers to introduction are also present, particularly, because each application must be proven to be not only technically feasible, but cost effective as well. Thus, although the use of costly continuous fibers has been considered (Folgar 1987), as of this writing, the primary metal-matrix composites of interest have been the more cost effective, discontinuously reinforced aluminum (DRA) composites. Opportunities for the use of discontinuously reinforced magnesium composites also exist, although these materials have been largely unexplored.

16.1.1 Chassis and Driveline Applications

Driveshafts in trucks and large passenger cars offer a particularly attractive application of DRA. Although the majority of automotive driveshafts are constructed from steel, aluminum driveshafts have been in use in selected applications for a number of years. Their primary appeal has been the significantly lower weight of the aluminum driveshaft, coupled with the ease with which it can be balanced. Current driveshafts, whether steel or aluminum, are constrained by the speed at which the shaft becomes dynamically unstable. The critical driveshaft speed, N_c, is given by

$$N_c = \tfrac{15\pi}{L^2}\sqrt{\left[\tfrac{E}{\rho}\right]g(R_o + R_i)^2} \qquad (16.1)$$

where: L is the driveshaft length, R_o the outer radius, R_i the inner radius, E the elastic modulus, ρ the density, and g is the gravitational acceleration.

As seen in Equation 16.1, the specific modulus $\tfrac{E}{\rho}$ is the only material property that affects the critical speed. Shaft length and tube diameter are geometric variables that control critical speed.

A number of current vehicle concepts require driveshafts that are longer than is feasible with aluminum or steel. Packaging and weight constraints require that tube diameter is not increased and, because the specific modulus of both steel and aluminum are quite close, 26.2×10^8 mm and 25.9×10^8 mm, respectively, critical speed limitations of these materials constrain driveshaft lengths. DRA driveshafts represent a cost-effective solution to this problem. Because the specific modulus of DRA composites is higher than that of either steel or aluminum, DRA offers a means of increasing driveshaft lengths without violating and altering packaging constraints and without increasing weight. As an example, a 20% Al_2O_3 reinforced aluminum DRA would exhibit a specific modulus of 35.3×10^8 mm. This would allow an 8% increase in driveshaft length for a constant cross-sectional area (Hoover 1991). Alternatively, the diameter of DRA driveshafts can be reduced relative to either aluminum or steel shafts, producing further packaging options and further decreasing weight.

Composite alloy choice for driveshafts is dictated by a number of considerations, including corrosion resistance and strength/ductility requirements. Al-Mg-Si, 6xxx series aluminum alloy composites are considered primary candidates for this application. Manufacturing technologies that are critical to the successful implementation of driveshafts in high-volume automotive applications include consistent material quality, high tolerance tube extrusion, and composite to aluminum fusion welding technology. These technologies appear to be well developed and have been used to manufacture driveshafts such as that shown in Figure 16.4. These driveshafts have been subjected to component testing and have met expectations (Hoover 1991).

Reduction of overall vehicle weight is important for improving fuel economy. Therefore, the application of DRA to brake components, especially disk brake rotors, has been receiving considerable worldwide attention. Examples of DRA brake rotors are shown in Figure 16.5. Because the brake rotors represent unsprung, rotating weights, reduction in their mass is especially beneficial and can also increase vehicle dynamics and acceleration. A brake rotor weight savings of approximately 60% may be possible if DRA can be substituted for the cast iron normally found in this application. If this can be applied to brake systems on all four wheels of a typical passenger car, resultant fuel economy savings, as high as 0.25 mpg have been projected. The high thermal conductivity of aluminum reinforced with SiC provides additional advantages for the thermal management of brake systems.

Figure 16.5. A variety of different MMC brake rotors *(Courtesy of Ford Motor Company).*

Because of the complex shape of brake rotors, casting is the primary manufacturing process. Casting procedures for DRA brake rotors are under development and appear to be feasible. The materials of choice for this application would be cast aluminum alloys reinforced with 15% to 30% SiC_p (Cole 1992). The function of a brake system is complex and the use of cast iron and rotor designs has evolved over many years. Thus, substitution of a material such as DRA requires significant development and testing. The development of new friction materials for brake pads, and the impact of high temperatures and thermal cycles experienced under extreme braking situations, are durability issues of prime importance. Brake rotors manufactured from 20% to 30% SiC_p have performed well in racing applications and vehicle tests.

In addition to brake rotors, exploratory developments are also being conducted on brake calipers to improve braking performance by making use of the improved elevated temperature modulus of DRA composites. Aluminum alloys and casting processes have been developed for use in suspension components and are currently in production in high performance vehicles (Komatsu et al. 1991). These components are often stiffness limited, thus providing an opportunity for the use of DRA. Of concern, these components are safety critical, making ductility and fracture toughness especially important. Therefore, cast composites with higher ductilities and impact toughness must be developed before these applications can be seriously considered.

Figure 16.4. MMC driveshaft on vehicle *(Courtesy of Duralcan USA).*

16.1.2 Engine Applications

A number of potential engine applications have been identified for metal-matrix composites. However, the benefits of composite use to these applications are more difficult to quantify and there are a wide range of alternative materials/processes competing for these same applications. Although DRA engine components have been, and continue to be, developed, implementation of these components should be viewed as more long term in nature.

All major automobile companies have developed or are developing aluminum engine blocks in a move to reduce gross vehicle weight and therefore fuel consumption. The demonstrated reduction in overall engine weight varies from 15 kg to 35 kg. To ease casting and reduce machining costs, aluminum alloys with moderate silicon levels are typically used for these blocks. These alloys, however, cannot withstand the harsh conditions existing in the cylinder region and thus many manufacturers have inserted cylinder liners made from cast iron. The cylinder area is subjected to extremes of temperature, wear, and combustion loading, making the properties of wear and scuff resistance, fatigue resistance, and creep resistance important. Thermal expansion and thermal conductivity are also important characteristics.

Alternative approaches, including metal-matrix composites, offer the possibility of eliminating the cast iron liner and provide further weight savings of 3 kg to 4.5 kg. In addition, these approaches offer superior thermal conductivity and, potentially, durability. The three primary approaches for replacing cast iron cylinder liners are metal-matrix composite/high silicon aluminum alloy cylinder liners, coatings, and "parent bore," blocks (Cole 1992). DRA composite liners and liners manufactured from high silicon alloys such as 390 Al or 3HA are also under consideration for liners. These liners could

Figure 16.6. Honda Prelude engine block with MMC cylinder liners *(Courtesy of Honda Motor Ltd.)*.

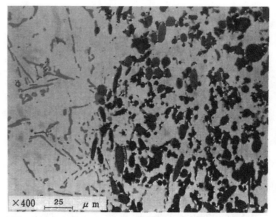

Figure 16.7. Microstructure of aluminum MMC cylinder liners *(Courtesy of Honda Motor Ltd.)*.

be manufactured by casting or by extrusion. A variant on the liner concept, made possible by the use of a selectively reinforced composite, incorporates Al_2O_3/graphite preform "liners" into the block via novel medium pressure die-casting procedures (Ebisawa et al. 1991). This process is being used commercially to produce engine blocks for the Honda Prelude (Figure 16.6), although its long-term cost effectiveness and reliability remains to be demonstrated. The reinforcement is a combination of graphite and alumina dispersed in an aluminum matrix (Figure 16.7). Honda has made a noteworthy commitment to discontinuous fiber reinforced automotive applications in which the process and the component design were key features. Potential cylinder coatings include thick (25 μm to 50 μm) coatings of Cr, Ni, Ni-P, and their composites, produced either by electrodeposition or by thermal spray techniques. Thin anodization coatings are also under consideration. Coatings represent a potential cost savings over cast iron liners if production efficiencies can be improved and if the coatings prove capable of meeting the liner requirements. "Parent bore" blocks are engine blocks that are fabricated entirely from an aluminum alloy that is capable of meeting the requirements of the cylinder area. Al 390 and 3HA are typical of this class of materials, although cast DRA has also been considered. The parent bore approach is costly though, because of the special care that must be taken during block casting and because of the difficulty encountered when machining these high-silicon alloys.

Metal-matrix composites have been widely considered for use in lightweight connecting rods and pistons. Reducing the weight of these components has a number of benefits. Four-cylinder engines are desirable for their superior fuel economy; however, when the engine displacement is larger than about two liters, the reciprocating masses of the connecting rod/piston assembly produce unbalanced secondary shaking forces that are objectionable. Thus, a principal benefit of using a lightweight connecting rod is the reduction of these secondary forces, which will allow consideration of larger four-cylinder engines. There are also engine designs that utilize higher speeds to provide increased fuel efficiency and higher power densities. Again, the reciprocating forces can become objectionable and even destructive at these higher speeds. Reducing the mass of the connecting rod will mitigate this problem and will enable consideration of higher engine speeds. In addition, lower reciprocating loads should lead to lower loads on the crank shaft and lower friction losses, which can decrease fuel consumption and/or boost performance.

Almost every major automotive company has considered the use of alternate materials for connecting rods. Investigations of aluminum composites reinforced with either continuous fibers (Folgar et al. 1987) or particles, forged titanium, powder metallurgy titanium, and rapidly solidified aluminum alloys have been made. Of these, the most promising candidates appear to be either DRA or a recently developed "free-machining" titanium alloy (Kimura et al. 1991). The primary design property for the connecting rod is high cycle fatigue at 150°C to 180°C. The steels used for connecting rods have axial fatigue strengths at 10^7 cycles of 240 MPa to 310 MPa. The DRA of choice for this application appears to be a 2xxx alloy reinforced with particles of SiC,

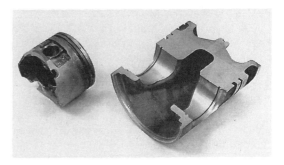

Figure 16.9. MMC pistons for heavy-duty diesel engines (right) and passenger cars (left) *(Courtesy of Ford Motor Company).*

produced by powder metallurgy. An example of a connecting rod forged from such a composite is shown in Figure 16.8. One such DRA, X2080/20/SiCp, possesses a fatigue strength at engine temperatures of approximately 170 MPa, which, although below that of the currently used steel, may be adequate for certain applications. Additional properties that are also important are coefficient of thermal expansion, modulus of elasticity, and wear resistance. From a technical and a cost effective perspective, the connecting rod is a challenging application for DRA.

The first automotive production usage of DRA was the selective reinforcement of the piston ring groove using a hybrid alumina/graphite preform infiltrated by squeeze casting aluminum (Kubo et al. 1988; Hamajima et al. 1990). The cost effectiveness of this application remains to be demonstrated. Squeeze cast, selectively reinforced aluminum pistons are widely used to reinforce pistons in heavy-duty diesel engines. Examples of metal-matrix composite pistons for both diesel and spark ignition engines are shown in Figure 16.9. More recently, forged DRA pistons have been evaluated in engines of racing motorcycles, in which use was made of the controlled coefficient of thermal expansion of composite materials to produce a "zero" clearance piston (Harrigan 1992). This allowed the reduction of piston clearances from .051 mm to 0.005 mm at the top of the piston skirt, and reduced ring seal leakage or "blow by," thus enhancing performance.

An important consideration of piston design is the awareness that metal temperature is known to vary strongly with location. The top of the piston experiences a temperature of approximately 250°C to 300°C under high-speed conditions, while in the pin region the temperatures are as low as 120°C to 150°C. The piston must withstand fatigue loading, thermal cycling, and resistance to combustion gases. It must also have acceptable wear resistance and a coefficient of thermal expansion that is well matched to the surrounding cylinder. In

Figure 16.8. A forged MMC connecting rod for a passenger car engine *(Courtesy of Ford Motor Company).*

addition, thermal conductivity close to that of aluminum is important for heat conduction. There are several possible composite systems that could be used for this application, including aluminum and magnesium matrices reinforced either fully, by particles, or selectively, using fiber preforms. Selective reinforced composite pistons would be produced via squeeze casting. Particle reinforced composite pistons could be produced by gravity casting, squeeze casting, semisolid forming, or potentially, die casting. In addition to engine components, MMCs have been considered for other wear resistant automotive applications (Figure 16.10).

16.1.3 Technological and Infrastructural Challenges

Despite the fact that significant opportunities for metal-matrix composites have been identified in the automotive industry, there are also a number of barriers that exist that must be surmounted prior to introduction of these materials in high production volume automobiles. These can be divided into "technological" barriers and "infrastructural" barriers.

16.1.3.1 Technological Considerations

Primary technological challenges that must be met include development of rapid and inexpensive machining technology, near-net shape forming technology, inexpensive raw materials, and the development of recycling technology. For most metal-matrix composites, final parts costs will be higher than those of existing components because of the higher costs of raw materials, shape fabrication, and machining. Although in some of the cases listed, detailed cost/benefit analyses indicate that the application of these materials is justified, technological developments that further reduce their costs will lead more quickly to the acceptance of these materials. In this context, the development of forming processes that are faster and provide high yields, and the development of machining techniques that are faster and exhibit improved tool lives, are important. An additional technical deficiency limiting widespread application of MMCs is a lack of design data, particularly in the so-called "secondary" properties such as fatigue, creep, wear, and corrosion. This problem is compounded by the large number of new materials that are currently being developed. Also, the lack of information on failure modes that can occur in composites, and the manner in which these failure modes differ from existing materials, can limit acceptance by the automobile design community. This is especially true when design criteria have been established over the course of time by experience and extensive durability testing.

16.1.3.2 Infrastructural Considerations

Despite the significant technological developments that either have taken place or will take place, the ultimate acceptance of MMCs will be limited by a number of "infrastructural" issues. This is especially true in North America, where the singular orientation of the major automobile manufacturers and their suppliers is to high production volume automobiles. Typical production levels for individual car lines range from 200,000 to 500,000 vehicles per year. This means that the economic risks for substituting new materials are perceived as high, naturally leading to conservatism. This perception is compounded by the fact that materials such as metal-matrix composites are embryonic in nature and large-scale production has not been generally demonstrated. In this context, it is important to recognize the establishment of a large scale DRA production facility by the Duralcan USA division of Alcan Aluminum Corporation. This facility has an annual capacity in excess of 11,000 metric tons. Such developments and investments are critical to the successful introduction of metal-matrix composites in the automotive industry.

Figure 16.10. MMC aluminum applications *(Courtesy of Kobe Steel Ltd.)*.

16.2 Aerospace Applications

16.2.1 Metal-Matrix Composites: Aerospace Requirements

Motivation for metal-matrix composites has been driven by defense applications, although current utilization of MMCs in aerospace engines may become more difficult based on cost considerations. The substitution of MMC and CMC for high thrust-to-weight engine applications must be evaluated in light of cost effectiveness and long-term reliability (Figure 16.11) (Driver 1989). Materials substitution in high-performance engine applications will mean a reduction in nickel, iron, and titanium alloys, and a projected increase in MMC, CMC, and carbon-carbon composites. The range of temperatures and specific property goals are detailed for structural engine applications in terms of rotating, stationary, and heat-resistant components versus conventional alloys (2024 Al, Hastealloy X, and Ti-6Al-4V) (Figures 16.12 and 16.13).

With the improvement in titanium alloys in the areas of use temperatures, casting technology, alloy chemistry, and cleanliness, the competitive application of metal-matrix composites to aerospace application has become more dominated by performance/cost considerations. Aerospace applications for metal-matrix composites require materials with high specific strengths and moduli, impact strength, and heat and erosion resistances. For satellite and space structures, high specific stiffness, controlled thermal expansion coefficients, and thermal cycling stability are the more important considerations. For aircraft structures, high specific strength is necessary for fan and compression turbine blades coupled with heat resistance. Of prime importance is the need to develop improved continuous-fiber reinforced materials for use at elevated temperatures in engines and high-speed vehicle structures in the following temperature ranges: (1) aluminum alloys to 500°C, (2) titanium alloys to 1000°C, and (3) intermetallic and nickel base alloys to 1200°C.

For ambient temperature applications, both high strength and stiffness considerations are considered necessary for struts, access doors, wing, frame, stiffener, and floor beams applications.

Helicopters have greater need for weight savings and, therefore, the utilization of composites in their construction has increased. For skin structures, the temperatures are significantly lower, vis a vis, high performance aircraft; as a result, polymer-matrix composites are considered the material of choice. Nevertheless, rotor applications demand higher specific strength and stiffness properties. Thus, metal-matrix properties have been considered for transmission cases, transmission structures, swash plates, push rods, and trailing edges for tail and rotor blades.

Projections of materials for aerospace applications are varied based on performance and temperature considerations (Figures 16.11, 16.12, and 16.13). Because of the elevated operating temperature requirements for

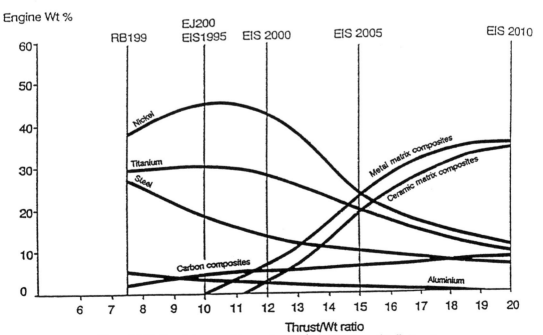

Figure 16.11. Trends in gas turbine engine materials *(Courtesy of Rolls Royce).*

308   APPLICATIONS

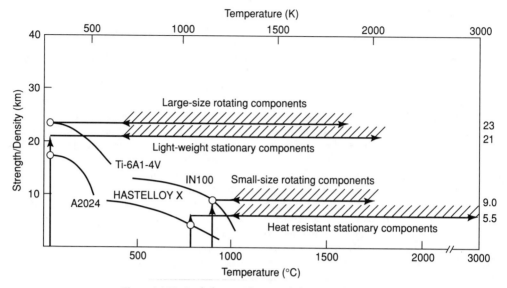

Figure 16.12. Goals for specific strength for engine materials.

turbine engines, MMCs have several applications at different combinations of operating stresses and temperatures (blades, vanes, and housing). See Figure 16.14. (Grisaffe 1991). With reinforcements of graphite, boron, silicon carbide, and metallic fibers, the potential utilization of MMCs are detailed in Table 16.5 (Kelly 1988). As the temperature increases, a degradation of fiber strength results. Consequently, the performance is fiber limited (Figure 16.15). In selecting the appropriate fiber, several material challenges remain, including fiber stability at elevated temperatures, the need for interface coatings, alloy optimization for fiber mechanical and thermodynamic compatibility, and interface stability, coupled with design considerations for long-term property data base (elevated temperature creep, fatigue, and toughness).

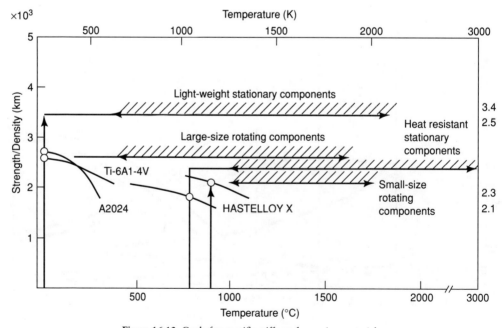

Figure 16.13. Goals for specific stiffness for engine materials.

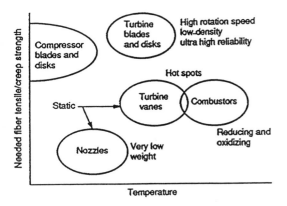

Figure 16.14. Strength temperature requirements for engine applications.

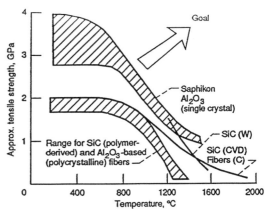

Figure 16.15. Tensile strengths of fibers as a function of temperature.

16.2.2 Aluminum Metal-Matrix Composites for Aerospace Applications

Aluminum metal-matrix composites have been widely considered for aerospace application in continuous fiber (C, SiC, B, and Al_2O_3) reinforcement in particulate and whisker (SiC, Al_2O_3), as well as in layered laminate structures (sandwich structure of aluminum and aramid or glass/epoxy such as ARALL or GLARE for high performance fatigue critical applications) (Bucci 1987).

Table 16.5. Potential Applications of Metal-Matrix Composites

Fiber	Matrix	Potential Applications
Graphite	Aluminum	Satellite, missile, helicopter structures
	Magnesium	Space and satellite structures
	Lead	Storage-battery plates
	Copper	Electrical contacts and bearings
Boron	Aluminum	Compressor blades and structural supports
	Magnesium	Antenna structures
	Titanium	Jet engine fan blades
Borsic	Aluminum	Jet engine fan blades
	Titanium	High-temperature structures and fan blades
Alumina	Aluminum	Transmission housings
	Magnesium	Helicopter transmission structures
Silicon carbide	Aluminum, titanium	High-temperature structures
	Superalloy (cobalt-based)	High-temperature engine components
Molybdenum, Tungsten	Superalloy	High-temperature engine components

Data from Kelly (1988).

16.2.2.1 Continuous Fiber Systems

Boron fiber systems add substantial strength and stiffness to metal-matrix systems at no penalty to density. However, the application of boron reinforced metal-matrix systems has been limited because of the high cost. Applications include B/Al tubes for space shuttle compression members (Figure 16.16). The mechanical properties of boron reinforced aluminum and silicon carbide reinforced titanium systems are detailed in Tables 16.6 and 16.7. The high temperature strength response of boron/aluminum composite indicates good strength retention to temperatures of 300°C and good stress rupture properties to 500°C.

Aluminum matrix alloy systems can also be reinforced with graphite and silicon carbide fibers. Although these fiber systems typically do not provide the high strength of boron reinforced systems, the potential, lower cost silicon carbide systems are competitive on a cost/performance basis because the fibers may be in the form of whiskers and fine diameter fiber tows such as Nicalon SiC (Nippon Carbon, Tokyo, Japan). The tensile properties of SiC reinforced aluminum demonstrate good strength retention up to 250°C, although the improvements in strength and modulus is at the expense of a loss of ductility. The trade off between strength/modulus versus ductility, toughness, and notch sensitivity is an important consideration in the application of metal-matrix composite systems. An important design parameter is notched compression strength as well as compression after impact, particularly with graphite fiber reinforced systems.

The graphite/aluminum system has also been widely investigated with several aluminum matrix systems (1100, 2024, 6061, etc.). The graphite/aluminum system is attractive because of projected improvements in specific strength and modulus. However, the thermal,

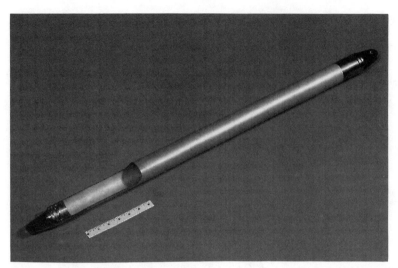

Figure 16.16. Boron/aluminum tube applications *(Courtesy of Textron)*.

chemical, and strength incompatibilities between the fiber and the matrix systems result in processing difficulties. The properties of graphite/aluminum composite systems show considerable variation for several combinations of aluminum matrix systems and PAN- and pitch-based carbon fiber systems compared with rule of mixtures strength predictions. The property fluctuations are a function of the varying processing conditions and the resulting degradation of the fiber properties. Graphite/aluminum and graphite/magnesium applications are considered optimum for satellite and space vehicle applications, in which weight saving, thermal management, and specific modulus are critical requirements.

Table 16.6. Unidirectional Boron/Aluminum Composite Mechanical Properties

Material Property		Value
Tensile strength	(ksi)	220
	(MPa)	1520
Tensile modulus	(msi)	31
	(MPa)	214
Compression strength	(ksi)	400
	(MPa)	2760
Compression modulus	(msi)	30
	(GPa)	207
Shear strength	(ksi)	23
	(MPa)	159
Shear modulus	(msi)	6
	(GPa)	41
Thermal expansion	PPM/°F	3.4
	PPM/°C	6.1

Data from Textron (1991).

16.2.2.2 Discontinuous Fiber Systems

Apart from continuous fiber reinforced MMC systems, dispersoid and particulate reinforced alloys have been introduced with the development of high-strength 7000 series alloys, aluminum lithium, high-temperature aluminum alloys, and dispersion strengthened elevated-temperature alloys. In addition, several process developments, such as Osprey (Alcan 1990), XD synthesis (Kumar 1991), direct metal oxidation (Lanxide 1991), preform infiltration (Harris 1988, 1990), and mechanical alloying (Benjamin 1970, 1977) have raised the performance limits of metal-matrix composite systems. It has been shown to be desirable to further strengthen these alloys by superimposing an additional constituent such as an oxide or carbide dispersoid, to improve elevated-temperature response, or by the addition of a high-volume fraction of a whisker or particulate, to increase the strength and modulus. To this end, several hybrid aluminum powder-metallurgy alloys have been

Table 16.7. Properties of SiC reinforced Aluminum and Titanium

Property	SiC/6061 Aluminum	SiC/Ti-6Al-4V
V_f	0.48	0.35
Density g/cc (lb/in³)	2.84 (0.103)	3.86 (0.14)
0° Tensile strength MPa (ksi)	1550 (225)	1725 (250)
0° Tensile modulus GPa (msi)	193 (28)	193 (28)
90° Tensile strength MPa (ksi)	83 (12)	415 (60)

Data from Textron (1991).

developed. These include mechanically alloyed elemental and pre-alloyed aluminum alloy powder systems; aluminum metal-matrix composites developed by mixing and consolidating aluminum powder and high modulus, low-density micron- and submicron-sized carbides (SiC, TiC, B_4C); and direct-spray deposition of aluminum powders with refractory reinforcements, such as Osprey (Alcan 1991; Harris 1988). In each of these processes, the additional constituents provide both desirable and detrimental changes to the alloy's mechanical and physical properties. With the addition of refractory SiC particulate and/or whiskers, the strength, modulus, and wear resistance increases with a reduction of ductility and fracture toughness. With regard to physical and electrical properties, a reduction is experienced with respect to the coefficient of thermal expansion, electrical conductivity, and an increase in thermal conductivity for SiC additions. A significant increase in the modulus can be achieved (50%) with the addition of high-volume fraction carbides in an aluminum matrix. However, for high-strength applications, the nominal upper limit of carbide reinforcement addition is 20 vol% in order to achieve a reasonable balance of strength, modulus, fracture toughness, and ductility. A comparison of the mechanical properties of several aluminum matrix alloys and their respective composites, processed by varying routes, is provided in Table 16.8 (Harris 1988).

The addition of oxide and carbide dispersions via mechanical alloying (MA) has been developed for 2000 and 5000 series aluminum alloys, and Novamet has introduced IN9021 (Benjamin 1970, 1977; Gilman et al. 1983). In the 5000 series alloys, mechanical alloying has produced Al-4 wt % Mg alloys designated as IN9051 and IN9052. The strengthening effects are hierarchical, namely combining multiple strengthening mechanisms (a microstructure that has an Mg solid solution, a submicron grain size, dispersions of MgO, Al_2O_3, and Al_4C_3, and low-angle textured microstructure) developed from attrition and extrusion processes. These alloys were developed from the ball milling of mixed elemental powders with the additions of process control agents. During the mechanical alloying process, the reactions of carbon and oxygen with the alloy

Table 16.8. Tensile Properties of Metal-Matrix Composites

Material	Fabrication Method and Form	Young's Modulus (GPa)	0.2% Ps (MPa)	UTS (MPa)	Elongation (%)	Fracture Toughness (MPa m)
Al-Cu	Squeeze cast	70.5	174	261	14.0	—
Al-Cu + Al_2O_3 (V_f = 0.2 fiber)	Squeeze cast	95.4	238	374	2.2	—
Al-Cu-Mg (T6) (2014)	Spray-formed sheet	73.8	432	482	10.2	—
Al-Cu-Mg (T6) (2014) (V_f = 0.1, 10 μm particle)	Spray-formed sheet	93.8	437	484	6.9	—
Al0-Cu-Mg (T4) (2124)	Powder-rolled plate	72.4	360	525	11.0	—
Al-Cu-Mg + SiC (T4) (V_{vf} = 0.17, 3μm particle)	Powder-rolled plate	99.3	420	610	8.0	18
Al-Cu-Mg (T6) (2124)	Powder-rolled plate	73.1	425	474	8.0	26
Al-Cu-Mg + SiC (T6) (V_{vf} = 0.17, 3μm particle)	Powder-rolled plate	99.6	510	590	4.0	17
Al-Si-Mg (T6) (6061)	Spray-rolled sheet	69.0	240	264	12.3	—
Al-Si-Mg + SiC (T6) (V_f = 0.1, 10μm particle)	Spray-rolled sheet	91.9	321	343	3.8	—
Al-Zn-Mg-Cu (T6) (7075)	Spray formed	71.7	617	659	11.3	—
Al-Zn-Mg-Cu + SiC (T6) (V_f = 0.12, 10μm particle)	Spray-formed extrusion	92.2	597	646	2.6	—
Al-Li-Cu-Mg (T6) (8090)	Spray-formed plate	79.5	420	505	6.5	38
Al-Li-Cu-Mg + SiC (T6) (V_f = 0.17, 3μm particle)	Spray-formed plate	104.5	510	550	2.0	—

Data from Harris (1988).

constituents facilitate the controlled oxidation and/or carburization and subsequent strengthening of the alloy. The benefits of this process include improved ambient strength, elevated temperature creep response, good levels of toughness and ductility, a stable microstructure, and stress corrosion resistance.

16.2.3 Elevated Temperature Metal-Matrix Applications

In general, aerospace applications can be divided into two categories: (1) subsonic or stratospheric applications, which require lower temperature, relatively low-performance metallic and composite systems, and (2) high-performance supersonic applications, operating at higher altitudes. Here, performance is severely materials limited. The latter category presents a formidable challenge in terms of re-entry temperatures and environments. The operational aircraft skin temperature is a function of both speed and altitude; conventional monolithic aluminum and polymer composites have a materials performance limitation of 120°C to 450°C. For high-speed transport systems, temperatures can reach over 1600° K at the nose, 1000° K to 1600° K in the lower body, and can fall below 700° K in the upper body upon re-entry. Other factors affecting materials performance include vacuum and radiation environments, acoustic fatigue, and a disassociating gas environment upon re-entry. SiC/titanium and SiC/aluminum composites (Figure 16.17a and b) are targeted for use in structural components for the national aerospace planes (NASP), for advanced fighters, and for gas turbine applications. The application for the NASP X-30 also includes titanium and intermetallic matrix systems (for example, B-21S reinforced with Textron SCS SiC fibers) (Textron 1991).

The improvement of aerospace engine thrust-to-weight is associated with a corresponding increase in engine pressure ratio, turbine entry temperatures, and a reduction in specific fuel consumption. See Table 16.9 (Driver 1989; Postans 1989). For defense-related engine applications, a goal of the Department of Defense, NASA Integrated High Performance Turbine Engine Technologies (IHPTET) is to increase the thrust-to-weight ratio from 10:1 to 20:1, with the application of silicon carbide reinforced titanium for shafts, rotors, and hollow fan blades. The weight savings are enhanced in rotating components in which the centrifugal forces are reduced. As a result, lower density components are

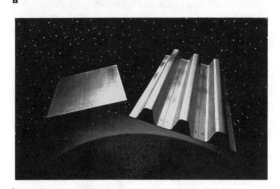

Figure 16.17. (a) Silicon carbide/aluminum aerospace applications. (b) Silicon carbide/titanium aerospace applications. *(Courtesy of Textron.)*

Table 16.9. Improvement in Gas Turbine Performance

	Civil		Military	
Year into service	1952	1988	1980	2000
Engine	Ghost	RB211-524H	RB199	—
Aircraft	Comet	Boeing 767	Tornado	—
Thrust	5,050 lb	60,000 lb	9,100 lb dry 16,400 lb reheat	7,500 lb (approx)
Thrust: Weight	2.3:1	6:1	4.2:1 dry 8:1 reheat	20:1
Overall pressure ratio	4.5:1	33:1	23.5:1	32:1
Turbine entry temperature	1100K	1700°K	1600K	2400°K target
Specific fuel consumption	1–2 lb/hr/lb	0.56 lb/hr/lb	0.65 lb/hr/lb without reheat	25% less than current technology

Data from Driver (1989).

utilized and the containment structures and bearings requirements such as stress and weight can be reduced (Dimiduk 1991).

16.2.4 Nickel-Based Dispersion Alloys for Aerospace Applications

The development of metal-matrix composites for elevated temperature applications (Anton et al. 1990) has been considered for particulate, fiber, filament, and to a limited extent, whisker reinforcement. The high processing temperatures (1500° C for nickel-base systems) can preclude utilization of fine whiskers of silicon carbide or alumina, which are not chemically and thermally compatible with the matrix system at these temperatures. The compatibility of fiber and matrix systems is based on elastic modulus, thermodynamic, and coefficient of thermal expansion considerations. Consequently, the optimum reinforcements are very limited in selection based on this criteria. As a result, continuous or long fiber reinforced intermetallics are not considered to be cost effective and are subject to rapid degradation at temperatures above 1000° C. An alternative approach is dispersion strengthening or discontinuous reinforcement, which has a variable volume fraction of up to 60 vol% and an isotropic structure.

The high-temperature matrix systems have included nickel-, niobium-, and cobalt-based systems as well as intermetallics. There has been significant interest in intermetallic matrix materials, which include nickel, titanium, niobium, tantalum, and iron aluminides. Considerable research activity has focused on molybdenum disilicide with less attention paid to chromium, tantalum, and niobium silicides. For elevated temperature composites, alternative process techniques have been considered including vapor deposition, hot pressing, reactive synthesis and consolidation, injection molding, mechanical alloying, in situ directional solidification, cryomilling, hot isostatic consolidation, gas liquid in situ reactions, XD exothermic dispersion synthesis, self-propagating synthesis, and powder cloth preform consolidation (Vedula 1991).

One of the early applications of reinforced alloys for elevated temperature application consisted of dispersion of oxides in nickel, copper, or aluminum alloys. Of these, the most successful development was by International Nickel Company, which developed nickel-based mechanically alloyed (MA) dispersion alloys (Benjamin et al. 1977). The mechanical alloying process combines an elemental powder metallurgy mix with a dispersion of an oxide or a carbide. An early alloy was MA754, which combined a Ni-20 wt % Cr solid solution alloy with a dispersion of 0.6 wt % Y_2O_3 for aircraft station bands and vanes. A second precipitation hardened MA alloy, MA6000, is Ni, 15 wt % Cr, 4 wt % W, 2 wt % Mo, 4.5 wt % Al, 2.5 wt % Ti, and 2.0 wt % Ta with a 1.1 wt % Y_2O_3 dispersion. This alloy has a major application for turbine vanes. A ferrous-based MA alloy, MA 956, containing 4 wt % Cr, 4.5 wt % Al, and 0.5 wt % Y_2O_3, is considered for combustor application. One of the benefits of MA alloys is the application of conventional thermal mechanical forming operations that permits fabrication of forged, rolled, and spun components with zone annealed directional microstructure. The utilization of the mechanical alloying process for the production of dispersion-hardened composite structure allows for a great deal of alloying freedom and the application to new alloy systems. Limitations include reduced ductility, secondary processing, and limitations to reinforcement levels. Nevertheless, it has proved to be a successful application of a metal-matrix dispersion alloy for elevated temperature aerospace engine applications.

In summary, there are several processing routes for elevated temperature metal-matrix composites processed via solidification or powder processing routes. Of particular interest are the dispersion or particulate systems such as XD synthesis (Kumar 1991), in which controlled variation in physical and mechanical properties can be achieved. In some cases, the reinforcements can weaken the structure and serve as sites for crack initiation. The area of development for metal-matrix composites for elevated temperature application (T > 800°C) is seen to be in a formative stage, with several areas of expertise to include reinforcement, deformation mechanisms, crystal structure, and fabrication where the science and processing has not been clearly established. Fracture toughness and embrittlement by interstitial oxygen, hydrogen, and nitrogen remain critical issues for these elevated temperature matrix systems.

16.3 Other Industrial Component Applications

Metal-matrix composites have a wide range of applications in defense, aerospace, and automotive industries because of their high specific properties as well as their ability to tailor other properties such as impact resistance, fatigue properties, and thermal coefficient of expansion. The materials utilized for these applications have been reviewed in earlier sections. Applications are increasingly being found for metal-matrix composites in other industrial components. The properties that make MMCs attractive for the aerospace and automotive industry, specific strength and modulus, controlled toughness, thermal expansion coefficient, hardness, and improved fatigue response, have made metal-matrix composites suitable for use in a variety of industrial component applications such as wear resistance parts, electrical and electronic components, thermal management, graded structures, acoustical damping composites, and sporting goods. These applications require a wide range of properties; wear resistant components require hard surfaces, electrical and electronic applications need good conductivity and a very low coefficient of thermal expansion, and damping composites require high attenuation of sound waves. With a suitable combination of the matrix material and the reinforcing phase, it is possible to tailor the physical and mechanical properties of the composite for the required application.

16.3.1 Wear-Resistant Materials

Wear resistant and hard materials have been used extensively in commercial applications such as cutting tool bits, wear resistant surface finishes, dies, and automotive tire studs. Carbides and nitrides have shown great promise as wear resistant materials because of their high temperature strength and hardness. The usefulness of carbides and nitrides in pure form as wear resistant materials is limited because of their brittle nature at ordinary temperatures. Hence, they are dispersed in a metallic matrix to increase the toughness of the wear component. Figure 16.18 (a) and (b) show hard materials developed by Kobe Steel Ltd. (Kobe, Japan) and Alcan for wear-resistant and tooling applications.

One of the earliest techniques to utilize the properties of the carbides and nitrides consisted of binding or "cementing" the particles with a binder, usually a metal/alloy that is a liquid at the sintering temperature. These type of materials are known as cemented carbides or hardmetals. The first cemented carbides were developed to produce dies for drawing tungsten wires. The cemented carbides are made up of a carbide phase, WC, TiC, TaC, NbC and/or HfC of sizes ranging from submicron to about 10 μm with a metal binder (Pastor 1987). Although Co has been the most preferred binder material for the carbides, cost considerations have motivated researchers to look at alternative binders such as stainless steel, Ni, Ni-Al, Fe-Ni-Co, Ni-Co, and other Co-based alloys. The binder has a volume fraction of about 2% to 25%. In general, the hardness and the wear resistance of the material increases with a finer grain size, and smaller amounts of binder. Conversely, a higher metal content and a coarse grain size increases the strength and toughness of the cemented carbide. With time, cemented carbides have found applications

Figure 16.18. (a) Cermets and cemented carbides for use in tooling materials *(Courtesy of Kobe Steel Ltd.)*. (b) Aluminum-matrix composite for wear, industrial, and automotive applications *(Courtesy of Alcan Ltd.)*.

outside machining in metal cutting, mining, rock drilling, metal forming tools, and abrasive grits. Some of the wear applications for the cemented carbides are listed in Table 16.10 (Stevenson 1984). The applications are classified into high, medium, and low abrasion resistance, corrosion resistance, and speciality applications.

Table 16.10. Applications of Cemented Carbides

Type	Applications
Light-impact, high-abrasion resistance applications	Blast nozzles, spray nozzles, reamers, gun drills, wear pads, powder compaction dies, glass cutters, plastic extrusion dies, guide rings, circuit board drills, paper slitters
Medium-impact, medium-abrasion resistance applications	Tape slitters, metal draw dies, aluminum extrusion dies, shear knives, guide rolls, snowplow blades
High-impact, low-abrasion resistance applications	Cold header dies, stamping dies, bar mill rolls
Specialty applications	Nuclear components, nonmagnetic guidance gyros, nonmagnetic dies for magnets

Data from Stevenson (1984).

In addition to cemented carbides, cermets are also extensively used as wear resistant materials. Cermets include a very large class of materials that includes oxide, nitride, and boride particulate reinforced metals. About 40% of the metal cutting tools used in Japan are titanium carbide or titanium carbonitride based cermets. Modern TiC based cermets are of the type TiC-Mo$_2$C-Ni/Mo or (Ti,Mo)C-Ni/Mo. They are used in machining steel and in ductile cast iron, and they display lower wear, longer life, and increased cutting speed. However, TiC based cermets are not suitable for machining high temperature alloys and superalloys. The success of TiC cermets is due to their high hardness and thermal conductivity, as well as their lower coefficient of friction. Carbonitride cermets did not find many applications until the 1970s. These grades are based on Ti(C,N)-Ni/Mo, Ti(C,N)-Ni Ti(C,N)-Ni/Fe/Mo. Compared to the TiC, the carbonitrides are much more complicated. For instance, the influence of N/C ratio, hard phase/binder ratio, Ti/Mo or W ratio, solid solution of Ti(C,N) as opposed to a mixture of carbides and nitrides, and the effect of sintering atmospheres are not very well understood. The carbonitride cermets seem to be finding increasing success in Japan, where they are being widely employed as cutting tools in the automotive industry, exhibiting excellent cutting performance compared to the traditional cemented WC/Co carbides. In addition to carbides and carbonitrides, borides, boronitrides, and borocarbides are also being investigated for wear resistance applications. These metal-matrix composites have been developed over the years to optimize reinforcement size as well as matrix chemistry, which serves as the binder.

Cemented carbides and cermets can be thought of as traditional forms of metal-matrix composites used for wear resistant and cutting tool material applications. Recently, however, lightweight aluminum and magnesium-matrix composites are finding increasing use in wear resistance applications. These composites have potential for use in parts such as nozzles, sleeves, tubes, and other applications in which abrasion and erosion are primary failure mechanisms. These composites have been developed by P/M routes by Ceracon (Santa Barbara, CA), Cabot (Reading, PA), and Mitsubishi (Tokyo, Japan). Ceracon has developed a process in which the preheated canned powders are consolidated in a very short time (5 to 60 seconds) in a granular ceramic, carbonaceous, or other free flowing material also called pressure transmitting medium (PTM) (Metal Powder Report 1990). The result of this process is fabrication of a near-net-shaped, fully dense product. This process is also suitable to bond PC (polycrystalline) diamond and cemented carbides to metallic substrates like steel. Using this process, metal-matrix composites for wear resistance applications, such as silicon carbide reinforced aluminum, high Al$_2$O$_3$ content Fe composites, and intermetallic compounds, have been manufactured. This process has also been used to fabricate layered structures consisting of steel coated with cobalt, diamond, and WC/Co. Some of the drilling tools used for the oil field components and the microstructural aspects of the composites are shown in Figure 16.19. Also shown are the wear resistant WC/Co, PCD, and cobalt coatings on 1018 steel for wear resistant applications. The Ceracon process has also been used to manufacture rotary tri-cone drill bits (Reed Tool Co.), water well drill bits (Numa Tool Co.), heavy equipment, and ground engaging tools.

As opposed to Ceracon, which makes composites via P/M routes, Lanxide has two processes for making metal-matrix composite from liquid metal alloys (Ashley 1991): (1) DIMOX is a process that involves oxidation of liquid metal alloy such as Al-Si-Mg to produce a monolithic metal/ceramic matrix with fiber and/or particulate reinforcement, (2) PRIMEX is a process in which composites are made via pressureless infiltration of molten metal through a preform of the reinforcing phase. One of the products from the DIMOX process, 50% to 70% SiC reinforced alumina, toughened by about 10 to 20 vol% aluminum, exhibits exceptional wet wear resistance. Another commercial application is a corrosion-resistant composite consisting of tungsten

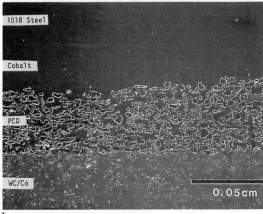

Figure 16.19. (a) Drilling tools. (b) Diamond and tungsten carbide-coated cobalt *(Courtesy of Ceracon).*

carbide particles reinforcing a high cobalt bronze alloy for valve applications. Other promising applications are products with superior corrosion resistance for use in the steel industry, such as glide gates to control the flow of molten steel, and tooling for polymer injection molding.

Composites made from self-propagating, high-temperature synthesis (SHS) can be used in a variety of applications. The SHS process has been shown to produce various carbides, borides, silicides, nitrides, hydride, and oxide reinforced metal-matrix composites as well as powders. The resultant products have applications as sheaths for thermocouples, protective coatings, refractory material, nozzles for metal spraying, and line equipment in chemical industry.

16.3.2 Electrical and Electronic Applications

Electrical and electronic applications, in addition to requiring higher specific strength and modulus for weight savings, require controlled electrical and thermal conductivity and lower coefficient of thermal expansion. As with wear products, the ability to combine different materials to tailor the properties make metal-matrix composites very attractive in any of these applications. Advances have been made in the use of metal-matrix composites for electrical and electronic connectors, storage battery plates, packaging material, transformer material, superconducting wires, and thermal management devices.

16.3.2.1 ODS Copper Composites

Oxide dispersion strengthening (ODS) is an effective method for increasing the strength of copper without significantly reducing the electrical and thermal conductivity. Dispersion strengthened copper, in addition to maintaining high electrical and thermal conductivity, maintains resistance to softening at high temperatures, which is important in electrical applications at elevated temperatures. Alumina DS copper maintains over 80% IACS conductivity at over 3 vol % of alumina, whereas the strength increases from 300 MPa to over 500 MPa. ODS copper composites are manufactured via techniques such as simple mechanical mixing, mechanical alloying, rapid solidification followed by consolidation, and selective internal oxidation. Internal oxidation has proved to be very useful because of high diffusivity of oxygen in copper. The most common dispersoids utilized are Al_2O_3, ZrO_2, Cr_2O_3, and ThO_2. The volume fraction of the dispersoids are of the order of 2% to 3%. Conductivities over 88% IACS have been achieved at temperatures up to 800°C with 3 v/o ThO_2. The typical applications for ODS copper are resistant welding electrodes, lead wires, commutators for helicopter starter motors, relay blades, electrical contact supports, and continuous casting molds (Kubozono and Mori 1989; Nakashima and Miyafuji 1989).

16.3.2.2 Superconducting Composite Wires

Superconductors have shown great potential for use in power transmission cables, communication cables, and magnetic field applications such as superconducting generators and MRI. Their practical use, however, has been limited because of the inherent brittleness of the intermetallic and oxide type of superconductors, and because of the low critical current densities, critical magnetic field, and ac losses. Some of these problems are overcome by use of multifilamentary composite wires (Figure 16.20). These wires are processed by filling a metal tube such as silver or bronze with a superconductor and drawing or rolling. Wires from low temperature superconductors NbTi, V_3Ga, and Nb_3Sn have been drawn with as many as 1057 superconducting filaments. Nb-46.5%Ti alloy wire with a 2.5 μm diameter filament in a bronze matrix have achieved a critical current density of 3.8×10^5 A/cm^2 at 4.2 K (Tanaka 1989). In addition to NbTi, V_3Ga and Nb_3Sn superconducting MMC wires have been fabricated with current densities in excess of 10^5 A/cm^2. To obtain high current densities, it is important to control processing conditions to achieve homogeneous composition and good bonding between matrix and wire without any interfacial chemical reaction. Since their invention, wires and tapes of high-temperature oxide superconductors in a silver matrix have been fabricated. Their critical current densities, however have been limited to about 10^4 A/cm^2.

16.3.2.3 Electronic Packaging

Electronic packaging materials are required to structurally support electronic components, provide protection from hostile environmental effects, and dissipate excess heat generated by electronic components. The mechanical, physical, and thermal properties important for electronic packaging applications include high stiffness, a high thermal conductivity for heat dissipation (and, therefore, a superior thermal management), a very low coefficient of thermal expansion, and a low density (Geiger and Jackson 1989). In addition, hermetic packages needed to protect electronic circuits from environmental hazards seals require glass to metal seals. In order to minimize residual stresses upon fabrication, the CTE of the housing should equal that of glass and the compression seal should have a CTE that is moderately higher than glass. Traditionally, low expansion Fe-Ni alloys (Kovar) have been used for packages, while copper and cold-rolled steels have been used for compression seals (Geiger and Jackson 1989). High densities of these alloys, however, make them unsuitable for avionic applications. Zweben (1992) has outlined the limitations and deficiencies of conventional metallic materials for packaging, circuit board, substructure, and support plates for electronic applications.

Many of the carbide, nitride, and oxide reinforcements have an extremely low CTE that, when combined with aluminum or magnesium, provide a material with a low CTE and a high thermal conductivity. At the time of this writing, aluminum and copper reinforced with high thermal conductivity carbon fibers and silicon carbide particle reinforced aluminum are most widely used among MMCs for electronic applications (Zweben 1992; Foster 1989). Figure 16.21 (a) and (b) shows the predicted and experimental data of the effect of SiC volume fraction on the CTE and the effect of thermal conductivity on SiC/Al composites with 6061, 6063, and Lanxide's SiC/Al composite produced via liquid metal infiltration (LMI) or PRIMEX process (Zweben 1992). It is clear that the CTE and the thermal conductivity can be tailored over a wide range. In addition, boron fiber/aluminum and particulate BeO/aluminum composites have also been recently commercialized for applications as heat sinks (Packard 1984). Composite systems such as continuous PAN C/Al, C/Cu, laminated Cu/Mo/Cu, diamond/Al, diamond/Mg, and AlN/Al systems show potential for use as packaging material (Zweben 1992).

MMCs have been finding increasing use in electronic packaging, especially in avionics, where the higher costs can be justified by the weight savings. A 40% SiC/Al carrier produced via P/M techniques is 65% lighter than the same part made from Kovar (Zweben 1992). The EPX series of composites, high volume fraction (greater than 60%) SiC reinforced aluminum composites developed by Lanxide by their PRIMEX process, are typical examples of MMCs suitable for applications in electronic industry. Some of the electronic components made by Lanxide such as packaging material and carrier plates, are shown in Figure 16.22. The composite has a coefficient of thermal expansion lower than 8.5×10^{-6} K; this low CTE minimizes the thermal residual stresses that arise during bonding of electronic components to the packaging material. Also, the thermal conductivity of 160 W/m-K of EPX-100 is seven times that

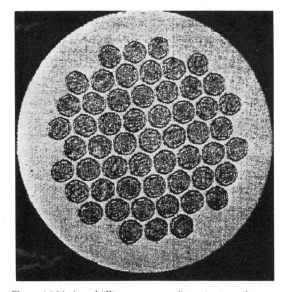

Figure 16.20. A multifiliment supercooling wire in a silver matrix *(Courtesy of Kobe Steel Ltd.)*.

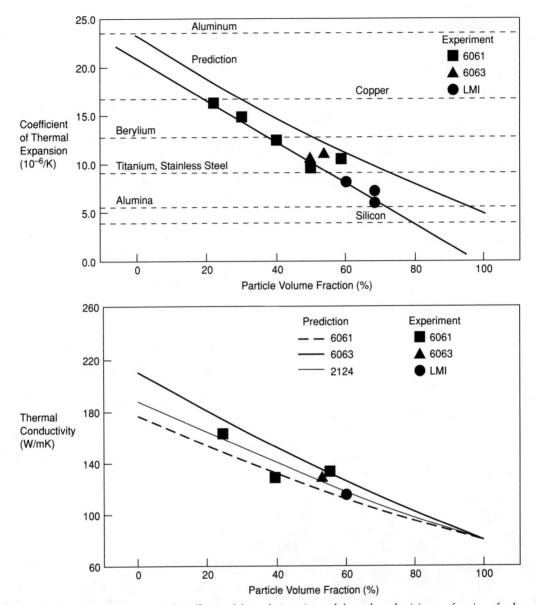

Figure 16.21. Predicted and experimental coefficient of thermal expansion and thermal conductivity as a function of volume fraction of SiC in aluminum MMCs (*Zweben 1992*.)

of conventional materials widely used now as packaging material. Because of their ability for near-net shaped manufacturing (which minimizes the need for machining and brazing), and their improved electrical, thermal and specific properties, these composites are expected to be competitive on a cost basis. Alcoa has produced a multichip module, in which the unreinforced aluminum pins are provided for better thermal management and 65 to 75 v/o SiC/Al composite substrate with exact CTE is present for mounting active devices (Premkumar 1992). In addition, MMCs are being used for applications in avionic racks, microwave packages, heat sinks, and enclosures (Zweben 1992; Premkumar et al. 1992).

Figure 16.22. (a) Thermal management composites by Lanxide. *(Courtesy of Du Pont Lanxide Composites, Newark, DE.)* (b) Electronic packaging material made from the PRIMEX process. *(Courtesy of Lanxide Electronic Components, LP, Newark, DE).*

16.3.3 Miscellaneous Industrial Applications

16.3.3.1 Graded Structures

The functionally gradient materials (FGM) composite program was started as a national program in Japan in 1987 to develop materials with effective compositional and microstructural control. The specific goal of this program is to develop materials for heat control for space planes, engine walls, combustion chambers, and turbine blades. These components are heated to a temperature of about 2000 K with a 1000 K difference in a thin layer toward the inner portion of the material. The purpose of designing a compositionally graded material is to withstand thermomechanical loadings by having a heat-resistant, antioxidation surface layer with a mechanically tough material on the inside (Watanabe et al. 1991). A transition in the material properties such as E, α, K is designed through the thickness. In addition, thermal stress relaxation is sought throughout the material. FGMs are designed to have heat-resistant ceramics on their high temperature side and tough metal in their low temperature side, with a gradual compositional change. The material is synthesized under controlled conditions to make a gradient distribution in internal composition and microstructure. The fabrication of the graded structure has been accomplished via powder metallurgy techniques and deposition processes such as plasma spraying, physical vapor deposition (PVD), chemical vapor deposition (CVD), electroforming, and self-propagation, high-temperature synthesis (SHS) (Sunakawa and Sakamoto 1991). Compositional gradient is achieved by varying the chemistry of the powders and the reactive gases with time in P/M and PVD and CVD processes, respectively. Sumitomo Electric (Osaka, Japan) has produced a material having a TiC/Ti gradient on a C/C composite substrate on one side, and an SiC/Si gradient on the other side by physical vapor deposition for thermal shock applications. A functionally gradient TiB_2-Cu composite, produced via the SHS process, was used to design a rocket thrust chamber that was exposed to a temperature difference of 1200 K throughout the wall thickness. This material showed no stress concentration or stress discontinuities internally, as seen in the layered structures (Sunakawa and Sakamoto 1991). Also, functionally gradient metal-matrix composites such as ZrO_2/Ni, $ZrO_2/0.8\%Y_2O_3/Ni.20\%Cr$, TiB_2/Cu, $MoSi_2/TiAl$, and SiC/C have been synthesized by several groups, including Nippon Kokan (Tokyo, Japan), Nippon Steel (Kawasaki, Japan), and GIRI (Tohoku, Japan).

16.3.3.2 Damping Composites

High damping materials (HIDAMATS) are used in applications in which elevated damping levels must be accompanied by superior mechanical properties. The most well-known example of composite material in damping application is gray cast iron, which is used extensively in heavy machinery (Ritchie and Pan 1991). Silicon and titanium carbide particulate reinforced aluminum composites in aerospace applications also show superior damping properties when compared with monolithic metals.

Vibrational damping composite steels have been widely used in automobiles, building materials, and household appliances (Taniuchi et al. 1991). Vibrational damping composite steel sheets with copolymers of vinyl acetate, butyl acrylate, and acrylic acid show a loss factor of 0.5 at 20°C. Damping properties of industrial components are important for regulating noise and vibration. ARALL and GLARE (Alcoa, Pittsburgh,

PA), a new family of metal/polymer laminates for aerospace structural applications, show damping response two to three times better than that of monolithic aluminum (Bucci et al. 1987).

16.3.3.3 Other Applications

Metal-matrix composites are finding a wide range of applications in the nuclear, biotechnology, and sporting goods industries. Titanium containing yttrium oxide dispersion strengthened ferritic alloys show promise for use in fast breeder reactors because of their high creep strength and neutron irradiation resistance. Initial studies of SiC/Al composites for structural applications in nuclear industry have shown encouraging results (Kano et al. 1989). Although these composites show a degradation in strength when exposed to elevated temperatures, increases in modulus and strength were observed after irradiation of the preform wires in the fast neutron spectrum (Okuda et al. 1989).

Nicalon SiC/Al composites are also being used in the manufacture of golf heads. In addition, SiC/Al composite tubes were considered for use in the catamaran "Stars and Stripes" to replace an aluminum tube (Harrigan 1992). Particulate reinforced aluminum has also been used to make the frame for the world's lightest bicycle, fabricated by Kobe Steel Ltd. Figure 16.23 shows a bicycle frame and a golf club head manufactured from aluminum-matrix composites.

Figure 16.23. Recreational application of Duralcan aluminum MMC. (a) A lightweight bicycle made from particulate reinforced aluminum. (b) Aluminum MMC driver head *(Courtesy of Alcan)*.

Efforts are being made in the biomedical industry to use reinforcements in existing cobalt-, titanium-, tantalum-, and molybdenum-based alloys for use in implants such as fracture plates, screws, pins, and dental implants. The major hindrance to MMC use in body implants has been the chemical instability of the metals in the corrosive human body environment.

Summary

The applications of metal-matrix composites have been reviewed for ground vehicle, aerospace, industrial, and recreational components. Several trends are noteworthy: (1) the application of MMCs to automotive engine applications (the commercial application in diesel pistons and cylinder liners by Toyota and Honda, respectively) coupled with the possible near-term application to automotive brakes, (2) the large-scale production facility by Alcan for the production of particulate MMCs, primarily for nonaerospace applications, (3) the utilization of MMCs for electrical, wear, and thermal management applications, and (4) the achievement of high performance properties in powder metallurgy and continuous fiber systems. The first two achievements, if successful, can lead to the increased utilization of MMCs apart from the limited production requirements for aerospace, wear, and electrical applications. These factors signify that the development of MMC composites has reached a watershed period. Future application or decline of this technology depends on successful near-term applications, particularly for automotive components. Several manufacturers are currently poised to produce metal-matrix composites by low-cost solidification and casting approaches in bulk and near-net shapes. Additionally, powder and Osprey approaches can produce high-performance structural shapes. Market forces, coupled with materials cost and reliability concerns, will be critical factors for future materials utilization. Several interesting processes and concepts requiring evaluation include exothermic dispersion, self-propagating synthesis, in situ processing, Osprey and plasma spray deposition, nanocomposite structures, functionally graded, hybrid-layered composites, and microstructurally toughened materials.

The development of processes and commercial applications for MMCs are worldwide. Efforts in Japan are particularly significant, especially because of their focus and implementation of processing for application-driven commercial products (Table 16.11). Corporate involvement includes many industries, including automotive (Honda, Toyota, Nissan), heavy (Kawasaki, Fuji, Mitsubishi), electronic component (Sumitomo Electric, Toshiba, Mitsubishi Electric), traditional metal ferrous

Table 16.11. List of Companies in Japan Undertaking R&D Work on Metal-Matrix Composites

Company	Main Areas of Research
Art Metal Manufacturing Co.	• Manufacture of automotive components
Fuji Heavy Industries	• Solid-state processing, including roll diffusion bonding for aerospace application
Hitachi Ltd.	• Copper-carbon composites for electrical applications
Honda	• Pressure-cast aluminum engine block featuring alumina and carbon short fiber reinforced cylinder liners (mass produced for the Honda Prelude) • Powder metallurgy Al alloy/alumina-zirconia-silica particle retainer ring (limited production) • Pressure cast stainless steel filament reinforced aluminum connecting rod (limited production) • Various MMC prototypes: Al-MMC piston pin, Mg-MMC piston. • Squeeze-cast carbon fiber reinforced Mg
Ishikawajima-Harima Heavy Industries	• Manufacturing methods and forming/metal working techniques for fiber reinforced metals (W-fiber reinforced superalloy turbine blades by hot isostatic pressing, Ti-matrix composites), continuous B/Al matrix composites
Kawasaki Heavy Industries	• Alumina fiber reinforced aluminum turbo jet engine impeller
Kobe Steel	• SiC whisker and particle reinforced aluminum by powder forging, HIP or pressurecasting, and deformation processing, for parts including automotive engines, bicycle frames, and optical telescope supports. • SiC filament reinforced Ti by P/M processing • Spray deposition of SiC/Al
Mitsubishi Aluminum	• Pressure infiltrated and extruded SiC whisker/Al • Particle reinforced aluminum rear shock absorber cylinder for a Yamaha motorcycle • Reinforced aluminum alloy for internal combustion engine part
Mitsubishi Chemical Industries	• Infiltration of fiber reinforced aluminum
Mitsubishi Electric	• Squeeze-cast SiC whisker-reinforced Al for space applications, short fiber C/Al composites • Laser melting/roll diffusion bonding of SiC fiber/Al
Mitsubishi Heavy Industries	• Pressure-infiltrated and extruded SiC whisker/Al alloys • Consolidation of C/Al and Nicalon/Al preform wires manufactured by Toray and Nippon Carbon, respectively • Ti matrix composites by HIPing
Mitsubishi Materials	• Wear, cutting tool applications, and aluminum PM composites
Mitsui Mining Company, Ltd.	• Alumina fiber reinforced aluminum parts for various applications
Nippon Carbon Company	• Infiltrated SiC fiber reinforced aluminum tows • Face insert for golf clubs (commercial production)
Nippondenso	• Whisker-reinforced Cu • Tribological material: $SiC/WS_2/Al$
NKK*	• Cast whisker and short fiber reinforced Al
Nippon Light Metals	• Agglomerated fiber reinforced aluminum
Nippon Steel*	• Squeeze casting of fiber reinforced aluminum • Compocasting of particle reinforced aluminum • Micromechanics of fiber reinforced metals • Functionally graded $NiCr/ZrO_2$ composites by low-pressure plasma spraying
Nissan	• SiC/Al Connecting Rods, Directed Metal Oxidation
Shikoku Chemicals Corporation*	• Al-B-O whisker reinforced Al
Showa Aluminum	• Mechanical alloyed SiC/Al and alumina/Al with particle sizes < 1 μm
Showa Denko	• Alumina, graphite, or silicon nitride particle reinforced Al alloys by pressing and sintering, including hypereutectic Al-Si

Table 16.11. List of Companies in Japan Undertaking R&D Work on Metal-Matrix Composites (*continued*)

Company	Main Areas of Research
Sumitomo Metals Industries Ltd.	• SiC whisker reinforced 2014 by squeeze casting or powder metallurgy • Ti-based MMC coatings • Ceramic particle reinforced superalloys
Suzuki Motor Co. (in collaboration with Agency of Industrial Science and Technology of Tokyo)	• Stir-casting process for making hypereutectic Al-Si alloy matrix composites
Toho Beslon	• Ion-plated C/Al tape (possibly discontinued)
Tokai Carbon Company	• P/M and squeeze cast SiC whisker/Al
Toray*	• C fiber-reinforced Al by squeeze casting • C/Al preform wires by the TiB process • C fiber-reinforced Cu and Sn • Casting of MMCs
Toshiba	• Fiber and whisker reinforced 6061 • Fabrication of a whisker reinforced aluminum bolt • W fiber reinforced FeCrAlY
Toshiba Machine Co. Ltd.	• Stir casting composite materials using an electromagnetic stirring apparatus
Toyoda Loom Company	• Oriented fiber preform processing • Pressure-cast oriented short fiber and whisker reinforced aluminum
Toyota Central R&D Laboratories	• Squeeze-cast fiber reinforced aluminum • Invention and development of hybrid fiber reinforced metals • NDE of MMCs

Data courtesy of A. Mortensen (1992).
* = private communication.

and nonferrous companies (Nippon Steel, Mitsubishi Materials, Sumitomo Light Metals) and fiber producers (Nippon Carbon, Toray). Japan's effort is notable, building on its strong base of fiber and reinforcement developments and an attitude that promotes materials performance and application-driven technological progress (Mortensen and Koczak 1993; Koczak 1989, 1992).

The European research and development programs are noted for their product and team-driven programs sponsored by the European Community (EURAM BRITE and EUREKA). Several programs focus on the development of metal-matrix composites (Table 16.12). The EURAM BRITE and EUREKA programs strengthen the competitive position of the EC by developing its technological base and providing new services and products. In many cases, the programs are precompetitive and are market-oriented. As a result, the research efforts generally involve corporate and academic participation, as well a product focus. The applications in these MMC programs include automotive pistons, connecting rod applications, primary aerospace structures and gear boxes, spray-deposited molds, and graded structures and cermets (Table 16.12).

The individual national programs of the United States and Europe are summarized as follows (Bader and Koczak 1991).

Austria. There is interest in both continuous- and short-fiber/particulate reinforced systems. Significant research effort is devoted to

- Liquid metal infiltration processes for continuous short fiber and particulate systems using aluminum- and magnesium-based alloys.
- Machinery and process development for extrusion and pressure die casting.
- Casting research for Duralcan-type materials.

Denmark. The emphasis is on the more academic aspects of research, primarily the understanding of structure/property relationships, including:

- Internal stress analysis and creep.
- General theories for two-phase materials.
- Microstructural effects—recrystallization.
- Interfaces and coatings.

Table 16.12. Recent European Community Sponsored Programs in Metal-Matrix Composites Sponsored by BRITE-EURAM (1990-1991)

Topic	Partners (*Prime Investigator)
Workability of metals and metal-matrix composites in particular Al-Li alloys and aluminium-based metal-matrix composites	University of Reading* SGM National Technical University of Athens University of Bordeaux I/Ensam
Optimization of ceramic fiber—reinforced aluminium alloys	Renault SA* University of Surrey University of Porto Senter for Industriforskning Hydro Aluminum
Prepregs and composite materials made of aluminium alloys reinforced with continuous fibers	DLR* University of Bordeaux Aerospatiale, Les Mureaux
Development of fiber reinforced aluminium metal-matrix composites for applications in aerospace primary components using powder metallurgy techniques	Agusta SPA* Aluminia SPA Novara Technical University of Delft Carborundum, Sale
Novel metal-matrix composites based on hypereutectic aluminium/silicon alloys	University of Sheffield* Lucas Automotive Ltd. Osprey Metals Ltd. Ceit (San Sebastian) Ederlan Cooperative University of Ruhr
Development of novel automotive piston/rod components and aerospace gearboxes from long fiber/metal-matrix composites	ICI Advanced Materials* Didier Werke AG Kolbenschmidt AG Fraunhofer Gesellschaft Imperial College Ricardo Consulting Engineers Ltd. VW AG Ray Advanced Materials Ltd. Agusta SPA VAW AG
Innovative manufacturing, design, and assessment of aluminium-matrix composites for high-temperature performance	University of Dublin* Ecole Centrale de Paris Politecnico Universidad Madrid University of Oxford University of Birmingham
A new approach to high-performance reinforced aluminium components using fibers with predetermined orientation	Simbi SPA* Mogan Materials Alpan HF Chalmers Industriteknik Magma GMBH Icetec
MMC sheets and sheet structures made from particulate reinforced aluminium alloys	MBB* Aerospatiale Alusuisse-Lonza Raufoss RWTH Aachen IMMG
Development of advanced carbon-magnesium metal-matrix composites by applying the semi-liquid phase infiltration	ONERA* Deutsche Forschung fur Luft Und Raumfahrt Aerospatiale

Table 16.12. Recent European Community Sponsored Programs in Metal-Matrix Composites Sponsored by BRITE-EURAM (1990–1991) (*continued*)

Topic	Partners (*Prime Investigator)
Squeeze casting of light alloys and metal-matrix composites—Mechanical property evaluation	Hi-Tec Metals R&D Ltd.* Agusta SPA National Aerospace Laboratory NLR University of Southampton Airbus Industrie Raufoss A/S
Improved aluminium alloy matrix composites through microstructural control of the processing and fabrication routes	University of Strathclyde* Laboratorio Nacional de Engenharia E Tecnologia Industrial (LNETI)
Processing and microstructural modeling for development of advanced materials by rapid solidification technology	University of Sheffield* Technical University of Denmark University of Oxford Swiss Federal Institute of Technology
Cast light alloy matrix composites—Assessment of a rheocasting route	Pechiney Centre de Recherches* Instituto Nacional de Tecnica Aeroespacial Delft University of Technology
Assessment of semi-solid-state forming of aluminium metal-matrix composites	Delft University of Technology* Institut National Polytechnique de Grenoble
Fabrication and joining of graded cermets by a technique of metal infiltration	British Ceramic Research Ltd.* Hamburger Institut fuer Technologiefoerderung
Nondestructive characterization of damage in particle reinforced aluminium-matrix composites	Shell Research Arnhem* Biosonic Sarl, Famars Fraunhofer Institut, Saarbrucken Catholic University of Leuven University of Surrey
Centrifugal casting of metal-matrix composites for the production of reinforced near-net shape components	Les Bronzes D'Industrie* SGM Tekniker INPG
Shape casting of particulate reinforced aluminium alloy feedstock	AEA Technology-Harwell* Foseco International Ltd. Hydro Aluminium A/S Renault Automobiles Vereinigte Aluminium-Werke AG
Development of new boride-based cermets and ceramics	Bonastre SA* Sandvik Hard Materials Ceit
Manufacture of tools and dies using spray-forming techniques (MUST)	Sprayforming Developments Ltd.* Danish Technological Institute Universit of Edinburgh Magma GMBH Grundfos International A/S

Data from BRITE-EURAM (1991).

France. There are several active groups in both industry and academia. Overall, more MMC research is conducted here than in any other European country. Their emphasis is on the development of viable industrial systems. This research includes:

- High- and low-pressure preform infiltration processes.
- Studies of wetting, interfaces, and coatings to support infiltration process development.
- Mechanical property evaluation and modeling, as well as development of test methods.
- Studies of fiber/matrix interactions; degradation at the interface, and metallurgical effects such as aging.
- Semisolid processing routes.
- Co-spray and related technologies.
- New fiber development: SiC/SiN (Sicar) (Rhone Poulenc).
- Materials systems that include Ti reinforced with B and SiC-based fibers, and Al and Mg alloys with C, SiC, and Al_2O_3.

Germany. The predominant interest shown here is in short-fiber and particulate systems, although there is also significant research effort directed toward continuous-fiber materials for advanced aerospace projects. Joint funding schemes encourage collaboration between industry and academia:

- Piston production using infiltration and powder technology.
- Machinability, friction, and wear studies associated with automotive applications.
- Powder-based fabrication routes: process and materials optimization.
- Interface and coating studies.
- Mechanical property studies, including creep and elevated temperature performance.
- Materials of interest: Al and Mg alloys reinforced with particulate and short-fiber (SiC, Al_2O_3); Al and Ti reinforced with continuous fibers, and metallic wire reinforcement.

Great Britain. Research is conducted by industry, research associations, and universities. Significant programs are directed toward low-cost automotive and general engineering applications, as well as toward high-performance aerospace requirements.

- Mechanical testing and modeling, and thermochemical modeling for process development.
- Process development: high- and low-pressure preform infiltration, co-spray and related processes, powder metallurgy, and foundry-based processes.
- Fiber development: Al_2O_3 (ICI), SiC (BP).
- Interface characterization, matrix optimization, coating technology.
- Materials system development. Al and Mg with continuous and discontinuous reinforcement, Ti/SiC coupled with superplastic forming and diffusion bonding (SPFDB) techniques.

Netherlands/Belgium. Research includes:

- Development of GLARE, ARALL, laminate composites (AKZO).
- PM Aluminum metal-matrix composites.

Sweden/Norway. Primary research focus includes:

- Processing, microstructure, and properties of Al and Mg alloy reinforced with SiC and Al_2O_3 (short-fiber, particulate), produced by foundry and powder routes.
- Tungsten fiber (wire)-reinforced superalloys for gas turbine applications.
- Hybrid composites (Ranfoss).

United States. Research focus is on large industrial involvement, together with several university groups and research institutions.

- Large companies are involved with the development of commodity MMC based on particulate systems (Alcan, Alcoa). Many smaller companies are developing defense-driven, high-performance materials for military and aerospace applications.
- Fiber development: (Textron, DuPont, Corning, Amoco): B, SiC, C, and ceramic.
- Process development for particulate and continuous systems: foundry- and powder-based routes.
- Mechanical property studies, creep, fatigue, and elevated-temperature performance.
- Chemical synthesis techniques, intermetallic based systems (Ti_xAl_y).
- Interfacial studies, wetting, alloy optimization.
- Large-scale production casting (Alcan, Duralcan).
- Numerical modeling for performance and processing.
- Interfacial study development.

The process of materials and process developments has been traditionally spurred by the aerospace and defense industries. Their legacy has been the development of high-performance, continuous metal-matrix composites, which has served as a catalyst for the accelerated development of whisker and particulate reinforced MMC composites.

The limitations of the development of metal-matrix composites include cost, process, scientific, and developmental concerns. Research issues focus on interface understanding and control, high-performance, high-temperature fibers and property models, and fracture toughness enhancement. Developmental and economic concerns include low-cost processing and reinforcement approaches, life prediction, design properties and techniques for fabricating multiaxial composites, nondestructive evaluation, and quality control for component reliability.

Application-driven research and development of MMC reinforcements, processes, and applications is actively being pursued worldwide.

Acknowledgements

M.J.K. and S.K. would like to acknowledge ONR, Alcoa, and NAWC for program support for metal-matrix composites. We are also grateful to Textron, Ford, Lanxide, Alcan, Ceracon, and Kobe for their cooperation. J.E.A. would like to acknowledge many fruitful collaborations on MMCs with a number of employees of the Ford Motor company, with special thanks to G. Cole and J. Lasecki.

References

Alcan. 1990. Product Information. Banbury, UK.

Alcan. 1991. Product Brochure. San Diego, CA.

Allison, J.E., Sherman, A.M., and R. Bapna. 1987. *J. Metals.* 39:14.

Anton, D.L., Martin, P.L., Miracle, D.B., and R. McMeeking. 1990. Intermetallic Matrix Composites. Vol. 194. Pittsburg, PA: Materials Research Society.

Arnhold, V., and D. Mueller-Schwelling. 1991. SAE Paper 910156. Warrendale, PA: SAE.

Ashley, S. 1991. *Mech. Eng.* 44.

Bader, M.G., and M.J. Koczak. 1991. *ESNIB.* 3:18.

Benjamin, J.S., and M.J. Bamford. 1977. *Metall. Trans.* 8A:1301.

Benjamin, J.S. 1970. *Metall. Trans.* 6:2943.

Bucci, R.J., Mueller, L., Schultz, R., and J. Prohaska. 1987. Presented at the 32nd International SAMPE Conference. Anaheim, CA.

Cole, G. 1992. Unpublished research.

Dimiduk, D., Miracle, D.B., and C.H. Ward. 1991. *In:* High Temperature Intermetallics. (C. Hippsely, I. Jones, and M.J. Koczak, eds.), London, The Institute of Materials.

Dowling, W.E., Allison J.E., and A.M. Sherman. 1992. *In:* Titanium '92. Proceedings, 1992 World Titanium Congress, Warrendale, PA: TMS. In press.

Driver, D. 1989. *In:* The Materials Evolution Through the 90's. 7th International Conference. Oxford: BNF Materials Technology Center.

Ebisawa, M., Hara, T., Hayashi, T., and H. Ushio. 1991. SAE Paper 910835, Warrendale, PA: SAE.

Folgar, F., Widrig, J.E., and J.W. Hunt. 1987. SAE Paper 870406, Warrendale, PA: SAE.

Foster, D.A. 1989. *SAMPE Quarterly.* 58.

Funatani, K. 1986. *J. Jap. Soc. of Mech. Eng.* 89:241.

Geiger, A.L., and M. Jackson. 1989. *Adv. Mater. Proc.* 139: 23.

Gilman, P.S., and J.S. Benjamin. 1983. *Ann. Rev. Mater. Sci.* 13:279.

Grissafe. Japan Technology Evaluation Center. 1991. *Advanced Composites in Japan.* NTIS Report # PB 90, 215–740.

Hamajima, K., Tanaka, A., and T. Suganuma. 1990. *JSAE Review.* 11:80–84.

Harrigan, W.A. 1992. Private communication.

Harrigan, W.A. 1992. To be published.

Harris, S.J. 1990. *In:* New Light Alloys. AGARD Lecture Series. No. 174. Specialized Printing Services, Ltd.

Harris, S.J. 1988. *In:* New Light Alloys. AGARD Lecture Series. No. 444, Paper 25.

Honda Motor Co. 1991. Brochure. Tokyo, Japan.

Hoover, W. 1991. 12th Risø International Symposium. 387–392. (N. Hansen et al. eds.), Roskilde, Denmark: Risø National Laboratory.

Hughes, D. 1988. *Aviation Week and Space Technology.* November 28.

Kano, S., Koyama, M., and S. Nomura. 1989. *In:* Proceedings of 1st Japan International SAMPE Symposium. 1610. (N. Igata et al., eds.) Chiba, Japan.

Kelly, A. 1988. *In:* Cast Reinforced Metal Matrix Composites. Proceedings of ASM International Conference. (S.G. Fishman and A.K. Dhingra, eds.), Chicago.

Kimura, A., Nakamura, S., Isogawa, S., et al. 1991. SAE Paper 910425. Warrendale, PA: SAE.

Kobe Steel Materials Laboratory. 1991. Brochure.

Koczak, M.J. 1992. Materials Research in Europe. *ESNIB.* 1.

Koczak, M.J., Prewo, K., Mortensen, A., et al. 1989. ONRFE-M7.

Komatsu, Y., Arai, T., Abe, H., et al. 1991. SAE Paper 910554. Warrendale, PA: SAE.

Kubo, M., Tanaka, A., and T. Kato. 1988. *JSAE Review.* 9:56–61.

Kubozono, K., and T. Mori. 1989. *In:* Proceedings of 1st Japan International Conference. p. 387. (N. Igata et al., eds.) Chiba, Japan.

Kumar, K.S., and J.D. Whittenberger. 1991. *In:* High Temperature Intermetallics. (C. Hippsely, I. Jones, and M.J. Koczak, eds.), London: The Institute of Materials.

Lanxide. 1991. Brochure.

Metal Powder Report. 1990. Advances in Powder Consolidation at Ceracon. 45:246.

Minoda, Y. 1989. *In:* Proceedings of 6th Symposium on Basic Technologies for Future Industries: Metal and Composite Materials. 193. Tokyo, Japan.

Minoda, Y. 1986. *In:* Proceedings of Japan-U.S. CCM-III. (K. Kawata, S. Umekawa, and A. Kobayashi, eds.), Tokyo, Japan: Japan Society of Composite Materials.

Mortensen, A. 1992. Private communications.

Mortensen, A., and M.J. Koczak. 1993. *JOM* 45(3):10–18.

Nakashima, Y., and M. Miyafuji. 1989. *In:* Proceedings of 1st Japan International Conference. Chiba, Japan (N. Igata et al., eds.). p. 387.

Okuda, T., Nomura, S., Nakanishi, M. et al. 1989. *In:* Proceedings of 1st Japan International SAMPE Symposium. Chiba, Japan (N. Igata et al., eds.). p. 1616.

Packard, D.C. 1984. *SAMPE Quarterly.*

Pastor, H. 1987. *R&HM.* p. 196.

Postans, P.J. 1989. *In:* Metals Fight Back Conference and Exhibition. Shephard London Conferences.

Premkumar, M.K., Hunt, W.H., Jr., and R.R. Sawtell. 1992. *J. Metals.* 44:24.

Ritchie, I.G., and Z.L. Pan. 1991. *Metall. Trans. A.* 22A:607.

Stevenson, R.W. 1984. *ASM Metals Handbook.* 7:773.

Sunakawa, M., and A. Sakamoto. 1991. American Institute of Aeronautics and Astronautics, Inc.

Tanaka, Y. 1989. Proceedings of 1st Japan International Conference. Chiba, Japan. (N. Igata, ed.) 447.

Taniuchi, M., Takatsuka, K., Fujiwara H., and K. Korida. 1991. *Metall. Trans. A.* 22A:629.

Textron. 1991. Product Brochure. Lowell, MA.

Toyota. 1991. Toyota AXV-IV Brochure. Toyota City, Japan.

Vedula, K. 1991. *In:* High Temperature Intermetallics. (C. Hippsely, I. Jones, and M.J. Koczak, eds.). London: The Institute of Materials.

Watanabe, R., Kawasaki, A., and H. Takahashi. 1991. To be published.

Zweben, C. 1992. *J. Metals.* 44:15.

Index

Page numbers followed by t and f denote table and figure, respectively.

ABAQUS finite element code, 188
Ab-initio calculations, 102–103
Accelerated aging, in reinforced metals, 120–121, 121f
    crystal plasticity models of, 146
Acoustic emission, 62
Aerospace applications, 307–313
    aluminum-based composites for, 309–312
    reinforcement materials for, 300
Age hardening. *See* Precipitation hardening
Aging
    accelerated, in reinforced metals, 120–121, 121f
        crystal plasticity models of, 146
    and crack closure, 285, 285f
    and cyclic stress-strain response, 270, 270f
    and dislocation substructure, 271, 271f
    and fatigue-life behavior, 282, 282f
Aging characteristics, of reinforced metals, 119–136
    basics of, 119–120, 120f
    experimental methods, 126–127
    fundamental and practical significance of, 121
    general observations on, 121–127, 122t–123t
    mechanisms and theoretical models, 131–135
    thermomechanical and processing variables in, 127–131
Aging temperature
    effects of, 127–129
    peak aging time as function of, 128, 128f
Aging time
    effects of, 127–129, 128f
    as function of aging temperature, 128, 128f
    microhardness variation as function of, 120–121, 121f
Agitation devices, dry blending with, 26
Alcan Aluminum Corporation, 298, 314
Aligned reinforcements
    continuous, 159, 161–163, 162f
    discontinuous, 159–161, 160f–161f
        versus randomly oriented reinforcements, 163, 163f
    of elastic-strain hardening matrices, 165–168, 166f–167f
Alloying, mechanical, 25, 27–28
    aerospace applications of, 310, 313
Alloying additions, effect on wetting and bonding, 46–47, 47f
Alloy solidification, 15
Alumina fibers
    applications of, 3, 5, 300, 309t
    infiltration of, 5, 8, 10f
Aluminum
    coefficient of thermal contraction, 110
    punching distances, 111
    silicon carbide-reinforced, properties of, 310t

    wetting and bonding of, effect of oxygen on, 48, 48f
Aluminum alloys
    aerospace applications of, 307
    ductile failure in, 234
    particle-reinforced, 18
    precipitation hardening of, 119
Aluminum-matrix composites
    for aerospace applications, 309–312
    for automotive applications, 302–306, 306f
    bicycle frame constructed from, 320, 320f
    development of, commercial motivation for, 301f, 301–302
    for electronic packaging materials, 317
    grain refinement methods for, 31–32
    stress-strain curves of, 114, 115f
    tensile properties of, 311t
    as wear resistant materials, 315
Aluminum oxide, characteristics of, 299t
Aluminum wires, production of molten-metal drops from, 13
Anisotropic shape change, during thermal cycling, 206–207
Annealing
    of glass, 61
    static, grain refinement with, 32
Applied pressure, during solidification, 9
ARALL, 309, 319–320
Aramid, characteristics of, 299t
Asperity contact, 284–285, 285f
ASTM Standards, for hole drilling, 66–67
Atom columns
    location of, determination of, 87, 88f
    number of, determination of, 85–87, 87f
Atomic Resolution Microscope (ARM), 89
Atomistic structure
    of commensurate regions at interfaces, 93–94, 94f–97f
    of misfit dislocations, 91, 92f
Austria, research and development programs in, 322
Automobile, cost of, relative fiber and materials costs compared to, 298t
Automotive industry, applications in, 302–306
Averaging models, for polycrystals, 140
Avionics, electronic packaging materials in, 317
Axial stress, predicted evolution of, as function of time and unit cell position, 202, 203f
Axial Young's modulus, 176, 181, 182t, 182
Axisymmetric cell model
    of interfacial decohesion, 247–248, 248f
    of void growth along metal/ceramic interface, 239, 239f

Bauschinger effect, 273, 273f
Belgium, research and development programs in, 325
BF imaging. *See* Bright-field imaging
Bicycle frames, particulate reinforced aluminum, 320, 320f

328 INDEX

Bimodal grain size distribution, in isothermally rolled composite, 32, 32f
Binary systems, interfaces in, 99
Blending, in primary solid-state synthesis, 26–27, 27f
Bonding. *See also specific type*
  chemical, 82
Bonding, *continued*
  between metal and ceramics, 81–82
  physical, 82
  physicochemistry of, 45–50
Bonding energy, level of, 82
Boron, potential applications of, 309t
Boron/aluminum composites
  aerospace applications of, 309–310, 310f
  mechanical properties of, 310t
Boron carbide, wetting of, 46, 47t
Borosilicate glass sphere, matrix punched by, prismatic dislocation loops in, 125, 125f
Borsic, potential applications of, 309t
Bottom-mixing process, 10
Boundary conditions
  and calculation of residual stresses, 75
  in infiltration processes, 8
Bragg reflections, 64–65, 98
Brake rotors, 303, 303f
Bright-field imaging, 83–84, 84f
  of misfit dislocations, 91–93, 92f–93f
BRITE-EURAM, research and development programs sponsored by, 323t–324t
Brittle debonding, in presence of plastic flow, 229–231, 229f–231f
Brittle failure, of reinforcement, 233, 248
Brittle reinforcement, and aging kinetics, 127
Burgers vector, 91, 92f, 103

Cabot, 315
Capillary phenomena, 42–58
Carbides
  as wear-resistant materials, 314
  wetting of, 46, 47t
Carbon, characteristics of, 299t
Carbon-carbon composites, strength versus temperature for, 300f
Carbon fiber aluminum composite, precipitate-free zones adjacent to, 245–246, 245f–246f
Carbon fiber-reinforced polymers, strength versus temperature for, 300f
Carrier gas, injection of particles under melt with, 10
Casting. *See also specific method*
  composites produced by
    automotive application of, 303
    fatigue crack initiation in, 279–280
  continuous, 18
  cooling rate of, 110
  and microstructure, 109–110
Cast iron cylinder liners, replacement of, 304
Cauchy stress, 141
Cauchy stress tensor, Jaumann rate of, 235
Cell model problem, for calculation of residual stresses, 75
Cemented carbides, 314
  applications of, 314f, 314–315, 315t
Centrifugal casting methods, 5
Ceracon, 315
Ceramic. *See also* Metal/ceramic interfaces
  brazing alloys for, 49
  nonreactive, wetting of, 46, 47t
  three-dimensional network of, for liquid-metal infiltration, 17, 18f
Ceramic-matrix composites
  coefficient of thermal contraction, 110
  strength versus temperature for, 300f
Cermets, 314f, 315
CFAL. *See* Carbon fiber aluminum composite
Chassis applications, 302–303
Chemical bonding, 82
Chemical reactions. *See also* Interfacial reactions
  in in-situ processes, 17
  in mechanical alloying, prevention of, 27–28
  at metal/ceramic interfaces, 99–102
    experimental observations of, 100–101, 101f
      in nonreactive systems, 100–101, 101f
      in reactive systems, 101–102, 102f
    theoretical considerations, 99–100, 100f
Chemical vapor deposition, 52, 319
CKIR testing, 290
Closed-die forging process, 38
Coarse-grain microstructure, with rolling and reheating, 32
Cobalt bronze alloy, for valve applications, 315–316
Coble creep, 171
Coefficient of thermal expansion, 23
  determination of, in ideally plastic matrix composite, 186
  effects of thermal cycling on, 37–38
  of electronic packaging materials, 317–318, 318f
  mismatch of, 192t, 204
    stresses caused by, 62, 68, 75, 120, 204
Coextrusion, 24
Coincidence site lattice, 82–83, 83f
Cold deformation microstructures, 111–112
Cold-rolling texture, 114
Cold work, and precipitation kinetics, 130–131, 130f–131f
Commensurate regions, at interfaces
  ab-initio modeling of, 102–106
  atomistic structure of, 93–94, 94f–97f
Complementary diffraction investigations, 88
Compocasting, 11
Composite creep rate
  formulation for, 199
  under transverse loading, prediction of, 193–194
Composite strain, recovery of, 192
Composite yielding, influence of residual stress on, 168, 169f–171f
Compression, creep behavior in, 196
Compressive yielding, 73
Compressive yield strength, effect of CTE mismatch on, 204
Computer simulation, in quantitative HREM, 93
Concentration factor tensors, 176
Conductivity, 23
Conical particle, and average residual strain, 74, 74f
Connecting rods, 304–305, 305f
Consolidation
  foil-fiber, 24, 25f
  high-rate, 25
  powder, 23–24, 28
  in primary solid-state synthesis, 26–27, 27f
Constant scattering angle, in neutron diffraction, 76
Constant wavelength, in neutron diffraction, 76
Contact angle
  and alloying additions, 46, 47f
  experimental values of, for nonreactive metals, 45–46, 45t–47t
  and interfacial energy, relation between, 82, 82f
  measurement of, 43

Index   329

in nonreactive systems, with reactive solute additions, 50, 50f–51f
in reactive systems, 48–49
Contamination, in mechanical alloying, prevention of, 27–28
Continuous casting, 18
Continuous fiber-reinforced composites, 23, 174–190
   aerospace applications of, 309–310, 310f, 310t
   aligned, 159, 161–163, 162f
   background of, 174–175
   behavior of, under thermal cycling, 204–206
   deformation of
      elastic, 175–182
         analytic/semi-analytic methods for, 178–179
         comparison of models, 180–182, 182t
         numerical models of, 179–180, 179f–181f
         thermal residual stresses in, 184–185
      plastic, 182–184, 183f–185f
         thermal residual stresses in, 186f, 186–187
   interfacial debonding of, 187–188, 188f
   isothermal creep in, 191–194
   longitudinal properties of, derivation of, 176
   summary of, 188–189
   thermal residual stresses in, 184–187
   transverse strengthening of, 159, 161–163, 162f
Continuous fiber tows, 299t, 300
Continuum mechanics boundary value problem, calculation of residual stress from, 74, 74f
Continuum models. *See also specific model*
   of deformation, 158–190
   of infiltration processes, 6, 6f
Continuum theories, 139
Contrast transfer function, of high-resolution electron microscopy, 84–85, 86f, 89
Conventional transmission electron microscopy, 83–84, 84f
Cooling, residual stresses generated during, 68–69
Copper-matrix composites, for electronic packaging materials, 317
Copper-type rolling texture, 114
Corotational rate, of Cauchy stress tensor, 235
Correspondence Principle, 193
Corrosion, resistance to, 53
Co-spray deposition products, microstructure of, 110
Cost(s)
   and aerospace applications, 307
   of reinforcement materials, 300
Covalent bonding, 82
C&Q process, 149–151, 153
Crack(s). *See also* Fatigue crack(s)
   blunting, 253–255, 254f–255f
   closure, 284–286, 285f–286f
   deflection, by fiber/matrix sliding, 253, 254f
   growth
      in aluminum alloy-based composites, 282f, 283–284, 284f
      catastrophic, 253, 254f, 256
      and debond resistance, 226–231
      in long fiber reinforced composites, 254–256, 255f–256f
      longitudinal, 255
      under monotonic loading, 263, 263f–264f
   short, behavior of, 291, 291f
   tensile, propagation of, 255
   trapping, 287
Crack bridging, 227, 227t, 227f
   and background plasticity, 227–228, 228f
   in discontinuously reinforced composites, 264
   in long fiber reinforced composites, 256, 256f

and reinforcement, 287
strength, steady-state shielding ratio as function of, 228, 228f
Cracking
   edge, during rolling operation, 30
   interfacial, testing of, 45
   of reinforcement, 116
   surface, during forging process, 38
   during thermal cycling, 205
   thin-film, 220, 220f
Crack tip
   blunted, stress distribution around, 225, 225f
   plasticity, 224–226, 224f–227f
      constrained, 225–226, 225f–226f
   processes, 284–287
      in discontinuously reinforced composites, 263, 263f
      in long fiber reinforced composites, 253, 254f
Crack-tip fields, crystal plasticity models of, 140
Creep
   accelerated, during thermal cycling, 204
   Coble, 171
   deformation, time-dependent, by internal and external stresses, 209
   diffusional, 40
   in in-situ composites, 194, 194f
   isothermal, 191–204
      in discontinuous reinforcements, 194–204
         analytic studies of, 199–204, 200f–203f
      summary of, 210–211
   Nabarro-Herring, 171
   off-axis, 193f, 193–194
   power law, 199–201
      steady-state, 170–171
      and temperature dependence of composite strain rate, 193
   process, 28
   rate
      composite
         formulation for, 199
         under transverse loading, prediction of, 193–194
      effect of reinforcement aspect ratio on, 201
   reinforcement against, 170–173
   steady-state, 191–192
      in off-axis loading, 193, 193f
      shear lag analysis of, 199
      temperature-compensated, as function of applied stress, 195, 195f
   strain
      composite, derivation of, 192
      plastic, predicted evolution of, as function of time and unit cell position, 202, 203f
   strength, 23, 172f, 172–173
   tertiary, 193, 200, 202
   thermal cycle, effect of microstructure on, 208–209, 209f
   transverse, 193
   uniaxial, 191–193
Creeping fibers, 199–200
   with sliding interface, model of, 201
Creeping matrix, stresses from, redistribution of, 202, 202f
Creep-resistant reinforcements, 210
Crossed triangle quadrilateral, 143
Crossply configurations, creep properties of, 194
Crystalline materials
   embedded in organic composites, measurement of residual stresses with, 62–63
   HREM analysis of, 85, 87f
Crystal plasticity models, 139–157
   description of, 143–146

Crystal plasticity models, *continued*
  discussion of, 152–156
  versus phenomenological models, 154–156, 155f
  results of, 146–152
  review of, 140–141
  theory and numerics of, 140–143
CSL. *See* Coincidence site lattice
CTE. *See* Coefficient of thermal expansion
CTEM. *See* Conventional transmission electron microscopy
CTF. *See* Contrast transfer function
Cutting tool bits, 314–315
CVD. *See* Chemical vapor deposition
Cyclic deformation, of discontinuously reinforced composites, 269–273, 270f–273f, 291
Cyclic hardening/softening behavior, 269–270
Cyclic loading
  fracture during
    in discontinuously reinforced composites, 264
    in long fiber reinforced composites, 255–256, 256f
  and plastic deformation, 271–272
Cylinder liners
  cast iron, replacement of, 304
  metal-matrix composite, 3, 301f, 301–302, 304, 304f
Cylindrical particles
  aligned, and strengthening ratio, 159–161, 161f
  and average residual strain, 74, 74f

Damping composites, 319–320
D'Arcy's law, 7
Dark-field imaging, 83–84, 84f
  of dislocations, in whisker- and particle-containing materials, 109f, 115f–116f
Debond driving force, 218–220, 219f–220f
Debonded interface, creep properties of, 200–201, 201f
Debonding, 52–53, 217–232, 233
  brittle, in presence of plastic flow, 229–231, 229f–231f
  by fiber/matrix sliding, 253, 254f
  interfacial, 248
    of continuous fiber-reinforced composites, 187–188, 188f
    during thermal cycling, 205
  on whisker filament, during tensile deformation, 245, 245f
Debond resistance, 220–221
  growing cracks and, 226–231
  inelastic processes accompanying, 220, 221f
  measurement of, 221, 222f
Decohesion front, in network of pre-existing dislocations, 229, 229f
Defect structure, of interface, 103, 104f
Defense-related engine applications, 312–313
Deformation. *See also* Elastic deformation; Plastic deformation
  cold, 111–112
  continuum models for, 158–190. *See also specific model*
  creep, time-dependent, by internal and external stresses, 209
  cyclic, of discontinuously reinforced composites, 269–273, 270f–273f, 291
  hot, 112
  inelastic, in interface fracture, 218
  isothermal, in polycrystalline composites, 148, 149f–150f
  processing and post-processing, in polycrystalline particulate composites, 149–152, 150f–151f
  and recrystallization, 112–114, 113f
  superplastic, 34–38, 35f–37f
  theory, infinitesimal, 72
  under thermal cycling, 208–209

Deformation gradient, 140–141
Deformation processing, 28–40
Deformation zones, around whiskers and particles, 112f, 112–114, 116
Denmark, research and development programs in, 322
Deposition
  chemical vapor, 52
  co-spray, and microstructure, 110
  processes, 24
  sputter, 52
  transport phenomena after, 14–15
DF imaging. *See* Dark-field imaging
Diamond-coated drilling tools, 136f, 315
Die(s), 314
Die-casting, particle-reinforced aluminum alloy ingots for, 18
Diesel engines, metal-matrix composite pistons for, 3, 304–305, 305f
Differential scanning calorimetry, determination of aging characteristics with, 121, 122t–123t, 126–127, 131
  in cold rolling, 130–131, 131f
Diffraction, 62–66, 64t
  electron, of deformation structures near shear band interfaces, 140
  investigations, complementary, 88
  neutron, 62, 65–66, 66f, 76
    measurement of plastic deformation with, 70, 70f–71f
    measurement of thermal stresses with, 68–69, 69f, 117
  x-ray, 62, 64–65
Diffraction contrast, with misfit dislocations, 92
Diffusion, 116
Diffusional relaxation, of reinforcement, 171–173, 172f
Diffusion barrier coating, of reinforcements, 52
Diffusion bonding, 24–25, 25f–26f, 40
  in metal/ceramic interfaces, 100–102, 101f–102f
DIMOX process, 15, 315
Direct lattice imaging, 85–87, 87f, 90, 90f
Direct metal oxidation, 310
Discontinuously reinforced aluminum composites
  automotive applications of, 302–305, 303f, 305f
  large-scale production facility for, 306
Discontinuously reinforced metal-matrix composites, 158–173
  aerospace applications of, 310–312
  aligned, 159–161, 160f–161f
    versus randomly oriented reinforcements, 163, 163f
  behavior of, under thermal cycling, 204, 206–210
  cyclic deformation of, 269–273, 270f–273f, 291
  deformation processing of, 28–40
  ductile failure in, 234–240
  fatigue behavior of, 269–294
  fracture of, 258–264
    during cyclic loading, 264
    damage mechanics, 262–264
    summary of, 264–265
    tensile, 258–265
  isothermal creep in, 194–204, 210
    analytic studies of, 199–204, 200f–203f
  primary solid-state processing of, 26–28, 27f
Disc-shaped particles, as strengthening agents, 159–160, 160f–161f
Dislocation(s)
  core, 91, 92f
  density
    and CTE mismatch, 204
    enhanced, 124–126, 124f–126f, 130
    measurement of, 111, 116

precipitate nucleation as function of, 132, 133f
  and growth rate, 134f, 134–135
with differential thermal contraction, 110f, 110–111
emission, at fiber ends, 126
glide, 116
helical, 124, 124f
misfit, 90–93, 91f–92f
motion, barriers to, 116
in particle- and whisker-containing material, 114–116, 115f–116f
and particles, attractive force between, 196
pre-existing, network of, decohesion front in, 229, 229f
prismatic loops, 125f, 125–126
substructure, after cyclic deformation, 271, 271f
Dislocation-climb mechanism, 196
Dislocation creep mechanism maps, 196
Dispersion alloys, nickel-based, for aerospace applications, 313
Dispersion processes, 3, 9–12
  description of, 9–11, 10f
  free energy change in, 43
  governing phenomena, 11–12
  optimization of, 17–18
DRA composites. *See* Discontinuously reinforced aluminum composites
Drainage/imbibition curves, 44–45
Drilling, hole, 66–68, 67f, 77–78
Drilling tools, 136f, 315
Driveshafts, 302–303, 303f
DRMMCs. *See* Discontinuously reinforced metal-matrix composites
Dry blending, 26
DSC. *See* Differential scanning calorimetry
Ductile failure, 233–240
Ductile matrix, crack blunting in, 253–254, 254f
Ductility
  effect of reinforcement shape on, 235–236, 236f, 259–260
  measurements, of fiber-reinforced composites and unreinforced matrix alloys, 252t
Duralcan aluminum-matrix composite, 320, 320f
Duralcan USA, 306

Edge cracking, during rolling operation, 30
EDS. *See* Energy dispersive spectroscopy
E-Glass, characteristics of, 299t
Eigenvalue problem, 221–224
Elastic cell, 229–230, 230f
Elastic constants, 67, 175
Elastic deformation
  of continuous fiber-reinforced composites, 175–182
    analytic/semi-analytic methods for, 178–179
    comparison of models, 180–182, 182t
    numerical models of, 179–180, 179f–181f
    thermal residual stresses in, 184–185
  and particle fracture, 241–242, 242f
Elasticity coefficients, 77
Elastic moduli, tensor of, 72, 74
Elastic strain, 72
  residual, with CTE mismatch, 204
Elastic-strain hardening matrices, aligned reinforcement of, 165–168, 166f–167f
Elastic stress analysis, of complex stress intensity factor, 223–224
Elastic unloading solution, 73

Electrical applications, 316–318
Electrical conductivity measurements, determination of aging characteristics with, 121, 122t–123t, 127
Electromagnetic body forces, infiltration with, 5, 18
Electron back-scattering, 116
Electron diffraction studies, of deformation structures near shear band interfaces, 140
Electronic applications, 316–318
Electronic packaging materials, 298, 317–318, 318f–319f
Electron microscopy. *See* High-resolution electron microscopy; Scanning electron microscopy; Transmission electron microscopy
Elevated temperature metal-matrix applications, in aerospace industry, 312f, 312–313
Ellipsoidal inclusion, 125, 126f
Ellipsoidal particles
  aligned, and strengthening ratio, 159–160, 160f
  and average residual strain, 74, 74f
Elongated particles, as strengthening agents, 159–161, 160f–161f
Energy balance model, of interface fractures, 218–221, 219f–222f
Energy dispersive spectroscopy, 84, 84f
Energy loss mechanisms, in composite fabrication processes, 44
Energy loss near edge structure (ELNES), 84
Engine applications, 304–306
Enthalpy, analysis of, 126–127
EPX series, 317
Equiaxed particles, aligned reinforcement by, and strengthening ratio, 166–168, 167f
Eshelby's equivalent inclusion method, 125–126, 177–178, 201–202, 241
European research and development programs, 322–325, 323t–324t
Eutectic systems, 15
Extended energy loss fine structure studies (EXELFS), 84
Extrusion, 29, 29f–30f
  and microstructure, 109–110, 112

Fabrication techniques, classification of, 3
Failure, 234–240
  basic mechanisms of, 233–234, 248
  brittle, of reinforcement, 233, 248
  of discontinuously reinforced composites, damage mechanics of, 262–264
  interfacial, 243–248
    microstructural observations of, 243–246
    modeling of, 247–248, 248f
  micromechanics of, 233–250
  in polycrystalline composites, 152–154, 153f
Far-field uniform stress, 175
Fatigue behavior
  of discontinuously reinforced composites, 269–294
    and process improvements, 273–274, 274f
    and reinforcement characteristics, 280–282, 281f–282f
    versus strain, 276–277, 277f–278f
    versus stress, 274–276, 275f–276f
    summary of, 291–292
  phenomenology of, 274–279
Fatigue crack(s)
  closure, 284–286, 285f–286f
  growth of, 255, 256f, 282–291
    early, 273–282
    effect of particle size and volume fraction on, 264

Fatigue crack(s), *continued*
  growth of, *continued*
    midrange rates of, 290–291
    overview of, 282f, 282–284, 284f
    initiation of, 273–282
    mechanisms of, 279f, 279–280
FGM composite program. *See* Functionally graded materials composite program
Fiber damage, during thermal cycling, 205
Fiber distribution, and plastic response, 182–184, 183f–185f
  effect of thermal residual stresses on, 186f, 187
Fiber flanks, geometric imperfections on, 246, 246f
Fiber fracture, single, 253–254
Fiber geometry, and interfacial failure, 244–245
Fiber-matrix interface, fracture of, 257–258, 258f, 265
Fiber packing, and transverse tension stress-strain response, in continuous fiber composite, 188, 188f
Fiber packing models, for continuous fiber composites, 178–180, 179f–180f, 182t, 183
Fiber-reinforced composites. *See also* Continuous fiber-reinforced composites; Long fiber reinforced metal-matrix composites
  deformation processing of, 39, 39f
  fracture properties of, 251, 252t
Fiber-reinforcement materials, characteristics of, 299t
Finite element analysis, 140–141, 143
  of isothermal creep, 202–204f203f
Finite element meshes, deformed, 248, 248f
First-order interaction parameter, 48–49
Flame spraying, of aluminum wires, 13
Flow strength
  of composites with nonhardening matrices, 159–164
  and plastic constraint, 184
  reduction in, after tension deformation, 273
Flow stress, and microstructure, 116–117
Fluid flow
  in dispersion processes, 11–12
  in infiltration processes, 6–7
Foams, for preform fabrication, 17, 18f
Foil-fiber consolidation process, 24, 25f
Foil processing, 34, 35f
Forging, 38–39, 38f–39f
  and microstructure, 109–110
Forscheimer equation, 6–7
Fourier filtering technique, 93
Fourier transformation, 84
Fracture
  crystal plasticity models of, 140
  interface, 233
    mechanics of, 217–218
    metal/ceramic, models for, 217–232
  particle, 240–243
    experimental observations of, 240–241, 241f–242f
    modeling of, 241–243, 242f–243f
  resistance, 226–231
  single fiber, 253–254
  strengths, of fiber reinforced composites, 253, 253t
  surface
    of long fiber reinforced composites, 252, 253f
    roughness of, and crack closure, 286
  tensile
    in discontinuously reinforced composites, 258–265
    in long fiber reinforced composites, 252–255, 254f–255f, 258
    patterns of void initiation in, 244, 244f

Fracture behavior, 251–268
  of discontinuously reinforced composites, 258–264
  of long fiber reinforced composites, 252–258
  summary of, 264–265
Fracture resistance curve, 231, 231f
  resulting from background plasticity shielding, 228, 228f
France, research and development programs in, 324
Free energy change, 43, 45, 49
Free-machining titanium alloy connecting rods, 305
Fresnel diffraction, 246
Friction stress, 179, 199
Functionally graded materials composite program, 319

Gas
  carrier, injection of particles under melt with, 10
  reaction of molten metal with, to form in-situ composites, 15
Gas turbine engine
  materials, trends in, 307, 307f
  performance, improvement of, 312t
Geometric interface model, 82–83
Geometric softening, 140–141, 153
Germany, research and development programs in, 325
Gibbs free-energy change, 45, 49
Gibbs' phase rule, 99
GIRI, 319
GIXS. *See* Gracing incidence x-ray scattering
GLARE, 309, 319–320
Glass-ceramic matrix composites, strength versus temperature for, 300f
Glass-matrix composites, strength versus temperature for, 300f
Glass sphere, matrix punched by, prismatic dislocation loops in, 125, 125f
Golf heads, 320
GP zones. *See* Guinier-Preston zones
Gracing incidence x-ray scattering, 95–99
  description of, 95
  of metal/ceramic interfaces
    results of, 96t, 96–99, 98f–99f
    techniques for, 95–96
Graded structures, 319
Grain boundaries
  formation of, 116
  high-angle, 111, 113
  sliding of, during thermal cycling, 205
Grain coarsening, during rolling and reheating, 32
Grain interaction, constraint caused by, 153
Grain refinement, 31–34, 32f, 40
Grain size
  bimodal distribution of, in isothermally rolled composite, 32, 32f
  and flow stress, calculation of, 116
  and silicon carbide particle content, 33, 33f
Grain structure, factors affecting, 40
Graphite
  characteristics of, 299t
  potential applications of, 309t
Graphite-aluminum, 23
  aerospace applications of, 309–310
Great Britain, research and development programs in, 325
Griffith energy, 226, 230
Ground vehicle applications, 297–326, 298t
Guinier-Preston zones, 119–120, 126
  formation of, 128
Gurson model, of matrix failure, 234–235

Hardening index, computed dependence of reference stress on, 166, 166f
Hardening matrices, elastic-strain, aligned reinforcement of, 165–168, 166f–167f
Hardening/softening behavior, cyclic, 269–270
Hardmetals, 314, 314f
Hardness advancement, 152
Hardness increments, 148, 152
Hardness techniques. *See also* Macrohardness; Microhardness
  for estimating aging response, 122t–123t, 127
Hard reinforcing particles, and hot deformation structures, 112
Hard-to-deform metals, high-rate consolidation of, 25
Hashin-Shtrikman bounds, 176–177, 182t
Heat transfer
  in infiltration processes, 7
  in spray processes, 14
Heat treatment
  stresses caused by, 62
  studies using, errors in, 121
Helical dislocations, 124, 124f
Helicopters, 307
Helmholtz potential energy, 142
Heteropolar bonding, 82
Hexagonal packing model, 178
High damping materials (HIDAMATS), 319
High Performance Turbine Engine Technologies, 312–313
High-rate consolidation, 25
High-resolution electron microscopy, 83–95
  contrast transfer function of, 84–85, 86f
  description of, 83
  of dislocation density, 125
  of geometric imperfections on fiber flanks, 246, 246f
  lattice imaging by, 85–87, 87f, 90, 90f
  limitations of, 94–95, 96f
  of misfit dislocations, 90–91, 91f
  quantitative, 83–88
    atomistic configurations obtained by, 93–94
    flow chart for, 87, 88f
  resolution of, 85–87, 87f
  structural studies by, examples of, 88–90, 102–106, 104f–106f
High-temperature matrix systems, aerospace applications of, 312f, 312–313
High voltage electro-microscopy, dislocation density measurement with, 111, 116
Hill connections, 177
Hill's self-consistent method, for continuous fiber composites, 177–178, 182t
HIP. *See* Hot isostatic pressing
Hole drilling, 66–68, 67f, 77–78
Homopolar bonding, 82
Honda Prelude engine, with MMC cylinder liners, 3, 304, 340f
Hot deformation, 112
Hot isostatic pressing, 28
Hot pressing, 28, 245–246
  liquid-phase, 18
HPTET. *See* High Performance Turbine Engine Technologies
HREM. *See* High-resolution electron microscopy
HVEM. *See* High voltage electro-microscopy
Hydridization, for preform fabrication, 17
Hydrostatic extrusion, 29
Hydrostatic pressure, and stress-strain curve of P/M composite, 36–37, 37f
Hydrostatic stresses, 72–73, 152, 153f, 154
  distribution of, and matrix failure, 236, 237f
  predicted evolution of, as function of time and unit cell position, 202, 203f

IACS. *See* International Annealed Copper Standard
Ideal microstructures, 139
Inclusion, ellipsoidal, 125, 126f
Inclusion aspect ratio, punching distance variation as function of, 125, 126f
Industrial applications, 313–320
Inelastic deformation, in interface fracture, 218
Inelastic processes, accompanying debond, 220, 221f
Infiltration processes, 3–9
  advantages and disadvantages of, 5
  description of, 3–5, 4f
  free energy change in, 43
  interfacial bond promotion in, 51–52
  and microstructure, 109–110
  models of, 5–6, 6f
  with no external force (spontaneous), 4f, 4–5
  optimization of, 17, 18f
  preform, 310
  pressure-driven, 4f, 5
  solidification in, 8–9
  transport phenomena in, 6–8
  vacuum-driven, 5
  wetting during, 51
Infinitesimal deformation theory, 72
In-plan radial scans, 98, 98f
In-situ composites, 15
  aging of, 16–17
  behavior of, under thermal cycling, 204–206
  creep properties of, 194, 194f
  reinforcement in, 17
In situ plastic properties, of matrix material, 158
In-situ processes, 3, 15–17
  advantages and disadvantages of, 15
  chemical reactions in, 17
  description of, 15, 16f
  governing phenomena, 15–16
  and microstructure, 109–110
  optimization of, 18
  solidification in, 16–17
Interdiffusion zone, 218
Interface(s). *See also* Metal/ceramic interfaces
  behavior, prediction of, 63
  commensurate regions at
    ab-initio modeling of, 102–106
    atomistic structure of, 93–94, 94f–97f
  decohesion, 116
  defect structure of, 103, 104f
  energies, 82, 82f
  engineering, 50–53
  fiber-matrix, fracture of, 257–258, 258f, 265
  formation
    governing parameters of, 42–45
    mechanics of, 44f, 44–45
    thermodynamics of, 42–44
  fracture, 233
    mechanics of, 217–218
  geometric model of, 82–83
  tailoring
    for composite performance, 52–53
    for processing, 50–52

Interface(s), *continued*
  work-hardened, strengthening effect of, 201
Interfacial bonding, 42–58
  enhancement of, 51–52
  forms of, 42–43
  mechanical features of, 44f, 44–45
  summary of, 53–54
  testing of, 52–53
  and transverse tension stress-strain response, in continuous fiber composite, 188, 188f
Interfacial debonding, 248
  of continuous fiber-reinforced composites, 187–188, 188f
  during thermal cycling, 205
Interfacial failure, 243–248
  microstructural observations of, 243–246
  modeling of, 247–248, 248f
  reinforcement and, 233–250
Interfacial reactions, 43–44, 52
  control of, 52
  and reactive wetting, 48–49
  studies of, in selected composite systems, 56–58
Interfacial zone, with creep characteristics, 192–193
Interfiber spacing, in infiltration processes, 8–9, 10f
Intermetallics, 300
Internal oxidation process, metal/ceramic interfaces produced by, 88
Internal stresses
  and flow stress, 116
  superplasticity of, 197
  during temperature changes, 204
International Annealed Copper Standard, 127
International Nickel Company, 313
Intrinsic threshold, and near-threshold behavior, 290
Investment casting, particle-reinforced aluminum alloy ingots for, 18
Isothermal creep, 191–204
  summary of, 210–211
Isothermal deformations, in polycrystalline composites, 148, 149f–150f
Isothermal rolling, 30–31, 31f–32f

Jacobian determinant, of deformation gradient, 141
Japan, research and development programs in, 320–322, 321t–322t
Jaumann rate, of Cauchy stress tensor, 235
J_2 flow theory, versus crystal plasticity theory, 154–156, 155f
J-integral, 219
Jisedai project, goal of FRM in, 299t

Kirchhoff stress, 141, 153f
Kobe Steel Ltd., 314, 320
Kovar, 317
Kronecker delta, 74

Lagrangian strain, 142
Laminated composites, 24
  residual stresses in, 62
Lanxide Corporation, 4, 298, 315, 317, 319f
Laplace transformation, 193
Large-diameter fibers, applications of, 300
Large-scale contact, of interface crack faces, 223
Large strain behavior, 166
Laser holography, 68
Laser melting, 18
Lattice imaging, direct, 85–87, 87f, 90, 90f

Lattice plane spacings, incommensurate, 93, 94f
Lattice reorientation
  and composite strengthening, 152–153
  in shear bands, 140
Lattice strain, measurement of, 76
Levy-Von Mises equations, 208–209
LFMMC. *See* Long fiber reinforced metal-matrix composites
Limit flow strength, and nonuniform spatial distribution, 164
Limiting strains, during forging, 38–39, 38f–39f
Limit-yield stress, for aligned reinforcements, 159, 161
Liquid metal, viscosity of, 11–12, 12f
Liquid-metal infiltration, 4
Liquid-state processing, 3–22
  development of, 18
  energy loss mechanisms in, 44
  methods of, 3–17
    classification of, 3
  and microstructure, 109–110
  optimization of, 17–18
  summary of, 18–19
Loading
  cyclic
    fracture during
      in discontinuously reinforced composites, 264
      in long fiber reinforced composites, 255–256, 256f
    and plastic deformation, 271–272
  monotonic
    crack propagation under, 263, 263f–264f
    and plastic deformation, 271–272
  stresses due to, 72–73
  transverse, matrix stresses and composite creep rate under, prediction of, 193–194
Load ratio effects, and near-threshold behavior, 290
Long fiber reinforced metal-matrix composites, fracture of, 252–258
  during cyclic loading, 255–256, 256f
  interfacial, 257–258, 258f, 258t
  summary of, 264–265
  surfaces in, 252, 253f
  tensile, 252–255
Lorentz polarization, 64

Macrohardness, measurement of, 121, 122t–123t, 127
Macroscopic shape change, during thermal cycling, 205–206, 206f
Magnesium-matrix composites, as wear resistant materials, 315
Magnetic measurements, of residual stresses, 62
Martin Marietta Laboratories, 15
Mass transfer, in infiltration processes, 7
Mass transport
  by diffusion, 171–173
  sliding without, and creep strength, 172, 172f
Matrix alloys, mechanical and physical properties of, 191, 192t
MBE grown Nb films on sapphire (α-Al_2O_3)
  gracing incidence x-ray scattering of, 95–99, 96t, 98f–99f
  high-resolution electron microscopy of, 88–90
Mean-field theory approximations, calculation of residual stresses with, 75
Mechanical alloying, 25, 27–28
  aerospace applications of, 310, 313
Mechanical interfacial bond promotion, 51–52
Mechanical phenomena, 42
  and interfacial zone formation, 44, 44f
Mechanical properties
  and CTE mismatch, 204

after superplastic forming, 37
    and thermal cycling, 206
Melting, production of molten-metal drops by, 13
Metal/ceramic interfaces, 81–108
    case study of, 102–106
    chemical processes at, 99–102
        experimental observations of
            in nonreactive systems, 100–101, 101f
            in reactive systems, 101–102, 102f
        theoretical considerations, 99–100, 100f
    fracture of, 257–258, 258f, 265
    models for, 217–232
    fundamentals of, 81–83
    structure determination
        by gracing incidence x-ray scattering, 95–99
        by HREM, 83–95
        example of, 88–90
Metal cohesive energy, experimental values of, for nonreactive metals, 45–46, 45t–47t
Metal foil, bonded between ceramic substrates, subjected to stress intensity factor, 226, 226f
Metallic wires, 299t, 300
Metal-matrix composites
    applications of, 297–326, 298t–299t
        in aerospace industry, 307–313
        in automotive industry, 302–306
        industrial, 313–320
        summary of, 320–325, 321t–324t
    benefits of, 23
    compared to monolithic materials, 299t
    processing of, 3. See also specific process
    strength versus modulus for, 300, 301f
    strength versus temperature for, 300f
    tensile properties of, 311t
Metal-reinforcement interface, importance of, 40
Microbands, 111–112
Microhardness
    measurement of
        aging response represented by, 121, 122t–123t, 127, 131
        as function of aging time, 128–129, 128f–129f
        variation of, as function of aging time, 120–121, 121f, 127–128, 128f
Micromechanical models
    of continuous fiber composites, 178
    of isothermal creep, 199–200
    of thermal cycling, 209
Micromechanics, of failure processes, 233–250
Microstrain effects, 61
Microstructure
    cold deformation, 111–112
    and crack closure, 285
    and cyclic stress-strain response, 270, 270f
    damage to, during thermal cycling, 205
    and differential thermal contraction, 110f, 110–111
    and fatigue life behavior, 280–282, 281f–282f
    flow stress and, 116–117
    hot deformation, 112
    ideal, 139
    inhomogeneities in, 111–112
    and interfacial failure, 243–246
    of interfacial zone, 44, 44f
    and near-threshold behavior, 289f, 289–290
    and processing route, 109–110
    recrystallization, 112–114, 113f

and reinforcement content, 129, 129f
and thermal cycle creep, 208–209, 209f
after thermomechanical processing, 33f, 33–34, 36, 36f
of whisker- and particle-containing materials, 109–118
Midrange crack growth rates, 290–291
Minimum potential energy, 176
Mises stress, macroscopic, 234
Mises yield criterion, 72, 74, 159, 182
Misfit dislocations, 90–93, 91f–93f
MMCs. See Metal-matrix composites
Mode mixity, 221–224
Modulus, versus strength, for metal-matrix composites, 300, 301f
Moiré contrast, in misfit dislocations, 91–93, 92f
Moiré fringe displacement measurement, 62, 68
Molybdenum
    characteristics of, 299t
    potential applications of, 309t
Monochromatic neutrons, 65
Monofilaments, 299t, 300
Monolithic materials, compared to composites, 299t
Monotapes
    development of, net shape parts fabricated by, 40
    produced by deposition processes, hollow turbine blades fabricated with, 24–25, 27f
Monotonic loading
    crack propagation under, 263, 263f–264f
    and plastic deformation, 271–272
Motorcycles, DRA pistons for, 305
Multiaxial stress state, components of, 193
Multifilament supercooling wire, in silver matrix, 317f

Nabarro-Herring creep, 171
National aerospace planes (NASP), 312
National Center for Electron Microscopy (NCEM), 84, 86f, 89
National Synchrotron Light Source, 95–96
Near-threshold behavior, 288–290
    effect of particle size and volume fraction on, 288–289, 288f–289f
    and load ratio effects, 290
    and microstructure, 289f, 289–290
Netherlands, research and development programs in, 325
Neutron(s)
    monochromatic, 65
    penetration depth of, for difference materials, 64t
    polychromatic, 65
Neutron diffraction, 62, 65–66, 66f, 76
    measurement of plastic deformation with, 70, 70f–71f
    measurement of thermal stresses with, 68–69, 69f, 117
Nicalon, 300, 320
Nickel-based alloys, 4
    for aerospace applications, 313
Nippon Kokan, 319
Nippon Steel, 319
Nitrides, as wear resistant materials, 314
Nonhardening matrices, composites with, flow strength of, 159–164
Nonreactive systems
    metal/ceramic interfaces in, 100–101
    reactive solute additions to, 49–50, 50f–51f
    wetting and bonding in, 45–48
Nonuniformity
    in continuous fiber-reinforced metals, 180, 180f
    effect of, 164, 165f

Norway, research and development programs in, 325
Nucleation
  dislocation
    with differential thermal contraction, 110, 110f
    in particle- and whisker-containing material, 114–116, 115f–116f
      accelerated aging caused by, 130
    of Guinier-Preston zones, 128
    particle-stimulated, 113
      weakening of recrystallization texture by, 114
    precipitate, as function of dislocation density, 132, 133f
    and growth rate, 134f, 134–135
    recrystallization, 112–114, 113f

ODS copper composites, 316
Off-axis creep, 193f, 193–194
Orowan bowing, 196
Orowan equation, 116
Orthotropic materials, measurement of residual stresses with, 77–78
Osprey process, 13, 310
Overaging, grain refinement with, 31
Oxidation process, internal, metal/ceramic interfaces produced by, 88
Oxide dispersion strengthening, 316
Oxide reinforcements, 45–46
  wetting and bonding of, influence of oxygen on, 48, 48f
Oxygen, influence of, on wetting and bonding, 47–48, 48f

Parent bore blocks, 304
Particle(s). *See also specific type*
  deformation zones around, 112, 112f
  and dislocations, attractive force between, 196
  fracture, 240–243
    experimental observations of, 240–241, 241f–242f
    modeling of, 241–243, 242f–243f
    and reinforcement, 287, 288f
  incorporation of, 11
  macroscopic effect of, 112
  migration of, 12
  shape
    and average residual strain, 74, 74f
    and cyclic deformation, 272–273
    and fatigue life behavior, 281–282
    and strengthening ratio, 159–161, 160f–161f, 166–168, 167f
  size
    and crack closure levels, 286, 286f
    and creep strength, 172f, 173
    and cyclic deformation, 272–273
    and fatigue crack growth, 264
    and fatigue life behavior, 280–281, 281f
    and near-threshold behavior, 288–289, 288f–289f
  spacing, and flow stress, 116
  spherical
    nonuniform distribution of, 164
    and stress-strain curve, 168, 169f
  suspended in liquid, behavior of, 11–12
  textural weakening caused by, 114
Particle-reinforced composites, 17–19
  microstructural evolution in, 109–118
  stress-strain behavior of, 114–116, 115f–116f
  toughness measurements of, 252t
Particle-stimulated nucleation, 113
  weakening of recrystallization texture by, 114

Particulates, and semisolid metal, mixing of, 11
Particulate systems, 299t
  applications of, 300, 313
Performance, of composite
  and aerospace applications, 307–308, 308f
  interface tailoring for, 52–53
Periodic hexagonal array (PHA) model, 178–179, 179f, 182t, 183–184, 187–188
Periodic square array, cell models based on, 178–179, 180f
Petch-Hall relationship, 116
PFZs. *See* Precipitate-free zone
Phase alignment, and aniostropic distribution, 109
Phase transformations, stresses caused by, 62
Phenomenological models
  of creep properties in discontinuous reinforcements, 196
  versus crystal plasticity models, 154–156, 155f
  of debond resistance, 220f, 221
  of interface fracture mechanics, 217–218, 249
Phenomenology, of fatigue behavior, 274–279
Physical bonding, 82
Physical modeling approach, for metal-matrix composite analysis, 139–140. *See also* Crystal plasticity models
Physical vapor deposition, 319
Physicochemistry, of wetting and bonding, 45–50
Pinning effect theory, 33
Piola-Kirchoff stress, 142
Pistons
  metal-matrix composite, 3, 304–305, 305f
  zero clearance, 305
Plasma torches, production of molten-metal drops by, 13
Plastic constraint, 152
  and composite toughening, 255
  and flow strength and strain hardening, 184
  and void growth, 239, 239f
Plastic deformation, 139
  of continuous fiber-reinforced composites, 182–184, 183f–185f
    thermal residual stresses in, 186f, 186–187
  localized, and cyclic and monotonic loading, 271–272
  and particle fracture, 242–243, 243f
  residual stresses caused by, 62–63, 69–71, 70f–71f
Plastic flow
  brittle debonding in presence of, 229–231, 229f–231f
  and calculation of residual stresses, 75
  and composite strengthening, 152
Plasticity
  background, and crack-bridging mechanisms, 227–228, 228f
  crack tip, 224–226, 224f–227f
    constrained, 225–226, 225f–226f
  matrix, and composite toughening, 255
Plastic relaxation, of thermal residual stresses, 124–125, 130
Plastic strain
  effective, 236, 237f–238f
  rate of, 74, 235
  total, 204
Plastic strain-life behavior, 278f, 278–279
Plastic zone, 72–73
  size of, 224–225
PM. *See* Powder metallurgy
P/M composites, forging limits of, 38, 38f
P/M process, 34–36, 35f
Poisson ratio, 177
Polychromatic neutrons, 65

Polycrystal(s), averaging models for, 140
Polycrystalline particulate composites, 144–146, 145f
   internal stresses in, during temperature changes, 204
   isothermal deformations in, 148, 149f–150f
   versus phenomenological models, 154, 155f
   processing and post-processing deformations in, 149–152, 150f–151f
   strengthening and failure in, 152–154, 153f
Porosity, of matrix produced by liquid-state processing, 18
Post-processing deformation, in polycrystalline particulate composites, 149–152, 150f–151f
Powder consolidation, 23–24, 28
Powder diffractometers, 65
Powder metallurgy, 23–28, 245
   composites produced by
     aerospace applications of, 310–311
     imperfections in, 280
Power law creep, 191, 199–201
   steady-state, 170–171
   and temperature dependence of composite strain rate, 193
Precipitate-free zones, 245f, 245–246
Precipitate nucleation, as function of dislocation density, 132, 133f
   and growth rate, 134f, 134–135
Precipitate phases, in interfacial failure, 245, 245f
Precipitation-hardenable alloys, 119
Precipitation hardening, 119–120, 120f
   size effect in, 158
Preform fabrication, optimization of, 17, 18f
Preform infiltration, 310
Pressure casting, 18
Pressure-driven infiltration, 4f, 5
Pressure transmitting medium, 315
Prestrain, imposed, 73, 73f
PRIMEX process, 4–5, 51, 315, 317
Prismatic dislocation loops, 125f, 125–126
Process improvements, and fatigue life behavior, 273–274, 274f
Processing deformations, in polycrystalline particulate composites, 149–152, 150f–151f
Processing route. *See also specific route*
   and microstructure, 109–110
   significance of, in crystal plasticity models, 139–140
Processing variables, in aging response, 127–131
Production capacity, large-scale, 298, 306
Progressively cavitating material model, of matrix failure, 234–235
Pseudoplastic behavior, of composite slurries, 12
PTM. *See* Pressure transmitting medium
Punching distances, 110–111
   variation of, as function of inclusion aspect ratio, 125, 126f
PVD. *See* Physical vapor deposition

Q process, 149–150
Quantitative HREM, 83–88
   atomistic configurations obtained by, 93–94
   flow chart for, 87, 88f
Quenching, stresses caused by, 62

Raman spectroscopy, measurement of residual stresses with, 62
Ramberg-Osgood relation, 165–168, 167f
Randomly oriented reinforcements, versus aligned discontinuous reinforcements, 163, 163f

Randomly placed fiber model, 180, 181f
Random texture components, 114
Rapidly solidified metals, high-rate consolidation of, 25
Rate tangent algorithm, 142–143
Reactive solute additions, to nonreactive systems, 49–50, 50f–51f
Reactive systems
   metal/ceramic interfaces in, 100–101
   wetting and bonding in, 48–50
Reactivity, 42–58
Reciprocal coincidence density, 82
Recovery processes, during thermal cycling, 210, 210f
Recrystallization, 112–114, 113f
   thermomechanical processing for, 31–34
Recrystallization texture, 114
Recrystallized materials, stress-strain curves of, 114, 115f
Rectangular unit cell, for periodic hexagonal array model, 179, 179f
Reinforced metals
   accelerated aging in, 120–121, 121f
   aging characteristics of, 119–136
     basics of, 119–120, 120f
     experimental methods, 126–127
     fundamental and practical significance of, 121
     general observations on, 121–127, 122t–123t
     mechanisms and theoretical models, 131–135
     thermomechanical and processing variables in, 127–131
   behavior of, during deformation. *See* Crystal plasticity models
   enhanced dislocation density in, 124–126, 124f–126f
Reinforcement(s). *See also specific type*
   advantages of, 23
   aligned
     continuous, 159, 161–163, 162f
     discontinuous, 159–161, 160f–161f
       versus randomly oriented reinforcements, 163, 163f
     of elastic-strain hardening matrices, 165–168, 166f–167f
   aspect ratio, and creep rate, 201
   brittle, and aging kinetics, 127
   continuous. *See* Continuous fiber-reinforced composites
   and crack closure, 286, 286f
   cracking of, 116
   and crack trapping/bridging, 287
   creep-resistant, 170–173, 210
   diffusional relaxation of, 171–173, 172f
   discontinuous. *See* Discontinuously reinforced metal-matrix composites
   inert diffusion barrier coating of, 52
   in infiltration processes, 8, 9f
   and interfacial failure, 233–250
   in liquid-state processing, 17–18, 18f
   materials
     characteristics of, 299t
     mechanical and physical properties of, 191, 192t
     and microstructure, 129, 129f
     selection of, 300
   modification of, to promote wetting, 51
   and particle fracture, 287, 288f
   and precipitation kinetics, 130–131, 130f–131f
   randomly oriented, versus aligned discontinuous reinforcements, 163, 163f
   shape
     and fatigue life behavior, 281–282
     and matrix failure, 235–236, 236f–239f

Reinforcement(s), *continued*
   size, and fatigue life behavior, 280–281, 281f
   spatial distribution of, 164, 165f
   strengthening effect of, in polycrystalline composites, 152–154, 153f
Relaxation
   diffusional, of reinforcement, 171–173, 172f
   plastic, of thermal residual stresses, 124–125, 130
   processes, 62
      during thermal cycling, 210, 210f
Research and development programs
   American, 325
   European, 322–325, 323t–324t
   Japanese, 320–322, 321t–322t
Residual elastic strain, with CTE mismatch, 204
Residual expansion, 73, 73f
Residual stress(es), 61–80. *See also* Thermal residual stresses
   classification of, 61
   and composite yielding, 168, 169f–171f
   definition of, 61
   effects of, 61
   macroscopic, measurement of, 116–117
   measurement of, 62
      reasons for, 63
      techniques for, 63–68, 76–78
   microscopic, measurement of, 116
   numerical simulations of, 71–75
   origins of, 61–62, 68–71
Reuss-type approximation, 176
Reynolds number, 6
Rheocasting, 11
Rice's J-integral, 219
Rigid-lattice model, 103, 105f
Rod profiles, 98, 98f
Roll bonding, 24, 26f
Rolling, 30–31, 30f–31f
   grain refinement with, 31–32, 32f
Rolling texture, 114
Rotated cube orientation, 114, 116f
Roughness-induced closure, 286
RS metals. *See* Rapidly solidified metals
Rule of mixtures, 176, 182t

SAD. *See* Selected area diffraction
Safimax, applications of, 300
Saffil, 8, 9f, 18
   applications of, 300
Sapphire, MBE grown Nb films on
   gracing incidence x-ray scattering of, 95–99, 96t, 98f–99f
   high-resolution electron microscopy of, 88–90
Scanning electron microscopy
   of chemical processes at metal/ceramic interfaces, 100
   of crack propagation, 263, 263f–264f
   measurement of residual stresses with, 62
   of particle fracture, 241
   probe range in, 84
Scherzer focus, 85, 87, 90
SCS-6 silicon carbide fiber reinforced Ti-6Al-4V alloy, 24, 26f
Secondary processing, 111–114
Second-phase particles, 111, 113–114, 116
Selected area diffraction, 83–84, 84f
Self-consistent method, for continuous fiber composites, 177–178, 182t
Self-propagating high-temperature synthesis, 316, 319
SEM. *See* Scanning electron microscopy

Semisolid metal, and particulates, mixing of, 11
Sessile-drop experiment, 43, 43f
Shape change
   anisotropic, during thermal cycling, 206–207
   macroscopic, during thermal cycling, 205–206, 206f
Shear, 120
   creep behavior in, 196
Shear bands, 111–113
   formation of, crystal plasticity models of, 140–141
Shear flow, 140
Shear free sliding, and creep strength, 172, 172f
Shear-lag analysis, of isothermal creep, 199, 200f, 202
Shear-lag model, 126
Shear modulus, transverse, 177, 181, 182t
Shear rate, 141
   influence on viscosity, 12, 12f
Shear stress, 140–141
   interfacial, measurement of, 257–258, 258f, 258t
   threshold, and creep behavior, 195–196
Shear-stress strain, analysis of, in crystal plasticity models, 145–146
Shielding ratio
   computed, 231, 231f
   in elastic cell, 230
   steady-state, as function of bridging strength to yield strength, 228, 228f
Short crack behavior, 291, 291f
SHS. *See* Self-propagating high-temperature synthesis
Silicon carbide
   characteristics of, 299t
   coefficient of thermal contraction, 110
   potential applications of, 309t
   wetting of, 46, 47t, 49
Silicon carbide/aluminum
   aerospace applications of, 312, 312f
   for electronic packaging materials, 317, 318f
Silicon carbide whiskers, stress-strain behavior of, 114
Silicon nitride, characteristics of, 299t
Silver chloride, punching distances, 111
Simple interface, 42
Single crystal model
   with embedded fibers, 143–144, 144f
      results of, 147f, 147–148
   hydrostatic stress in, 152
Single crystal theory, 140–141
Single fiber fracture, 253–254
Sintering, 18, 23–28
Sliding
   effect of particle size on, 173
   fiber/matrix, debonding and crack deflection by, 253, 254f
   with no mass transport, and creep strength, 172, 172f
   shear free, and creep strength, 172, 172f
Sliding interface, creeping fibers with, model of, 201
Slip marking, surface, during thermal cycling, 205
Slip system hardness, 141, 145
Slow oscillation, in stress field, 222–223
Small-scale contact, of interface crack faces, 223
Small-strain theory, 176
Sol-gel techniques, 52
Solidification
   applied pressure during, 9
   in infiltration processes, 8–9
   in in-situ processes, 16–17
   particle migration caused by, 12
   in spray processes, 15

Solid-state phase change, internal stresses during, 204
Solid-state processing, 3, 23–41
  of discontinuously reinforced composites, 26–28, 27f
  energy loss mechanisms in, 44
  free energy change in, 43
  methods of, 23–25. See also specific method
  and microstructure, 109–110
  summary of, 40
Source shortening stress, 179
Spallation neutron sources, 65
SPATE. See Stress-pattern analysis by measurement of thermal emission
Spatial distribution, of reinforcement, 164, 165f
Spectroscopy, 84, 84f
  energy dispersive, 84, 84f
  Raman, measurement of residual stresses with, 62
Sphere reinforcement, 236, 237f
Spherical particles
  nonuniform distribution of, effect of, 164
  and strengthening ratio, 159–161, 161f
  and stress-strain curve, 168, 169f
Spherical shell, pressurized, residual deformation measurement in, 72–73, 73f
Spontaneous infiltration, 4f, 4–5
Spray processes, 3, 13–15
  advantages and disadvantages of, 13
  applications of, 298
  cooling rate of, 110
  description of, 13, 14f
  free energy change in, 43
  governing phenomena, 13–14
  and microstructure, 109–110
  optimization of, 18
  solidification in, 15
  transport phenomena, 14
    after deposition, 14–15
  wetting during, promotion of, 51
Sputter deposition, 52
Square packing, 178–179, 180f, 182t
  diagonal, 179, 180f, 182t
  edge, 179, 180f, 182t
  zero-degree, 179
Squeeze-cast composites
  aligned reinforcement of, stress-strain curves of, 166–168, 167f
  automotive applications of, 305
Squeeze casting, 5, 18
  thermal cycling during, 206–207
Static annealing, grain refinement with, 32
Steady-state creep, 191–192
  in off-axis loading, 193, 193f
  power law, 170–171
  shear lag analysis of, 199
  temperature-compensated, as function of applied stress, 195, 195f
Steady-state shielding ratio, as function of bridging strength to yield strength, 228, 228f
Steady-state strain rate, and creep behavior in shear, 196
Steel
  high-carbon, characteristics of, 299t
  titanium-carbide-reinforced, 4
Step-function decay, in yield strength, model system with, 230, 230f
Stiffness, of aircraft engine materials, goals for, 308, 308f
Stir-casting, 45

Strain. See also Residual stress(es)
  composite, recovery of, 192
  elastic, 72
    residual, with CTE mismatch, 204
  Lagrangian, 142
  lattice, measurement of, 76
  limiting, during forging, 38–39, 38f–39f
  per thermal cycle, influence of thermal cycle amplitude on, 208, 209f
  plastic
    effective, 236, 237f–238f
    rate of, 74, 235
    total, 204
  stress-shear, analysis of, in crystal plasticity models, 145–146
  thermal mismatch, 40, 209
  uniform
    in elastic deformation, 175
    rate of, 164
Strain energy density function, 218
Strain-gauge rosette, for hole drilling, 66, 67f, 77
Strain hardening properties
  of crystal plasticity models, 145–148, 152
    versus phenomenological models, 154
  and plastic constraint, 184
Strain-life behavior
  plastic, 278f, 278–279
  total, 276–277, 277f–278f
Strain rate
  in extrusion, 30
  as function of stress, during thermal cycling, 208, 209f
  for P/M composite, 36, 36f
    effect of hydrostatic pressure on, 36–37, 37f
  in rolling, 29
  sensitivity, 29
  steady-state, and creep behavior in shear, 196
  superplasticity, 197
  in thermomechanical processing, 32
  uniform, 164
Stratospheric applications, 312
Strength
  of aircraft engine materials, goals for, 308, 308f
  versus modulus, for metal-matrix composites, 300, 301f
  versus temperature, for metal- and ceramic-matrix composites, 300, 300f
Strengthening
  in continuous fiber-reinforced composites, classification of, 178–179
  transverse, of continuous fiber-reinforced composites, 159, 161–163, 162f
Strengthening effect
  of reinforcement, in polycrystalline composites, 152–154, 153f
  of work-hardened interface, 201
Strengthening ratio
  for aligned reinforcements
    continuous, 161–163, 162f
    discontinuous, 159–161, 160f–161f
  cell model computation for, results of, 159, 160f
  as function of nonuniformity, 164, 165f
  for plastic matrix reinforced with aligned cylindrical particles, 159–161, 161f
  for randomly oriented reinforcements, 163, 163f
Stress(es). See also Flow stress; Residual stress(es)
  applied
    absence of, thermal cycling in, 206–207

Stress(es), *continued*
  applied, *continued*
    temperature-compensated steady-state creep as function of, 195, 195f
    thermal cycling in, 207–208, 208f–209f
  axial, 202, 203f
  Cauchy, 141
  distribution, around blunted crack tip, 225, 225f
  external, time-dependent creep deformation by, 209
  far-field uniform, 175
  flow, and microstructure, 116–117
  friction, 179, 199
  hydrostatic, 152, 153f, 154
    predicted evolution of, 202, 203f
  internal
    and flow stress, 116
    during temperature changes, 204
    time-dependent creep deformation by, 209
  Kirchhoff, 141, 153f
  limit-yield, for aligned reinforcements, 159, 161
  operating, and aerospace applications, 308, 308f
  Piola-Kirchoff, 142
  redistribution of, from creeping matrix, 202, 202f
  shear, 140–141
    interfacial, measurement of, 257–258, 258f, 258t
    threshold, and creep behavior, 195–196
  source shortening, 179
  state, multiaxial, components of, 193
  strain rate as function of, during thermal cycling, 208, 209f
  tensile, during deformation processing, 40
  under transverse loading, prediction of, 193–194
Stress field, slow oscillation in, 222–223
Stress intensity factor
  complex, 221–222
    elastic stress analysis of, 223–224
  remote, metal foil bonded between ceramic substrates subjected to, 226, 226f
Stress intensity range, 282–283
  and crack closure, 284–285, 285f
Stress-life behavior, 274–276, 275f–276f
Stress-pattern analysis by measurement of thermal emission, 62
Stress-shear strain, analysis of, in crystal plasticity models, 145–146
Stress-strain curve
  of continuous fiber-reinforced composite, and thermal residual stress, 187, 187f
  of P/M composite, 36, 36f
    and hydrostatic pressure, 36–37, 37f
  of polycrystalline composites, 148, 149f, 151, 151f, 152
  versus phenomenological models, 154–155, 155f
  and reinforcement shape, 235–236, 236f
  of residual stress models, 168–169, 169f–171f
  of single crystal composites, 147f, 147–148
  tensile
    of aligned reinforcement elastic-strain hardening matrices, 166, 167f
    of isotropic matrix materials, 165
    of whisker- and particle-containing materials, 114, 115f
    of tensile deformation along fiber axis, in continuous fiber composite, 182–183, 183f
  transverse-tension, and interfacial bonding, 188, 188f
  of whisker- and particle-containing materials, 114–117, 115f–116f

Stress-strain response, cyclic, in discontinuously reinforced composites, 269–270, 270f
Subgrain formation, in whisker- and particle-containing materials, 116, 116f
Subgrain size, measurement of, 116
Subsonic applications, 312
Sumitomo Electric, 319
Superconducting composite wires, 316–317, 317f
Superplastic deformation, 34–38, 35f–37f
Superplastic forging, 38
Superplasticity, 197–199, 198t, 210
  internal stress, 197
Surface adsorption, and wetting, 46–47, 47f
Surface cracking, during forging process, 38
Surface-relaxation techniques, measurement of residual stresses with, 62
Surface roughening, during thermal cycling, 205
Surface roughness, and crack closure, 286
Surface slip markings, during thermal cycling, 205
Sweden, research and development programs in, 325

Taylor factor, 154
Taylor M-factor, 116
TEM. *See* Transmission electron microscopy
Temperature
  aging, effects of, 127–129, 128f
  considerations
    in aerospace applications, 307–309, 309f
    in automotive applications, 305–306
  and creep properties of discontinuous reinforcements, 195, 195f
  initial, in infiltration processes, 8, 9f
  metal, in dispersion processes, 11–12
  in powder consolidation, 24
  in rolling operation, 30–31, 31f–32f
  and strain hardening properties, in crystal plasticity models, 145–146
  versus strength, for metal- and ceramic-matrix composites, 300, 300f
Tensile fracture
  in discontinuously reinforced composites, 258–265
  in long fiber reinforced composites, 252–255, 254f–255f
  propagation of, 255
  void initiation in, patterns of, 244f, 244–245
Tensile properties, of metal-matrix composites, 311t
Tensile stresses, during deformation processing, 40
Tensile stress-strain curves
  of aligned reinforcement elastic-strain hardening matrices, 166, 167f
  of isotropic matrix materials, 165
  of whisker- and particle-containing materials, 114, 115f
Tensile yield strength, and CTE mismatch, 204
Tension, creep behavior in, 196
Ternary systems, interfaces in, 99–100, 100f
Tertiary creep, 193, 200, 202
Texture, 114
Thermal conductivity, of electronic packaging materials, 317–318, 318f
Thermal contraction coefficients, differential, effects of, 110f, 110–111
Thermal cycle
  amplitude, and strain per thermal cycle, 208, 209f
  creep, effect of microstructure on, 208–209, 209f
Thermal cycling, 204–210
  in absence of applied stress, 206–207

characteristics of, variation in, effects of, 206, 208
of continuously reinforced composites, 204–206
of discontinuous reinforcements, 204, 206–210
microstructural damage during, 205
in presence of applied stress, 207–208, 208f–209f
and residual stresses, 69
summary of, 210–211
Thermal expansion coefficient. *See* Coefficient of thermal expansion
Thermal expansion tensor, 74
Thermal management composites, 317, 319f
Thermal mismatch strains, 40, 209
Thermal ratcheting, prediction of, 204–205
Thermal residual stresses, 68–69, 69f, 120
calculation of, 74–75
in continuous fiber-reinforced composites, 184–187
plastic relaxation of, 124–125, 130
yielding caused by, 168–169, 169f–171f
Thermomechanical processing, 31–34, 32f–34f
microstructure after, 33f, 33–34, 36, 36f
in polycrystalline composite model, 153
Thermomechanical variables, in aging response, 127–131
Thin-film cracking, 220, 220f
Thixomolding, 11
Thixotropic nature, of composite slurries, 12
Three-dimensional network, for preform fabrication, 17, 18f
Three-phase damage model, of elastic deformation in particle fracture, 241–242, 242f
Threshold. *See also* Near-threshold behavior
intrinsic, 290
Tire studs, 314
Titanium alloys
aerospace applications of, 307, 309–310, 310t, 312
applications of, 320
properties of, 310t
Tool guide, for hole drilling, 66–67, 67f
Toughness measurements
of fiber-reinforced composites and unreinforced matrix alloys, 252t
of particle- and whisker-reinforced composites, 252t
Toyota Motor Corporation, 3, 298
Transition bands, 111–112
Translation state, 104–106, 106f
Transmission electron microscopy
of aging response, 121, 122t–123t, 126, 129, 131–132, 132f
of chemical processes at metal/ceramic interfaces, 100
conventional, 83–84, 84f
of crack propagation, 263, 263f–264f
of dislocation density, 124f, 124–126
of matrix failure, 234
objective lens of, geometric beam path through, 84, 85f
subgrain size measurement with, 116
of thermal cycle strain, 210, 210f
Transmission function, 84
Transport phenomena
in infiltration processes, 5–8
in spray processes, 14
after deposition, 14–15
Transverse creep, 193
Transverse loading, matrix stresses and composite creep rate under, prediction of, 193–194
Transverse shear modulus, 177, 181, 182t
Transverse strengthening, of continuous fiber-reinforced composites, 159, 161–163, 162f

Transverse Young's modulus, 176, 180–181, 182t
Tresca-type yield criterion, 199
Triangle packing model, 179f, 183
Tungsten
applications of, 136f, 309t, 315–316
characteristics of, 299t
Turbine blades, hollow, fabrication of, with monotapes, 24–25, 27f
Tyranno, 300

Ultrasonics, measurement of residual stresses with, 62
Uniaxial creep, 191–193
Uniform strain
in elastic deformation, 175
rate of, 164
Unit cell model
of continuous fiber composites, 178–179, 179f–181f, 182t, 183
of interfacial decohesion, 247–248, 248f
of particle fracture, 242–243
of polycrystalline particulate composites, 144–145, 145f
of residual stress, 168
of single crystal composites, 143–144, 144f
Unit density, strength and stiffness per, 23
United States, research and development programs in, 325
Unloading solution, elastic, 73
Unreinforced matrix alloys, fracture properties of, 251, 252t
UTS, 252–253

Vacuum-driven infiltration, 5
wetting during, promotion of, 51
V-blender, 26
Vertical clustering, of reinforcement, and stress-strain curve and ductility, 236f, 236–239
Vibrational damping composite steels, 319
Vickers diamond pyramid indentor microhardness, aging response represented by, 121, 122t–123t, 127
as function of aging time, 127–129, 128f–129f
Viscosity, of liquid metals, 11–12, 12f
Void nucleation, growth, and coalescence, 239–240, 244f, 244–245
in long fiber reinforced composites, 258–262, 259f–262f
and plastic constraint, 239, 239f
Void volume fraction, 236, 237f–238f
Voigt-type approximation, 176
Volume fraction
and crack closure levels, 286, 286f
and fatigue crack growth, 264
and fatigue life behavior, 280–282, 281f–282f
and near-threshold behavior, 288–289, 288f–289f
Vortex method, 10

Wagner's first-order interaction parameter, 48–49
W-based alloys, 300
Weak-beam imaging, of misfit dislocations, 92–93
Wear-resistant composites, 23, 298, 314f, 314–316, 316f
Weibull distribution, 202, 202f
Wetting
enhancement of, 51–52
nonreactive, 45–48
reactive, 48–50
Whisker(s), 299t, 300. *See also specific type*
cracked, 244–245, 245f
deformation zones around, 112, 112f
macroscopic effect of, 112
textural weakening caused by, 114

Whisker-reinforced composites
   fatigue life behavior of, 281–282
   microstructural evolution in, 109–118
   strengthening in, and plastic constraint, 184
   stress-strain behavior of, 114–116, 115f–116f
   toughness measurements of, 252t
Williams's singularity, 221–224
Work-hardened interface, strengthening effect of, 201
Work hardening, cyclic, 272
Work of adhesion, 43, 81
   determination of, 82
   experimental values of, for nonreactive metals, 45–46, 45t–47t
   influence of alloying additions on, 46, 47f
Work of immersion, 43

XD synthesis, 15, 17, 310, 313
X-ray(s), penetration depth of, for difference materials, 64t
X-ray diffraction, 62, 64–65
X-ray reflectivity curves, 99, 99f

Yield criterion
   Mises, 72, 74, 159, 182
   Tresca-type, 199
Yielding, composite, influence of residual stress on, 168, 169f–171f
Yield strength
   effect of CTE mismatch on, 204
   in elastic cell, 230, 230f
   steady-state shielding ratio as function of, 228, 228f
   step-function decay in, model system with, 230, 230f
Yield zone, development of, in matrix punched by glass sphere, 125, 125f
Young-Dupré equation, 43, 82, 82f
Young's modulus, 72, 165, 230, 240–242
   axial, 176, 181, 182t, 182
   transverse, 176, 180–181, 182t

Z-axis surface scattering spectrometer, 95
Zener's pinning effect theory, 33
Zero clearance piston, 305
Zero-degree square packing, 179

TA 481 .F87 1993

Fundamentals of metal-matrix composites